ISNM
International Series of
Numerical Mathematics
Vol. 123

# General Inequalities 7

7th International Conference at Oberwolfach, November 13–18, 1995

Edited by

C. Bandle
W.N. Everitt
L. Losonczi
W. Walter

Springer Basel AG

Editors:

Catherine Bandle
Institute of Mathematics
University of Basel
Rheinsprung 21
4051 Basel
Switzerland
e-mail: bandle@math.unibas.ch

William N. Everitt
School of Mathematics and Statistics
University of Birmingham
Edgbaston
Birmingham B15 2TT
England, U.K.
e-mail: w.n.everitt@bham.ac.uk

Laszlo Losonczi
Department of Mathematics and Computer Science
Kuwait University
P.O. Box 5969
13060 Safat
Kuwait
e-mail: LOSONCZI@MATH-1.sci.kuniv.edu.kw

Wolfgang Walter
Institute of Mathematics I
University of Karlsruhe
Kaiserstr. 12
76128 Karlsruhe
Germany
e-mail: walter@ma1ws1.mathematik.uni-karlsruhe.de

1991 Mathematics Subject Classification 26A51, 26DXX, 39B72, 35P15, 52A41, 15A42

A CIP catalogue record for this book is available from the Library of Congress, Washington D.C., USA

Deutsche Bibliothek Cataloging-in-Publication Data
**General inequalities 7** / 7th International Conference on
General Inequalities, Oberwolfach, November 13–18, 1995. Ed.
by C. Bandle ... - Basel ; Boston ; Berlin : Birkhäuser, 1997
(International series of numerical mathematics ; Vol. 123)
ISBN 978-3-0348-9837-9 ISBN 978-3-0348-8942-1 (eBook)
DOI 10.1007/978-3-0348-8942-1
NE: Bandle, Catherine [Hrsg.]; Internationale Tagung über Allgemeine
Ungleichungen <7, 1995, Oberwolfach>; GT


Originally published by Birkhäuser Verlag in 1997
Softcover reprint of the hardcover 1st edition 1997
Printed on acid-free paper produced from chlorine-free pulp. TCF ∞
Cover design: Heinz Hiltbrunner, Basel

ISBN 978-3-0348-9837-9

9 8 7 6 5 4 3 2 1

# Preface

Inequalities continue to play an essential role in mathematics. The subject is perhaps the last field that is comprehended and used by mathematicians working in all the areas of the discipline of mathematics. Since the seminal work *Inequalities* (1934) of Hardy, Littlewood and Pólya mathematicians have laboured to extend and sharpen the earlier classical inequalities. New inequalities are discovered every year, some for their intrinsic interest whilst others flow from results obtained in various branches of mathematics. So extensive are these developments that a new mathematical periodical devoted exclusively to inequalities will soon appear; this is the Journal of Inequalities and Applications, to be edited by R.P. Agarwal. Nowadays it is difficult to follow all these developments and because of lack of communication between different groups of specialists many results are often rediscovered several times. Surveys of the present state of the art are therefore indispensable not only to mathematicians but to the scientific community at large.

The study of inequalities reflects the many and various aspects of mathematics. There is on the one hand the systematic search for the basic principles and the study of inequalities for their own sake. On the other hand the subject is a source of ingenious ideas and methods that give rise to seemingly elementary but nevertheless serious and challenging problems. There are many applications in a wide variety of fields from mathematical physics to biology and economics.

The main goal of the *General Inequalities* meetings held at Oberwolfach is to bring together research workers from many parts of the globe, working in different and seemingly unconnected areas but interested in the many different aspects of inequalities. These meetings have now a long tradition extending over more than twenty years; the first meeting took place in 1976 inspired by the efforts of J. Aczél, G. Aumann, E.F. Beckenbach, R. Bellman and W. Walter.

This volume contains the proceedings of the *General Inequalities* 7 meeting held at Oberwolfach in 1995. It is conceived in the spirit of the preceding volumes in the sense that it not only contains the latest results presented by the participants, but is also a useful reference book for both lecturers and research workers. The material is ordered in seven chapters from *Inequalities in Analysis* to *Applications*, and ends in the same way as the preceding volumes with a lively chapter entitled *Problems and Remarks*. The contributions here given reflect the ramifications of general inequalities into many areas of mathematics, and also present a synthesis of results in both theory and practice.

During the GI 7 meeting a reception was held in honour of Professor Wolfgang Walter to mark his retirement from the GI Committee, and to recognise his unique and dedicated service to the sequence of meetings at Oberwolfach. Particular mention was made of his tireless efforts in serving as Editor of the series of Proceedings for these meetings; these volumes, prepared with metriculous attention to detail, now form an invaluable collection of results in the general field of inequalities in mathematics.

The dedicated labours of Mrs. Irene Jendrasik have been essential to the production of this volume in its final form; with her expertise and professional skill she processed all the TEX files and so presented them in a uniform manner. The Editors thank her warmly for this excellent contribution.

The Editors are also indebted to Dr. Gerd Herzog, Dr. Wolfgang Reichel, and Dr. Ch. Schmoeger who assisted in the technical details and undertook all the necessary proof reading.

The inspiring atmosphere and excellent facilities of the Mathematical Research Institute at Oberwolfach contributed essentially to the success of the meeting.

November 1996

C. Bandle
W.N. Everitt
L. Losonczi
W. Walter

# Participants

J. Aczél, University of Waterloo, Waterloo, Ontario, Canada

R.P. Agarwal, National University of Singapore, Kent Ridge, Singapore

C.D. Ahlbrandt, University of Missouri, Columbia, Missouri, USA

H. Alzer, Waldbröl, Germany

M. S. Ashbaugh, University of Missouri, Columbia, Missouri, USA

R. Badora, Silesian University, Katowice, Poland

C. Bandle, Universität Basel, Basel, Switzerland

R. Benguria, Pontificia University Catolica, Santiago, Chile

C. Bennewitz, University of Lund, Lund, Sweden

M. Bohner, University of Ulm, Ulm, Germany

A. Brillard, University of Haute Alsace, Mulhouse, France

F. Brock, Universität zu Köln, Köln, Germany

B.M. Brown, University of Wales, Cardiff, Wales, U.K.

R.C. Brown, University of Alabama, Tuscaloosa, Alabama, USA

A. Clausing, Westfäl. Wilhelms-Universität, Münster, Germany

W. Eichhorn, Universität Karlsruhe, Kalsruhe, Germany

M. Essén, University of Uppsala, Uppsala, Sweden

W.N. Everitt, University of Birmingham, Birmingham, England

J. Fleckinger, University of Toulouse I, Toulouse, France

M. Flucher, Universität Basel, Basel, Switzerland

T. Furuta, Science University of Tokyo, Tokyo, Japan

R. Ger, Silesian University, Katowice, Poland

M. Goldberg, Israel Institute of Technology, Haifa, Israel

W. Hayman, Imperial College of Science and Technology, London, England

D.B. Hinton, University of Tennessee at Knoxville, Knoxville, Tennessee, USA

K. Kartak, Technical University of Prague, Prague, Czech Republic

B. KAWOHL, Universität zu Köln, Köln, Germany

H. KÖNIG, Universität des Saarlandes, Saarbrücken, Germany

H. KÖNIG, Universität Kiel, Kiel, Germany

A. KUFNER, Československá Akademie Véd, Praha, Czech Republic

L. LOSONCZI, Kuweit University, Kuwait

E.R. LOVE, University of Melbourne, Parkville, Victoria, Australia

M. MARCUS, Israel Institute of Technology, Haifa, Israel

J. MATKOWSKI, Technical University, Bielsko-Biala, Poland

G.V. MILOVANOVIC, University of Nis, Nis, Yugoslavia

R.N. MOHAPATRA, University of Central Florida, Orlando, Florida, USA

G. NASRI-ROUDSARI, Rhein.-Westf. Techn. Hochschule, Aachen, Germany

R.J. NESSEL, Rhein.-Westf. Techn. Hochschule, Aachen, Germany

ZS. PÁLES, Lajos Kossuth University, Debrecen, Hungary

C.E.M. PEARCE, University of Adelaide, Adelaide, Australia

J.E. PEČARIĆ, Faculty of Technology, Zagreb, Yugoslavia

M. PLUM, Universität Karlsruhe, Karlsruhe, Germany

TH.M. RASSIAS, National Technical University of Athens, Athens, Greece

J. RÄTZ, Universität Bern, Bern, Switzerland

W. REICHEL, Universität Karlsruhe, Karlsruhe, Germany

D.C. RUSSELL, Ventnor, England

S. SAITOH, Gunma University, Kiryu, Japan

G. TALENTI, Università degli Studi, Firenze, Italy

P. VOLKMANN, Universität Karlsruhe, Karlsruhe, Germany

W. WALTER, Universität Karlsruhe, Karlsruhe, Germany

A. ZETTL, Northern Illinois University, DeKalb, Illinois, USA

# Contents

## Inequalities for Eigenvalue Problems



## Inequalities for Differential Operators



## Convexity

## Problems and Remarks

# Inequalities in Analysis

International Series of Numerical Mathematics, Vol. 123, © 1997 Birkhäuser Verlag Basel

# Higher dimensional Hardy inequality

Pavel Drábek, Hans P. Heinig, and Alois Kufner

**Abstract**

Let $(Hf)(x) = \int_{B(0,|x|)} f(y)\,dy$, $x \in \mathbb{R}^n$, where $B(0,|x|)$ is the ball in $\mathbb{R}^n$ centered at zero with radius $|x|$. If $U$ and $W$ are weight functions, then it is shown that $H : L^p_U \to L^q_W$, $1 < p, q < \infty$, is bounded if and only if $U$, $W$ satisfy conditions which are the natural extensions to the one-dimensional conditions. For specific choices of the weights one obtains the $n$-dimensional Hardy inequalities whose constants are sharp and independent of $n$.

In addition a gradient inequality and an exponential integral inequality is characterized in terms of the weight functions.

## 1. Introduction

The characterization of weights for which the Hardy operator $(Hf)(x) = \int_0^x f(t)\,dt$ is bounded on weighted Lebesgue spaces is largely solved (cf. [OK]). For the higher dimensional case the situation is much different and only for the dimension $n = 2$ a characterization is known where the Hardy operator is of the form $(Hf)(x,y) = \int_0^x \int_0^y f(s,t)\,dsdt$ (cf., e.g., [S2] or [AK]). On the other hand, if the integral operator $H$ is of the form

$$(Hf)(x) = \int_{B(|x|)} f(y)\,dy, \tag{1.1}$$

where $B(|x|) = B(0,|x|)$ is the ball in $\mathbb{R}^n$ centered at the origin and of radius $|x|$, one has for $1 < p < \infty$

$$\left(\int_{\mathbb{R}^n} \left| \frac{1}{|B(|x|)|} \int_{B(|x|)} f(y)\,dy \right|^p dx\right)^{1/p} \le \frac{p}{p-1} \left(\int_{\mathbb{R}^n} |f(y)|^p dy\right)^{1/p}, \tag{1.2}$$

where $|B(|x|)| = \int_{|y|\le|x|} dy = \int_{\Sigma_{n-1}} \left(\int_0^{|x|} t^{n-1} dt\right) d\sigma = \frac{|x|^n |\Sigma_{n-1}|}{n}$ is the volume of $B(|x|)$ (cf. [CG]). Here $\Sigma_{n-1}$ is the surface of the unit ball in $\mathbb{R}^n$ and $|\Sigma_{n-1}|$ its area.

The aim of this note is to extend (1.2) to general weights and indices $p, q$ with $1 < p, q < \infty$, i.e., to derive inequalities of the form

$$\left(\int_{\mathbb{R}^n} W(x)|Hf(x)|^q dx\right)^{1/q} \leq C\left(\int_{\mathbb{R}^n} U(x)|f(x)|^p dx\right)^{1/p}. \tag{1.3}$$

The weights $W$ and $U$ are non-negative locally integrable functions on $\mathbb{R}^n$ and without loss of generality we assume $f(x) \geq 0$, $x \in \mathbb{R}^n$. The conjugate index of $r > 1$ is denoted by $r'$, i.e., $r' = \frac{r-1}{r}$.

As a consequence of our result (Theorem 2.1) we obtain the inequality

$$\int_{\mathbb{R}^n} |B(|x|)|^{-r-1}\left(\int_{B(|x|)} f(y)\,dy\right)^p dx \leq \left(\frac{p}{r}\right)^p \int_{\mathbb{R}^n} |B(|x|)|^{p-r-1} f^p(x)\,dx \tag{1.4}$$

for $1 < p < \infty$ and $r > 0$, with constant $(p/r)^p$ being the best possible (see Remark 2.1 below).

## 2. The Hardy inequality

To prove our first result we require the following continuous version of Minkowski's inequality: For $g(x) \geq 0$, $h(x) \geq 0$ on $(0, \infty)$ and for $r \geq 1$

$$\left\{\int_0^\infty g(x)\left(\int_0^x h(y)\,dy\right)^r dx\right\} \leq \left\{\int_0^\infty h(y)\left(\int_y^\infty g(x)\,dx\right)^{1/r} dy\right\}^r. \tag{2.1}$$

Our first characterization is given by the following.

**Theorem 2.1.** *Let $W$ and $U$ be weight functions on $\mathbb{R}^n$. For $1 < p \leq q < \infty$, the inequality*

$$\left\{\int_{\mathbb{R}^n} W(x)\left[\int_{B(|x|)} f(y)\,dy\right]^q dx\right\}^{1/q} \leq C\left\{\int_{\mathbb{R}^n} U(x) f^p(x)\,dx\right\}^{1/p} \tag{2.2}$$

*holds for $f \geq 0$ if and only if*

$$A := \sup_{\alpha>0}\left\{\int_{|x|\geq\alpha} W(x)\,dx\right\}^{1/q}\left\{\int_{|x|\leq\alpha} U^{1-p'}(x)\,dx\right\}^{1/p'} < \infty. \tag{2.3}$$

*Moreover, if $C$ is the smallest constant for which (2.2) holds, then*

$$A \leq C \leq A(p')^{1/p'} p^{1/q}. \tag{2.4}$$

*Proof.* Observe first that for any $\alpha > 0$, (2.3) can be written in the polar coordinates form:

$$\left\{\int_\alpha^\infty t^{n-1}\int_{\Sigma_{n-1}} W(t\tau)\,d\tau dt\right\}^{1/q}\left\{\int_0^\alpha t^{n-1}\int_{\Sigma_{n-1}} U^{1-p'}(t\tau)\,d\tau dt\right\}^{1/p'} \le A. \quad (2.5)$$

Moreover, the left side of (2.2) has the form

$$\int_{\mathbb{R}^n} W(x)\left[\int_{B(|x|)} f(y)\,dy\right]^q dx \quad (2.6)$$
$$= \int_0^\infty \int_{\Sigma_{n-1}} t^{n-1}W(t\tau)\left[\int_0^t\int_{\Sigma_{n-1}} s^{n-1}f(s\sigma)\,d\sigma\,ds\right]^q d\tau\,dt.$$

If we define $h$ by

$$h(t) = \left\{\int_{\Sigma_{n-1}}\int_0^t s^{n-1}U^{1-p'}(s\sigma)\,ds\,d\sigma\right\}^{1/(pp')},$$

then by Hölder's inequality

$$\begin{aligned}\int_0^t\int_{\Sigma_{n-1}} s^{n-1}f(s\sigma)\,d\sigma ds &= \int_{\Sigma_{n-1}}\int_0^t s^{(n-1)/p}f(s\sigma)U^{1/p}(s\sigma)h(s)s^{(n-1)/p'}\\ &\quad\times\left[U^{1/p}(s\sigma)h(s)\right]^{-1} ds\,d\sigma\\ &\le \left\{\int_{\Sigma_{n-1}}\int_0^t s^{n-1}\left[f(s\sigma)U^{1/p}(s\sigma)h(s)\right]^p ds\,d\sigma\right\}^{1/p}\\ &\quad\times\left\{\int_{\Sigma_{n-1}}\int_0^t s^{n-1}\left[U^{1/p}(s\sigma)h(s)\right]^{-p'} ds\,d\sigma\right\}^{1/p'},\end{aligned}$$

and if we define $\widetilde{W}_n$, $F_n$, $H_n$ by

$$\widetilde{W}_n(t) = \int_{\Sigma_{n-1}} t^{n-1}W(t\tau)\,d\tau, \qquad F_n(s) = \int_{\Sigma_{n-1}} s^{n-1}\left[f(s\sigma)U^{1/p}(s\sigma)h(s)\right]^p d\sigma,$$

$$H_n(t) = \int_0^t\int_{\Sigma_{n-1}} s^{n-1}\left[U^{1/p}(s\sigma)h(s)\right]^{-p'} d\sigma\,ds,$$

$t, s > 0$, then we obtain from (2.6)

$$\int_{\mathbb{R}^n} W(x)\left[\int_{B(|x|)} f(y)\,dy\right]^q dx \le \int_0^\infty \widetilde{W}_n(t)\left[\int_0^t F_n(s)\,ds\right]^{q/p} H_n^{q/p'}(t)\,dt. \quad (2.7)$$

Applying now (2.1) with $r = q/p \geq 1$, the right side of (2.7) is not larger than

$$\left\{\int_0^\infty F_n(s)\left[\int_s^\infty \widetilde{W}_n(t) H_n^{q/p'}(t)\,dt\right]^{p/q} ds\right\}^{q/p}. \tag{2.8}$$

Using the notation $U_n(s) = \int_{\Sigma_{n-1}} s^{n-1} U^{1-p'}(s\sigma)d\sigma$, we obtain by integration and by inequality (2.5) that

$$\begin{aligned}
H_n(t) &= \int_{\Sigma_{n-1}}\int_0^t s^{n-1}U^{1-p'}(s\sigma)\left[\int_0^s\int_{\Sigma_{n-1}} z^{n-1}U^{1-p'}(z\tau)\,d\tau dz\right]^{-1/p} ds d\sigma \\
&= \int_0^t U_n(s)\Big(\int_0^s U_n(z)\,dz\Big)^{-1/p} ds = p'\int_0^t \frac{d}{ds}\Big(\int_0^s U_n(z)\,dz\Big)^{1/p'} ds \\
&= p'\Big(\int_0^t U_n(s)\,ds\Big)^{1/p'} = p'\Big(\int_0^t\int_{\Sigma_{n-1}} s^{n-1}U^{1-p'}(s\sigma)\,d\sigma ds\Big)^{1/p'} \\
&\leq p'A\Big\{\int_t^\infty s^{n-1}\int_{\Sigma_{n-1}} W(s\tau)\,d\tau ds\Big\}^{-1/q} = p'A\Big\{\int_t^\infty \widetilde{W}_n(s)\,ds\Big\}^{-1/q}.
\end{aligned}$$

Similarly, again using integration and inequality (2.5), we can show that

$$\begin{aligned}
&\int_s^\infty \widetilde{W}_n(t)\Big(\int_t^\infty \widetilde{W}_n(s)\,ds\Big)^{-1/p'} dt \\
&= p\Big(\int_s^\infty \widetilde{W}_n(t)\,dt\Big)^{1/p} \\
&= p\Big(\int_s^\infty\int_{\Sigma_{n-1}} t^{n-1}W(t\tau)\,d\tau dt\Big)^{1/p} \\
&\leq pA^{q/p}\Big\{\int_0^s t^{n-1}\int_{\Sigma_{n-1}} U^{1-p'}(t\tau)\,d\tau dt\Big\}^{-q/(p'p)} \\
&= pA^{q/p}h^{-q}(s).
\end{aligned}$$

Hence the integral (2.8) can be estimated from above by

$$\begin{aligned}
&(p'A)^{q/p'}pA^{q/p}\Big\{\int_0^\infty F_n(s)h^{-p}(s)\,ds\Big\}^{q/p} \\
&= A^q(p')^{q/p'}p\Big\{\int_0^\infty\int_{\Sigma_{n-1}} s^{n-1}f^p(s\sigma)U(s\sigma)\,d\sigma\,ds\Big\}^{q/p} \\
&= A^q(p')^{q/p'}p\Big\{\int_{\mathbb{R}^n} U(x)f^p(x)\,dx\Big\}^{1/q},
\end{aligned}$$

and due to (2.7), it follows that (2.2) holds with $C = A(p')^{1/p'}p^{1/q}$.

The necessity of condition (2.3) follows by taking $f(x) = U^{1-p'}(x)\chi_{(0,\alpha)}(|x|)$ in (2.2) with $\alpha > 0$. Hence

$$\begin{aligned} C\left\{\int_{\mathbb{R}^n} U(x) f^p(x)\,dx\right\}^{1/p} &= C\left\{\int_{|x|\le\alpha} U^{1-p'}(x)\,dx\right\}^{1/p} \\ &\ge \left\{\int_{|x|\ge\alpha} W(x)\left[\int_{|y|\le|x|} f(y)\,dy\right]^q dx\right\}^{1/q} \\ &= \left\{\int_{|x|\ge\alpha} W(x)\,dx\right\}^{1/q}\left\{\int_{|y|\le\alpha} U^{1-p'}(y)\,dy\right\} \end{aligned}$$

and (2.3) follows on dividing by $\{\int_{|x|\le\alpha} U^{1-p'}(x)\,dx\}^{1/p}$ with $C \ge A$.

**Remark 2.1.** For $p = q > 1$, $W(x) = |B(|x|)|^{-r-1}$, $U(x) = |B(|x|)|^{p-r-1}$ with $r > 1$, we have

$$\begin{aligned} A &= \left(\int_{|x|\ge\alpha} |B(|x|)|^{-r-1} dx\right)^{1/p}\left(\int_{|x|\le\alpha} |B(|x|)|^{(p-r-1)(1-p')} dx\right)^{1/p'} \\ &= \left(\int_{\Sigma_{n-1}} d\sigma \int_\alpha^\infty t^{n-1}\left(\frac{t^n|\Sigma_{n-1}|}{n}\right)^{-r-1} dt\right)^{1/p} \\ &\quad\times\left(\int_{\Sigma_{n-1}} d\sigma \int_0^\alpha t^{n-1}\left(\frac{t^n|\Sigma_{n-1}|}{n}\right)^{(p-r-1)(1-p')} dt\right)^{1/p'} \\ &= |\Sigma_{n-1}|^{-r/p} n^{(r+1)/p}\left(\int_\alpha^\infty t^{-1-nr} dt\right)^{1/p} |\Sigma_{n-1}|^{r/p} n^{(p-r-1)/p} \\ &\quad\times\left(\int_0^\alpha t^{-1+nr(p'-1)} dt\right)^{1/p'} \\ &= n\left(\frac{\alpha^{-nr}}{nr}\right)^{1/p}\left(\frac{\alpha^{nr(p'-1)}}{nr(p'-1)}\right)^{1/p'} = \frac{1}{r}\,(p'-1)^{-1/p'} = \frac{1}{r}\left(\frac{p}{p'}\right)^{1/p'}, \end{aligned}$$

and consequently in this case the constant $C$ in (2.2) has the form $A(p')^{q/p'}p^{1/p} = p/r$, which is independent of $n$. Therefore we obtained the inequality (1.4). Moreover, the constant $(p/r)^p$ is best possible (cf. [HLP, Theorem 330]).

For $r = p - 1$ we obtain inequality (1.2) with sharp constant. This result was also obtained in [CG].

Our next result concerns the case $q < p$:

**Theorem 2.2.** *Let $W$ and $U$ be weight functions on $\mathbb{R}^n$. For $1 < q < p < \infty$, the* *inequality* (2.2) *holds for* $f \geq 0$ *if and only if*

$$\int_{\mathbb{R}^n} \left[ \int_{\mathbb{R}^n \setminus B(|x|)} W(y)\, dy \right]^{r/q} \left[ \int_{B(|x|)} U^{1-p'}(y)\, dy \right]^{r/q'} U^{1-p'}(x)\, dx < \infty, \tag{2.9}$$

*where* $\frac{1}{r} = \frac{1}{q} - \frac{1}{p}$.

*Proof.* Let us use polar coordinates. Write $S(x) = U^{1-p'}(x)$, $G_n(s) = \int_{\Sigma_{n-1}} s^{n-1} g(s\sigma) S(s\sigma) d\sigma$ for $g \geq 0$, and define $\widetilde{W}_n(t)$ as in the proof of Theorem 2.1. Then

$$\begin{aligned}
&\int_{\mathbb{R}^n} W(x) \left[ \int_{B(|x|)} g(y) S(y)\, dy \right]^q dx \\
&= \int_0^\infty \int_{\Sigma_{n-1}} t^{n-1} W(t\tau) d\tau \left[ \int_0^t \int_{\Sigma_{n-1}} s^{n-1} g(s\sigma) S(s\sigma)\, d\sigma\, ds \right]^q dt \\
&= \int_0^\infty \widetilde{W}_n(t) \left( \int_0^t G_n(s)\, ds \right)^q dt \\
&= q \int_0^\infty G_n(s) \left( \int_0^s G_n(t)\, dt \right)^{q-1} \left( \int_0^s \widetilde{W}_n(t)\, dt \right) ds \\
&= q \int_{\Sigma_{n-1}} \int_0^\infty s^{n-1} g(s\sigma) S(s\sigma) \left( \int_0^s \int_{\Sigma_{n-1}} t^{n-1} g(t\tau) S(t\tau)\, d\tau\, dt \right)^{q-1} \\
&\qquad \times \left( \int_s^\infty \widetilde{W}_n(t)\, dt \right) ds\, d\sigma \\
&= q \int_{\Sigma_{n-1}} \int_0^\infty s^{n-1} g(s\sigma) S^{\frac{1}{p} + \frac{q-1}{p} + \frac{p-q}{p}}(s\sigma) \left[ \frac{\int_{\Sigma_{n-1}} \int_0^s t^{n-1} g(t\tau) S(t\tau)\, dt\, d\tau}{\int_{\Sigma_{n-1}} \int_0^s t^{n-1} S(t\tau)\, dt\, d\tau} \right]^{q-1} \\
&\qquad \times \left[ \left( \int_{\Sigma_{n-1}} \int_0^s t^{n-1} S(t\tau)\, dt\, d\tau \right)^{q-1} \left( \int_s^\infty \widetilde{W}_n(t)\, dt \right) \right] ds\, d\sigma.
\end{aligned}$$

Since $\frac{1}{p} + \frac{q-1}{p} + \frac{p-q}{p} = 1$ we can apply Hölder's inequality (with three factors) and obtain that

$$\int_{\mathbb{R}^n} W(x) \left[ \int_{B(|x|)} g(y) S(y)\, dy \right]^q dx \leq q I_1 I_2 I_3$$

where

$$\begin{aligned}
I_1 &= \left( \int_{\Sigma_{n-1}} \int_0^\infty s^{n-1} g^p(s\sigma) S(s\sigma)\, ds d\sigma \right)^{1/p} = \left( \int_{\mathbb{R}^n} g^p(x) S(x)\, dx \right)^{1/p}, \\
I_2 &= \left\{ \int_{\Sigma_{n-1}} \int_0^\infty s^{n-1} S(s\sigma) \left[ \frac{\int_{\Sigma_{n-1}} \int_0^s t^{n-1} g(t\tau) S(t\tau)\, dt\, d\tau}{\int_{\Sigma_{n-1}} \int_0^s t^{n-1} S(t\tau)\, dt\, d\tau} \right]^p ds\, d\sigma \right\}^{\frac{q-1}{p}},
\end{aligned}$$

and

$$I_3 = \left\{\int_{\Sigma_{n-1}}\int_0^\infty s^{n-1}S(s\sigma)\left[\int_{\Sigma_{n-1}}\int_0^s t^{n-1}S(t\tau)\,dt\,d\tau\right]^{\frac{q-1}{q}\frac{pq}{p-q}} \times\left[\int_s^\infty \widetilde{W}_n(t)\,dt\right]^{\frac{pq}{p-q}\frac{1}{q}} ds\,d\sigma\right\}^{\frac{p-q}{p}}.$$

Now $\frac{1}{r} = \frac{1}{q} - \frac{1}{p} = \frac{p-q}{pq}$, so that

$$\begin{aligned} I_3 &= \left\{\int_0^\infty\int_{\Sigma^{n-1}}\left(\int_s^\infty \widetilde{W}_n(t)\,dt\right)^{r/q}\left(\int_{\Sigma_{n-1}}\int_0^t t^{n-1}S(t\tau)\,dt\,d\tau\right)^{r/q'} \times s^{n-1}S(s\sigma)\,d\sigma ds\right\}^{\frac{p-q}{p}} \\ &= \left\{\int_{\mathbb{R}^n}\left(\int_{\mathbb{R}^n\setminus B(|x|)} W(y)\,dy\right)^{r/q}\left(\int_{B(|x|)} S(y)\,dy\right)^{r/q'} S(x)\,dx\right\}^{\frac{p-q}{p}} \\ &= C_0 < \infty \end{aligned}$$

by (2.9) (since $S = U^{1-p'}$).

Next,

$$I_2 = \left\{\int_{\mathbb{R}^n} S(x)\left[\frac{1}{\int_{B(|x|)} S(y)\,dy}\int_{B(|x|)} S(y)g(y)\,dy\right]^p dx\right\}^{\frac{q-1}{p}}$$

and using Theorem 2.1 for $p = q$ with $W(x) = S(x)(\int_{B(|x|)} S(y)dy)^{-p}$, $U(x) = S^{1-p}(x)$, and $f(x) = S(x)g(x)$, we have

$$I_2 \le C\left\{\int_{\mathbb{R}^n} S^{1-p+p}(x)g^p(x)dx\right\}^{\frac{q-1}{p}} = C\left\{\int_{\mathbb{R}^n} g^p(x)S(x)\,dx\right\}^{\frac{q-1}{p}},$$

provided

$$A = \left(\int_{|x|\ge\alpha} S(x)\left(\int_{B(|x|)} S(y)\,dy\right)^{-p} dx\right)^{1/p}\left(\int_{|x|\le\alpha} S(x)\,dx\right)^{1/p'} < \infty.$$

But this condition is satisfied since with $S_n(s) = \int_{\Sigma_{n-1}} s^{n-1}S(s\tau)d\tau$ we have

$$\begin{aligned} A &= \left(\int_{\Sigma_{n-1}}\int_\alpha^\infty t^{n-1}S(t\tau)\left(\int_0^t S_n(s)\,ds\right)^{-p} dt\,d\tau\right)^{1/p}\left(\int_0^\alpha S_n(s)\,ds\right)^{1/p'} \\ &= \left\{\int_\alpha^\infty\left(\int_0^t S_n(s)\,ds\right)^{-p} S_n(t)\,dt\right\}^{1/p}\left\{\int_0^\alpha S_n(s)\,ds\right\}^{1/p'} \\ &\le \left\{\frac{1}{p-1}\left(\int_0^\alpha S_n(s)\,ds\right)^{1-p}\right\}^{1/p}\left\{\int_0^\alpha S_n(s)\,ds\right\}^{1/p'} = (p-1)^{-1/p} < \infty. \end{aligned}$$

Hence we have shown that

$$\int_{\mathbb{R}^n} W(x)\left[\int_{B(|x|)} g(y)S(y)\,dy\right]^q dx \le CC_0\left\{\int_{\mathbb{R}^n} g^p(x)S(x)\,dx\right\}^{\frac{1}{p}+\frac{q-1}{p}}$$

from which (2.2) follows on taking $f = gS$ since

$$g^pS = \left(\frac{f}{S}\right)^p S = f^pS^{1-p} = Uf^p.$$

Thus the condition (2.9) is sufficient for (2.2) to hold.

For the necessity, we put for $f(x)$ in (2.2) the functions

$$\begin{aligned} f_n(x) &= \left(\int_{|y|\ge x} W(y)\,dy\right)^{r/(pq)} \left(\int_{a_n\le|y|\le|x|} U^{1-p'}(y)\,dy\right)^{r/(pq')} \\ &\qquad \times U^{1-p'}(x)\chi_{(a_n,b_n)}(|x|), \end{aligned}$$

$n = 1, 2, \dots,$ where $0 < a_n < b_n$ with $a_n \searrow 0$, $b_n \nearrow \infty$ for $n \to \infty$ (cf. [OK, Lemma 3.9]).

**Remark 2.2.** Obviously, the proofs of the foregoing results are based on their one-dimensional analogues (cf. [OK] for Theorem 2.1 and [S1] for Theorem 2.2). Thus, it is also clear how the results can be modified for the cases $p, q = 1, \infty$.

Similarly, in analogue with the adjoint one-dimensional Hardy operator $(H^*f)(x) = \int_x^\infty f(t)\,dt$, we can show that the inequality

$$\left\{\int_{\mathbb{R}^n} W(x)\left[\int_{\mathbb{R}^n\setminus B(|x|)} f(y)\,dy\right]^q dx\right\}^{1/q} \le C\left\{\int_{\mathbb{R}^n} W(x)f^p(x)\,dx\right\}^{1/p} \tag{2.10}$$

holds if and only if either

$$A := \sup_{\alpha>0}\left(\int_{|x|\le\alpha} W(x)\,dx\right)^{1/q}\left(\int_{|x|\ge\alpha} U^{1-p'}(x)\,dx\right)^{1/p'} < \infty \tag{2.11}$$

for $1 < p \le q < \infty$ or

$$\int_{\mathbb{R}^n}\left[\int_{B(|x|)} W(y)\,dy\right]^{r/q}\left[\int_{\mathbb{R}^n\setminus B(|x|)} U^{1-p'}(y)\,dy\right]^{r/q'} U^{1-p'}(x)\,dx < \infty \tag{2.12}$$

for $1 < q < p < \infty$ with $\frac{1}{r} = \frac{1}{q} - \frac{1}{p}$.

## 3. A weighted gradient inequality

The Hardy inequality appears also in a form containing differential operators. Here, we will use Theorem 2.1 for deriving an $n$-dimensional analogue, namely, the inequality

$$\Big(\int_{\mathbb{R}^n} W(x)|F(x)|^q dx\Big)^{1/q} \le C\Big(\int_{\mathbb{R}^n} U(x)\Big|\frac{x\cdot\nabla F(|x|)}{|x|}\Big|^p dx\Big)^{1/p}. \tag{3.1}$$

**Theorem 3.1.** *Let $U$, $W$ be weight functions on $\mathbb{R}^n$. For $1 < p \le q < \infty$, the inequality* (3.1) *holds for radial functions $F$ such that $F(0) = 0$ if and only if*

$$A := \sup_{\alpha>0}\Big(\int_{|x|\ge\alpha} W(x)\,dx\Big)^{1/q}\Big(\int_0^\alpha\Big[\int_{\Sigma_{n-1}} t^{n-1}U(t\tau)\,d\tau\Big]^{1-p'}dt\Big)^{1/p'} < \infty. \tag{3.2}$$

*Proof.* Let us write $F$ in the form

$$F(x) = \int_{B(|x|)} f(y)\,dy = \int_{\Sigma_{n-1}}\int_0^{|x|} s^{n-1}f(s\sigma)\,ds\,d\sigma, \tag{3.3}$$

and denote $\widetilde{F}(t) = F(x)$ for $|x| = t$. Then we have

$$F'(|x|) = \widetilde{F}'(t) = t^{n-1}\int_{\Sigma_{n-1}} f(t\tau)\,d\tau$$

and

$$\begin{aligned}
&\Big\{\int_{\mathbb{R}^n} W(x)|F(x)|^q dx\Big\}^{1/q}\\
&\quad= \Big\{\int_{\Sigma_{n-1}}\int_0^\infty t^{n-1}W(t\tau)|\widetilde{F}(t)|^q dt\,d\tau\Big\}^{1/q}\\
&\quad= \Big\{\int_0^\infty\int_{\Sigma_{n-1}} t^{n-1}W(t\tau)\Big|\int_0^t\int_{\Sigma_{n-1}} s^{n-1}f(s\sigma)\,d\sigma\,ds\Big|^q d\tau\,dt\Big\}^{1/q}\\
&\quad= \Big\{\int_0^\infty \widetilde{W}_n(t)\Big|\int_0^t \widetilde{F}_n(s)\,ds\Big|^q dt\Big\}^{1/q}
\end{aligned}$$

where we denoted

$$\widetilde{W}_n(t) = \int_{\Sigma_{n-1}} t^{n-1}W(t\tau)\,d\tau, \qquad \widetilde{F}_n(s) = \int_{\Sigma_{n-1}} s^{n-1}f(s\sigma)d\sigma.$$

Denoting further $\widetilde{U}_n(t) = \int_{\Sigma_{n-1}} t^{n-1}U(t\tau)\,d\tau$, then the (one-dimensional) Hardy inequality

$$\left(\int_0^\infty \widetilde{W}_n(t)\left|\int_0^t \widetilde{F}_n(t)\,dt\right|^q dt\right)^{1/q} \le C\left(\int_0^\infty \widetilde{U}_n(t)\left|\widetilde{F}_n(t)\right|^p dt\right)^{1/p} \tag{3.4}$$

holds (see Theorem 2.1) if and only if

$$A := \sup_{\alpha>0}\left(\int_\alpha^\infty \widetilde{W}_n(t)\,dt\right)^{1/q}\left(\int_0^\alpha \widetilde{U}_n^{1-p'}(t)\,dt\right)^{1/p'} < \infty. \tag{3.5}$$

Now

$$\begin{aligned}
\int_0^\infty \widetilde{U}_n(t)\widetilde{F}_n^p(t)\,dt &= \int_0^\infty \widetilde{U}_n(t)\left[\int_{\Sigma_{n-1}} t^{n-1}f(t\tau)\,d\tau\right]^p dt \\
&= \int_0^\infty \widetilde{U}_n(t)[\widetilde{F}'(t)]^p dt = \int_{\Sigma_{n-1}}\int_0^\infty t^{n-1}U(t\tau)[\widetilde{F}'(t)]^p dt d\tau \\
&= \int_{\mathbb{R}^n} U(x)[F'(|x|)]^p dx,
\end{aligned}$$

and (3.5) can be written in the form

$$A := \sup_{\alpha>0}\left(\int_{|x|\ge\alpha} W(x)\,dx\right)^{1/q}\left(\int_0^\alpha\left[\int_{\Sigma_{n-1}} t^{n-1}U(t\tau)\,d\tau\right]^{1-p'} dt\right)^{1/p'} < \infty. \tag{3.6}$$

So, we have shown that for differentiable radial functions $F$ satisfying $F(0) = 0$ the inequality

$$\left(\int_{\mathbb{R}^n} W(x)F^q(x)\,dx\right)^{1/q} \le C\left(\int_{\mathbb{R}^n} U(x)[F'(|x|)]^p dx\right)^{1/p} \tag{3.7}$$

holds if and only if (3.6) is satisfied.

Now, since $F'(|x|) = \frac{x\cdot\nabla F(|x|)}{|x|}$, the assertion follows.

These and similar results for a gradient inequality (weighted Friedrichs/Poincaré inequality) were also obtained in [Si].

## 4. A weighted exponential inequality

Let

$$(Pf)(x) = \frac{1}{|B(x)|}\int_{B(x)} f(y)\,dy,$$

then

$$\lim_{p\to\infty}\left[(Pf^{1/p})(x)\right]^p = \exp(P\log f)(x).$$

It is natural to expect weighted estimates also for this special exponential operator whose one-dimensional analogue was discussed in [HKK].

The final result concerns the extension of one of the results in [HKK].

**Theorem 4.1.** *Let $f$, $U$, $V$ be positive functions defined on $\mathbb{R}^n$. Define $W$ by*

$$W(x) = U(x)\exp\left(P\log\frac{1}{V}\right)(x), \tag{4.1}$$

*where $P$ is the average operator defined above. Then the inequality*

$$\int_{\mathbb{R}^n} U(x)\exp(P\log f)(x)\,dx \le C\int_{\mathbb{R}^n} V(x)f(x)\,dx \tag{4.2}$$

*holds if and only if*

$$A := \sup_{t>0} t^n \int_{|x|\ge t} \frac{W(x)}{|x|^{2n}}\,dx < \infty. \tag{4.3}$$

*Proof.* Writing $Vf = g$ then (4.2) is equivalent to

$$\int_{\mathbb{R}^n} W(x)\exp(H\log g)(x)\,dx \le C_1\int_{\mathbb{R}^n} g(x)\,dx. \tag{4.4}$$

Substituting $y = |x|\xi$, $dy = |x|^n d\xi$, we can write the left side as

$$\begin{aligned}&\int_{\mathbb{R}^n} W(x)\exp\left[\frac{1}{B(|x|)}\int_{|y|\le|x|}\log g(y)\,dy\right]dx \\ &= \int_{\mathbb{R}^n} W(x)\exp\left[\frac{1}{|x|^n B(1)}\int_{B(1)}\log g(|x|\xi)|x|^n d\xi\right]dx\end{aligned} \tag{4.5}$$

and since

$$\int_{B(1)}\log(|\xi|^n)\,d\xi = -|B(1)|,$$

the integral in (4.5) becomes

$$\begin{aligned}&\int_{\mathbb{R}^n} W(x)\exp\left[\frac{1}{B(1)}\int_{B(1)}\log(|\xi|^n)g(|x|\xi)\,d\xi + 1\right]dx \\ &= e\int_{\mathbb{R}^n} W(x)\exp\left[\frac{1}{B(1)}\int_{B(1)}\log(|\xi|^n)g(|x|\xi)\,d\xi\right]dx \\ &\le \frac{e}{B(1)}\int_{\mathbb{R}^n} W(x)\int_{B(1)}|\xi|^n g(|x|\xi)\,d\xi dx,\end{aligned}$$

where the last estimate follows from Jensen's inequality. Now in polar coordinates the last integral is

$$\int_{\Sigma_{n-1}}\int_{\Sigma_{n-1}}\int_0^\infty t^{n-1}W(t\tau)\int_0^1 s^{2n-1}g(ts\sigma)\,ds\,dt\,d\sigma\,d\tau. \tag{4.6}$$

An interchange of order of integration and substitution $ts=r$ yields

$$\begin{aligned}
&\int_0^\infty t^{n-1}W(t\tau)\int_0^1 s^{2n-1}g(ts\sigma)\,ds\,dt\\
&\quad=\int_0^1 s^{2n-1}\int_0^\infty W\left(\frac{r}{s}\tau\right)\left(\frac{r}{s}\right)^{n-1}g(r\sigma)\frac{dr}{s}\,ds\\
&\quad=\int_0^1 s^{n-1}\int_0^\infty W\left(\frac{r}{s}\,\tau\right)r^{n-1}g(r\sigma)\,dr\,ds\\
&\quad=\int_0^\infty r^{n-1}g(r\sigma)\left(\int_0^1 s^{n-1}W\left(\frac{r\tau}{s}\right)ds\right)dr\\
&\quad=\left(\text{substituting } \tfrac{r}{s}=z,\ ds=-\tfrac{r}{z^2}\,dz\right)\\
&\quad\int_0^\infty r^{n-1}g(r\sigma)\left(\int_r^\infty\left(\frac{r}{z}\right)^{n-1}W(z\tau)r\frac{dz}{z^2}\right)dr\\
&\quad=\int_0^\infty r^{2n-1}g(r\sigma)\left(\int_r^\infty z^{-n-1}W(z\tau)dz\right)dr.
\end{aligned}$$

Substituting into (4.6) shows that

$$\begin{aligned}
&\int_{\mathbb{R}^n}W(x)\exp\left[\frac{1}{B(|x|)}\int_{B(|x|)}\log g(y)\,dy\right]dx\\
&\leq \frac{e}{B(1)}\int_{\Sigma_{n-1}}\int_{\Sigma_{n-1}}\int_0^\infty r^{n-1}g(r\sigma)r^n\left(\int_r^\infty\frac{z^{n-1}W(z\tau)\,dz}{z^{2n}}\right)d\tau\,d\sigma\\
&= \frac{e}{B(1)}\int_{\mathbb{R}^n}g(x)\left(|x|^n\int_{|y|\geq|x|}\frac{W(y)}{|y|^{2n}}\,dy\right)dx\\
&\leq A\frac{e}{B(1)}\int_{\mathbb{R}^n}g(x)\,dx,
\end{aligned}$$

where the last estimate follows from (4.3).

Conversely, if (4.2) or equivalently (4.4) is satisfied, then taking

$$g(x)=t^{-n}\chi_{(0,t)}(|x|)+e^{-2n}|x|^{-2n}t^n\chi_{(t,\infty)}(|x|)$$

with $t>0$ fixed and $x \in \mathbb{R}^n$ in (4.4) shows that the right side of (4.4) is ($g$ is radial!)

$$C_1 \int_{\Sigma_{n-1}} \int_0^\infty s^{n-1} g(s)\, ds d\sigma = C_1 |\Sigma_{n-1}| \left\{ \int_0^t s^{n-1} t^{-n} ds + \int_t^\infty e^{-2n} s^{n-1} t^n s^{-2n} ds \right\}$$
$$= C_1 |\Sigma_{n-1}| \left\{ \frac{1}{n} + e^{-2n}\, \frac{t^n t^{-n}}{n} \right\} = C(n).$$

Hence

$$\begin{aligned}
C(n) &\geq \int_{\mathbb{R}^n} W(x) \exp(P \log g)(x)\, dx \\
&= \int_{\Sigma_{n-1}} \int_0^\infty s^{n-1} W(s\sigma) \exp\left\{ \frac{1}{B(s)} \int_{\Sigma_{n-1}} \int_0^s r^{n-1} \log g(r)\, dr d\tau \right\} ds d\sigma \\
&= \int_{\Sigma_{n-1}} \int_0^\infty s^{n-1} W(s\sigma) \exp\left[ \frac{n}{s^n} \int_0^s r^{n-1} \log g(r)\, dr \right] ds\, d\sigma \\
&= \int_{\Sigma_{n-1}} \left\{ \int_0^t s^{n-1} W(s\sigma) \exp\left[ \frac{n}{S^n} \int_0^s r^{n-1} \log g(r)\, dr \right] ds\, d\sigma \right. \\
&\quad + \int_t^\infty s^{n-1} W(s\sigma) \exp\left[ \frac{n}{s^n} \left( \int_0^t r^{n-1} \log t^{-n} dr \right.\right. \\
&\quad \left.\left.\left. + \int_t^s r^{n-1} \log(e^{-2n} r^{-2n} t^n) dr \right) dy \right] ds \right\} d\sigma \\
&= \int_{\Sigma_{n-1}} (I_1 + I_2) d\sigma, \text{ respectively.}
\end{aligned}$$

Now $I_1 \geq 0$ so it suffices to estimate $\int_{\Sigma_{n-1}} I_2 d\sigma$ from below. We have

$$\begin{aligned}
I_2 &= \int_t^\infty s^{n-1} W(s\sigma) \exp\left\{ \frac{n}{s^n} \left[ \int_0^t r^{n-1} \log t^{-n} dr \right.\right. \\
&\quad \left.\left. + \int_t^s r^{n-1} \log e^{-2n} dr - 2n \int_t^s r^{n-1} \log r\, dr + \int_t^s r^{n-1} \log t^n dr \right] \right\} ds \\
&= \int_t^\infty s^{n-1} W(s\sigma) \exp\left\{ \frac{n}{s^n} \left[ \frac{t^n \log t^{-n}}{n} - 2n \frac{s^n - t^n}{n} \right.\right. \\
&\quad \left.\left. -2n \left( \frac{r^n \log r}{n} \Big|_t^s - \frac{r^n}{n^2} \Big|_t^s \right) + \frac{s^n - t^n}{n} \log t^n \right] \right\} ds \\
&= \int_t^\infty s^{n-1} W(s\sigma) \exp\left[ -2n + \frac{2nt^n}{s^n} - 2n \log s + 2 - \frac{2t^n}{s^n} + n \log t \right] ds \\
&\geq e^{2-2n} \int_t^\infty s^{n-1} W(s\sigma) \frac{t^n}{s^{2n}}\, ds.
\end{aligned}$$

Hence $\displaystyle\int_{\Sigma_{n-1}} I_2 d\sigma \geq e^{2-2n} t^n \int_{|x| \geq t} \frac{W(x)}{|x|^{2n}}\, dx$ and this implies (4.3).

**Acknowledgement.** The results contained in the paper formed a part of the talk held by the third author at the Oberwolfach meeting General Inequalities VII, November 1995. The first version of this paper was written during the second author's visit at the Department of Mathematics, University of West Bohemia, Pilsen, where he was supported by the Grant No. 0065 of the Czech Ministry of Education. The research of the first and third authors was supported by the Grants No. 201/94/0008 and No. 201/94/1066, respectively, of the Grant Agency of Czech Republic.

## References

[AK] J. Appell and A. Kufner, *On the two-dimensional Hardy operator in Lebesgue spaces with mixed norms.* Analysis **15** (1995), 91–98.

[CG] M. Christ and L. Grafakos, *Best constants for two nonconvolution inequalities.* Proc. AMS 6, **123** (1995), 1687–1693.

[HKK] H.P. Heinig, R. Kerman, M. Krbec, *Weighted exponential inequalities.* Preprint No. 79 (1992), Math. Inst. Acad. Sci. Prague, 30 pp.

[HLP] G.H. Hardy, J.E. Littlewood and G. Pólya, *Inequalities.* University Press, Cambridge 1959.

[OK] B. Opic and A. Kufner, *Hardy type inequalities.* Pitman Series **219**, Harlow (1990).

[S1] E.T. Sawyer, *Weighted Lebesgue and Lorentz norm inequalities for the Hardy operator.* Trans. Amer. Math. Soc. **281**(1) (1984), 329–337.

[S2] E.T. Sawyer, *Weighted inequalities for the two-dimensional Hardy operator.* Studia Math. **82** (1985), 1–16.

[Si] G. Sinnamon, *A weighted gradient inequality.* Proc. Royal Soc. Edinburgh **111** A (1989), 329–335.

Pavel Drábek, Department of Mathematics, University of West Bohemia, P.O. Box 314, 306 14 Plzeň, Czech Republic.

Hans P. Heinig, Department of Mathematics & Statistics, McMaster University, 1280 Main Street West, Hamilton, Ontario, Canada L8S 4K1.

Alois Kufner, Mathematical Institute, Czech Academy of Sciences, Žitná 25, 115 67 Praha 1, Czech Republic.

International Series of Numerical Mathematics, Vol. 123, © 1997 Birkhäuser Verlag Basel

# Integral inequalities for algebraic polynomials

Gradimir V. Milovanović

**Abstract**

In this paper we consider two extremal problems for algebraic polynomials in $L^2$ metrics.

(1) Let $\mathcal{P}_n$ be the class of all algebraic polynomials $P(x) = \sum_{k=0}^{n} a_k x^k$ of degree at most $n$ and $\|P\|_{d\sigma} = (\int_{\mathbb{R}} |P(x)|^2 d\sigma(x))^{1/2}$, where $d\sigma(x)$ is a nonnegative measure on $\mathbb{R}$. We determine the best constant in the inequality $|a_k| \le C_{n,k}(d\sigma)\|P\|_{d\sigma}$, for $k = 0, 1, \ldots, n$, when $P \in \mathcal{P}_n$ and such that $P(\xi_k) = 0$, $k = 1, \ldots, m$. The cases $C_{n,n}(d\sigma)$ and $C_{n,n-1}(d\sigma)$ were studied by Milovanović and Guessab [5], and only for the Legendre measure by Tariq [9].

(2) Let $\hat{\mathcal{P}}_N$ be the set of all monic algebraic polynomials of degree $N$ and $\varepsilon_s$ be $M$th roots of unity, i.e., $\varepsilon_s = \exp(i2\pi s/M)$, $s = 0, 1, \ldots, M-1$. Polynomials orthogonal on the radial rays in the complex plane with respect to the inner product

$$(f, g) = \int_0^a \left( \sum_{s=0}^{M-1} f(x\varepsilon_s)\overline{g(x\varepsilon_s)} \right) w(x)\, dx$$

have been introduced and studied recently in [3]. Here, $w$ is a weight function and $0 < a \le +\infty$. We consider the extremal problem

$$\inf_{P \in \hat{\mathcal{P}}_N} \int_0^a \left( \sum_{s=0}^{M-1} |P(x\varepsilon_s)|^2 \right) w(x)\, dx,$$

as well as some inequalities for coefficients of polynomials under some restrictions of the polynomial class.

## 1. Introduction

Let $\mathcal{P}_n$ be the class of algebraic polynomials $P(x) = \sum\limits_{k=0}^{n} a_k x^k$ of degree at most $n$. The first inequality of the form $|a_k| \le C_{n,k}\|P\|$ was given by V.A. Markov [2]. Namely, if

$$\|P\| = \|P\|_\infty = \max_{x \in [-1,1]} |P(x)|$$

and $T_n(x) = \sum_{\nu=0}^{n} t_{n,\nu} x^\nu$ denotes the $n$-th Chebyshev polynomial of the first kind, Markov proved that

$$|a_k| \le \begin{cases} |t_{n,k}| \cdot \|P\|_\infty & \text{if } n-k \text{ is even,} \\ |t_{n-1,k}| \cdot \|P\|_\infty & \text{if } n-k \text{ is odd.} \end{cases}$$

Precisely,

$$|a_k| \le \frac{2^{k-1} D_{n,k}}{k!} \|P\|_\infty, \tag{1.1}$$

where

$$D_{n,k} = \begin{cases} n \dfrac{\left(\frac{n+k-2}{2}\right)!}{\left(\frac{n-k}{2}\right)!} & \text{if } n-k \text{ is even,} \\[2ex] (n-1) \dfrac{\left(\frac{n+k-3}{2}\right)!}{\left(\frac{n-k-1}{2}\right)!} & \text{if } n-k \text{ is odd.} \end{cases}$$

For $k = n$ (1.1) reduces to the well-known Chebyshev inequality

$$|a_n| \le 2^{n-1} \|P\|_\infty. \tag{1.2}$$

Using a restriction on the polynomial class, inequality (1.1) could be improved. For example, taking $P(1) = 0$ or $P(-1) = 0$, Schur [8] found the following improvement of (1.2)

$$|a_n| \le 2^{n-1} \left( \cos \frac{\pi}{4n} \right)^{2n} \|P\|_\infty.$$

This result was extended later by Rahman and Schmeisser [7] for polynomials with real coefficients, which have at most $n-1$ distinct zeros in $(-1, 1)$.

Our interest are the corresponding inequalities in $L^2$ norm. Such one result was obtained by Labelle [1]

$$|a_k| \le \frac{1 \cdot 3 \cdot 5 \cdots (2k-1)}{k!} \left( k + \frac{1}{2} \right)^{1/2} \binom{[(n-k)/2] + k + 1/2}{[(n-k)/2]} \|P\|_2, \tag{1.3}$$

where $P \in \mathcal{P}_n$, $0 \le k \le n$,

$$\|P\| = \|P\|_2 = \left( \int_{-1}^{1} |P(x)|^2 \, dx \right)^{1/2},$$

and the symbol $[x]$ denotes the integral part of $x$. Equality in this case is attained only for the constant multiples of the polynomial

$$\sum_{\nu=0}^{[(n-k)/2]} (-1)^\nu (4\nu + 2k + 1) \binom{k + \nu - 1/2}{\nu} P_{k+2\nu}(x),$$

where $P_m(x)$ denotes the Legendre polynomial of degree $m$.

This result was improved by Tariq [9]. Under restriction $P(1) = 0$, he proved that

$$|a_n| \leq \frac{n}{n+1} \cdot \frac{(2n)!}{2^n (n!)^2} \left( \frac{2n+1}{2} \right)^{1/2} \|P\|_2, \tag{1.4}$$

with equality case

$$P(x) = P_n(x) - \frac{1}{n^2} \sum_{\nu=0}^{n-1} (2\nu+1) P_\nu(x).$$

In comparing with (1.4), we note that in inequality (1.3) for $k = n$ the factor $n/(n+1)$ does not exist. Tariq also obtained that

$$|a_{n-1}| \leq \frac{(n^2+2)^{1/2}}{n+1} \cdot \frac{(2n-2)!}{2^{n-1}((n-1)!)^2} \left( \frac{2n-1}{2} \right)^{1/2} \|P\|_2, \tag{1.5}$$

with equality case

$$P(x) = \frac{2n+1}{n^2+2} P_n(x) - P_{n-1}(x) + \frac{1}{n^2+2} \sum_{\nu=0}^{n-2} (2\nu+1) P_\nu(x).$$

In the absence of the hypothesis $P(1) = 0$ the factor $(n^2+2)^{1/2}/(n+1)$ appearing on the right-hand side of (1.5) is to be dropped.

Recently, Milovanović and Guessab [5] have extended this result to polynomials with real coefficients, which have $m$ zeros on real line.

In this paper we consider more general problem including $L^2$ norm of polynomials with respect to a nonnegative measure on the real line $\mathbb{R}$ and we give estimates for all coefficients. Also, we consider extremal problems for polynomials with respect to an inner product defined on the radial rays in the complex plane. Polynomials orthogonal with respect to such a product have been introduced and studied in [3].

## 2. Estimates for coefficients

For $\xi_k \in \mathbb{C}$, $k = 1, \ldots, m$, we consider a restricted polynomial class

$$\mathcal{P}_n(\xi_1, \ldots, \xi_m) = \{P \in \mathcal{P}_n \mid P(\xi_k) = 0,\ k = 1, \ldots, m\} \qquad (0 \leq m \leq n).$$

In the case $m = 0$ this class of polynomials reduces to $\mathcal{P}_n$. If $\xi_1 = \cdots = \xi_k = \xi$ $(1 \leq k \leq m)$ then the restriction on polynomials at the point $x = \xi$ becomes $P(\xi) = P'(\xi) = \cdots = P^{(k-1)}(\xi) = 0$.

Let

$$\prod_{i=1}^{m}(x-\xi_i) = x^m - s_1x^{m-1} + \cdots + (-1)^{m-1}s_{m-1}x + (-1)^m s_m$$

where $s_k$ denotes elementary symmetric functions of $\xi_1, \ldots, \xi_m$, i.e.,

$$s_k = \sum \xi_1 \cdots \xi_k \quad for \quad k = 1, \ldots, m. \tag{2.1}$$

For $k = 0$ we have $s_0 = 1$, and $s_k = 0$ for $k > m$ or $k < 0$.

Let $d\sigma(x)$ be a given nonnegative measure on the real line $\mathbb{R}$, with compact or infinite support, for which all moments $\mu_k = \int_{\mathbb{R}} x^k\, d\sigma(x)$, $k = 0, 1, \ldots$, exist and are finite, and $\mu_0 > 0$. In that case, there exist a unique set of orthonormal polynomials $\pi_n(\cdot) = \pi_n(\cdot; d\sigma)$, $n = 0, 1, \ldots$, defined by

$$\pi_n(x) = b_n^{(n)}(d\sigma)x^n + b_{n-1}^{(n)}(d\sigma)x^{n-1} + \cdots + b_0^{(n)}(d\sigma), \qquad b_n^{(n)}(d\sigma) > 0,$$
$$(\pi_n, \pi_m) = \delta_{nm}, \qquad n, m \geq 0,$$

where

$$(f, g) = \int_{\mathbb{R}} f(x)\overline{g(x)}\, d\sigma(x) \qquad (f, g \in L^2(\mathbb{R})). \tag{2.2}$$

Also, we put

$$\|P\|_{d\sigma} = \sqrt{(P, P)} = \left(\int_{\mathbb{R}} |P(x)|^2\, d\sigma(x)\right)^{1/2}. \tag{2.3}$$

We mention first that every polynomial $P(x) = \sum_{\nu=0}^{n} a_\nu x^\nu \in \mathcal{P}_n$ can be represented in the form

$$P(x) = \sum_{\nu=0}^{n} \alpha_\nu \pi_\nu(x; d\sigma),$$

where $\alpha_\nu = (P, \pi_\nu)$, $\nu = 0, 1, \ldots, n$, and the inner product is given by (2.2). Then we have

$$a_{n-k} = \sum_{i=0}^{k} \alpha_{n-i} b_{n-k}^{(n-i)}(d\sigma) = \left(P, \sum_{i=0}^{k} b_{n-k}^{(n-i)}(d\sigma)\pi_{n-i}\right), \quad k = 0, 1, \ldots, n, \tag{2.4}$$

where $\pi_\nu(\cdot) = \pi_\nu(\cdot; d\sigma)$. If we suppose that $P \in \mathcal{P}_n(\xi_1, \ldots, \xi_m)$, then we have

$$P(x) = Q(x)\prod_{k=1}^{m}(x - \xi_k), \tag{2.5}$$

where $Q(x) = a'_{n-m}x^{n-m} + a'_{n-m-1}x^{n-m-1} + \ldots + a'_0 \in \mathcal{P}_{n-m}$. Also, we have

$$\prod_{i=1}^{m}(x-\xi_i) = x^m - s_1x^{m-1} + \cdots + (-1)^{m-1}s_{m-1}x + (-1)^m s_m$$

where $s_k$, $k = 0, 1, \ldots, m$, denotes elementary symmetric functions (2.1). Using (2.5) we find that

$$a_{n-k} = \sum_{i=0}^{k} a'_{n-m-i}(-1)^{k-i} s_{k-i}, \qquad k = 0, 1, \ldots, n,$$

and $a'_k = 0$ for $k < 0$ and $k > n - m$. Now, the corresponding equalities (2.4) for polynomial $Q$ in the measure $d\hat{\sigma}(x)$, given by

$$d\hat{\sigma}(x) = \prod_{k=1}^{m} |x - \xi_k|^2 d\sigma(x) \tag{2.6}$$

become

$$a'_{n-m-i} = \Big(Q, \sum_{j=0}^{i} \hat{b}_{n-m-i}^{(n-m-j)} \hat{\pi}_{n-m-j}\Big), \quad i = 0, 1, \ldots, n - m,$$

where $\hat{\pi}_\nu(\cdot) = \pi_\nu(\cdot; d\hat{\sigma})$. According to (2.5), we have

$$a_{n-k} = \sum_{i=0}^{k} (-1)^{k-i} s_{k-i} \left( Q, \sum_{j=0}^{i} \hat{b}_{n-m-i}^{(n-m-j)} \hat{\pi}_{n-m-j} \right) = (Q, W_{n-m}) \tag{2.7}$$

where

$$\begin{aligned} W_{n-m}(x) &= \sum_{i=0}^{k} (-1)^{k-i} s_{k-i} \sum_{j=0}^{i} \hat{b}_{n-m-i}^{(n-m-j)} \hat{\pi}_{n-m-j}(x) \\ &= \sum_{j=0}^{k} \hat{\pi}_{n-m-j}(x) \sum_{i=j}^{k} (-1)^{k-i} s_{k-i} \hat{b}_{n-m-i}^{(n-m-j)} \end{aligned}$$

and $\hat{b}_\nu^{(\mu)} = 0$ for $\nu < 0$.

Now, we can prove the following result:

**Theorem 2.1.** *Let $P \in \mathcal{P}_n(\xi_1, \ldots, \xi_m)$ and $s_1, \ldots, s_m$ be given by (2.1). If the measure $d\hat{\sigma}(x)$ is given by (2.6) and $\|P\|_{d\sigma}$ is defined by (2.3), then*

$$|a_{n-k}| \le \left( \sum_{j=0}^{k} \left( \sum_{i=j}^{k} (-1)^{k-i} s_{k-i} \hat{b}_{n-m-i}^{(n-m-j)} \right)^2 \right)^{1/2} \|P\|_{d\sigma}, \tag{2.8}$$

*for $k = 0, 1, \ldots, n$, where $\hat{b}_\nu^\mu = b_\nu^\mu(d\hat{\sigma})$, $\nu = 0, 1, \ldots, \mu$, are the coefficients in the orthonormal polynomial $\hat{\pi}_\mu(\cdot) = \pi_\mu(\cdot; d\hat{\sigma})$. Inequality (2.8) is sharp and becomes* *an equality if and only if $P(x)$ is a constant multiple of the polynomial*

$$\left(\sum_{j=0}^{k}\hat{\pi}_{n-m-j}(x)\sum_{i=j}^{k}(-1)^{k-i}s_{k-i}\hat{b}_{n-m-i}^{(n-m-j)}\right)\prod_{k=1}^{m}(x-\xi_k).$$

*Proof.* Using Cauchy inequality from (2.7) we get

$$|a_{n-k}| \le C_{n,n-k}\|Q\|_{d\hat{\sigma}}$$

where

$$C_{n,n-k} = \|W_{n-m}\|_{d\hat{\sigma}} = \left(\sum_{j=0}^{k}\left(\sum_{i=j}^{k}(-1)^{k-i}s_{k-i}\hat{b}_{n-m-i}^{(n-m-j)}\right)^2\right)^{1/2}.$$

Since

$$\|Q\|_{d\hat{\sigma}}^2 = \int_{\mathbb{R}}|Q(x)|^2 d\hat{\sigma}(x) = \int_{\mathbb{R}}|P(x)|^2 d\sigma(x) = \|P\|_{d\sigma}^2$$

we obtain inequality (2.8). The extremal polynomial is

$$x \mapsto W_{n-m}(x)\prod_{k=1}^{m}(x-\xi_k). \qquad \square$$

For $k = 0$ and $k = 1$ this theorem gives the results obtained by Milovanović and Guessab [5] (see also [6], pp. 432–439). In the simplest case when $d\sigma(t) = dt$ on $(-1, 1)$, these results reduce to the Tariq's inequalities (1.4) and (1.5).

## 3. Extremal problems on the radial rays

Let $\hat{\mathcal{P}}_N$ be the set of all monic algebraic polynomials of degree $N$ and $\varepsilon_s$ be $M$th roots of unity, i.e., $\varepsilon_s = \exp(i2\pi s/M)$, $s = 0, 1, \ldots, M-1$. Polynomials orthogonal on the radial rays with respect to the inner product

$$(f, g) = \int_0^a \left(\sum_{s=0}^{M-1} f(x\varepsilon_s)\overline{g(x\varepsilon_s)}\right) w(x)\, dx \tag{3.1}$$

have been introduced and studied recently in [3]. Here, $w$ is a weight function and $0 < a \le +\infty$. The case $a = +\infty$ with an exponential weight gives the generalized Hermite polynomials on the radial rays (see [4]).

In this section we consider the following extremal problem

$$\inf_{P\in\hat{\mathcal{P}}_N} \int_0^a \left(\sum_{s=0}^{M-1} |P(x\varepsilon_s)|^2\right) w(x)\, dx. \tag{3.2}$$

**Theorem 3.1.** *Let $0 < a \le +\infty$. For each $P \in \hat{\mathcal{P}}_N$ we have*

$$\int_0^a \left(\sum_{s=0}^{M-1} |P(x\varepsilon_s)|^2\right) w(x)\, dx \ge \|\pi_N\|^2, \tag{3.3}$$

*where $(\pi_N)$ are monic orthogonal polynomials with respect to the inner product* (3.1). *In* (3.3) *equality holds if and only if $P(z) = \pi_N(z)$.*

*Proof.* Let $P \in \hat{\mathcal{P}}_N$ and $(\pi_k)_{k=0}^{+\infty}$ be a sequence of the monic polynomials orthogonal with respect to the inner product (3.1), where $0 < a \le +\infty$. Then $P(z)$ can be expressed in the form

$$P(z) = \sum_{k=0}^{N} c_k \pi_k(z),$$

where $c_k = (P, \pi_k)/\|\pi_k\|^2$, $k = 0, 1, \ldots, N$. Notice that $c_N = 1$. Since

$$\|P\|^2 = \sum_{k=0}^{N} |c_k|^2 \|\pi_k\|^2 \ge |c_N|^2 \|\pi_N\|^2 = \|\pi_N\|^2,$$

we obtain (3.3), with equality if and only if $P(z) = \pi_N(z)$. □

Using an explicit form of the norm in a particular case when $M = 4$ (see [3], Remark 5.4) we obtain the following result:

**Corollary 3.2.** *Let $P \in \hat{\mathcal{P}}_N$, $N = 4n + \nu$, $n = [N/4]$, and $\nu \in \{0, 1, 2, 3\}$. Then*

$$\int_0^1 \left(|P(x)|^2 + |P(ix)|^2 + |P(-x)|^2 + |P(-ix)|^2\right) dx \ge L_N,$$

*where*

$$L_N = \frac{4}{2N+1} \left(\prod_{k=n}^{2n-1} \frac{4(k-\nu+1)}{4k+2\nu+1}\right)^2$$

*for $N \ge 4$, and $L_N = 4/(2N+1)$ if $N \le 3$.*

Taking an arbitrary polynomial $P$ of degree $N$, i.e., $P \in \mathcal{P}_N$, and using Theorem 3.1 we can obtain an estimate for its leading coefficient $a_N$ like (1.2) or (1.3) without the factor $n/(n+1)$.

**Corollary 3.3.** *Let $0 < a \le +\infty$. For each $P_N(z) = \sum_{k=0}^{N} a_k z^k \in \mathcal{P}_N$ we have*

$$|a_N| \le \frac{\|P\|}{\|\pi_N\|}, \tag{3.4}$$

*where $(\pi_N)$ are monic orthogonal polynomials with respect to the inner product (3.1). In (3.4) equality holds if and only if $P(z)$ is a constant multiple of the polynomial $\pi_N(z)$.*

We consider now a restricted class of polynomials of degree at most $N$ such that $P(z)$ has a multiple zero of the order $m$ in the origin $z = 0$. Denote this class by $\mathcal{P}_N^m$. Evidently, each polynomial $P \in \mathcal{P}_N^m$ can be expressed in the form $P(z) = z^m Q(z)$, where $Q \in \mathcal{P}_{N-m}$.

Introduce the weight function $x \mapsto w_m(x) = x^m w(x)$ on $(0, a)$, where $m \in \mathbb{N}$, and let $(\pi_{k,m})_{k=0}^{+\infty}$ be the set of monic polynomials orthogonal with respect to the inner product (3.1) with the weight function $w_m$ instead of $w$. Then Corollary 3.3 can be interpreted in the following way:

**Corollary 3.4.** *Let $0 < a \le +\infty$ and $P$ be an arbitrary polynomial in $\mathcal{P}_N^m$. Then for its leading coefficient the following inequality*

$$|a_N| \le \frac{\|P\|}{\|\pi_{N-m,m}\|}$$

*holds, with equality if and only if $P(z)$ is a constant multiple of the polynomial $z^m \pi_{N-m,m}(z)$.*

## References

[1] G. Labelle, *Concerning polynomials on the unit interval.* Proc. Amer. Math. Soc. **20** (1969), 321–326.

[2] V.A. Markov, *On functions deviating least from zero in a given interval.* Izdat. Imp. Akad. Nauk, St. Petersburg 1892 (Russian) [German transl. Math. Ann. **77** (1916), 218–258].

[3] G.V. Milovanović, *A class of orthogonal polynomials on the radial rays in the complex plane.* J. Math. Anal. Appl., to appear.

[4] G.V. Milovanović, *Generalized Hermite polynomials on the radial rays in the complex plane.* In: Theory of Functions and Applications (H.B. Nersessian, ed.), Collection of Works Dedicated to the Memory of M.M. Djrbashian, Louys Publishing House, Yerevan, 1995, pp. 125–129.

[5] G.V. Milovanović and A. Guessab, *An estimate for coefficients of polynomials in $L^2$ norm.* Proc. Amer. Math. Soc. **120** (1994), 165–171.

[6] G.V. Milovanović, D. S. Mitrinović and Th. M. Rassias, *Topics in Polynomials: Extremal Problems, Inequalities, Zeros.* World Scientific, Singapore – New Jersey – London – Hong Kong, 1994.

[7] Q.I. Rahman and G. Schmeisser, *Inequalities for polynomials on the unit interval.* Trans. Amer. Math. Soc. **231** (1977), 93–100.

[8] I. Schur, *Über das Maximum des absoluten Betrages eines Polynoms in einem gegebenen Intervall.* Math. Z. **4** (1919), 217–287.

[9] Q.M. Tariq, *Concerning polynomials on the unit interval.* Proc. Amer. Math. Soc. **99** (1987), 293–296.

Gradimir V. Milovanović, Department of Mathematics, University of Niš, Faculty of Electronic Engineering, P.O. Box 73, 18000 Niš, Yugoslavia.

International Series of Numerical Mathematics, Vol. 123, © 1997 Birkhäuser Verlag Basel

# Inequalities of Gauß-Minkowski type

C.E.M. Pearce, J. Pečarić, and S. Varošanec

**Abstract**

Some new inequalities of Gauß-Minkowski type are derived. As our principal tools we employ some inequalities of Hölder type.

## 1. Introduction

The result known as Gauß's inequality and stated by him without proof in 1821 [3] is as follows.

*Suppose $f$ is a nonnegative, nonincreasing function. Then provided the integrals involved exist,*

$$\left[\int_0^\infty x^2 f(x)\, dx\right]^2 \leq \frac{5}{9}\int_0^\infty f(x)\, dx \int_0^\infty x^4 f(x)\, dx. \tag{1}$$

This inequality has a nice probabilistic interpretation. Write $\sigma$ for the probability measure with nonnegative support given by

$$d\sigma(x) = \frac{f(x)dx}{\int_0^\infty f(x)dx}$$

and denote by $\nu_r$ the $r$-th absolute moment $E_\sigma(X^r)$ of a random variable $X$ whose distribution is prescribed by $\sigma$. Then (1) becomes

$$\nu_2^2 \leq \frac{5}{9}\nu_4.$$

Gauß's inequality is the special case $n = 2$, $r = 4$ of the Gauß-Winckler inequality

$$[(n+1)\nu_n]^{1/n} \leq [(r+1)\nu_r]^{1/r} \quad (0 \leq n \leq r)$$

stated by Winckler [13] in 1866 (with an erroneous proof) and established by Faber [2] in 1922.

These ideas have proved remarkably seminal. In 1984 a long review paper of Beesack [13] included versions of more than 20 inequalities of Winckler type. In another direction, Pólya gave [7, I p.83] that for $f$ nonnegative and nondecreasing and $a, b \geq 0$,

$$\left[\int_0^1 x^{a+b} f(x)\, dx\right]^2 \geq \left[1 - \left(\frac{a-b}{a+b-1}\right)^2\right] \int_0^1 x^{2a} f(x)\, dx \int_0^1 x^{2b} f(x)\, dx.$$

This last result has itself given rise to several lines of development. Thus the following extension, derived by Varošanec and Pečarić [10] using the Hölder and Popoviciu inequalities, includes an additional monotone function $g$ and two sets of parameters. See [5], [6], [9], [11] and [12] for related results.

(i) *Let $g : [a, b] \to \mathbb{R}$ (with $a > 0$) be a nonnegative, nondecreasing, differentiable function and $f : [a, b] \to \mathbb{R}$ a nonnegative function such that $f/g'$ is nondecreasing. Further, let $p_1, \ldots, p_n$ be positive real numbers such that $\sum_{i=1}^n 1/p_i = 1$ and $a_1, \ldots, a_n$ real numbers such that $a_i > -1/p_i$. Then*

$$\int_a^b g(x)^{a_1+\ldots+a_n} f(x)\, dx \geq \frac{\prod_{i=1}^n (a_i p_i + 1)}{1 + \sum_{i=1}^n a_i} \prod_{i=1}^n \left[\int_a^b g(x)^{a_i p_i} f(x)\, dx\right]^{1/p_i}.$$

(ii) *If $g(a) = 0$ and $f/g'$ is nonincreasing but the other conditions above are unchanged, then the inequality in is reversed.*

A second line includes some results of Minkowski type. The following result was proved in [8].

(a) *Let $\varphi : [a, b] \to \mathbb{R}$ be a nonnegative, nondecreasing function and $x_i : [a, b] \to \mathbb{R}$ $(i = 1, \ldots, n)$ nonnegative, nondecreasing functions with continuous first derivatives. If $p > 1$, then*

$$\left(\int_a^b \left(\sum_{i=1}^n x_i(t)^p\right)' \varphi(t) dt\right)^{1/p} \geq \sum_{i=1}^n \left(\int_a^b (x_i^r(t)'\varphi(t) dt\right)^{1/p}.$$

(b) *Under the same conditions but with $\varphi$ nonincreasing and $x_i(a) = 0$ for all $i = 1, \ldots, n$, the inequality is reversed.*

In particular, we have in the special case $n = 2$, $p = 2$, $x_1(t) = \sqrt{f(t)}$, $x_2(t) = \sqrt{g(t)}$ that

$$\begin{aligned} &\int_a^b \left(\left(\sqrt{f(x)} + \sqrt{g(x)}\right)^2\right)' \varphi(x) dx \\ &\quad \geq \left(\sqrt{\int_a^b f'(x)\varphi(x) dx} + \sqrt{\int_a^b g'(x)\varphi(x) dx}\right)^2. \end{aligned} \tag{2}$$

If the squares are expanded and common terms cancelled, we derive

$$\begin{aligned} &\int_a^b \left(\sqrt{g(x)h(x)}\right)' \varphi(x)dx \\ &\qquad \geq \left(\int_a^b g'(x)\varphi(x)dx\right)^{1/2} \left(\int_a^b h'(x)\varphi(x)dx\right)^{1/2}. \end{aligned} \tag{3}$$

This is in turn the special case $n = 2$, $p_1 = p_2 = 2$, $x_1(t) = f(t)$, $x_2(t) = g(t)$ of the following result established in [5].

*Let $f : [a,b] \to \mathbb{R}$ be nonnegative and increasing and let $x_i : [a,b] \to \mathbb{R}$ $(1 \leq i \leq n)$ be nonnegative and increasing functions with a continuous first derivative. If $p_i$ $(1 \leq i \leq n)$ are positive real numbers such that $\sum_{i=1}^{n} 1/p_i = 1$ then*

$$\int_a^b \left(\prod_{i=1}^{n} (x_i(t))^{1/p_i}\right)' f(t)\, dt \geq \prod_{i=1}^{n} \left(\int_a^b x_i'(t) f(t)\, dt\right)^{1/p_i}.$$

In a reverse manner to the derivation of (3) from (2), we may deduce the following from (3).

*Under the conditions* (a) *above, we have*

$$\begin{aligned} &\int_a^b \left(\left(\sqrt{f(x)} - \sqrt{g(x)}\right)^2\right)' \varphi(x)dx \\ &\qquad \leq \left(\sqrt{\int_a^b f'(x)\varphi(x)dx} - \sqrt{\int_a^b g'(x)\varphi(x)dx}\right)^2. \end{aligned} \tag{4}$$

That is, under the assumptions (a), the inequality (2) is reversed if the two plus signs are both replaced with minus signs.

In this paper we derive some general theorems of Gauß-Minkowski type of which this last result appears as a special case. In Section 2 we consider results involving a first derivative and in Section 3 results involving higher derivatives. Our tools are some inequalities of Hölder type [4] which have already found interesting application in other problems.

Finally, in Section 4, we note briefly some further related results, involving positive signs only, which complement theorems derived recently in [8].

## 2. Basic results

**Theorem 1** *Let $f, g, \varphi : [a,b] \to \mathbb{R}$ be nonnegative functions with $\varphi$ nondecreasing and possessing a continuous first derivative. Further let $p$, $q$ real numbers satisfying $p + q = 1$.*

(a) *If $\varphi$ is nondecreasing and $p, q > 0$ then*

$$\begin{aligned}\int_a^b [(f^p(x) \pm g^p(x))]\,[(f^q(x) \pm g^q(x))]'\,\varphi(x)\,dx \\ \gtreqless \left[\left(\int_a^b f'(x)\varphi(x)\,dx\right)^p \pm \left(\int_a^b g'(x)\varphi(x)\,dx\right)^p\right] \\ \times \left[\left(\int_a^b f'(x)\varphi(x)\,dx\right)^q \pm \left(\int_a^b g'(x)\varphi(x)\,dx\right)^q\right].\end{aligned}$$

*Here the convention is that the greater than or equal possibility is associated with taking the plus throughout and the less than or equal with minus throughout.*

(b) *If $\varphi$ is a nonincreasing function, $p, q > 0$ and $f(a) = g(a) = 0$, the inequality is reversed.*

(c) *If $\varphi$ is nondecreasing and $pq < 0$, the inequality is reversed.*

*Proof.* Integration by parts gives

$$\begin{aligned}\int_a^b [(f^p(x) \pm g^p(x))\,(f^q(x) \pm g^q(x))]'\,\varphi(x)dx \\ = (f^p(x) \pm g^p(x))\,(f^q(x) \pm g^q(x))\,\varphi(x)|_a^b \\ - \int_a^b (f^p(x) \pm g^p(x))\,(f^q(x) \pm g^q(x))\,d\varphi(x).\end{aligned} \tag{5}$$

We now invoke two results from [4]. For $\varphi$ nondecreasing, an easy extension of Theorem 2.2 of [4] gives that when $f$, $g$ and $\varphi$ are nonnegative and $p, q > 0$, $p + q = 1$,

$$\begin{aligned}\left[\left(\int_a^b f(x)\,d\varphi\right)^p - \left(\int_a^b g(x)\,d\varphi\right)^p\right]\left[\left(\int_a^b f(x)\,d\varphi\right)^q - \left(\int_a^b g(x)d\varphi\right)^q\right] \\ \le \int_a^b (f^p(x) - g^p(x))\,(f^q(x) - g^q(x))\,\,d\varphi(x)\end{aligned}$$

whenever the relevant integrals exist. Similarly from Theorem 3.2 of [4], a reversed inequality arises if plus replaces minus throughout. Combining these results provides

$$\begin{aligned}\left[\left(\int_a^b f(x)d\varphi\right)^p \pm \left(\int_a^b g(x)d\varphi\right)^p\right]\left[\left(\int_a^b f(x)d\varphi\right)^q \pm \left(\int_a^b g(x)d\varphi\right)^q\right] \\ \gtreqless \int_a^b (f^p(x) \pm g^p(x))\,(f^q(x) \pm g^q(x))\,\,d\varphi(x).\end{aligned} \tag{6}$$

Taking (5) with (6) gives

$$\begin{aligned}
&\int_a^b [(f^p(x) \pm g^p(x))\,(f^q(x) \pm g^q(x))]'\,\varphi(x)\,dx\\
&\quad \gtreqless (f^p(b) \pm g^p(b))\,(f^q(b) \pm g^q(b))\,\varphi(b) - (f^p(a) \pm g^p(a))\,(f^q(a) \pm g^q(a))\,\varphi(a)\\
&\qquad - \left[\left(\int_a^b f(x)\,d\varphi(x)\right)^p \pm \left(\int_a^b g(x)\,d\varphi(x)\right)^p\right]\\
&\qquad\qquad \times \left[\left(\int_a^b f(x) d\varphi(x)\right)^q \pm \left(\int_a^b g(x)\,d\varphi(x)\right)^q\right].
\end{aligned} \tag{7}$$

Similarly we have the following from Theorem 4.1 of [4] *mutatis mutandis.*

*Let* $w_k,\ a_k,\ b_k > 0\ (1 \le k \le n)$, $p + q = 1$ *and* $w_1 a_1 - \sum_{i=2}^n w_i a_i > 0$, $w_1 b_1 - \sum_{i=2}^n w_i b_i > 0$. *Then*

$$\begin{aligned}
&[(w_1 a_1 - \textstyle\sum_{i=2}^n w_i a_i)^p \pm (w_1 b_1 - \sum_{i=2}^n w_i b_i)^p]\\
&\quad \times [(w_1 a_1 - \textstyle\sum_{i=2}^n w_i a_i)^q \pm (w_1 b_1 - \sum_{i=2}^n w_i b_i)^q]\\
&\qquad \lesseqgtr w_1 (a_1^p \pm b_1^p)(a_1^q \pm b_1^q) - \textstyle\sum_{i=2}^n w_i (a_i^p \pm b_i^p)(a_i^q \pm b_i^q).
\end{aligned}$$

Application of this inequality to the right-hand side of (7) provides

$$\begin{aligned}
&\int_a^b [(f^p(x) \pm g^p(x))(f^q(x) \pm g^q(x))]'\,\varphi(x)\,dx\\
&\quad \gtreqless \left[\left((\varphi(b) f(b) - \varphi(a) f(a) - \int_a^b f(x)\,d\varphi(x)\right)^p \pm \left(\varphi(b) g(b) - \int_a^b g(x)\,d\varphi(x)\right)^p\right]\\
&\quad \times \left[\left((\varphi(b) f(b) - \varphi(a) f(a) - \int_a^b f(x)\,d\varphi(x)\right)^q \pm \left(\varphi(b) g(b) - \int_a^b g(x)\,d\varphi(x)\right)^q\right]\\
&\quad = \left[\left(\int_a^b f'(x)\varphi(x)\,dx\right)^p \pm \left(\int_a^b g'(x)\varphi(x)\,dx\right)^p\right]\\
&\quad \times \left[\left(\int_a^b f'(x)\varphi(x)\,dx\right)^q \pm \left(\int_a^b g'(x)\varphi(x)\,dx\right)^q\right],
\end{aligned}$$

and we have (a).

For the second part, let $\varphi$ be a nonincreasing function and $f(a) = g(a) = 0$. Then

$$\begin{aligned}
&\int_a^b [(f^p(x) \pm g^p(x))\,(f^q(x) \pm g^q(x))]'\,\varphi(x)\,dx\\
&\quad = (f^p(b) \pm g^n(b))\,(f^q(b) \pm g^q(b)) + \int_a^b (f^p(x) \pm g^p(x))\,(f^q(x) \pm g^q(x))\,\,d\overline{\varphi}(x),
\end{aligned}$$

where $\overline{\varphi} = -\varphi$. By (6) we have

$$\begin{aligned}
&\int_a^b \left[(f^p(x) \pm g^p(x))\,(f^q(x) \pm g^q(x))\right]' \varphi(x) \\
&\quad \lesseqgtr (f^p(b) \pm g^p(b))\,(f^q(b) \pm g^q(b))\,\varphi(b) \\
&\qquad + \left[\left(\int_a^b f(x)d\overline{\varphi}(x)\right)^p \pm \left(\int_a^b g(x)\, d\overline{\varphi}(x)\right)^p\right] \\
&\qquad\qquad \times \left[\left(\int_a^b f(x)d\overline{\varphi}(x)\right)^q \pm \left(\int_a^b g(x)\, d\overline{\varphi}(x)\right)^q\right].
\end{aligned} \tag{8}$$

We now invoke the appropriate modification of two further results from [4]. Theorems 2.1 and 3.1 together provide the result below.

*Let $w_k, a_k, b_k > 0$ for $k = 1, \ldots, n$ and $p, q > 0, p + q = 1$. Then*

$$\begin{aligned}
&\left[\left(\sum_{k=1}^n w_k a_k\right)^p \pm \left(\sum_{k=1}^n w_k b_k\right)^p\right]\left[\left(\sum_{k=1}^n w_k a_k\right)^q \pm \left(\sum_{k=1}^n w_k b_k\right)^q\right] \\
&\qquad \gtreqless \sum_{k=1}^n w_k (a_k^p \pm b_k^p)(a_k^q \pm b_k^q).
\end{aligned} \tag{9}$$

Application of (9) to the right-hand side of (8) gives

$$\begin{aligned}
&\int_a^b \left[(f^p(x) \pm g^p(x))\,(f^q(x) \pm g^q(x))\right]' d\varphi(x) \\
&\quad \lesseqgtr \left[\left(f(b)\varphi(b) + \int_a^b f(x)d\overline{\varphi}(x)\right)^p \pm \left(g(b)\varphi(b) + \int_a^b g(x)d\overline{\varphi}(x)\right)^p\right] \\
&\qquad \times \left[\left(f(b)\varphi(b) + \int_a^b f(x)d\overline{\varphi}(x)\right)^q \pm \left(g(b)\varphi(b) + \int_a^b g(x)d\overline{\varphi}(x)\right)^q\right] \\
&\quad = \left[\left(f(b)\varphi(b) + \int_a^b f(x)d\varphi(x)\right)^p \pm \left(g(b)\varphi(b) - \int_a^b g(x)d\varphi(x)\right)^p\right] \\
&\qquad \times \left[\left(f(b)\varphi(b) - \int_a^b f(x)d\varphi(x)\right)^q \pm \left(g(b)\varphi(b) - \int_a^b g(x)d\varphi(x)\right)^q\right] \\
&\quad = \left[\left(\int_a^b f'(x)\varphi(x)dx\right)^p \pm \left(\int_a^b g'(x)\varphi(x)dx\right)^p\right] \\
&\qquad \times \left[\left(\int_a^b f'(x)\varphi(x)dx\right)^q \pm \left(\int_a^b g'(x)\varphi(x)dx\right)^q\right].
\end{aligned}$$

□

**Remark.** For $p = q = 1/2$ we recover both (2) and the inequality (4) noted at the end of the introduction.

## 3. Higher derivatives

**Theorem 2** *Let $p$ and $q$ be real numbers such that $p+q=1$ and let $\varphi, f, g : [a,b] \to \mathbb{R}$ be nonnegative functions with continuous derivatives of $n$-th order and properties (1)–(4) below.*

1. $(-1)^n \varphi^{(n)} < 0,\ f^{(n)} > 0,\ g^{(n)} > 0,\ (f^p \pm g^p)(f^q \pm g^q)^{(n)} > 0;$
2. $(-1)^k \varphi^{(k)}(b) < 0$ *for* $k = 1, 2, \ldots, n-1,\ \varphi(b) > 0;$
3. $f^{(k)}(a) = g^{(k)}(a) = 0$ *for* $k = 0, 1, \ldots, n-1$ *and for* $n \geq 2;$
4. $f^{(k)}(b) > 0,\ g^{(k)}(b) > 0$ *for* $k = 0, 1, \ldots n-1.$

*We have the following.*

(a) *If $p$ and $q$ are positive numbers, then*

$$\begin{aligned}
&\textstyle\int_a^b \left((f^p(x) \pm g^p(x))(f^q(x) \pm g^q(x))\right)^{(n)} \varphi(x)\, dx \\
&\gtreqless \Delta + \left[\left(\textstyle\int_a^b f^{(n)}(x)\varphi(x)dx\right)^p \pm \left(\textstyle\int_a^b g^{(n)}(x)\varphi(x)\, dx\right)^p\right] \\
&\qquad \times \left[\left(\textstyle\int_a^b f^{(n)}(x)\varphi(x)\, dx\right)^q \pm \left(\textstyle\int_a^b f^{(n)}(x)\varphi(x)\, dx\right)^q\right],
\end{aligned} \tag{10}$$

*where*

$$\Delta = \begin{cases} 0 & \text{for } n = 1 \\ \sum_{k=0}^{n-1} (-1)^k \varphi^{(k)}(b) \left[ ((f^p \pm g^p)(f^q \pm g^q))^{(n-k-1)}(b) - \right. \\ \quad ((f^{(n-k-1)}(b))^p \pm (g^{(n-k-1)}(b))^p) \\ \left. \times ((f^{(n-k-1)}(b))^q \pm (g^{(n-k-1)}(b))^q) \right] & \text{for } n \geq 2. \end{cases}$$

(b) *If $f^{(k)}(b) = g^{(k)}(b)$ for all $k = 0, 1, \ldots, n-1$ then $\Delta = 0$.*

(c) *If $pq < 0$, then the sign in (10) is reversed.*

*Proof.* We proceed in a similar fashion to the proof of the previous theorem, employing integration by parts and results from [4]. We have

$$
\begin{aligned}
&\int_a^b ((f^p(x) \pm g^p(x))(f^q(x) \pm g^q(x)))^{(n)} \varphi(x)dx \\
&= \sum_{k=0}^{n-1} (-1)^k \varphi^{(k)}(b)((f^p \pm g^p)(f^q \pm g^q))^{(n-k-1)}(b) \\
&\qquad - \int_a^b (f^p(x) \pm g^p(x))(f^q(x) \pm g^q(x))(-1)^{n-1}\varphi^{(n)}(x)dx \\
&\gtreqless \sum_{k=0}^{n-1} (-1)^k \varphi^{(k)}(b)((f^p \pm g^p)(f^q \pm g^q))^{(n-k-1)}(b) \\
&\qquad - \left[\left(\int_a^b f(x)(-1)^{n-1}\varphi^{(n)}(x)dx\right)^p \pm \left(\int_a^b g(x)(-1)^{n-1}\varphi^{(n)}(x)dx\right)^p\right] \\
&\qquad\qquad \times \left[\left(\int_a^b f(x)(-1)^{n-1}\varphi^{(n)}(x)dx\right)^q \right.\\
&\qquad\qquad\qquad \left. \pm \left(\int_a^b g(x)(-1)^{n-1}\varphi^{(n)}(x)dx\right)^q\right] \\
&= \Delta + \sum_{k=0}^{n-1} (-1)^k \varphi^{(k)}(b) \left((f^{(n-k-1)}(b))^p \pm (g^{(n-k-1)}(b))^p\right) \\
&\qquad \times \left((f^{(n-k-1)}(b))^q \pm (g^{(n-k-1)}(b))^q\right) \\
&\qquad - \left[\left(\int_a^b f(x)(-1)^{n-1}\varphi^{(n)}(x)dx\right)^p \pm \left(\int_a^b g(x)(-1)^{n-1}\varphi^{(n)}(x)dx\right)^p\right] \\
&\qquad\qquad \times \left[\left(\int_a^b f(x)(-1)^{n-1}\varphi^{(n)}(x)dx\right)^q \right.\\
&\qquad\qquad\qquad \left. \pm \left(\int_a^b g(x)(-1)^{n-1}\varphi^{(n)}(x)dx\right)^q\right] \\
&\gtreqless \Delta + \left[\left(\sum_{k=0}^{n-1} (-1)^k \varphi^{(k)}(b) f^{(n-k-1)}(b) - \int_a^b f(x)(-1)^{n-1}\varphi^{(n)}(x)dx\right)^p \right.\\
&\qquad \left. \pm \left(\sum_{k=0}^{n-1} (-1)^k \varphi^{(k)}(b) g^{(n-k-1)}(b) - \int_a^b g(x)(-1)^{n-1}\varphi^{(n)}(x)dx\right)^p\right] \\
&\qquad\qquad \times \left[\left(\sum_{k=0}^{n-1} (-1)^k \varphi^{(k)}(b) f^{(n-k-1)}(b) \right.\right.\\
&\qquad\qquad\qquad \left. - \int_a^b f(x)(-1)^{n-1}\varphi^{(n)}(x)dx\right)^q \\
&\qquad \left. \pm \left(\sum_{k=0}^{n-1} (-1)^k \varphi^{(k)}(b) g^{(n-k-1)}(b) - \int_a^b g(x)(-1)^{n-1}\varphi^{(n)}(x)dx\right)^q\right] \\
&= \Delta + \left[\left(\int_a^b f^{(n)}(x)\varphi(x)dx\right)^p \pm \left(\int_a^b g^{(n)}(x)\varphi(x)dx\right)^p\right] \\
&\qquad \times \left[\left(\int_a^b f^{(n)}(x)\varphi(x)dx\right)^q \pm \left(\int_a^b f^{(n)}(x)\varphi(x)dx\right)^q\right],
\end{aligned}
$$

giving the first part.

For the second, suppose that $f^{(k)}(b) = g^{(k)}(b) = B_k$ for all $k = 0, 1, \ldots, n-1$. By induction by $k$ we can prove that $(f^p g^q)^{(k)}(b) = B_k$ and $(f^q g^p)^{(k)}(b) = B_k$ (cf. [11]).

So we have

$$
\begin{aligned}
&((f^p \pm g^p)(f^q \pm g^q))^{(k)}(b) \\
&\qquad = f^{(k)}(b) + g^{(k)}(b) \pm (f^p g^q)^{(k)}(b) \pm (f^q g^p)^{(k)}(b) \\
&\qquad = 4B_k.
\end{aligned}
$$

It follows readily that $\Delta = 0$.

The final part is proved much as before. □

A similar theorem applies for the case when the signs of the derivatives of $\varphi$ are the reverses of those in Theorem 3.

**Theorem 3** *Let $p$ and $q$ be real numbers such that $p+q=1$ and let $\varphi, f, g : [a,b] \to \mathbb{R}$ be nonnegative functions with continuous derivatives of $n$-th order possessing properties (1)–(4) below.*

1. $(-1)^n \varphi^{(n)} > 0$, $f^{(n)} > 0$, $g^{(n)} > 0$, $(f^p \pm g^p)(f^q \pm g^q)^{(n)} > 0$;
2. $(-1)^k \varphi^{(k)}(b) \geq 0$ *for* $k = 0, 1, 2, \ldots, n-1$;
3. $f^{(k)}(a) = g^{(k)}(a) = 0$ *for* $k = 0, 1, \ldots, n-1$;
4. $f^{(k)}(b) > 0$, $g^{(k)}(b) > 0$ *for* $k = 0, 1, \ldots n-1$.

*We have the following.*

(a) *If $p$ and $q$ are positive numbers, then (10) holds with the inequality reversed;*

(b) *if $pq < 0$, then (10) holds.*

## 4. Further results

**Theorem 4** *Let $f : [a,b] \to \mathbb{R}$ be nonnegative and suppose $x_i : [a,b] \to \mathbb{R}$ ($i = 1, \ldots, n$) are nonnegative and nondecreasing functions with continuous first derivatives. Suppose $p < 1$.*

(a) *If $f$ is nondecreasing, then*

$$
\left( \int_b^a \left( \left( \sum_{i=1}^n x_i(t) \right)^p \right)' f(t)\, dt \right)^{1/p} \leq \sum_{i=1}^n \left( \int_a^b (x_i^p(t))' f(t) dt \right)^{1/p}.
$$

(b) *If $f$ is a nonincreasing function and $x_i(a) = 0$ for all $i = 1, \ldots, n$, then the inequality is reversed.*

The corresponding situations for $p > 1$ have been considered in [8]. The proof of the present result is the same as that of Theorem 4 in [8] except that in place of Bellman's and Minkowski's inequalities for $p > 1$ we use their analogues for $p < 1$.

Our final result similarly takes $p < 1$ in the situation treated for $p > 1$ in Theorem 5 of [8]. The proof is modified as for the theorem above.

**Theorem 5** *Let $f,\ x_i : [a,b] \to \mathbb{R}$ $(i = 1,\ldots,m)$ be nonnegative functions with continuous derivatives of order $n$ which satisfy the conditions (1)–(4) below.*

1. $(-1)^k f^{(k)}(b) \geq 0$ *for* $k = 1,\ldots,n-1$;
2. $x_i^{(k)}(a) = 0$ *for* $k = 0,\ldots,n-1,\ i = 1,\ldots,m$ *and* $\left(\sum_{i=1}^m x_i^p(b)\right)^{(k)} \geq 0$ *for* $k = 1,\ldots,n-1$;
3. $(x_i^p(t))^{(n)} \geq 0$ *for* $t \in [a,b]$ *and* $i = 1,\ldots,m$;
4. $(-1)^n f^{(n)}(t) \geq 0$ *for* $t \in [a,b]$.

(a) *If $p < 0$ and*

$$x_i^{(k)}(b) = x_j^{(k)}(b) \text{ for all } i,j \in \{1,2,\ldots,m\} \text{ and } k = 0,1,\ldots,n-1, \tag{11}$$

*then*

$$\int_a^b \left(\left(\sum_{i=1}^m x_i(t)\right)^p\right)^{(n)} f(t)\,dt \leq \left(\sum_{i=1}^m \left(\int_a^b (x_i(t)^p)^{(n)} f(t)\,dt\right)^{1/p}\right)^p. \tag{12}$$

(b) *If $p < 0$ but (11) is not satisfied, then*

$$\int_a^b \left(\left(\sum_{i=1}^m x_i(t)\right)^p\right)^{(n)} f(t)\,dt \leq \Delta + \left(\sum_{i=1}^m \left(\int_a^b (x_i^p(t))^{(n)} f(t)\,dt\right)^{1/p}\right)^p, \tag{13}$$

*where*

$$\Delta = \sum_{k=0}^{n-1}(-1)^k f^{(k)}(t)\left(\left(\left(\sum_{i=1}^m x_i(t)\right)^p\right)^{(n-k-1)} - \left(\sum_{i=1}^m \left((x_i^p(t))^{(n-k-1)}\right)^{1/p}\right)^p\right)\Bigg|_{t=b}.$$

(c) *If the directions of the inequalities in conditions (1) and (4) are reversed, then the inequalities in (12) and (13) are also reversed.*

(d) *If $0 < p < 1$, then the inequalities in (12) and (13) are reversed.*

**References**

[1] P.R. Beesack, *Inequalities for absolute moments of a distribution: from Laplace to von Mises.* J. Math. Anal. Appl. **98** (1984), 435–457.

[2] G. Faber, *Bemerkungen zu Sätzen der Gaußschen theoria combinationis observationum.* Heft 1, Sitzungber. Bayer Akad. Wiss. (1922), 7–21.

[3] C.F. Gauß, *Theoria combinationis observationum.* 1821; translated into German in: "Abhandlungen zur Methode der kleinsten Quadrate", 1887, pp. 9, 12, Neudruck, Würzburg, 1964.

[4] B. Guljaš, C.E.M. Pearce and J. Pečarić, *Some inequalities related to Hölder's inequality.* To appear in Revue Roumaine de Mathématiques Pures et Appliquées.

[5] J. Pečarić and S. Varošanec, *A Generalization of Pólya's Inequalities.* Inequalities and Applications, World Scientific Publishing Company, Singapore 1994, 501–504.

[6] J. Pečarić and S. Varošanec, *Remarks on Gauß-Winckler's and Stolarsky's Inequalities.* Utilitas Mathematica **48** (1995), 233–241.

[7] G. Pólya and G. Szegő, *Aufgaben und Lehrsätze aus der Analysis I, II*, Berlin (1925).

[8] S. Varošanec, *Inequalities of Minkowski type.* Real Analysis Exchange **20** (1995), 250–255.

[9] S. Varošanec and J. Pečarić, *On Pólya's Inequality.* Functional Analysis IV, Proc. of the Postgraduate School and Conference in Dubrovnik, November 10–17, 1993, Aarhus Univ. (1994), 275–278.

[10] S. Varošanec and J. Pečarić, *Gauß' and related inequalities.* Zeit. für Anal. und ihre Anwendungen **14** (1995), 175–183.

[11] S. Varošanec and J. Pečarić, *Some integral inequalities with bounds for moments of distributions.* To appear in J. Austral. Math. Soc., Ser. B.

[12] S. Varošanec, J. Pečarić and J. Šunce, *On Gauß-Polyá's inequality.* Preprint.

[13] A. Winckler, *Allgemeine Sätze zur Theorie der unregelmässigen Beobachtungsfehler.* Sitzungsber. Math.-Natur. Kl. K. Akad. Wiss. Wien, Zweite Abt. **53** (1866), 6–41.

C.E.M. Pearce, Department of Applied Mathematics, The University of Adelaide, Adelaide, Australia SA 5005.

J. Pečarić, Faculty of Textile Technology, University of Zagreb, Pierottijeva 6, 41000 Zagreb, Croatia.

S. Varošanec, Department of Mathematics, University of Zagreb, Bijenička cesta 30, 41000 Zagreb, Croatia.

International Series of Numerical Mathematics, Vol. 123, © 1997 Birkhäuser Verlag Basel

# Natural norm inequalities in nonlinear transforms

Saburou Saitoh

**Abstract**

We shall show the existence of natural norm inequalities in some general nonlinear transforms of reproducing kernel Hilbert spaces and as its applications we shall derive typical concrete norm inequalities in the nonlinear transforms.

## 1. General principles

Let $E$ be an arbitrary nonvoid abstract set and let $H_K(E)$ be a Hilbert (possibly finite-dimensional) space admitting a reproducing kernel $K(p,q)$ on $E$. Then, the Hilbert space $H_K(E)$ is composed of complex-valued functions $f(p)$ on $E$ such that

$$K(\cdot,q) \in H_K(E) \quad \text{for any fixed } q \in E \tag{1.1}$$

and, for any member $f$ of $H_K(E)$ and for any fixed point $q$ of $E$,

$$(f(\cdot), K(\cdot,q))_{H_K} = f(q). \tag{1.2}$$

In general, a reproducing kernel $K(p,q)$ on $E$ satisfying (1.1) and (1.2) is uniquely determined by the Hilbert space $H_K(E)$ and is a positive matrix in the sense that for any points $\{p_j\}_j$ of $E$ and for any complex numbers $\{C_j\}_j$

$$\sum_{j,j'} C_j \overline{C_{j'}} K(p_{j'}, p_j) \geq 0. \tag{1.3}$$

Conversely, a positive matrix $K(p,q)$ on $E$ satisfying (1.3) determines uniquely a functional Hilbert space (reproducing kernel Hilbert space $\equiv$ RKHS) $H_K(E)$ satisfying (1.1) and (1.2).

We shall consider the RKHS $H_K(E)$ as an input function space of the following nonlinear transform

$$\varphi : \quad f \in H_K(E) \longrightarrow \sum_{n=0}^{\infty} d_n(p) f(p)^n, \tag{1.4}$$

where $\{d_n(p)\}$ are any complex-valued functions on $E$. In this nonlinear transform $\varphi$, we shall show that the images $\varphi(f)$, $f \in H_K(E)$, belong to a Hilbert space $\mathbb{H}$ which is naturally determined by the nonlinear transform $\varphi$ and there exists a natural norm inequality between the two norms $\|\varphi(f)\|_{\mathbb{H}}$ and $\|f\|_{H_K}$.

In order to show these facts we need the three basic ideas of Aronszajn [1] for reproducing kernels; that is, sums, products and restrictions of reproducing kernels.

For two positive matrices $K_1(p,q)$ and $K_2(p,q)$ on $E$, the sum $K_3(p,q) = K_1(p,q)+K_2(p,q)$ is, of course, a positive matrix on $E$. The RKHS $H_{K_3}$ admitting the reproducing kernel $K_3(p,q)$ on $E$ is composed of all functions

$$f = f_1 + f_2 \quad (f_j \in H_{K_j}) \tag{1.5}$$

and the norm in $H_{K_3}$ is given by

$$\|f\|^2_{H_{K_3}} = \min\{\|f_1\|^2_{H_{K_1}} + \|f_2\|^2_{H_{K_2}}\}, \tag{1.6}$$

where the minimum is taken over all the expressions (1.5) for $f$.

The product $K_4(p_1,p_2;q_1,q_2) = K_1(p_1,q_1)K_2(p_2,q_2)$ on $(E\times E)\times(E\times E)$ is, of course, a positive matrix on $E \times E$. The RKHS $H_{K_4}$ admitting the reproducing kernel $K_4(p_1,p_2;q_1,q_2)$ on $E \times E$ is a tensor product $H_{K_1} \otimes H_{K_2}$ and the Hilbert space is composed of all functions

$$f(p_1,p_2) = \sum_{n=1}^{\infty} f_{1,n}(p_1)f_{2,n}(p_2) \quad (f_{j,n} \in H_{K_j}) \tag{1.7}$$

having finite norms

$$\|f\|^2_{H_{K_4}} = \sum_{n=1}^{\infty} \|f_{1,n}\|^2_{H_{K_1}} \|f_{2,n}\|^2_{H_{K_2}} < \infty. \tag{1.8}$$

The restriction $K_5(p,q) = K_4(p,p;q,q)$ to the diagonal set $E$ of $E \times E$ is a positive matrix and the RKHS $H_{K_5}$ admitting the reproducing kernel $K_5(p,q)$ on $E$ is composed of all functions $f(p) \equiv f(p,p)$ in (1.7) satisfying (1.8). The norm in $H_{K_5}$ is given by

$$\|f\|^2_{H_{K_5}} = \min \sum_{n=1}^{\infty} \|f_{1,n}\|^2_{H_{K_1}} \|f_{2,n}\|^2_{H_{K_2}}, \tag{1.9}$$

where the minimum is taken over all the expressions (1.7) satisfying (1.8) for $f(p) = f(p,p)$ on $E$.

For a RKHS $H_K$ on $E$ and for any complex-valued function $s(p)$ on $E$,

$$K_s(p,q) = s(p)\overline{s(q)}K(p,q) \quad \text{on} \quad E \times E$$

is the reproducing kernel for the Hilbert space $H_{K_s}$ comprising of all functions $f_s(p)$ on $E$ which are expressible in the form

$$f_s(p) = f(p)s(p) \text{ on } E \text{ for } f \in H_K$$

and we have the inequality

$$\|f_s\|_{H_{K_s}} \leq \|f\|_{H_K},$$

as we see from (1.9).

For $n$-times sum and $n$-times product, the circumstances are similar. Hence, we have, in particular, for any $f_j \in H_{K_j}$ $(j = 1, 2, \ldots, N)$

$$\|\sum_{j=1}^{N} f_j\|^2_{H_{\left(\sum_{j=1}^{N} K_j\right)}} \leq \sum_{j=1}^{N} \|f_j\|^2_{H_{K_j}} \tag{1.10}$$

and

$$\|f^n\|^2_{H_{K^n}} \leq \|f\|^{2n}_{H_K}. \tag{1.11}$$

Hence, we obtain, in general

**Theorem.** *If*

$$\sum_{n=0}^{\infty} |h_n(p)|^2 K(p,p)^n < \infty \quad \textit{on } E$$

*and if, for $f \in H_K$*

$$\sum_{n=0}^{\infty} (\|f\|_{H_K})^{2n} < \infty,$$

*then, for the nonlinear transform $\varphi(f)$ in* (1.4),

$$\varphi(f) = \sum_{n=0}^{\infty} d_n(p) f(p)^n$$

*converges absolutely on $E$, and*

$$\varphi(f) \in H_{\mathbb{K}_d} \tag{1.12}$$

*and*

$$\|\varphi(f)\|^2_{H_{\mathbb{K}_d}} \leq \sum_{n=0}^{\infty} (\|f\|_{H_K})^{2n}, \tag{1.13}$$

*where $H_{\mathbb{K}_d}$ is the* RKHS *admitting the reproducing kernel*

$$\mathbb{K}_d(p,q) = \sum_{n=0}^{\infty} d_n(p)\overline{d_n(q)} K(p,q)^n \quad \textit{on } E,$$

*which converges absolutely on $E \times E$.*

In the Theorem, the concrete realization of the norm in the RKHS $H_{\mathbb{K}_{\rm d}}$ is, in general, involved. See, Hejhal [4] and Saitoh [6] for a profound result for $\hat{K}(z,\bar{u})^2$ in the case of the classical Szegő reproducing kernel $\hat{K}(z,\bar{u})$ and for a prototype result of the Theorem, respectively.

At this moment, recall the fact that if

$$\mathbb{K}_{\rm d}(p,q) \ll \tilde{K}(p,q) \quad \text{on} \quad E; \tag{1.14}$$

that is,

$$\tilde{K}(p,q) - \mathbb{K}_{\rm d}(p,q)$$

is a positive matrix on $E$, then we have

$$H_{\mathbb{K}_{\rm d}} \subset H_{\tilde{K}}$$

(as the classes of functions) and, as we see from (1.5) and (1.6)

$$\|f\|_{H_{\tilde{K}}} \le \|f\|_{H_{\mathbb{K}_{\rm d}}} \quad \text{for all} \quad f \in H_{\mathbb{K}_{\rm d}}.$$

Hence, for some suitable reproducing kernel $\tilde{K}(p,q)$ satisfying (1.14) whose norm can be determined, in a reasonable way, we can obtain the inequality

$$\|\varphi(f)\|^2_{H_{\tilde{K}}} \le \sum_{n=0}^{\infty} (\|f\|_{H_K})^{2n},$$

in the Theorem.

When all the coefficients $\{d_n\}$ are constants, as a typical reproducing kernel $\tilde{K}(p,q)$ satisfying (1.14) and a large reproducing kernel for $K(p,q)$, we can consider the exponential of $K(p,q)$

$$\exp K(p,q) = 1 + K(p,q) + \frac{K(p,q)^2}{2!} + \cdots$$

which is a positive matrix on $E$. Note here that the constants $(n!)^{-1}$ are not essential in our arguments. Then, we have, in particular

**Corollary 1.1.** *If $\{d_n(p)\}$ are all constants, we have, in the Theorem*

$$\|\varphi(f)\|^2_{H_{\exp K}} \le \sum_{n=0}^{\infty} |d_n|^2 n! (\|f\|_{H_K})^{2n},$$

*if the right hand side converges.*

**Corollary 1.2.** *Let*

$$N(z) = \sum_{n=0}^{\infty} a_n z^n$$

*be analytic around $z = 0$ which converges on the disc $\{|z| < R\}$. We define the analytic function $N^+(z)$ on $\{|z| < R\}$ by*

$$N^+(z) = \sum_{n=0}^{\infty} |a_n| z^n.$$

*We assume that for a reproducing kernel $K(p,q)$ on $E$,*

$$K(p,p) < R.$$

*Then, $N^+(K(p,q))$ converges absolutely on $E \times E$ and is a positive matrix on $E$. For the* RKHS *$H_{N^+(K)}$ admitting the reproducing kernel $N^+(K(p,q))$ and for a function $f$ in $H_K$ satisfying*

$$N^+(\|f\|^2_{H_K}) < \infty,$$

*we have the norm inequality*

$$\|N(f)\|^2_{H_{N^+(K)}} \leq N^+(\|f\|^2_{H_K}).$$

In the theory of nonlinear partial differential equations, we meet to nonlinear transforms, for example, for $u(x,t)$

$$u \longrightarrow u_t + 6uu_x + u_{xxx}$$

and

$$u \longrightarrow u_{tt} - u_{xx} + m^2 \sin u \quad (m > 0; \ \text{constant}).$$

For such nonlinear transforms we shall show that similar results are valid as in our Theorem.

In order simply to state the result, we shall assume that $E$ is an open interval on $\mathbb{R}$. Then, for the smoothness of a RKHS $H_K(E)$, note that if

$$\frac{\partial^{(j+j')} K(x,y)}{\partial x^j \partial y^{j'}} \quad (j, j' \leq n)$$

are continuously differentiable on $E \times E$, then for any member $f$ of $H_K(E)$, $f^{(j)}$ $(j \leq n)$ are also continuously differentiable on $E$ (Kreĭn [5]), and we have

$$f^{(n)} \in H_{K^{n,n}}$$

and

$$\|f^{(n)}\|_{K^{n,n}} \leq \|f\|_{H_K},$$

for the RKHS $H_{K^{n,n}}$ admitting the reproducing kernel

$$K^{n,n}(x,y) = \frac{\partial^{2n} K(x,y)}{\partial x^n \partial y^n} \quad \text{on } E$$

(Saitoh [14]). Hence, for example, in the nonlinear transform

$$\psi : f \in H_K(E) \longrightarrow h_1(x) f''(x) + h_2(x) f'(x)^2 + h_3(x)|f(x)|^2 \tag{1.15}$$

for any complex-valued functions $\{h_j(x)\}$ on $E$, the images $\psi(f)$ belong to the RKHS $H_{\psi^+(K)}$ admitting the reproducing kernel

$$\psi^+(K(x,y)) = h_1(x)\overline{h_1(y)} K^{2,2}(x,y) \tag{1.16}$$

$$+ h_2(x)\overline{h_2(y)} K^{1,1}(x,y)^2 + h_3(x)\overline{h_3(y)} K(x,y)\overline{K(x,y)},$$

and we obtain the inequality

$$\|\psi(f)\|^2_{H_{\psi^+(K)}} \leq \|f\|^2_{H_K}(1 + 2\|f\|^2_{H_K}). \tag{1.17}$$

In some general linear transform of Hilbert spaces we could get essentially isometrical identities between the input and the output function spaces (see [9] and [13]), but in our nonlinear transforms we get norm inequalities, essentially and to determine the cases making the equalities hold in the inequalities is, in general, involved, in even the case of a finite dimensional RKHS $H_K$ and we need case by case arguments to determine the cases. See, for example, [11], [10] and [6]. However, for many cases (not always), for the reproducing kernels $f(p) = K(p,q)(q \in E)$ equalities hold in our inequalities. See [10], for example.

## 2. Concrete examples

Subsequently we shall derive typical norm inequalities in nonlinear transforms by applying our Theorem to several typical reproducing kernels.

### 2.1 $\ell^2$ space

The Kronecker delta $\delta_{jj'}$ $(j, j' \geq 1$; integers) is apparently the reproducing kernel for the usual $\ell^2$ space.

We shall consider the nonlinear transform of $\ell^2$

$$\varphi(\mathbf{c}) = \sum_{n=0}^{\infty} d_n \mathbf{c}^n \quad (d_n : \ \text{constants}) \tag{2.1.1}$$

in the sense of Hadamard's product

$$\mathbf{c}^n = \{c_j^n\}_{j=1}^{\infty}.$$

We consider the positive matrix

$$\Delta_{jj'} = \sum_{n=0}^{\infty} p_n (\delta_{jj'})^n \quad (p_n > 0; \;\; \text{constants}),$$

with

$$p = \sum_{n=1}^{\infty} p_n < \infty.$$

Then, by our Theorem we obtain the inequality

$$\frac{1}{p_0}|d_0|^2 + \frac{1}{p}\sum_{j=1}^{\infty} | \sum_{n=1}^{\infty} d_n c_j^n |^2 \leq \sum_{n=0}^{\infty} \frac{|d_n|^2}{p_n} (\sum_{j=1}^{\infty} |c_j|^2)^n,$$

if the right hand side converges.

## 2.2 Positive-definite Hermitian matrices

Recall first that to every positive-definite Hermitian matrix, there corresponds a uniquely determined inner product space admitting a reproducing kernel. See Saitoh [8,13] for this idea.

For a nonlinear transform of $\mathbb{C}^N$

$$\varphi(\mathbf{x}) = \sum_{n=0}^{\infty} d_n \mathbf{x}^n \quad (d_n : \;\; \text{constants}) \tag{2.2.1}$$

in the sense of Hadamard's product such that

$$\mathbf{x}^n = (x_1^n, x_2^n, \ldots, x_N^n)^T,$$

and for the corresponding positive matrix on $E$

$$\mathbb{K}(\nu, \mu) = \sum_{n=0}^{\infty} p_n (\tilde{a}_{\nu\mu})^n \quad (p_n > 0; \;\; \text{constants}) \tag{2.2.2}$$

with

$$\sum_{n=0}^{\infty} p_n |(\tilde{a}_{\nu\nu})^n| < \infty \quad \text{on} \quad E,$$

we have the inequality, by our Theorem

$$\left(\sum_{n=0}^{\infty} d_n \mathbf{x}^n\right)^* \left(\sum_{n=0}^{\infty} p_n \tilde{A}^n\right)^{\sim} \left(\sum_{n=0}^{\infty} d_n \mathbf{x}^n\right) \leq \sum_{n=0}^{\infty} \frac{|d_n|^2}{p_n} (\mathbf{x}^* A \mathbf{x})^n, \tag{2.2.3}$$

if the right hand side converges. Here, $\tilde{A}^n = \|\tilde{a}_{\nu\mu}^n\|$.

### 2.3 The reproducing kernel $\log \dfrac{\min(x,y)}{a}$

As we can see directly

$$K(x,y) = \log \frac{\min(x,y)}{a} \quad (0 < a \le x, y \le b \le \infty) \tag{2.3.1}$$

is the reproducing kernel for the Hilbert space $H(a,b)$ comprising of all real-valued and absolutely continuous functions $f(x)$ on $[a,b)$ such that $f(a) = 0$ and $f$ have finite norms

$$\left\{ \int_a^b f'(x)^2 x dx \right\}^{\frac{1}{2}} < \infty. \tag{2.3.2}$$

Meanwhile,

$$\mathbb{K}(x,y) = \frac{\min(x,y)}{a} - 1 \quad (0 < a \le x, y \le b \le \infty) \tag{2.3.3}$$

is the reproducing kernel for the Hilbert space $H_{\mathbb{K}}$ comprising of all real-valued and absolutely continuous functions $f(x)$ on $[a,b)$ such that $f(0) = 0$ and $f$ have finite norms

$$\left\{ a \int_a^b f'(x)^2 dx \right\}^{\frac{1}{2}} < \infty. \tag{2.3.4}$$

Then we have the identity

$$\exp K(x,y) = \mathbb{K}(x,y) + 1. \tag{2.3.5}$$

Since

$$H_{\mathbb{K}} \cap \mathbb{C} = \{0\},$$

the RKHS admitting the reproducing kernel $\mathbb{K}(x,y) + 1$ is the direct sum of $H_{\mathbb{K}}$ and $\mathbb{C}$. Hence, for the nonlinear transform of $f \in H(a,b)$

$$\varphi(f) = \sum_{n=0}^{\infty} d_n f(x)^n \quad (d_n : \ \text{constants}),$$

we have the inequality

$$|d_0|^2 + a \int_a^b \Big(\sum_{n=1}^{\infty} d_n f(x)^n\Big)'^2 dx \le \sum_{n=0}^{\infty} |d_n|^2 n! \left( \int_a^b f'(x)^2 x dx \right)^n, \tag{2.3.6}$$

if the right hand side converges.

In particular, we have

$$1 + a \int_a^b (\exp f(x))'^2 dx \le \exp\left\{ \int_a^b f'(x)^2 x dx \right\}$$

(Saitoh [16]).

## 2.4 The reproducing kernel $\min(x, y)$

Note that

$$K(x,y) = \min(x,y) \quad (0 \leq x, y)$$

is the reproducing kernel for the Hilbert space $H_0$ comprising of all real-valued and absolutely continuous functions $f(x)$ on $[0,\infty)$ such that $f(0) = 0$ and $f$ have finite norms

$$\left\{\int_0^\infty f'(x)^2 dx\right\}^{\frac{1}{2}} < \infty.$$

Then,

$$\mathbb{K}_N(x,y) = \sum_{n=1}^{N} K(x,y)^n \tag{2.4.1}$$

$$= \min\left\{\frac{x(1-x^N)}{1-x}, \frac{y(1-y^N)}{1-y}\right\}$$

is the reproducing kernel for the Hilbert space $H_{0,N}$ comprising of all real-valued and absolutely continuous functions $f(x)$ on $[0,\infty)$ such that $f(0) = 0$ and $f$ have finite norms

$$\left\{\int_0^\infty f'(x)^2 \left(\frac{x(1-x^N)}{1-x}\right)'^{-1} dx\right\}^{\frac{1}{2}} < \infty. \tag{2.4.2}$$

Hence, for the nonlinear transform of $f \in H_0$

$$\varphi(f) = \sum_{n=1}^{N} f(x)^n = \frac{f(x)(1-f(x)^N)}{1-f(x)},$$

we have the inequality

$$\int_0^\infty \left(\frac{f(x)(1-f(x)^N)}{1-f(x)}\right)'^2 \left(\frac{x(1-x^N)}{1-x}\right)'^{-1} dx$$

$$\leq \frac{\int_0^\infty f'(x)^2 dx \left\{1 - \left(\int_0^\infty f'(x)^2 dx\right)^N\right\}}{1 - \int_0^\infty f'(x)^2\, dx}.$$

In particular, for $f \in H_0$ satisfying

$$0 < \int_0^1 f'(x)^2 dx < 1,$$

we have the inequality, by letting $N \to \infty$

$$\int_0^1 \left(\frac{f(x)}{1-f(x)}\right)'^2 (1-x)^2\, dx \leq \frac{\int_0^1 f'(x)^2\, dx}{1 - \int_0^1 f'(x)^2\, dx}. \tag{2.4.4}$$

### 2.5 The reproducing kernel $\exp\{\min(x,y)\}$

From the identity

$$\exp\{\min(x,y)\} = \{\min(e^x, e^y) - 1\} + 1 \tag{2.5.1}$$

and from the arguments in (2.3), for the nonlinear transform of $f \in H_0$ (the RKHS in (2.4))

$$\varphi(f) = \sum_{n=0}^{\infty} d_n f(x)^n \quad (d_n: \ \text{constants}),$$

we have the inequality

$$|d_0|^2 + \int_0^\infty \left(\sum_{n=1}^{\infty} d_n f(x)^n\right)'^2 e^{-x}\, dx \tag{2.5.2}$$

$$\leq \sum_{n=0}^{\infty} |d_n|^2 n! \left\{\int_0^\infty f'(x)^2\, dx\right\}^n.$$

In particular, we have

$$1 + \int_0^\infty (e^{f(x)})'^2 e^{-x} dx \leq e^{\int_0^\infty f'(x)^2\, dx}.$$

### 2.6 The reproducing kernel $\log \dfrac{1}{(1-\overline{u}z)^2}$

Note that

$$K(z,\overline{u}) = \log \frac{1}{(1-\overline{u}z)^2} \quad \text{(principal value)} \tag{2.6.1}$$

is the reproducing kernel for the Hilbert space $H_D$ comprising of all analytic functions $f(z)$ on the unit disc $U = \{|z| < 1\}$ such that $f(0) = 0$ and $f(z)$ have finite norms

$$\left\{\frac{1}{2\pi}\iint_U |f'(z)|^2 dxdy\right\}^{\frac{1}{2}} < \infty. \tag{2.6.2}$$

Meanwhile,

$$\frac{1}{(1-\overline{u}z)^2} = \exp K(z,\overline{u}) \tag{2.6.3}$$

is the reproducing (Bergman) kernel for the Hilbert space comprising of all analytic functions $f(z)$ on $U$ having finite norms

$$\left\{\frac{1}{\pi}\iint_U |f(z)|^2\, dxdy\right\}^{\frac{1}{2}} < \infty. \tag{2.6.4}$$

Hence, for the nonlinear transform for $f \in H_D$

$$\varphi(f) = \sum_{n=0}^{\infty} d_n f(x)^n \quad (d_n : \ \text{constants}),$$

we have the inequality

$$|d_0|^2 + \frac{1}{\pi} \iint_U \left| \sum_{n=1}^{\infty} d_n f(z)^n \right|^2 dx\, dy \\ \leq \sum_{n=0}^{\infty} |d_n|^2 n! \left\{ \frac{1}{2\pi} \iint_U |f'(z)|^2 \, dx\, dy \right\}^n.$$

In particular, we have

$$\frac{1}{\pi} \iint_U |e^{f(z)}|^2 \, dx\, dy \leq \exp\left\{ \frac{1}{2\pi} \iint_U |f'(z)|^2 \, dx\, dy \right\}$$

(Saitoh [7]). For more general reproducing kernels of exponential type, see Saitoh [12,13]. See also [2] in connection with the theory of univalent functions on $U$.

## 2.7 The Bergman-Selberg kernels

For $q > \frac{1}{2}$,

$$K_q(z, \overline{u}) = \frac{\Gamma(2q)}{(z + \overline{u})^{2q}}$$

is the Bergman-Selberg reproducing kernel on the half plane $R^+ = \{\mathrm{Re}\, z > 0\}$ comprising of all analytic functions $f(z)$ on $R^+$ with finite norms

$$\|f\|^2_{H_{K_q}} = \frac{1}{\pi \Gamma(2q-1)} \iint_{R^+} |f(z)|^2 [2\, \mathrm{Re}\, z]^{2q-2} \, dx\, dy. \tag{2.7.1}$$

For $q = \frac{1}{2}$, $K_{1/2}(z, \overline{u})$ is the Szegő reproducing kernel on $R^+$ comprising of all analytic functions $f(z)$ on $R^+$ with finite norms

$$\|f\|^2_{H_{K_{1/2}}} = \frac{1}{2\pi} \sup_{x>0} \int_{-\infty}^{\infty} |f(x+iy)|^2 \, dy. \tag{2.7.2}$$

Then, a member $f(z)$ of $H_{K_{1/2}}$ has nontangential boundary values on the imaginary axis belonging to $L_2$ and we have

$$\|f\|^2_{H_{K_{1/2}}} = \frac{1}{2\pi} \int_{-\infty}^{\infty} |f(iy)|^2 \, dy$$

(cf. [3]). Note that

$$\frac{\partial^2 K_1(z,\overline{u})}{\partial z \partial \overline{u}} = \frac{6}{(z+\overline{u})^4}, \frac{\partial^4 K_1(z,\overline{u})}{\partial z^2 \partial \overline{u}^2} = \frac{120}{(z+\overline{u})^6},$$

$$\frac{\partial^2 K_{1/2}(z,\overline{u})}{\partial z \partial \overline{u}} = \frac{2}{(z+\overline{u})^3}, \quad \text{and} \quad \frac{\partial^4 K_{1/2}(z,\overline{u})}{\partial z^2 \partial \overline{u}^2} = \frac{24}{(z+\overline{u})^5}.$$

Hence, as in (1.16) and (1.17), for the nonlinear transforms of $f \in H_{K_1}$

$$\begin{gathered} d_1 f' + d_2 f^2, \\ d_1 f'' + d_2 f' f + d_3 f^3, \\ d_1 f f'' + d_2 (f')^2 + d_3 f^4, \\ d_1 f' f'' + d_2 f (f')^2 + d_3 f^5, \end{gathered}$$

and

$$d_1 (f'')^2 + d_2 (f')^3 + d_3 (f')^2 f^2 + d_4 f^6,$$

we have the specially simple norm inequalities. We have, for example,

$$\frac{6}{7} \|d_1 f' + d_2 f^2\|^2_{H_{K_2}} \leq \|f\|^2_{H_{K_1}} (\frac{1}{6}|d_1|^2 + |d_2|^2 \|f\|^2_{H_{K_1}}),$$

and

$$\frac{120}{127} \|d_1 f'' + d_2 f' f + d_3 f^3\|^2_{H_{K_3}}$$
$$\leq \|f\|^2_{H_{K_1}} \left( \frac{1}{120}|d_1|^2 + \frac{1}{6}|d_2|^2 \|f\|^2_{H_{K_1}} + |d_3|^2 \|f\|^4_{H_{K_1}} \right).$$

For the nonlinear transforms of $f \in H_{K_{1/2}}$

$$d_1 f' + d_2 f^3, \; d_1 f f' + d_2 f^4, \; d_1 (f')^2 + d_2 f' f^3 + d_3 f^6, \; d_1 f'' + d_2 f' f^2 + d_3 f^5,$$

and

$$d_1 f f'' + d_2 (f')^2 + d_3 f^6,$$

we have the corresponding and specially simple norm inequalities.

### 2.8 A transform with nonconstant coefficients

As we see directly by using the Taylor expansion,

$$K(z,\overline{u}) = \frac{1}{\overline{u}z} \log \frac{1}{1-\overline{u}z}$$

is the reproducing kernel for the Hilbert space $H_K$ comprising of all analytic functions $f(z)$ on $U$ with finite norms

$$\left\{ \frac{1}{\pi} \iint_U |f'(z)|^2 dx\, dy + \frac{1}{2\pi} \int_{\partial U} |f(z)|^2 |dz| \right\}^{\frac{1}{2}} < \infty,$$

([3]). Hence, from the identity

$$\frac{1}{1-\overline{u}z} = e^{z\overline{u}K(z,\overline{u})},$$

in the transform of $H_K$ functions

$$f \longrightarrow e^{zf},$$

we have the inequality

$$\frac{1}{2\pi}\int_{\partial U} |e^{zf(z)}|^2 |dz| \le \exp\left[\frac{1}{\pi}\iint_U |f'(z)|^2 dx\, dy + \frac{1}{2\pi}\int_{\partial U} |f(z)|^2 |dz|\right].$$

**Acknowledgements.** This research was partially supported by the Japanese Ministry of Education, Science, Sports and Culture; Grant-in-Aid Scientific Research, General Research (C) 06640201. Furthermore, the travel expenses for participating to the 7th International Conference on General Inequalities (November 12–18, Oberwolfach) were supported by the Ministry.

**References**

1. N. Aronszajn, *Theory of reproducing kernels.* Trans. Amer. Math. Soc. 68 (1950), 337–405.

2. L. de Branges, *Underlying concepts in the proof of the Bieberbach conjecture.* Proc. of International Congress of Math., Berkeley, California 1986, pp. 25–42.

3. J. Burbea, *Total positivity of certain reproducing kernels.* Pacific J. Math. **67** (1976), 101–130.

4. D.A. Hejhal, *Theta Functions, Kernel Functions and Abel Integrals.* Memoirs of Amer. Math. Soc. **129** (1972).

5. M.G. Kreĭn, *Hermitian positive kernels on homogeneous spaces.* Amer. Math. Soc. Transl. (2) **34** (1963), 69–108.

6. S. Saitoh, *The Bergman norm and the Szegő norm.* Trans. Amer. Math. Soc. **249** (1979), 261–279.

7. S. Saitoh, *Some inequalities for analytic functions with a finite Dirichlet integral on the unit disc.* Math. Ann. **246** (1979), 69–77.

8. S. Saitoh, *Positive definite Hermitian matrices and reproducing kernels.* Linear Algebra Appl. **48** (1982), 119–130.

9. S. Saitoh, *Integral transforms in Hilbert spaces.* Proc. Japan Acad. **58** (1982), 361–364.

10. S. Saitoh, *Reproducing kernels of the direct product of two Hilbert spaces.* Riazi J. Kar. Math. Assoc. **4** (1982), 1–20.

11. S. Saitoh, *Quadratic norm inequalities deduced from the theory of reproducing kernels.* Linear Algebra Appl. **93** (1987), 171–178.

12. S. Saitoh, *Inequalities of exponential type for analytic functions with finite Dirichlet integrals.* Complex Variables, **10** (1988), 123–139.

13. S. Saitoh, *Theory of Reproducing Kernels and its Applications.* Pitman Res. Notes in Math. Series 189 (1988), Longman Scientific & Technical, England.

14. S. Saitoh, *Decreasing principles in transforms of reproducing kernel spaces.* Mathematica Montisnigri **1** (1993), 99–109.

15. S. Saitoh, *Inequalities in the most simple Sobolev space and convolutions of $L_2$ functions with weights.* Proc. Amer. Math. Soc. **118** (1993), 515–520.

16. S. Saitoh, *An integral inequality of exponential type for real-valued functions.* Inequalities and Applications **3** (1994), 537–541.

Saburou Saitoh, Department of Mathematics, Faculty of Engineering, Gunma University, Kiryu 376, Japan.

E-mail address: ssaitoh@eg.gunma-u.ac.jp

Inequalities for Matrices and Discrete Problems Inequalities for Matrices and Discrete Problems Inequalities for Matrices and Discrete Problems Inequalities for Matrices and Discrete Problems Inequalities for Matrices and Discrete Problems Inequalities for Matrices and Discrete Problems Inequalities for Matrices and Discrete Problems Inequalities for Matrices and Discrete Problems Inequalities for Matrices and Discrete Problems Inequalities for Matrices and Discrete Problems Inequalities for Matrices and Discrete Problems Inequalities for Matrices and Discrete Problems Inequalities for Matrices and Discrete Problems Inequalities for Matrices and Discrete Problems Inequalities for Matrices and Discrete Problems Inequalities for Matrices and Discrete Problems Inequalities for Matrices and Discrete Problems Inequalities for Matrices and Discrete Problems Inequal-

ities f atrices

and D Prob-

lems alities

for M s and

Discre blems

# Inequalities for Matrices and Discrete Problems

Inequalities for Matrices and Discrete Problems Inequalities for Matrices and Discrete Problems Inequalities for Matrices and Discrete Problems Inequalities for Matrices and Discrete Problems Inequalities for Matrices and Discrete Problems Inequalities for Matrices and Discrete Problems Inequalities for Matrices and Discrete Problems Inequalities for Matrices and Discrete Problems Inequalities for Matrices and Discrete Problems Inequalities for Matrices and Discrete Problems Inequalities for Matrices and Discrete Problems Inequalities for Matrices and Discrete Problems Inequalities for Matrices and Discrete Problems Inequalities for Matrices and Discrete Problems Inequalities for Matrices and Discrete Problems Inequalities for Matrices and Discrete Problems Inequalities for Matrices and Discrete Problems Inequalities for Matrices and Discrete Problems Inequalities for Matrices and Discrete Problems Inequalities for Matrices and Discrete Problems Inequalities for Matrices and Discrete Problems Inequalities for Matrices and Discrete Problems Inequalities for Matrices and Discrete Problems Inequalities for Matrices and Discrete Problems Inequalities for Matrices and Discrete Problems Inequalities for Matrices and Discrete Problems Inequalities for Matrices and Discrete Problems Inequalities for Matrices and Discrete Problems Inequalities for Matrices and Discrete Problems Inequalities for Matrices and Discrete

International Series of Numerical Mathematics, Vol. 123, © 1997 Birkhäuser Verlag Basel

# Positive definiteness of discrete quadratic functionals

Martin Bohner

**Abstract**

This summary of the author's talk presented in Oberwolfach motivates the importance of certain discrete quadratic functionals and characterizes so-called positive definiteness of them via a condition on the principal solution of a related linear Hamiltonian difference system.

## 1. Motivation: Discrete variational problems

Given a function $L : \mathbb{Z} \times \mathbb{R}^2 \to \mathbb{R}$, two integers $M, N \in \mathbb{Z}$ with $M \le N$, and two reals $\alpha, \beta \in \mathbb{R}$, one might be interested in minimizing the functional

$$\mathcal{L}(y) = \sum_{k=M}^{N} L\left(k, y_{k+1}, \Delta y_k\right) \quad \text{with} \quad \Delta y_k = y_{k+1} - y_k$$

subject to the boundary conditions $y_M = \alpha$ and $y_{N+1} = \beta$, i.e., in finding an "optimal sequence" $y^* = \left(y_M^*, \ldots, y_{N+1}^*\right)$ with $y_M^* = \alpha$ and $y_{N+1}^* = \beta$ such that $\mathcal{L}(y^*) \le \mathcal{L}(y)$ for all sequences $y = (y_M, \ldots, y_{N+1})$ with $y_M = \alpha$ and $y_{N+1} = \beta$ satisfying $\max_{M \le k \le N+1} |y_k - y_k^*| < \delta$ for some $\delta > 0$.

For a so-called admissible variation, i.e., a function $\eta : \mathbb{Z} \cap [M, N+1] \to \mathbb{R}$ with $\eta_M = \eta_{N+1} = 0$, and an admissible (i.e., as above) $y^*$, we define a function $\phi : \mathbb{R} \to \mathbb{R}$ by $\phi(\varepsilon) := \mathcal{L}(y^* + \varepsilon\eta)$. We assume that $L$ is a $C^2$-function in the last two variables. Then the following necessary and sufficient conditions for a local (i.e., as above) minimum are easy to prove (see for example [5, Chapter 8]):

**Theorem 1.** *If $y^*$ is a local minimum, then $\dot{\phi}(0) = 0$ and $\ddot{\phi}(0) \ge 0$ (for all admissible variations).*

**Theorem 2.** *If $y_M^* = \alpha$, $y_{N+1}^* = \beta$, $\dot{\phi}(0) = 0$, and $\ddot{\phi}(0) > 0$ (for all admissible variations), then $y^*$ is a (proper) local minimum.*

At this point, of course, we are interested in how $\ddot{\phi}(0)$ looks like. An easy calculation yields

$$\ddot{\phi}(0) = \mathcal{F}(\eta) := \sum_{k=M}^{N} \left\{ L_{uu}(\cdot)\eta_{k+1}^2 + 2L_{uv}(\cdot)\eta_{k+1}\Delta\eta_k + L_{vv}(\cdot)\left(\Delta\eta_k\right)^2 \right\},$$

where $(\cdot)$ denotes $(k, y_{k+1}^*, \Delta y_k^*)$, and $L_u$, $L_v$ are partial derivatives with respect to the second and third entry of $L$, respectively. Positive definiteness of $\mathcal{F}$ now means

$$\mathcal{F}(\eta) > 0 \quad \text{for all nontrivial admissible variations } \eta.$$

It is the goal of this survey to generalize this concept and to give an easy characterization of positive definiteness. For a more general discussion of the motivation we refer to [3].

## 2. From Sturm-Liouville difference equations to linear Hamiltonian difference systems

We start this section with looking at the following discrete quadratic functional:

$$\mathcal{F}(y) := \sum_{k=M}^{N} \left\{ q_k y_{k+1}^2 + p_k \left(\Delta y_k\right)^2 \right\} \quad \text{with} \quad p_k, q_k \in \mathbb{R},\ p_k \neq 0.$$

According to the introductory section, it is important to know about positive definiteness of $\mathcal{F}$, i.e., to answer the question when

$$\mathcal{F}(y) > 0 \text{ for all } y \neq 0 \text{ with } y_M = y_{N+1} = 0, \text{ we write } \mathcal{F} > 0,$$

holds. This is done by the following (at least for $p_k > 0$) well-known result.

**Theorem 3.** $\mathcal{F} > 0$ *iff* $\tilde{y}$ *satisfies* $\tilde{y}_k \tilde{y}_{k+1} p_k > 0$ *for all* $M + 1 \leq k \leq N$.

Here, $\tilde{y}$ is the (unique) solution of the Sturm-Liouville difference equation

$$\Delta\left\{p_k \Delta y_k\right\} = q_k y_{k+1}$$

satisfying the initial conditions $\tilde{y}_M = 0$ and $\tilde{y}_{N+1} = \frac{1}{p_M}$. Of course, Theorem 3 provides an easy and useful characterization of positive definiteness, and it will follow from our more general result, Theorem 4 of the next section.

We now wish to generalize the present questioning in a certain sense. To do so, we introduce a state variable $x := y$ and a control variable $u := p\Delta y$. The above Sturm-Liouville difference equation now becomes a system of two difference equations

$$\Delta x_k = \frac{1}{p_k} u_k, \qquad \Delta u_k = q_k x_{k+1},$$

where the first "artificial" equation is called the equation of motion. The functional $\mathcal{F}$ now reads

$$\mathcal{F}(x,u)=\sum_{k=M}^{N}\left\{q_k x_{k+1}^2+\frac{1}{p_k}u_k^2\right\},$$

while the above $\tilde{y}$ corresponds to the solution $(\tilde{x},\tilde{u})$ of the system satisfying $\tilde{x}_M=0$ and $\tilde{u}_M=1$. Positive definiteness now means

$$\mathcal{F}(x,u)>0 \text{ for all admissible } (x,u) \text{ with } x\neq 0 \text{ and } x_M=x_{N+1}=0,$$

where an $(x,u)$ is called admissible if it satisfies the equation of motion.

This is a perfect point to introduce general linear Hamiltonian difference systems

$$\Delta x_k=A_k x_{k+1}+B_k u_k,\qquad \Delta u_k=C_k x_{k+1}-A_k^T u_k$$

and corresponding discrete quadratic functionals

$$\mathcal{F}(x,u)=\sum_{k=M}^{N}\left\{x_{k+1}^T C_k x_{k+1}+u_k^T B_k u_k\right\}.$$

We are now dealing with real $n\times n$-matrices $A_k, B_k, C_k$, $k\in\mathbb{Z}$, satisfying our general assumptions

$$I-A_k \text{ invertible and } B_k, C_k \text{ symmetric},$$

while solutions $(x,u)$ of the system are vectors $x_k, u_k\in\mathbb{R}^n$, $M\leq k\leq N+1$. An $(x,u)$ satisfying the system's first equation is called admissible, and $\mathcal{F}>0$ now reads as before

$$\mathcal{F}(x,u)>0 \text{ for all admissible } (x,u) \text{ with } x\neq 0 \text{ and } x_M=x_{N+1}=0.$$

In the next section we prove a characterization of $\mathcal{F}>0$ involving the so-called principal solution $(\tilde{X},\tilde{U})$ of the Hamiltonian system, i.e., the solution of

$$\Delta X_k=A_k X_{k+1}+B_k U_k,\qquad \Delta U_k=C_k X_{k+1}-A_k^T U_k$$

satisfying the initial conditions $\tilde{X}_M=0$ (the $n\times n$-matrix with only zero entries) and $\tilde{U}_M=I$ (the $n\times n$-identity-matrix). Due to our assumptions, $(\tilde{X},\tilde{U})$ is easily calculated by the formula

$$\begin{pmatrix}\tilde{X}_k\\ \tilde{U}_k\end{pmatrix}=S_{k-1}\cdot S_{k-2}\cdot\ldots\cdot S_M\begin{pmatrix}0\\ I\end{pmatrix},\qquad M+1\leq k\leq N+1$$

with the $2n\times 2n$-matrices (see also [1])

$$S_k=\begin{pmatrix}\tilde{A}_k & \tilde{A}_k B_k\\ C_k\tilde{A}_k & C_k\tilde{A}_k B_k+I-A_k^T\end{pmatrix},\qquad \text{where}\quad \tilde{A}_k:=(I-A_k)^{-1}.$$

Finally note that the case treated before indeed fits into this concept using $n=1$, $A_k=0$, $B_k=\frac{1}{p_k}$, and $C_k=q_k$.

## 3. The main result on positive definiteness

Using last section's terminology, our main result now reads as follows.

**Theorem 4.** *$\mathcal{F} > 0$ iff $(\tilde{X}, \tilde{U})$ satisfies*

$$\mathrm{Ker}\tilde{X}_{k+1} \subset \mathrm{Ker}\tilde{X}_k \text{ and } \tilde{X}_k\tilde{X}_{k+1}^\dagger \tilde{A}_k B_k \geq 0 \quad \text{for all} \quad M+1 \leq k \leq N.$$

Here, Ker and the dagger denote the nullspace and the Moore-Penrose inverse (i.e., $M^\dagger$ is the unique matrix satisfying $MM^\dagger M = M$ and $M^\dagger M M^\dagger = M^\dagger$ such that both $MM^\dagger$ and $M^\dagger M$ are symmetric), respectively, and "$\geq 0$" indicates positive semidefiniteness. Also, we will write Im for the image.

We devote the remainder of this paper to the proof of Theorem 5 below. This result then obviously implies our main result, Theorem 4 above.

**Theorem 5.** *We abbreviate $\tilde{X}_k\tilde{X}_{k+1}^\dagger \tilde{A}_k B_k$ by $D_k$, and for admissible $(x,u)$ we put $z_k := u_k - \tilde{U}_k\tilde{X}_k^\dagger x_k$. Let $m \in [M,N] \cap \mathbb{Z}$.*

(i) *Suppose $\mathrm{Ker}\tilde{X}_{m+1} \subset \mathrm{Ker}\tilde{X}_m$, and let $(x,u)$ be admissible with $x_m \in \mathrm{Im}\tilde{X}_m$. Then $D_m$ is symmetric, $x_{m+1} \in \mathrm{Im}\tilde{X}_{m+1}$, $x_m = \tilde{X}_m\tilde{X}_{m+1}^\dagger x_{m+1} - D_m z_m$, and*

$$x_{m+1}^T C_m x_{m+1} + u_m^T B_m u_m = \Delta\left\{x_m^T \tilde{U}_m \tilde{X}_m^\dagger x_m\right\} + z_m^T D_m z_m.$$

(ii) *Suppose $\mathrm{Ker}\tilde{X}_{k+1} \subset \mathrm{Ker}\tilde{X}_k$ for all $M \leq k < m$ but $\mathrm{Ker}\tilde{X}_{m+1} \not\subset \mathrm{Ker}\tilde{X}_m$. Pick $c \in \mathrm{Ker}\tilde{X}_{m+1} \setminus \mathrm{Ker}\tilde{X}_m$. Then $(x,u)$ defined by*

$$(x_k, u_k) := \begin{cases} \left(\tilde{X}_k, \tilde{U}_k\right)c & M \leq k \leq m \\ 0 & m < k \leq N+1 \end{cases}$$

*is admissible and satisfies $x \neq 0$, $x_m = x_{N+1} = 0$, and $\mathcal{F}(x,u) = 0$.*

(iii) *Suppose $\mathrm{Ker}\tilde{X}_{k+1} \subset \mathrm{Ker}\tilde{X}_k$ for all $M \leq k \leq N$ and $d^T D_m d < 0$ for some $d \in \mathbb{R}^n$. Put $c := -\tilde{X}_{m+1}^\dagger \tilde{A}_m B_m d$. Then $(x,u)$ defined by*

$$(x_k, u_k) := \begin{cases} \left(\tilde{X}_k, \tilde{U}_k\right)c & M \leq k < m \\ \left(\tilde{X}_m c, \left[\tilde{X}_m\tilde{X}_{m+1}^\dagger \tilde{A}_m\right]^T d\right) & k = m \\ 0 & m < k \leq N+1 \end{cases}$$

*is admissible and satisfies $x \neq 0$, $x_m = x_{N+1} = 0$, and $\mathcal{F}(x,u) < 0$.*

*Proof.* It is easy to see (e.g. as in [2, Lemma 4]) that $\mathrm{Ker}\tilde{X}_{m+1} \subset \mathrm{Ker}\tilde{X}_m$ implies

$$\tilde{X}_m\tilde{X}^{\dagger}_{m+1}\tilde{X}_{m+1} = \tilde{X}_m, \quad \tilde{X}^{\dagger}_{m+1}\tilde{X}_{m+1}\tilde{X}^{\dagger}_m = \tilde{X}^{\dagger}_m, \quad \tilde{X}_{m+1}\tilde{X}^{\dagger}_{m+1}\tilde{A}_mB_m = \tilde{A}_mB_m.$$

Furthermore, $\Delta\left\{\tilde{X}_k^T\tilde{U}_k - \tilde{U}_k^T\tilde{X}_k\right\} = 0$ and $\tilde{X}_M^T\tilde{U}_M = 0$ imply that $\tilde{X}_k^T\tilde{U}_k$ are symmetric so that

$$\begin{aligned} D_m &= \left\{(I - A_m)\tilde{X}_{m+1} - B_m\tilde{A}_m^T\left(\tilde{U}_{m+1} - C_m\tilde{X}_{m+1}\right)\right\}\tilde{X}^{\dagger}_{m+1}\tilde{A}_mB_m \\ &= B_m - B_m\tilde{A}_m^T\left(\tilde{X}^{\dagger}_{m+1}\right)^T\tilde{X}^T_{m+1}\left(\tilde{U}_{m+1} - C_m\tilde{X}_{m+1}\right)\tilde{X}^{\dagger}_{m+1}\tilde{A}_mB_m \end{aligned}$$

is symmetric. Now, $x_m = \tilde{X}_mc_m \in \mathrm{Im}\tilde{X}_m$ yields

$$\begin{aligned} x_{m+1} &= \tilde{A}_mx_m + \tilde{A}_mB_mu_m = \tilde{X}_{m+1}c_m + \tilde{A}_mB_m\left(u_m - \tilde{U}_mc_m\right) \\ &= \tilde{X}_{m+1}\left\{c_m + \tilde{X}^{\dagger}_{m+1}\tilde{A}_mB_m\left(u_m - \tilde{U}_mc_m\right)\right\} \in \mathrm{Im}\tilde{X}_{m+1}, \end{aligned}$$

$$\begin{aligned} \tilde{X}^{\dagger}_{m+1}\tilde{A}_mB_mz_m &= \tilde{X}^{\dagger}_{m+1}\left\{x_{m+1} - \tilde{A}_mx_m - \left(\tilde{X}_{m+1} - \tilde{A}_m\tilde{X}_m\right)\tilde{X}^{\dagger}_mx_m\right\} \\ &= \tilde{X}^{\dagger}_{m+1}x_{m+1} - \tilde{X}^{\dagger}_mx_m = \Delta\left(\tilde{X}^{\dagger}_mx_m\right), \end{aligned}$$

$$D_mz_m = \tilde{X}_m\tilde{X}^{\dagger}_{m+1}x_{m+1} - \tilde{X}_m\tilde{X}^{\dagger}_mx_m = \tilde{X}_m\tilde{X}^{\dagger}_{m+1}x_{m+1} - x_m,$$

$$\begin{aligned} \Delta\left(x_m^T\tilde{U}_m\right) &= (x_m + B_mu_m)^T\left(\tilde{A}_m^TC_m\tilde{X}_{m+1} + \tilde{U}_m\right) - x_m^T\tilde{U}_m \\ &= x^T_{m+1}C_m\tilde{X}_{m+1} + u_m^TB_m\tilde{U}_m \\ &= x^T_{m+1}C_m\tilde{X}_{m+1} + u_m^T(I - A_m)\tilde{X}_{m+1} - u_m^T\tilde{X}_m, \end{aligned}$$

and finally

$$\begin{aligned} &\left[\Delta\left(x_m^T\tilde{U}_m\right)\right]\tilde{X}^{\dagger}_{m+1}x_{m+1} = x^T_{m+1}C_mx_{m+1} + u_m^T(I - A_m)x_{m+1} - u_m^T\tilde{X}_m\tilde{X}^{\dagger}_{m+1}x_{m+1} \\ &= x^T_{m+1}C_mx_{m+1} + u_m^TB_mu_m - u_m^T\tilde{X}_m\left[\Delta\left(\tilde{X}^{\dagger}_mx_m\right)\right] \\ &= x^T_{m+1}C_mx_{m+1} + u_m^TB_mu_m - \left(x_m^T\tilde{U}_m + z_m^T\tilde{X}_m\right)\left[\Delta\left(\tilde{X}^{\dagger}_mx_m\right)\right] \\ &= x^T_{m+1}C_mx_{m+1} + u_m^TB_mu_m - z_m^TD_mz_m - x_m^T\tilde{U}_m\left[\Delta\left(\tilde{X}^{\dagger}_mx_m\right)\right]. \end{aligned}$$

This takes care of part (i). Now we look at $(x, u)$ defined in (ii). Admissibility follows from

$$\tilde{A}_m(x_m + B_mu_m) = \tilde{A}_m\left(\tilde{X}_m + B_m\tilde{U}_m\right)c = \tilde{X}_{m+1}c = 0 = x_{m+1},$$

while an application of part (i) yields

$$
\begin{aligned}
\mathcal{F}(x,u) &= \sum_{k=M}^{m-1} \left\{ x_{k+1}^T C_k x_{k+1} + u_k^T B_k u_k \right\} + x_{m+1}^T C_m x_{m+1} + u_m^T B_m u_m \\
&= x_m^T \tilde{U}_m \tilde{X}_m^\dagger x_m + \sum_{k=M}^{m-1} z_k^T D_k z_k + u_m^T B_m u_m \\
&= c^T \tilde{U}_m^T \tilde{X}_m c + c^T \tilde{U}_m^T B_m \tilde{U}_m c = c^T \tilde{U}_m^T (I - A_m) \tilde{X}_{m+1} c = 0.
\end{aligned}
$$

Finally, we remark that $(x,u)$ from (iii) is admissible because of

$$\tilde{A}_m \left( x_m + B_m u_m \right) = \tilde{A}_m \left( -D_m d + D_m^T d \right) = 0 = x_{m+1}.$$

Another application of part (i) yields

$$\mathcal{F}(x,u) = \sum_{k=M}^{N} z_k^T D_k z_k = z_m^T D_m z_m = d^T D_m d < 0$$

since $D_m z_m = \tilde{X}_m \tilde{X}_{m+1}^\dagger x_{m+1} - x_m = -x_m = -\tilde{X}_m c = D_m d$ holds. This finishes the proof of (iii) and hence yields all our desired results. □

For a deeper discussion of this topic we refer to [4]. To finish with, we wish to remark and to emphasize that so-called Sturm-Liouville difference equations of higher order may be equivalently rewritten as certain linear Hamiltonian difference systems satisfying our general assumptions so that Theorem 4 applies to those important objects also.

## References


[1] C.D. Ahlbrandt and A.Peterson, *The $(n,n)$-disconjugacy of a $2n^{\text{th}}$ order linear difference equation.* Comput. Math. Appl., 1994.

[2] M. Bohner, *Controllability and disconjugacy for linear Hamiltonian difference systems.* In “Conference Proceedings of the first international conference on difference equations”, pages 65–77, 1994.

[3] M. Bohner, *Inhomogeneous discrete variational problems.* In “Conference Proceedings of the second international conference on difference equations”, 1995. To appear.

[4] M. Bohner, *Linear Hamiltonian difference systems: disconjugacy and Jacobi-type conditions.* J. Math. Anal. Appl., 1995. To appear.

[5] W.G. Kelley and A.Peterson, *Difference Equations: An Introduction with Applications.* Academic Press, San Diego, 1991.



Martin Bohner, Universität Hohenheim, Institut 110, D-70593 Stuttgart, Germany.
E-mail: bohner@uni-hohenheim.de

International Series of Numerical Mathematics, Vol. 123, © 1997 Birkhäuser Verlag Basel

# Stable norms – Examples and remarks

Moshe Goldberg*

Let $\mathcal{A}$ be an associative algebra over a field $\mathbb{F}$, either $\mathbb{R}$ or $\mathbb{C}$. We shall say that a norm $N$ on $\mathcal{A}$ is *strongly stable* if

$$N(a^k) \leq N(a)^k, \qquad a \in \mathcal{A}, \quad k = 1, 2, 3, \ldots .$$

While often $N$ will not be strongly stable, it may have weaker, yet related properties. For example, we call $N$ *quadrative* if

$$N(a^2) \leq N(a)^2, \qquad a \in \mathcal{A}.$$

In certain instances, strong stability and quadrativity are equivalent. This is typical, for instance, of weighted sup norms on various algebras (e.g. [1–5]). In general, however, it would seem that quadrativity is definitely weaker than strong stability.

Given an integer $k \geq 2$, we shall follow [2] and call our norm *$k$-bounded* if

$$N(a^k) \leq N(a)^k, \qquad a \in \mathcal{A}.$$

Hence, $N$ is quadrative if it is 2-bounded, and strongly stable if it is $k$-bounded for all $k = 2, 3, 4, \ldots$.

Boundedness for a particular $k$ larger than 2 does not usually ensure strong stability, not even quadrativity. For example, let $W = (\omega_{ij})$ be a fixed $n \times n$ matrix of positive entries, and consider the *$W$-weighted $l_\infty$* norm on $\mathbb{C}_{n\times n}$, the algebra of $n \times n$ complex matrices:

$$\|A\|_{W,\infty} = \max_{ij} |\omega_{ij}\alpha_{ij}|, \qquad A = (\alpha_{ij}) \in \mathbb{C}_{n\times n}.$$

For this norm we can prove,

**Proposition 1** [2]. (a) $\|\cdot\|_{W,\infty}$ *is strongly stable if and only if it is quadrative.*

(b) *If $k \geq 3$, then there exists a weight matrix $W$ for which $\|\cdot\|_{W,\infty}$ is $k$-bounded but not strongly stable, not even quadrative.*

*Research sponsored in part by the Fund for the Promotion of Research at the Technion, Grant 100-013.

Relaxing our definition of strong stability, we shall say that our norm $N$ is *stable* if for some positive constant $\sigma$,

$$N(a^k) \leq \sigma N(a)^k, \qquad a \in \mathcal{A}, \quad k = 1, 2, 3, \ldots .$$

With this new definition we can sweeten the observation in Theorem 1(b) for all norms on $\mathbb{C}_{n\times n}$:

**Proposition 2.** *If $k \geq 2$ and $N$ is $k$-bounded on $\mathbb{C}_{n\times n}$, then $N$ is stable.*

*Proof.* By hypothesis, for any $A \in \mathbb{C}_{n\times n}$,

$$N(A^{k^j}) \leq N(A)^{k^j}, \qquad j = 1, 2, 3, \ldots .$$

Thus

$$N(A) \geq N(A^{k^j})^{1/k^j} \underset{j\to\infty}{\longrightarrow} \rho(A)$$

where $\rho(A)$ denotes the spectral radius of $A$. Consequently, $N$ is *spectrally dominant,* i.e.,

$$N(A) \geq \rho(A), \qquad A \in \mathbb{C}_{n\times n};$$

hence $N$ is stable by the celebrated Friedland-Zeger Theorem [6, Theorem 1]. □

Contrary to what one may expect, the converse of Proposition 2 is false; that is, stability does not imply $k$-boundedness for any $k \geq 2$. For instance, consider $\|\cdot\|_{W,\infty}$ on $\mathbb{C}_{2\times 2}$, where $W$ is the $2 \times 2$ weight matrix

$$W = \begin{pmatrix} \dfrac{2}{1+\theta} & \dfrac{2}{1-\theta} \\ \\ \dfrac{2}{1-\theta} & \dfrac{2}{1+\theta} \end{pmatrix}, \qquad \theta = \text{constant}, \quad 0 < \theta < 1.$$

For each $k = 2, 3, 4, \ldots$, put

$$\lambda_k \equiv \{\|A^k\|_{W,\infty}\colon A \in \mathbb{C}_{2\times 2}, \quad \|A\|_{W,\infty} = 1\}. \tag{1}$$

Then evidently, $\lambda_k$ is the best (least) constant satisfying

$$\|A^k\|_{W,\infty} \leq \lambda_k \|A\|^k_{W,\infty}, \qquad A \in \mathbb{C}_{2\times 2}.$$

It was shown in [2] that

$$\lambda_k = \frac{1-\theta^k}{1-\theta};$$

so $\lambda_k > 1$, whence $\|\cdot\|_{W,\infty}$ is not $k$-bounded for any $k \geq 2$. On the other hand, we get

$$\|A^k\|_{W,\infty} \leq \frac{1}{1-\theta}\|A\|^k_{W,\infty}, \qquad A \in \mathbb{C}_{2\times 2}, \quad k = 1, 2, 3, \ldots,$$

hence $\|\cdot\|_{W,\infty}$ is stable.

Following (1), we realize (compare [1, Theorem 1.2]) that a norm $N$ is $k$-bounded on an algebra $\mathcal{A}$ if and only if

$$\mu_k \equiv \sup\{N(a^k) \colon a \in \mathcal{A}, \quad N(a) = 1\} \leq 1. \tag{2}$$

Further, we observe that $N$ is stable if and only if the sequence $\{\mu_k\}_{k=2}^{\infty}$ is bounded, and strongly stable if and only if the bound is 1.

If $\mathcal{A}$ is finite dimensional, a simple compactness argument shows that $\mu_k$ in (2) is finite. It follows in this case that sufficiently large multiples of $N$ will always be $k$-bounded. More precisely, *if $\mu$ is a positive constant such that*

$$\mu \geq (\mu_k)^{1/(k-1)},$$

*then $N_\mu \equiv \mu N$ will be a $k$-bounded norm on $\mathcal{A}$.* For then,

$$N_\mu(a^k) = \mu N(a^k) \leq \mu \cdot \mu_k N(a)^k \leq \mu^k N(a)^k = N_\mu(a)^k, \qquad a \in \mathcal{A}.$$

The moral of this observation is that $N$ must be *'sufficiently large'* in order to be $k$-bounded. To illustrate this somewhat vague statement, consider again $\|\cdot\|_{W,\infty}$, the weighted $l_\infty$ norm on $\mathbb{C}_{n\times n}$, for which we can state:

**Proposition 3** [2; compare 3]. *$\|\cdot\|_{W,\infty}$ is $k$-bounded if and only if*

$$(W_{-1})^k \leq W_{-1}, \tag{3}$$

*where the matrix $W_{-1}$, the Hadamard inverse of $W$, is defined by*

$$W_{-1} \equiv (\omega_{ij}^{-1}),$$

*and the inequality in (3) is construed entrywise.*

Indeed, we see without difficulty that (3) puts limits on the size of the reciprocals $\omega_{ij}^{-1}$; thus ensuring that $W$ is large.

Another example which illustrates the idea that a norm must be 'sufficiently large' in order to be $k$-bounded, concerns arbitrary norms on $\mathcal{A} = \mathbb{C}$, the algebra of complex numbers over the reals:

**Proposition 4.** *If $k \geq 2$ and $N$ is $k$-bounded on the algebra $\mathcal{A} = \mathbb{C}$ over $\mathbb{R}$, then*

$$N(z) \geq |z|, \qquad z \in \mathbb{C}.$$

*Proof.* Suppose the theorem is false. Then $N(z_0) < 1$ for some $z_0$ with $|z_0| = 1$; so for some fixed $0 < \theta < 1$,

$$N(z) < \theta$$

on some arc $\mathcal{C}$ of the unit circle. By $k$-boundedness therefore,

$$N(z^k) \leq N(z)^k \leq \theta^k, \qquad z \in \mathcal{C};$$

on in other words,

$$N(z) \leq \theta^k \text{ on } \mathcal{C}^k \equiv \{z^k : z \in \mathcal{C}\}.$$

Continuing in this way we obtain

$$N(z) \leq \theta^m \text{ on } \mathcal{C}^m \equiv \{z^m : z \in \mathcal{C}\}, \text{ for all } m = k^j, \quad j = 1, 2, 3, \ldots .$$

Since at some stage, $\mathcal{C}^m$ is the entire unit circle, it follows that there exists $j_0$ such that

$$N(z) \leq \theta^{k^j} \text{ for all } |z| = 1, \quad j \geq j_0.$$

Whence $N$ vanishes everywhere, a contradiction. □

We conclude this note by posing two open problems of obvious relevance to the above discussion:

(a) Prove or disprove: 'Quadrativity implies strong stability (at least in the finite dimensional case)'.

(b) For a given algebra (say $\mathbb{C}_{n \times n}$), characterize all norms that are stable but not strongly stable.

## References

1. R. Arens and M. Goldberg, *A class of seminorms on function algebras.* J. Math. Anal. Appl. **162** (1991), 592–609.

2. R. Arens and M. Goldberg, *Weighted $l_\infty$ norms for matrices.* Linear Algebra Appl. **201** (1994), 155–163.

3. R. Arens and M. Goldberg, *Homotonic mappings.* J. Math. Anal. Appl. **194** (1995), 414–427.

4. R. Arens, M. Goldberg and W.A.J. Luxemburg, *Multiplicativity factors for function norms.* J. Math. Anal. Appl. **177** (1993), 368–385.

5. R. Arens, M. Goldberg and W.A.J. Luxemburg, *Multiplicativity factors for Orlicz space function norms.* J. Math. Anal. Appl. **177** (1993), 386–411.

6. S. Friedland and C. Zenger, *All spectral dominant norms are stable.* Linear Algebra Appl. **58** (1984), 97–107.

Moshe Goldberg, Department of Mathematics, Technion – Israel Institute of Technology, Haifa 32000, Israel.

International Series of Numerical Mathematics, Vol. 123, © 1997 Birkhäuser Verlag Basel

# Applications of order preserving operator inequalities to a generalized relative operator entropy

Takayuki Furuta

*Dedicated to Professor Hisaharu Umegaki on his seventieth birthday*

**Abstract**

We shall introduce a generalized relative operator entropy and we shall discuss order preserving property and monotonicity property of this new one as applications of oreder preserving operator inequalities.

## 1. Introduction

In what follows, a capital letter means a bounded linear operator on a complex Hilbert space $H$. An operator $T$ is said to be positive (in symbol: $T \geq 0$) if $(Tx, x) \geq 0$ for all $x \in H$. Also an operator $T$ is strictly positive (in symbol: $T > 0$) if $T$ is positive and invertible. As an extension of Löwner-Heinz theorem [17], [19], we established the following "order preserving operator inequalities".

**Theorem F** [10]. *If $A \geq B \geq 0$, then for each $r \geq 0$*

(i) $(B^r A^p B^r)^{1/q} \geq (B^r B^p B^r)^{1/q}$ *and*

(ii) $(A^r A^p A^r)^{1/q} \geq (A^r B^p A^r)^{1/q}$

*hold for $p \geq 0$ and $q \geq 1$ with $(1 + 2r)q \geq p + 2r$.*

Mean theoretic alternative proofs of Theorem F are given in [6], [11] and [18] and also an elementary one-page proof is shown in [12]. Inequalities (i) and (ii) of Theorem F hold for the domain surrounded by $p$, $q$ and $r$, which are the conditions of Theorem F, and it is proved in [21] that this domain is the best possible one (see Figure). We remark that Theorem F yields the Löwner-Heinz theorem when we put $r = 0$ in (i) or (ii) stated above: if $A \geq B \geq 0$ ensures $A^\alpha \geq B^\alpha$ for any $\alpha \in [0, 1]$.

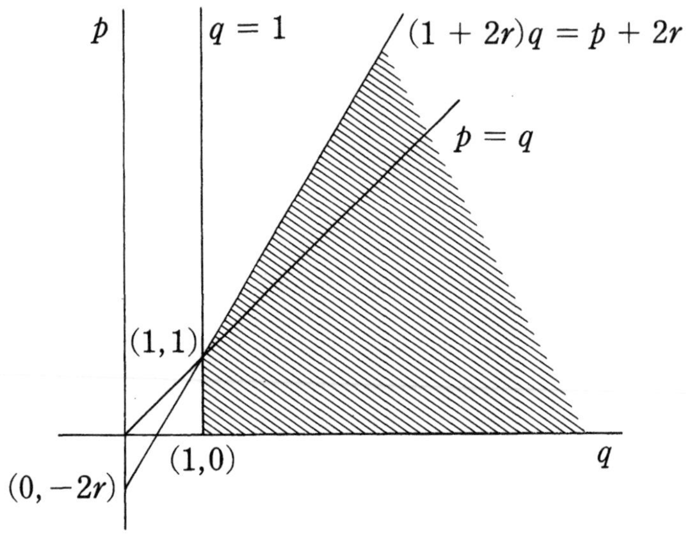


We write $A \gg B$ if $\log A \geq \log B$ for invertible operator $A$ and $B$, which is called the chaotic order [7], [8]. It is shown in [1] and well known as a useful result that $A \gg B$ if and only if $A^p \geq (A^{\frac{p}{2}} B^p A^{\frac{p}{2}})^{\frac{1}{2}}$ for all $p \geq 0$.

As an application of Theorem F, we obtain the following result.

**Theorem G** [16], [9]. *If $A \geq B \geq 0$ with $A > 0$, then for each $t \in [0, 1]$ and $p \geq 1$,*

$$F_{p,t}(A, B, r, s) = A^{\frac{-r}{2}} \{A^{\frac{r}{2}} (A^{\frac{-t}{2}} B^p A^{\frac{-t}{2}})^s A^{\frac{r}{2}}\}^{\frac{1-t+r}{(p-t)s+r}} A^{\frac{-r}{2}}$$

*is a decreasing function of both $r$ and $s$ such that $r \geq t$ and $s \geq 1$, and the following inequality holds: $F_{p,t}(A, A, r, s) \geq F_{p,t}(A, B, r, s)$, that is, for each $t \in [0, 1]$ and $p \geq 1$,*

$$A^{1-t+r} \geq \{A^{\frac{r}{2}} (A^{\frac{-t}{2}} B^p A^{\frac{-t}{2}})^s A^{\frac{r}{2}}\}^{\frac{1-t+r}{(p-t)s+r}}$$

*holds for any $s \geq 0$ and $r$ such that $r \geq t$.*

Recently Ando-Hiai [2] established various log majorization results to ensure excellent and useful inequalities for unitarily invariant norms. Theorem G interpolates this log-majorization by Ando-Hiai and Theorem F and also Theorem G extends results of [7], [8], [13] and [14].

In [4] and [5], the relative operator entropy $S(A|B)$ is defined by

$$S(A|B) = A^{\frac{1}{2}}(\log A^{\frac{-1}{2}} B A^{\frac{-1}{2}}) A^{\frac{1}{2}}$$

for invertible positive operators $A$ and $B$ and this relative operator entropy can be considered as an extension of the entropy by Nakamura and Umegaki [20] and the relative entropy by Umegaki [22]. We remark that $S(A|I) = -A\log A$ is the usual well-known operator entropy.

In this paper, we introduce a generalized relative operator entropy as an extension of $S(A|B)$ as follows. For positive invertible operators $A$ and $B$, ($s$-$\alpha$)-generalized relative operator entropy $W_{s,\alpha}(A|B)$ for each $\alpha \in [0,1]$ is defined by;

$$W_{s,\alpha}(A|B) = A^{\frac{1}{2}}\{\log A^{\frac{-1}{2}}(A^{\frac{-\alpha}{2}} B A^{\frac{-\alpha}{2}})^s A^{\frac{-1}{2}}\} A^{\frac{1}{2}} \tag{$*$}$$

for any $s \geq 1$. As easily seen by definition $(*)$, for each $\alpha \in [0,1]$ and any $s \geq 1$, $W_{s,\alpha}(A|A^{\alpha/2} B A^{\alpha/2}) = S(A|B^s)$. Especially $W_{s,\alpha}(A|B) = S(A|B)$ in the case $s = 1$ and $\alpha = 0$, that is, $W_{1,0}(A|B)$ coincides with the usual relative operator entropy $S(A|B)$.

Using Theorem G, we shall discuss monotonicity property of a generalized relative entropy $W_{s,\alpha}(A|B)$ and also we shall show order preserving property of $W_{s,\alpha}(A|B)$ as a continuation of the results in [14] and [15].

Finally we shall attempt to extend the differential geometric view in [3] between the geodesic from $A$ to $B$ and the relative operator entropy $S(A|B)$, that is, we shall show a relation between a generalized geodesic from $A$ to $B$ and a generalized relative entropy $W_{s,\alpha}(A|B)$ .

## 2 Statement of results

**Theorem 1.** *Let $A$ and $B$ be positive invertible operators. If $A \gg B$ (i.e., $\log A \geq \log B$), then the following* (I) *and* (II) *hold.*

(I) *For each $\alpha \in [0,1]$, and all $p \geq 0$ and $u \geq 0$,*

$$G_{p,u,\alpha}(A,B,s) = A^{\frac{-u}{2}}\{A^{\frac{u}{2}}(A^{\frac{u\alpha}{2}} B^p A^{\frac{u\alpha}{2}})^s A^{\frac{u}{2}}\}^{\frac{u(\alpha+1)}{(u\alpha+p)s+u}} A^{\frac{-u}{2}}$$

*is a decreasing function of $s$ such that $s \geq 1$.*

(II) *For each $\alpha \in [0,1]$, and all $p \geq 0$ and $u \geq 0$, $\dfrac{1}{(u\alpha+p)s+u} W_{s,\alpha}(A^{-u}|B^p)$ is a decreasing function of $s$ such that $s \geq 1$.*

**Theorem 2.** *Let $A$ and $B$ be positive invertible operators. Then the following assertions are mutually equivalent.*

(I) $A \gg B$ *(i.e.,* $\log A \geq \log B$*).*

(II$_1$) *For each* $\alpha \in [0,1]$, *and all* $p \geq 0$ *and all* $u \geq 0$,

$$A^{u(\alpha+1)} \geq \{A^{\frac{u}{2}}(A^{\frac{u\alpha}{2}} B^p A^{\frac{u\alpha}{2}})^s A^{\frac{u}{2}}\}^{\frac{u(\alpha+1)}{(u\alpha+p)s+u}}$$

*holds for any* $s \geq 1$.

(II$_2$) *For each* $\alpha \in [0,1]$, *a fixed positive number* $p_0$ *and a fixed number* $s_0$ *such that* $s_0 \geq 1$,

$$A^{u(\alpha+1)} \geq \{A^{\frac{u}{2}}(A^{\frac{u\alpha}{2}} B^{p_0} A^{\frac{u\alpha}{2}})^{s_0} A^{\frac{u}{2}}\}^{\frac{u(\alpha+1)}{(u\alpha+p_0)s_0+u}}$$

*holds for all* $u \in [0,u_0]$, *where* $u_0$ *is a fixed positive number.*

(III$_1$) *For each* $\alpha \in [0,1]$, *all* $p \geq 0$ *and all* $u \geq 0$,

$$\log A^{(u\alpha+p)s+u} \geq \log\{A^{\frac{u}{2}}(A^{\frac{u\alpha}{2}} B^p A^{\frac{u\alpha}{2}})^s A^{\frac{u}{2}}\}$$

*holds for any* $s \geq 1$.

(III$_2$) *For each* $\alpha \in [0,1]$, *a fixed positive number* $p_0$ *and a fixed number* $s_0$ *such that* $s_0 \geq 1$,

$$\log A^{(u\alpha+p_0)s_0+u} \geq \log\{A^{\frac{u}{2}}(A^{\frac{u\alpha}{2}} B^{p_0} A^{\frac{u\alpha}{2}})^{s_0} A^{\frac{u}{2}}\}$$

*holds for all* $u \in [0,u_0]$, *where* $u_0$ *is a fixed positive number.*

**Theorem 3.** *Let $A$, $B$ and $C$ be positive invertible operators. Then the following assertions are mutually equivalent.*

(I) $C \gg A \gg B$ *(i.e.,* $\log C \geq \log A \geq \log B$*).*

(II$_1$) *For each* $\alpha \in [0,1]$, *and all* $p \geq 0$ *and all* $u \geq 0$,

$$\begin{aligned}\{A^{\frac{u}{2}}(A^{\frac{u\alpha}{2}} C^p A^{\frac{u\alpha}{2}})^s A^{\frac{u}{2}}\}^{\frac{u(\alpha+1)}{(u\alpha+p)s+u}} &\geq A^{u(\alpha+1)} \\ &\geq \{A^{\frac{u}{2}}(A^{\frac{u\alpha}{2}} B^p A^{\frac{u\alpha}{2}})^s A^{\frac{u}{2}}\}^{\frac{u(\alpha+1)}{(u\alpha+p)s+u}}\end{aligned}$$

*holds for any* $s \geq 1$.

(II$_2$) *For each* $\alpha \in [0,1]$, *a fixed positive number* $p_0$ *and a fixed number* $s_0$ *such that* $s_0 \geq 1$,

$$\begin{aligned}&\{A^{\frac{u}{2}}(A^{\frac{u\alpha}{2}} C^{p_0} A^{\frac{u\alpha}{2}})^{s_0} A^{\frac{u}{2}}\}^{\frac{u(\alpha+1)}{(u\alpha+p_0)s_0+u}} \\ &\qquad \geq A^{u(\alpha+1)} \geq \{A^{\frac{u}{2}}(A^{\frac{u\alpha}{2}} B^{p_0} A^{\frac{u\alpha}{2}})^{s_0} A^{\frac{u}{2}}\}^{\frac{u(\alpha+1)}{(u\alpha+p_0)s_0+u}}\end{aligned}$$

*holds for all* $u \in [0,u_0]$, *where* $u_0$ *is a fixed positive number.*

(III$_1$) *For each $\alpha \in [0,1]$, all $p \geq 0$ and all $u \geq 0$,*

$$\begin{aligned}\log\{A^{\frac{u}{2}}(A^{\frac{u\alpha}{2}}C^pA^{\frac{u\alpha}{2}})^sA^{\frac{u}{2}}\} &\geq \log A^{(u\alpha+p)s+u}\\ &\geq \log\{A^{\frac{u}{2}}(A^{\frac{u\alpha}{2}}B^pA^{\frac{u\alpha}{2}})^sA^{\frac{u}{2}}\}\end{aligned}$$

*holds for any $s \geq 1$.*

(III$_2$) *For each $\alpha \in [0,1]$, a fixed positive number $p_0$ and a fixed number $s_0$ such that $s_0 \geq 1$,*

$$\begin{aligned}\log\{A^{\frac{u}{2}}(A^{\frac{u\alpha}{2}}C^{p_0}A^{\frac{u\alpha}{2}})^{s_0}A^{\frac{u}{2}}\} &\geq \log A^{(u\alpha+p_0)s_0+u}\\ &\geq \log\{A^{\frac{u}{2}}(A^{\frac{u\alpha}{2}}B^{p_0}A^{\frac{u\alpha}{2}})^{s_0}A^{\frac{u}{2}}\}\end{aligned}$$

*holds for all $u \in [0,u_0]$, where $u_0$ is a fixed positive number.*

(IV$_1$) *For each $\alpha \in [0,1]$, and all $p \geq 0$, and $u \geq 0$,*

$$W_{s,\alpha}(A^{-u}|C^p) \geq W_{s,\alpha}(A^{-u}|A^p) \geq W_{s,\alpha}(A^{-u}|B^p)$$

*holds for any $s \geq 1$.*

(IV$_2$) *For each $\alpha \in [0,1]$, and a fixed positive nymber $p_0$, and a fixed number $s_0$ such that $s_0 \geq 1$,*

$$W_{s_0,\alpha}(A^{-u}|C^{p_0}) \geq W_{s_0,\alpha}(A^{-u}|A^{p_0}) \geq W_{s_0,\alpha}(A^{-u}|B^{p_0})$$

*holds for all $u \in [0,u_0]$, where $u_0$ is a fixed positive number.*

We have the upper bound and the lower bound of $W_{s,\alpha}(A|B)$ as follows.

**Theorem 4.** *Let $A$ and $B$ be positive invertible operators. For each $\alpha \in [0,1]$ and any positive number $x_0$ and any $s \geq 1$, the following inequality holds;*

$$\begin{aligned}&(\log x_0 - 1)A + \tfrac{1}{x_0}(A^{-\alpha/2}BA^{-\alpha/2})^s\\ &\qquad \geq W_{s,\alpha}(A|B) \geq (1-\log x_0)A - \frac{1}{x_0}A(A^{\alpha/2}B^{-1}A^{\alpha/2})^sA,\end{aligned}$$

*especially, $W_{s,\alpha}(A|B) = 0$ holds if and only if $B^s = A^{1+\alpha s}$ holds.*

We have the following relation among $W_{s,\alpha}(A|B)$, $S(A|B)$ and usual operator entropy $-A\log A$.

**Theorem 5.** *Let $A$ and $B$ be positive invertible operators. Then the following* (I), (II), (III) *and* (IV) *hold:*

(I) *For each $\alpha \in [0,1]$, any $s \geq 1$, and any $r \geq 0$,*

$$W_{s,\alpha}(A|A^r) = ((r-\alpha)s-1)A\log A.$$

(II) *For each $\alpha \in [0,1]$ and any $s \geq 1$,*

$$W_{s,\alpha}(A|A^{\frac{\alpha}{2}}BA^{\frac{\alpha}{2}}) = S(A|B^s).$$

(III) *For each $\alpha \in [0,1]$ and any $s \geq 1$,*

$$W_{s,\alpha}(A|I) = -(\alpha s+1)A\log A.$$

(IV) *For each $\alpha \in [0,1]$, any $s \geq 1$ and any $t \in [0,1]$, $g_{s,\alpha}(t)$ is defined by*

$$g_{s,\alpha}(t) = A^{\frac{1}{2}}\{A^{\frac{-1}{2}}(A^{\frac{-\alpha}{2}}BA^{\frac{-\alpha}{2}})^s A^{\frac{-1}{2}}\}^t A^{\frac{1}{2}}.$$

*Then the following relation holds:*

$$W_{s,\alpha}(A|B) = s-\lim_{t\to+0}\frac{g_{s,\alpha}(t)-A}{t} = \frac{d}{dt}g_{s,\alpha}(t)\Big|_{t=0}.$$

**Remark 1.** We remark that differential geometric view between the geodesic from $A$ to $B$ and the relative operator entropy $S(A|B)$ is discussed in [3], that is, the geodesic $g(t)$ from $A$ to $B$ is defined by

$$g(t) = A^{\frac{1}{2}}(A^{\frac{-1}{2}}BA^{\frac{-1}{2}})^t A^{\frac{1}{2}}$$

and

$$S(A|B) = A^{\frac{1}{2}}(\log A^{\frac{-1}{2}}BA^{\frac{-1}{2}})A^{\frac{1}{2}} = \text{s}-\lim_{t\to+0}\frac{d}{dt}g(t)\Big|_{t=0}$$

and these relations just correspond to (IV) of Theorem 5 in the case $s=1$ and $\alpha=0$ since $g(t) = g_{1,0}(A|B)$ and $S(A|B) = W_{1,0}(A|B)$ by (II) of Theorem 5.

## 3. Corollaries of results in 2.

Theorem 2 implies the following Theorem A and also Theorem 3 yields Theorem B and Corollary 1 which is an extension of Corollary 2 .

**Theorem A** [14]. *Let $A$ and $B$ be positive invertible operators. Then the following assertions are mutually equivalent.*

(I) *$A \gg B$ (i.e., $\log A \geq \log B$).*

($\text{II}_1$) *$A^u \geq (A^{\frac{u}{2}}B^pA^{\frac{u}{2}})^{\frac{u}{p+u}}$ for all $p \geq 0$ and all $u \geq 0$,*

(II$_2$) $A^u \geq (A^{\frac{u}{2}} B^{p_0} A^{\frac{u}{2}})^{\frac{u}{p_0+u}}$ *for a fixed positive number* $p_0$ *and for all* $u$ *such that* $u \in [0, u_0]$, *where* $u_0$ *is a fixed positive number.*

(III$_1$) $\log A^{p+u} \geq \log(A^{\frac{u}{2}} B^p A^{\frac{u}{2}})$ *for all* $p \geq 0$ *and all* $u \geq 0$.

(III$_2$) $\log A^{u+p_0} \geq \log(A^{\frac{u}{2}} B^{p_0} A^{\frac{u}{2}})$ *for a fixed positive number* $p_0$ *and for all* $u$ *such that* $u \in [0, u_0]$, *where* $u_0$ *is a fixed positive number.*

**Theorem B** [14]. *Let* $A$, $B$ *and* $C$ *be positive invertible operators. Then the following assertions are mutually equivalent.*

(I) $C \gg A \gg B$ (*i.e.*, $\log C \geq \log A \geq \log B$).

(II$_1$) $(A^{\frac{u}{2}} C^p A^{\frac{u}{2}})^{\frac{u}{p+u}} \geq A^u \geq (A^{\frac{u}{2}} B^p A^{\frac{u}{2}})^{\frac{u}{p+u}}$ *for all* $p \geq 0$ *and all* $u \geq 0$.

(II$_2$) $(A^{\frac{u}{2}} C^{p_0} A^{\frac{u}{2}})^{\frac{u}{p_0+u}} \geq A^u \geq (A^{\frac{u}{2}} B^{p_0} A^{\frac{u}{2}})^{\frac{u}{p_0+u}}$ *for a fixed positive number* $p_0$ *and for all* $u$ *such that* $u \in [0, u_0]$, *where* $u_0$ *is a fixed positive number.*

(III$_1$) $\log(A^{\frac{u}{2}} C^p A^{\frac{u}{2}}) \geq \log A^{p+u} \geq \log(A^{\frac{u}{2}} B^p A^{\frac{u}{2}})$ *for all* $p \geq 0$ *and all* $u \geq 0$,

(III$_2$) $\log(A^{\frac{u}{2}} C^{p_0} A^{\frac{u}{2}}) \geq \log A^{p_0+u} \geq \log(A^{\frac{u}{2}} B^{p_0} A^{\frac{u}{2}})$ *for a fixed positive number* $p_0$ *and for all* $u$ *such that* $u \in [0, u_0]$, *where* $u_0$ *is a fixed positive number.*

(IV$_1$) $S(A^{-u}|C^p) \geq S(A^{-u}|A^p) \geq S(A^{-u}|B^p)$ *for all* $p \geq 0$, *and* $u \geq 0$.

(IV$_2$) $S(A^{-u}|C^{p_0}) \geq S(A^{-u}|A^{p_0}) \geq S(A^{-u}|B^{p_0})$ *for a fixed positive number* $p_0$, *and for all* $u$ *such that* $u \in [0, u_0]$, *where* $u_0$ *is a fixed positive number.*

**Corollary 1.** *Let* $A$, $B$ *and* $C$ *be positive invertible operators. If* $C \gg A^{-1} \gg B$, *then for for each* $\alpha \in [0, 1]$

$$W_{s,\alpha}(A|C) \geq -\{1 + (1+\alpha)s\} A \log A \geq W_{s,\alpha}(A|B)$$

*for any* $s \geq 1$.

**Corollary 2** [14]. *Let* $A$ *and* $B$ *be positive invertible operators. If* $C \gg A^{-1} \gg B$, *then* $S(A|C) \geq -2A \log A \geq S(A|B)$.

**Corollary 3** [15]. *Let* $A$ *and* $B$ *be positive invertible operators. For any positive number* $x_0$, *the following inequality holds;*

$$(\log x_0 - 1)A + \frac{1}{x_0} B \geq S(A|B) \geq (1 - \log x_0)A - \frac{1}{x_0} AB^{-1}A.$$

## 4. Proofs of results in 2. and 3.

*Proof of Theorem* 1. First of all, we recall the following (1) in [7], Theorem 1, and [14], Theorem 1.

$$\log A \geq \log B \text{ holds iff } A^u \geq (A^{\frac{u}{2}} B^p A^{\frac{u}{2}})^{\frac{u}{p+u}} \text{ holds for all } p \geq 0 \text{ and } u \geq 0. \quad (1)$$

We put $A_1 = A^u$ and $B_1 = (A^{\frac{u}{2}} B^p A^{\frac{u}{2}})^{\frac{u}{p+u}}$ in (1). The hypothesis $\log A \geq \log B$ implies $A_1 \geq B_1 \geq 0$ by (1) so that, by Theorem G, for $t \in [0,1]$, $r = 1$ and $p_1 \geq 1$,

$$F_{p_1,t}(A_1, B_1, 1, s) = A_1^{\frac{-1}{2}} \{A_1^{\frac{1}{2}} (A_1^{\frac{-t}{2}} B_1^{p_1} A_1^{\frac{-t}{2}})^s A_1^{\frac{1}{2}}\}^{\frac{2-t}{(p_1-t)s+1}} A_1^{\frac{-1}{2}} \quad (2)$$

is a decreasing function of $s$ such that $s \geq 1$, and the following inequality holds

$$A_1^{1-t} = F_{p_1,t}(A_1, A_1, 1, s) \geq F_{p_1,t}(A_1, B_1, 1, s) \quad (3)$$

for any $s \geq 1$. We may assume that $u > 0$ since in the case $u = 0$, the result is trivial. Put $p_1 = \dfrac{p+u}{u} \geq 1$ in (2), then $B_1^{p_1} = A^{\frac{u}{2}} B^p A^{\frac{u}{2}}$ and (2) implies

$$A^{\frac{-u}{2}} \{A^{\frac{u}{2}} (A^{\frac{u(1-t)}{2}} B^p A^{\frac{u(1-t)}{2}})^s A^{\frac{u}{2}}\}^{\frac{u(2-t)}{(p+u(1-t))s+u}} A^{\frac{-u}{2}} \quad (4)$$

is a decreasing function of $s$ such that $s \geq 1$.

Put $\alpha = 1 - t$. Then $\alpha \in [0,1]$ since $t \in [0,1]$. By (4) for each $\alpha \in [0,1]$, and all $p \geq 0$ and $u \geq 0$,

$$\begin{aligned} F_{p_1,t}(A_1, B_1, 1, s) &= G_{p,u,\alpha}(A, B, s) \\ &= A^{\frac{-u}{2}} \{A^{\frac{u}{2}} (A^{\frac{u\alpha}{2}} B^p A^{\frac{u\alpha}{2}})^s A^{\frac{u}{2}}\}^{\frac{u(\alpha+1)}{(u\alpha+p)s+u}} A^{\frac{-u}{2}} \end{aligned} \quad (5)$$

is a decreasing function of $s$ such that $s \geq 1$.

Hence the proof of (I) is complete.

(II). As $\log t$ is an operator monotone function, (I) implies that $A \gg B$ ensures

$$\frac{1}{(u\alpha+p)s+u} \log\{A^{\frac{u}{2}} (A^{\frac{u\alpha}{2}} B^p A^{\frac{u\alpha}{2}})^s A^{\frac{u}{2}}\}$$

is a decreasing function of $s$ such that $s \geq 1$, and for each $\alpha \in [0,1]$, and all $p \geq 0$ and $u \geq 0$, $\dfrac{1}{(u\alpha+p)s+u} W_{s,\alpha}(A^{-u}|B^p)$ is a decreasing function of $s$ such that $s \geq 1$.

Hence the proof of Theorem 1 is complete.

*Proof of Theorem* 2. (I) $\Longrightarrow$ ($\mathrm{II}_1$). $A \gg B$ ensures (3) in the proof of (I) in Theorem 1, so that by (3) and (5) for each $\alpha \in [0,1]$, and all $p \geq 0$ and $u \geq 0$, we have $A^{u\alpha} \geq G_{p,u,\alpha}(A,B,s)$ for any $s \geq 1$, that is, for each $\alpha \in [0,1]$, and all $p \geq 0$ and $u \geq 0$,

$$A^{u(\alpha+1)} \geq \{A^{\frac{u}{2}}(A^{\frac{u\alpha}{2}}B^pA^{\frac{u\alpha}{2}})^sA^{\frac{u}{2}}\}^{\frac{u(\alpha+1)}{(u\alpha+p)s+u}}$$

holds for any $s \geq 1$, so the proof of (I) $\Longrightarrow$ ($\mathrm{II}_1$) is complete.

($\mathrm{II}_1$) $\Longrightarrow$ ($\mathrm{II}_2$) and ($\mathrm{III}_1$) $\Longrightarrow$ ($\mathrm{III}_2$) are both trivial since we have only to put $p = p_0$ and $s = s_0$ in ($\mathrm{II}_1$) and ($\mathrm{III}_1$) respectively.

($\mathrm{II}_1$) $\Longrightarrow$ ($\mathrm{III}_1$) and ($\mathrm{II}_2$) $\Longrightarrow$ ($\mathrm{III}_2$) are both obvious since $\log t$ is an operator monotone function.

($\mathrm{III}_2$) $\Longrightarrow$ (I). Put $s_0 = 1$, $p_0 = 1$ and $u = 0$ in ($\mathrm{III}_2$), then $\log A \geq \log B$, so that we have (I).

Hence the proof of Theorem 2 is complete.

*Proof of Theorem* 3. (I) $\Longleftrightarrow$ ($\mathrm{II}_1$). The hopothesis $C \gg A$ in (I), that is, $A^{-1} \gg C^{-1}$ and by (I) and ($\mathrm{II}_1$) of Theorem 2, the latter condition is equivalent to the following; for each $\alpha \in [0,1]$, and all $p \geq 0$ and all $u \geq 0$,

$$A^{-u(\alpha+1)} \geq \{A^{\frac{-u}{2}}(A^{\frac{-u\alpha}{2}}C^{-p}A^{\frac{-u\alpha}{2}})^sA^{\frac{-u}{2}}\}^{\frac{u(\alpha+1)}{(u\alpha+p)s+u}}$$

holds for any $s \geq 1$. Taking inverses yields the following:

For each $\alpha \in [0,1]$, and all $p \geq 0$ and all $u \geq 0$,

$$\{A^{\frac{u}{2}}(A^{\frac{u\alpha}{2}}C^pA^{\frac{u\alpha}{2}})^sA^{\frac{u}{2}}\}^{\frac{u(\alpha+1)}{(u\alpha+p)s+u}} \geq A^{u(\alpha+1)}$$

and the rest of ($\mathrm{II}_1$) is already shown by (I) and ($\mathrm{II}_1$) of Theorem 2, so we have (I) $\Longrightarrow$ ($\mathrm{II}_1$). For the proof ($\mathrm{II}_1$) $\Longrightarrow$ (I), we have only to trace the reverse implication in the proof (I) $\Longleftrightarrow$ ($\mathrm{II}_1$). Hence the proof of equivalence relation between (I) and ($\mathrm{II}_1$) is complete. By the same method as in the proof of Theorem 2 and together with the same technique as in the proof (I) $\Longleftrightarrow$ ($\mathrm{II}_1$) in Theorem 3, we can easily show the equivalence relation among (I), ($\mathrm{II}_1$), ($\mathrm{II}_2$),($\mathrm{III}_1$) and ($\mathrm{III}_2$).

($\mathrm{III}_1$) $\Longleftrightarrow$ ($\mathrm{IV}_1$). ($\mathrm{III}_1$) is equivalent to the following inequalities: For each $\alpha \in [0,1]$, all $p \geq 0$ and all $u \geq 0$,

$$\begin{aligned} A^{\frac{-u}{2}}\log\{A^{\frac{u}{2}}(A^{\frac{u\alpha}{2}}C^pA^{\frac{u\alpha}{2}})^sA^{\frac{u}{2}}\}A^{\frac{-u}{2}} &\geq A^{\frac{-u}{2}}\log A^{(u\alpha+p)s+u}A^{\frac{-u}{2}} \\ &\geq A^{\frac{-u}{2}}\log\{A^{\frac{u}{2}}(A^{\frac{u\alpha}{2}}B^pA^{\frac{u\alpha}{2}})^sA^{\frac{u}{2}}\}A^{\frac{-u}{2}} \end{aligned}$$

holds for any $s \geq 1$, that is, for each $\alpha \in [0,1]$, and all $p \geq 0$, and $u \geq 0$,

$$W_{s,\alpha}(A^{-u}|C^p) \geq W_{s,\alpha}(A^{-u}|A^p) \geq W_{s,\alpha}(A^{-u}|B^p)$$

holds for any $s \geq 1$ and this is just ($\mathrm{IV}_1$).

Similarly we have $(\mathrm{III}_2) \Longleftrightarrow (\mathrm{IV}_2)$.

$(\mathrm{IV}_1) \Longrightarrow (\mathrm{IV}_2)$ is trivial since $(\mathrm{IV}_2)$ is a special case of $(\mathrm{IV}_1)$.

$(\mathrm{IV}_2) \Longrightarrow (\mathrm{I})$. Put $u = 0$, $p_0 = 1$ and $s_0 = 1$ in $(\mathrm{IV}_2)$. Then we have $C \gg A \gg B$.

Hence the proof of Theorem 3 is complete.

*Proof of Theorem* 4. First of all, we cite the following obvious inequality for any positive numbers $x$ and $x_0$

$$\log x_0 - 1 + \frac{x}{x_0} \geq \log x \geq 1 - \log x_0 - \frac{1}{x_0 x}. \tag{6}$$

We can interchange $x$ with positive operator $A^{-1/2}(A^{-\alpha/2}BA^{-\alpha/2})^s A^{-1/2}$ in (6), then

$$\begin{aligned} &A^{1/2}\{\log x_0 - 1 + \frac{1}{x_0} A^{-1/2}(A^{-\alpha/2}BA^{-\alpha/2})^s A^{-1/2}\}A^{1/2} \\ &\geq A^{1/2} \log\{A^{-1/2}(A^{-\alpha/2}BA^{-\alpha/2})^s A^{-1/2}\}A^{1/2} \\ &\geq A^{1/2}\{1 - \log x_0 - \frac{1}{x_0} A^{1/2}(A^{\alpha/2}B^{-1}A^{\alpha/2})^s A^{1/2}\}A^{1/2} \end{aligned}$$

that is,

$$\begin{aligned} &(\log x_0 - 1)A + \tfrac{1}{x_0}(A^{-\alpha/2}BA^{-\alpha/2})^s \\ &\geq W_{s,\alpha}(A|B) \geq (1 - \log x_0)A - \tfrac{1}{x_0}A(A^{\alpha/2}B^{-1}A^{\alpha/2})^s A. \end{aligned} \tag{7}$$

For proof of the latter part of Theorem 4, put $x_0 = 1$ and $W_{s,\alpha}(A|B) = 0$ in (7), then

$$-A + (A^{-\alpha/2}BA^{-\alpha/2})^s \geq 0 \geq A - A(A^{\alpha/2}B^{-1}A^{\alpha/2})^s A,$$

that is, $(A^{-\alpha/2}BA^{-\alpha/2})^s \geq A$ and $A(A^{\alpha/2}B^{-1}A^{\alpha/2})^s A \geq A$ and the latter inequality is equivalent to $A \geq (A^{-\alpha/2}BA^{-\alpha/2})^s$, so that $A = (A^{-\alpha/2}BA^{-\alpha/2})^s$, that is, $B^s = A^{1+\alpha s}$ holds. The reverse implication is trivial by the definition of $W_{s,\alpha}(A|B)$ in $(*)$.

Hence the proof of Theorem 4 is complete.

*Proof of Theorem* 5. We have only to give a proof of (IV) since (I), (II) and (III) are all immediate consequence of the definition $(*)$ of $W_{s,\alpha}(A|B)$. By using the following well-known formula $\lim_{t\to+0}\frac{x^t-1}{t} = \log x$, for each $\alpha \in [0,1]$ and any $s \geq 1$,

$$\begin{aligned}
\text{s-}\lim_{t\to+0}\frac{g_{s,\alpha}(t)-A}{t} &= \text{s-}\lim_{t\to+0}\frac{A^{\frac{1}{2}}\{A^{\frac{-1}{2}}(A^{\frac{-\alpha}{2}}BA^{\frac{-\alpha}{2}})^sA^{\frac{-1}{2}}\}^tA^{\frac{1}{2}}-A}{t} \\
&= A^{\frac{1}{2}}\left(\text{s-}\lim_{t\to+0}\frac{\{A^{\frac{-1}{2}}(A^{\frac{-\alpha}{2}}BA^{\frac{-\alpha}{2}})^sA^{\frac{-1}{2}}\}^t-I}{t}\right)A^{\frac{1}{2}} \\
&= A^{\frac{1}{2}}\log\{A^{\frac{-1}{2}}(A^{\frac{-\alpha}{2}}BA^{\frac{-\alpha}{2}})^sA^{\frac{-1}{2}}\}A^{\frac{1}{2}} \\
&= W_{s,\alpha}(A|B),
\end{aligned}$$

so the proof of (IV) is complete.

*Proof of Theorem* A. Put $s=1$ and $\alpha=0$ in Theorem 2 and also replace $ps \geq 0$ by $p \geq 0$ replace $p_0s_0 \geq 0$ by $p_0 \geq 0$.

*Proof of Theorem* B. As $W_{1,0}(A|B) = S(A|B)$ holds, we have only to put $s=1$ and $\alpha=0$ and also replace $ps \geq 0$ by $p \geq 0$ and replace $p_0s_0 \geq 0$ by $p_0 \geq 0$ in Theorem 3.

*Proof of Corollary* 1. Put $u=1$ and $p=1$ and also replacing $A$ by $A^{-1}$ in ($\text{IV}_1$) of Theorem 3.

*Proof of Corollary* 2. Put $\alpha=0$ and $s=1$ in Corollary 1 since $W_{1,0}(A|B) = S(A|B)$ holds.

*Proof of Corollary* 3. Put $\alpha=0$ and $s=1$ in Theorem 4 since $W_{1,0}(A|B) = S(A|B)$ holds.

## References

1. T. Ando, *On some operator inequality.* Math. Ann., **279** (1987), 157–159.
2. T. Ando and F. Hiai, *Log majorization and complementary Golden-Thompson type inequalities.* Linear Alg. and Its Appl. **197**, **198** (1994), 113–131.
3. G. Corach, H. Porta and L. Recht, *Geodesics and operator means in the space of positive operators.* To appear in Int. J. Math.
4. J.I. Fujii and E. Kamei, *Relative operator entropy in noncommutative information theory.* Math. Japon. **34** (1989), 341–348.
5. J.I. Fujii and E. Kamei, *Uhlmann's interpolational method for operator means.* Math. Japon. **34** (1989), 541–547.

6. M. Fujii, *Furuta's inequality and its mean theoretic approach.* J. Operator Theory **23** (1990), 67–72.
7. M. Fujii, T. Furuta and E. Kamei, *Furuta's inequality and its application to Ando's theorem.* Linear Alg. and Its Appl. **179** (1993), 161–169.
8. M. Fujii, and E. Kamei, *Furuta's inequality and a generalization of Ando's Theorem.* Proc. Amer. Math. Soc. **115** (1992), 409–413.
9. M. Fujii, and E. Kamei, *Mean theoretic approach to the grand Furuta inequality.* To appear in Proc. Amer. Math. Soc.
10. T. Furuta, *$A \geq B \geq 0$ assures $(B^r A^p B^r)^{1/q} \geq B^{(p+2r)/q}$ for $r \geq 0$, $p \geq 0$, $q \geq 1$ with $(1+2r)q \geq p+2r$.* Proc. Amer. Math. Soc. **101** (1987), 85–88.
11. T. Furuta, *A proof via operator means of an order preserving inequality.* Linear Alg. and Its Appl. **113** (1989), 129–130.
12. T. Furuta, *Elementary proof of an order preserving inequality.* Proc. Japan Acad. **65** (1989), 126.
13. T. Furuta, *Two operator functions with monotone property.* Proc. Amer. Math. Soc. **111** (1991), 511–516.
14. T. Furuta, *Applications of order preserving operator inequalities.* Operator Theory: Advances and Applications **59** (1992), 180–190.
15. T. Furuta, *Furuta's inequality and its application to the relative operator entropy.* J. Operator Theory **30** (1993), 21–30.
16. T. Furuta, *Extension of the Furuta inequality and Ando-Hiai log majorization.* Linear Alg. and Its Appl. **219** (1995), 139–155.
17. E. Heinz, *Beiträge zur Störungstheorie der Spektralzerlegung.* Math. Ann. **123** (1951), 415–438.
18. E. Kamei, *A satellite to Furuta's inequality.* Math. Japon. **33** (1988), 883–886.
19. K. Löwner, *Über monotone Matrixfunktionen.* Math. Z. **38** (1934), 177–216.
20. M. Nakamura and H. Umegaki, *A note on the entropy for operator algebras.* Proc. Japan Acad. **37** (1961), 149–154.
21. K. Tanahashi, *Best possibility of the Furuta inequality.* To appear in Proc. Amer. Math. Soc.
22. H. Umegaki, *Conditional expectation in operator algebra IV* (entropy and information). Kodai Math. Sem. Rep. **14** (1962), 59–85.

Takayuki Furuta, Department of Applied Mathematics, Faculty of Science, Science University of Tokyo, Kagurazaka, Shinjuku 162, Tokyo, Japan.

International Series of Numerical Mathematics, Vol. 123, © 1997 Birkhäuser Verlag Basel

# The arithmetic mean – the geometric mean and related matrix inequalities

J. Pečarić and B. Mond

## 1. Introduction

The difficulty of establishing a (noncommutative) matrix inequality involving the geometric mean was discussed in 1978 by K.V. Bhagwat and R. Subramanian [9] who pointed out that the problem of defining a geometric mean for non-commutative operators "makes it difficult to establish the validity or otherwise of the classical inequalities involving the geometric mean". However, in a recent paper, Sagae and Tanabe [32] define a geometric mean and establish an AG-GM inequality for a finite number of positive definite matrices.

Here we make use of their definition of the geometric mean of positive definite matrices to establish matrix versions of numerous inequalities involving means. These include mixed mean matrix inequalities and conditions under which the matrix inequalities of [32] are reversed. Converses for the matrix convexity of the inverse function as well as some bounds for the exponential functions of matrices are also given.

## 2. General inequalities

Let $w_i$, $C_i$ be positive numbers with $\sum_{i=1}^{r} w_i = 1$. Then the power means are defined by

$$M_r^{[t]}(C;w) = (w_1 C_1^t + \ldots + w_r C_r^t)^{1/t}, \quad t \neq 0 \tag{2.1}$$

and

$$M_r^{[0]}(C;w) = C_1^{w_1} \ldots C_r^{w_r}. \tag{2.2}$$

It is well known that if $s \leq t$, then

$$M_r^{[s]}(C;w) \leq M_r^{[t]}(C;w). \tag{2.3}$$

Moreover, if $C_i, i = 1, \ldots, r$ are $n \times n$ positive definite matrices, we can use the definition of power means of order $t \neq 0$ given by (2.1). The geometric mean $M_r^{[0]} \equiv G_r$ cannot be given by (2.2) since the product given by (2.2) need not be a positive definite matric even if all $C_i$ are positive definite.

Instead we shall use the definition of the geometric mean of $r$ matrices recently given in [32]:

$$\begin{aligned} M_r^{[0]}(C; w) \equiv \ & (C_r^{1/2}(C_r^{-1/2}C_{r-1}^{1/2} \cdots (C_3^{-1/2}C_2^{1/2}(C_2^{-1/2}C_1C_2^{-1/2})^{u_1} \\ & C_2^{1/2}C_3^{-1/2})^{u_2} \cdots C_{r-1}^{1/2}C_r^{-1/2})^{u_{r-1}}C_r^{1/2}, \end{aligned} \tag{2.4}$$

where $u_i = 1 - w_{i+1} \Big/ \sum_{k=1}^{i+1} w_k$ for $i = 1, \ldots, r-1$. We shall use the notation $A_r$ for the arithmetic mean $M_r^{[1]}$ and $H_r$ for the harmonic mean $M_r^{[-1]}$.

Sagae and Tanabe [32] proved that the inequalities between the harmonic, geometric and arithmetic means also hold in the matrix case, i.e.,

**Theorem 2.1.**

$$H_r \leq G_r \leq A_r, \tag{2.5}$$

*where $A \geq B$ means that $A - B$ is a positive semi-definite matrix. Equalities hold in (2.5) iff $C_1 = \ldots = C_r$.*

If the weights $w_i$ are not all positive, we can give reverse results.

We begin with the following ([2]):

**Lemma 2.1.** *If $\alpha \in (0, 1)$, then*

$$G_2 \equiv C_2^{1/2}(C_2^{-1/2}C_1C_2^{-1/2})^{\alpha}C_2^{1/2} \leq \alpha C_1 + (1-\alpha)C_2 \equiv A_2, \tag{2.6}$$

*but if either $\alpha < 0$ or $\alpha > 1$, the reverse inequality, i.e.,*

$$G_2 \geq A_2 \tag{2.7}$$

*is valid.*

*Proof.* The following generalization of Bernoulli's inequality is well known (see, e.g., [18], p.34), or [19], p.65]).

For $-1 < x \neq 0$

$$(1+x)^{\alpha} > 1 + \alpha x \quad \text{if} \quad \alpha > 1 \quad \text{or} \quad \alpha < 0$$

and

$$(1+x)^\alpha < 1+\alpha x \quad \text{if} \quad 0<\alpha<1. \tag{2.8}$$

For $1+x=t$, this is equivalent to

$$t^\alpha > \alpha t + 1 - \alpha \quad \text{if} \quad \alpha > 1 \quad \text{or} \quad \alpha < 0 \tag{2.9}$$

and

$$t^\alpha < \alpha t + 1 - \alpha \quad \text{if} \quad 0 < \alpha < 1. \tag{2.10}$$

For $t = 1$, we have equality.

If the positive definite matrix $C$ has the representation $C = \Gamma D_\lambda \Gamma^*$ when $\Gamma$ is unitary and $D_\lambda = \text{diag}\,(\lambda_1, \ldots, \lambda_n)$, where $\lambda_1, \ldots, \lambda_n$ are characteristic roots of $C$, then it follows from (2.9) and (2.10) that

$$D_\lambda^\alpha > \alpha D_\lambda + (1-\alpha) I \quad \text{if} \quad \alpha > 1 \quad \text{or} \quad \alpha < 0$$

and

$$D_\lambda^\alpha < \alpha D_\lambda + (1-\alpha) I \quad \text{if} \quad 0 < \alpha < 1.$$

Pre- and post-multiplication by $\Gamma$ and $\Gamma^*$ respectively, yields

$$C^\alpha > \alpha C + (1-\alpha) I \quad \text{if} \quad \alpha > 1 \quad \text{or} \quad \alpha < 0$$

and

$$C^\alpha < \alpha C + (1-\alpha) I \quad \text{if} \quad 0 < \alpha < 1$$

with equalities iff $C = I$.

If we now set $C = C_2^{-1/2} C_1 C_2^{-1/2}$, we obtain

$$(C_2^{-1/2} C_1 C_2^{-1/2})^\alpha > \alpha C_2^{-1/2} C_1 C_2^{-1/2} + (1-\alpha) I$$

if $\alpha > 1$ or $\alpha < 0$, and

$$(C_2^{-1/2} C_1 C_2^{-1/2})^\alpha < \alpha C_2^{-1/2} C_1 C_2^{-1/2} + (1-\alpha) I$$

for $0 < \alpha < 1$.

Pre- and post-multiplication by $C_2^{1/2}$ now yields (2.7) and (2.6), respectively. Equality holds iff $C_2^{-1/2} C_1 C_2^{-1/2} = I$ or equivalently $C_1 = C_2$.

Sagae and Tanabe [32] used (2.6) and mathematical induction to prove the inequality $G_r \leq A_r$ for positive weights. Similarly, we shall use (2.7).

**Theorem 2.2.** *Let $w_i, i = 1, \ldots, r$ be real numbers such that*

$$w_1 > 0, \quad w_i < 0, \; i = 2, \ldots, r; \quad w_1 + \ldots + w_r = 1.$$

*Then*

$$A_r \leq G_r. \tag{2.11}$$

*If we also have* $w_1 C_1^{-1} + \cdots + w_r C_r^{-1} > 0$, *then*

$$G_r \leq H_r. \tag{2.12}$$

*Equality holds in* (2.11) *and* (2.12) *iff* $C_1 = \cdots = C_r$.

*Proof.* For $r = 2$, (2.11) is proved in Lemma 2.1, i.e., it is inequality (2.7). So suppose (2.11) holds for $r-1$.

Let $A_{r-1}$ and $G_{r-1}$ be weighted arithmetic and geometric means of matrices $C_1, \ldots, C_{r-1}$ with weights $\tilde{w}_i = w_i \Big/ \sum_{i=1}^{r-1} w_i$ for $i = 1, \ldots, r-1$.

Note that $\tilde{w}_1 > 0$, $\tilde{w}_i < 0$, $i = 2, \ldots, r-1$; $\tilde{w}_1 + \cdots + \tilde{w}_{r-1} = 1$, and $\tilde{u}_i = 1 - \tilde{w}_{i+1} \Big/ \sum_{j=1}^{i+1} \tilde{w}_j = 1 - w_{i+1} \Big/ \sum_{j=1}^{i+1} w_j = u_i$ for $i = 1, \ldots, r-2$ and $u_{r-1} = 1 - w_r (> 0)$. So we have

$$\begin{aligned} A_r &= \sum_{i=1}^{r-1} w_i C_i + w_r C_r = (1 - w_r) A_{r-1} + w_r C_r \\ &\leq (1 - w_r) G_{r-1} + w_r C_r \quad \text{(by the inductive hypothesis)} \\ &\leq C_r^{1/2} (C_r^{-1/2} G_{r-1} C_r^{-1/2})^{1-w_r} C_r^{1/r} = G_r \qquad \text{(by (2.7))}. \end{aligned}$$

The equality $A_r = G_r$ holds only when all the equalities are valid simultaneously. Equality in the first inequality holds if $A_{r-1} = G_{r-1}$, i.e., $C_1 = C_2 = \cdots = C_{r-1}$ by induction for $r-1$ and equality in the second inequality holds if $G_{r-1} = C_r$ by the conditions for equality for $r = 2$. Therefore, the equality $A_r = G_r$ holds iff $A_r = G_{r-1} = C_r$, i.e., $C_1 = C_2 = \cdots = C_r$.

Now by the substitutions $C_i^{-1} \to C_i$, $i = 1, \ldots, r$, we get (2.12) from (2.11).

**Remark.** It is interesting that (2.11), i.e., the reverse arithmetic-geometric inequality is stronger than the same arithmetic-geometric means inequality in the sense that we can obtain the second inequality in (2.5) from (2.11) (see [2]).

Power means inequality (2.3) for $t, s \neq 0$, in the case when $w_i = 1/r$ (in this case we shall write $M_r^{[t]}(C)$ for $M_r^{[t]}(C; w)$) was considered by Bhagwat and Subramanian [9]. They proved

1. If $s \leq t$, $s \notin (-1, 1)$, $t \notin (-1, 1)$, then

$$M_r^{[t]}(C) \geq M_r^{[s]}(C).$$

2. For $t \geq 1/2$,
$$M_r^{[2t]}(C) \geq M_r^{[t]}(C).$$

Moreover, it can be shown that the following results hold [20, 21]:

**Theorem 2.3.** *Inequality* (2.3) *holds if one of the following is valid:*

(a) $t \geq s$, $t \notin (-1,1)$, $s \notin (-1,1)$; *or*

(b) $t \geq 1 \geq s \geq 1/2$; *or*

(c) $s \leq -1 \leq t \leq -1/2$.

*Proof.* First we shall prove (2.3) if either (a); (d) $t \geq 1 \geq s \geq t/2$; or (e) $s \leq -1 \leq t \leq s/2$, holds. For this we need only Jensen's inequality for matrix convex functions

$$f\left(\sum_{k=1}^{r} w_k A_k\right) \leq \sum_{k=1}^{r} w_k f(A_k), \quad w_k \geq 0, \quad w_1 + \ldots + w_r = 1. \tag{2.13}$$

Further, it is known that the function $f(x) = x^p$ is matrix concave for $0 < p \leq 1$ and matrix convex for $1 \leq p \leq 2$ or $-1 \leq p < 0$; while the functions $g(x) = x^{1/t}$, for $t \geq 1$, and $h(x) = -x^{1/t}$, for $t \leq -1$ are matrix monotone. Using these facts and the substitutions $f(x) = x^{s/t}$, $A_i = C_i^t$ (or $f(x) = x^{t/s}$, $A_i = C_i^s$) in (2.13), we get (2.3) with condition (a), (d) or (e). We shall only prove the case for $t \geq s \geq 1$.

The function $f(x) = x^{s/t}$ is matrix concave since $0 < s/t \leq 1$. With the substitution $A_i = C_i^t$, (2.13) becomes

$$\left(\sum_{k=1}^{r} w_k C_k^t\right)^{s/t} \geq \sum_{k=1}^{r} w_k C_k^s. \tag{2.14}$$

Since the function $g(x) = x^{1/s}$ is matrix monotone, (2.3) follows from (2.14).

Further (d) for $t = 1$ gives (for $1 \geq s \geq 1/2$)

$$M_r^{[1]}(C; w) \geq M_r^{[s]}(C; w).$$

On the other hand (a) for $s = 1$ gives

$$M_r^{[t]}(C; w) \geq M_r^{[1]}(C; w).$$

These inequalities give (2.3) if (b) holds. Similarly, using (e) and (a), we can prove that (2.3) holds if (c) is valid.

## 3. Mixed means matrix inequalities

Recently, K. Kedlaya [16] proved the following conjecture of F. Holland [12]:

Let $x_1, x_2, \ldots, x_n$ be positive real numbers. Then

$$\left(\prod_{j=1}^{n} \frac{x_1 + x_2 + \ldots x_j}{j}\right)^{1/n} \geq \frac{1}{n}\sum_{i=1}^{n} \sqrt[i]{x_1, x_2, \ldots, x_i} \tag{3.1}$$

with equality iff $x_1 = x_2 = \cdots = x_n$.

We shall consider related results for positive definite Hermitian matrices. The open problem is to give a generalization for positive definite matrices for arbitrary number of matrices. The cases of two matrices were considered in [22, 28]. Let the arithmetic, geometric and harmonic means be defined respectively by

$$A\nabla_\lambda B = \lambda A + (1-\lambda)B$$

$$A\#_\lambda B = A^{1/2}(A^{-1/2}BA^{-1/2})^{1-\lambda}A^{1/2}$$

$$A!_\lambda B = (\lambda A^{-1} + (1-\lambda)B^{-1})^{-1}$$

where $\lambda \in (0,1)$.

The following are special cases of more general results obtained in [22]:

Let $\alpha, \lambda \in (0,1)$. Then

$$\begin{gathered} A\#_\alpha(A\nabla_\lambda B) \geq A\nabla_\lambda(A\#_\alpha B) \\ A\#_\alpha(A!_\lambda B) \leq A!_\lambda(A\#_\alpha B), \\ A!_\alpha(A\nabla_\lambda B) \geq A\nabla_\lambda(A!_\alpha B) \end{gathered} \tag{3.2}$$

Moreover using the idea of the method of proof from [16] (but instead of Hölder's inequality, the Minkowski type matrix inequality from [6]), the following mixed arithmetic-harmonic mean inequality was proved in [28]:

**Theorem 3.1.** *Let $A_1, \ldots, A_n$ be positive definite $m \times m$ Hermitian matrices. Then*

$$\left[\frac{1}{n}\sum_{j=1}^{n}\left(\frac{A_1 + A_2 + \ldots + A_j}{j}\right)^{-1}\right]^{-1} \geq \frac{1}{n}\sum_{i=1}^{n}\left(\frac{1}{i}\sum_{k=1}^{i} A_k^{-1}\right)^{-1} \tag{3.3}$$

## 4. Reverse forms of a convex matrix inequality

Let $A$ and $B$ be two complex $n \times n$ Hermitian positive definite matrices, and let $0 \le \lambda \le 1$. Then

$$[\lambda A + (1-\lambda)B]^{-1} \le \lambda A^{-1} + (1-\lambda)B^{-1}. \tag{4.1}$$

This result, i.e., matrix convexity of the inverse function is an old result that appears explicitly in the papers of J. Bendat and S. Sherman [8], C. Davis [10], M.H. Moore [29], I. Olkin and J. Pratt [30] and P. Whittle [33].

B. Mond and J. Pečarić [23] have proved the following reverse results:

**Theorem 4.1.** *Let $A$ and $B$ be two complex Hermitian positive definite matrices, and let $0 \le \lambda \le 1$. Then*

$$[\lambda A + (1-\lambda)B]^{-1} \ge K(\lambda A^{-1} + (1-\lambda)B^{-1}) \tag{4.2}$$

*where*

$$K = 4 \min_i \frac{\mu_i}{(1+\mu_i)^2}, \tag{4.3}$$

*and the $\mu_i$ are the solutions of the equation*

$$\det(B - \mu A) = 0. \tag{4.4}$$

*The constant $K$ is best possible.*

**Theorem 4.2.** *Let $A$ and $B$ be positive definite Hermitian matrices. Then*

$$[\lambda A + (1-\lambda)B]^{-1} - (\lambda A^{-1} + (1-\lambda)B^{-1}) \ge \tilde{K} A^{-1} \tag{4.5}$$

*where*

$$\tilde{K} = \min_i \frac{(\sqrt{\mu_i} - 1)^2}{-\mu_i} \tag{4.6}$$

*and the $\mu_i$ are the solutions of equation* (4.4). *The constant $\tilde{K}$ is best possible.*

Note that the matrix convexity of the Moore-Penrose (generalized) inverse was considered in the papers of D.G. Kaffes [14], A. Giovagnoli and H.P. Wynn [11] and D.G. Kaffe, T. Mathew, M.B. Rao and K. Subramanyam [15].

Let $A$ and $B$ be two complex Hermitian positive semi-definite matrices of the same order. The inequality

$$[\lambda A + (1-\lambda)B]^{+} \le \lambda A^{+} + (1-\lambda)B^{+} \tag{4.7}$$

for every $0 \le \lambda \le 1$ holds iff $R(A) = R(B)$ where $R(A)$ is the range of $A$.

Similar results to those in Theorems 4.1 and 4.2 can also be given [24]. A related result was also obtained by M. Alić, J. Pečarić and V. Volenec [4].

**Theorem 4.3.** *Let $A$ and $B$ be two positive definite Hermitian matrices and let $\alpha \in (0,1)$. Then*

$$A\nabla_\alpha B \leq K(A\#_\alpha B) \tag{4.8}$$

*where* $A\#_\alpha B = B^{1/2}(B^{-1/2}AB^{-1/2})^\alpha B^{1/2}$ *and*

$$K = \frac{(\beta-1)\beta^{1/(\beta-1)}}{e\log\beta} \tag{4.9}$$

*and* $\beta = \max\{\lambda_1, \frac{1}{\lambda_n}\}, \lambda_1 \geq \cdots \geq \lambda_n$ *are the eigenvalues of* $A^{-1/2}BA^{-1/2}$.

**Remark.** In the papers [23] and [24], we have direct proofs of Theorems 4.1 and 4.2. Note that such results can also be proved using the following general result [13]:

Suppose $X$ and $Y$ are Hermitian matrices of the same size, with $X$ positive definite and $Y$ positive semi-definite. Then $X - Y$ is positive semi-definite iff $\rho(YX^{-1}) \leq 1$, where $\rho$ is a spectral radius. For example, application of this result with $X = [\lambda A + (1-\lambda)B]^{-1} - (\lambda A^{-1} + (1-\lambda)B^{-1})$, $Y = \hat{K}A^{-1}$ gives that the optimal value of $\hat{K}$ is over the entire range of admissible $\lambda$'s of

$$\frac{1}{\lambda + (1-\lambda)\mu_i} - \lambda - (1-\lambda)\frac{1}{\mu_i}.$$

The minimum of this is

$$-(\sqrt{\mu_i} - 1)^2/\mu_i$$

and, the best overall value of $\tilde{K}$ is

$$\min_i \frac{(\sqrt{\mu_i}-1)^2}{-\mu_i},$$

i.e., the constant obtained in Theorem 4.2.

## 5. A matrix version of the Kantorovich inequality and related results

Let $w_i$, $a_i$ be positive numbers such that $0 < m \leq a_i \leq M$, $i = 1, \ldots, n$ and $\sum_{i=1}^n w_i = 1$. Then the well-known Kantorovich inequality holds

$$A_n(a,w) \leq \frac{(m+M)^2}{4mM} H_n(a,w). \tag{5.1}$$

There exist many matrix extensions of this inequality. Here we present some results of B. Mond and J. Pečarić [27].

Let $A_j$ $(j = 1, \ldots, k)$ be positive definite Hermitian matrices of order $n$, with eigenvalues contained in the interval $[m, M]$ where $0 < m < M$ and let $U_j$, $j = 1, \ldots, k$ be $t \times n$ matrices such that

$$\sum_{j=1}^{k} U_j U_j^* = I.$$

We now consider the means

$$M_k^{[r]}(A; U) = \left( \sum_{j=1}^{k} U_j A_j^r U_j^* \right)^{1/r} \qquad (r \neq 0).$$

The following results are valid:

$$M_k^{[-1]}(A; U)^{-1} \leq \frac{(m+M)^2}{4mM} M_k^{[1]}(A; U)^{-1}, \tag{5.2}$$

$$M_k^{[1]}(A; U) - M_k^{[-1]}(A; U) \leq (\sqrt{M} - \sqrt{m})^2 I, \tag{5.3}$$

$$M_k^{[2]}(A; U)^2 \leq \frac{(M+m)^2}{4mM} M_k^{[1]}(A; U)^2, \tag{5.4}$$

$$M_k^{[2]}(A; U) - M_k^{[1]}(A; U) \leq \frac{(M-m)}{4(M+m)} I. \tag{5.5}$$

For $k = 1$, (5.2) was proved by A.W. Marshall and I. Olkin [17]. (5.3)–(5.5) were proved by B. Mond and J. Pečarić [26]. Similarly we can prove

$$M_k^{[2]}(A; U)^2 - M_k^{[1]}(A; U)^2 \leq \frac{(M-m)^2}{4} I. \tag{5.6}$$

As in [26], we can start with

$$mMI \leq (m+M)A_j - A_j^2$$

from which we get

$$mM \sum_{j=1}^{k} U_j U_j^* \leq (m+M) \sum_{j=1}^{k} U_j A_j U_j^* - \sum_{j=1}^{k} U_j A_j^2 U_j^*,$$

i.e.,

$$mMI \leq (m+M) \sum_{j=1}^{k} U_j A_j U_j^* - \sum_{j=1}^{k} U_j A_j^2 U_j^*,$$

i.e.,

$$\begin{aligned}
&M_k^{[2]}(A;U)^2 - M_k^{[1]}(A;U)^2 \\
&\quad \le (m+M)M_k^{[1]}(A;U) - mMI - M_k^{[1]}(A;U)^2 \\
&\quad = \frac{(M-m)^2}{4}I - \{M_k^{[1]}(A;U) - \frac{M+m}{2}I\}^2 \\
&\quad \le \frac{(M-m)^2}{4}I.
\end{aligned}$$

Some further extensions are also obtained by B. Mond and J. Pečarić [27]. Namely the following two theorems are proved:

**Theorem 5.1.** *Let $r, s$ be non-zero numbers with $s > r$ and either $s \notin (-1,1)$ or $r \notin (-1,1)$. Then*

$$M_k^{[s]}(A;U) - M_k^{[r]}(A;U) \le \Delta I \tag{5.7}$$

*where*

$$\Delta = [\theta M^s + (1-\theta)m^s]^{1/s} - [\theta M^r + (1-\theta)m^r]^{1/r}$$

*and where $\theta$ is more precisely defined.*

**Theorem 5.2.** *Let the conditions of Theorem* 5.1 *be satisfied. Then*

$$M_k^{[s]}(A;U) \le \tilde{\Delta} M_k^{[r]}(A;U), \tag{5.8}$$

*where*

$$\tilde{\Delta} = \left\{\frac{r(\gamma^s - \gamma^r)}{(s-r)(\gamma^r - 1)}\right\}^{1/s} \left\{\frac{s(\gamma^r - \gamma^s)}{(r-s)(\gamma^s - 1)}\right\}^{-1/r}$$

*and $\gamma = M/m$.*

**Remark.** S. Liu and H. Neudecker and, independently, C-K Li and R. Mathias (see [27]) have noted that such matrix inequalities for several matrices are equivalent to the corresponding case of one matrix.

**Remark.** Extensions of (5.2)–(5.6) to the case of an $n \times n$ positive semi-definite symmetric matrix were obtained in the papers of J.K. Baksalary and S. Puntanen [7] and J. Pečarić, S. Puntanen and G.P.H. Styan [31].

## 6. Some improvements of the AG inequality

The following result was proved in [4]:

**Theorem 6.1.** *Let $A$ and $B$ be two positive definite Hermitian matrices and let $a$ be a real number. If $A > B, k \in \mathbb{N}$, then*

$$\begin{aligned} &\binom{a}{k+1}\{A\#_\alpha B - A\nabla_\alpha B\} \\ \geq &\binom{a}{k+1}\sum_{p=2}^{k}\binom{a}{p} B^{1/2}[B^{-1/2}(A-B)B^{-1/2}]^p B^{1/2} \end{aligned} \tag{6.1}$$

*but, if $A < B$, then*

$$\begin{aligned} &(-1)^{k+1}\binom{a}{k+1}\{A\#_\alpha B - A\nabla_\alpha B\} \\ \geq &(-1)^{k+1}\binom{a}{k+1}\sum_{p=2}^{k}\binom{a}{p} B^{1/2}[B^{-1/2}(A-B)B^{-1/2}]^p B^{1/2}. \end{aligned} \tag{6.2}$$

A special case, $k = 2$ is the following:

**Corollary 6.1.** *Let $A$ and $B$ be two positive definite Hermitian matrices and let $a$ be a real number and let $A \geq B$. If $a \in (0,1) \cup (2,\infty)$, then*

$$A\nabla_\alpha B - A\#_\alpha B \leq \frac{a(1-a)}{2} B^{1/2}[B^{-1/2}(A-B)B^{-1/2}]^2 B^{1/2}. \tag{6.3}$$

*If $a \in (-\infty, 0) \cup (1,2)$, the reverse inequality in (6.3) is valid.*

*If $A \leq B$, we have reverse results.*

Further generalizations are obtained in [1].

## 7. Some bounds involving exponential functions

Let $A$, $G$ and $H$ denote, respectively, the arithmetic, geometric and harmonic means of $n \times n$ positive definite matrices $C_i$. The following results were proved in [32]:

$$\begin{aligned} G\exp(G^{-1}A - I) &= \exp(AG^{-1} - I)G \\ &\geq (AG^{-1}A + G)/2 \\ &\geq A \geq G \geq H \\ &\geq 2(H^{-1}GH^{-1} + G^{-1})^{-1} \\ &\geq G\exp(I - H^{-1}G) = \exp(I - GH^{-1})G \end{aligned} \tag{7.1}$$

and

$$
\begin{array}{rcl}
H\exp(H^{-1}A-I) & = & \exp(AH^{-1}-I)H \\
& \geq & (AH^{-1}A+H)/2 \\
& \geq & A\geq G\geq H \\
& \geq & 2(H^{-1}AH^{-1}+A^{-1})^{-1} \\
& \geq & A\exp(I-H^{-1}A) \\
& = & \exp(I-AH^{-1})A,
\end{array}
\tag{7.2}
$$

where all the inequalities are strict unless $C_1=\cdots=C_r$, in which case all the equalities hold.

In [2], it was noted that these results are a consequence of some general inequalities for exponential functions and of (2.5).

Namely the following result was proved in [2]:

**Theorem 7.1**. *Let $C$ and $D$ be two positive definite Hermitian matrices such that $C\geq D$. Then for $r,k\in\mathbb{N}$, we have*

$$
\begin{aligned}
&D\exp(D^{-1}C-I)=\exp(CD^{-1}-I)D \\
&\geq D\sum_{i=0}^{r}(D^{-1}C-I)^i/i!\geq\cdots\geq\frac{1}{2}(CD^{-1}C+D)\geq C\geq D \\
&\geq 2(D^{-1}CD^{-1}+C^{-1})^{-1}\geq\cdots\geq(\sum_{i=0}^{r}(CD^{-1}-I)^i/i!)^{-1}C \\
&\geq C\exp(I-D^{-1}C)=\exp(I-CD^{-1})C;
\end{aligned}
\tag{7.3}
$$

$$
\begin{aligned}
C\sum_{i=0}^{2k-1}(C^{-1}D-I)^i/i! &\leq C\exp(C^{-1}D-I)=\exp(DC^{-1}-I)C \\
&\leq C\textstyle\sum_{i=0}^{2k}(C^{-1}D-I)^i/i!,
\end{aligned}
\tag{7.4}
$$

*which, for $k=1$, gives*

$$
D\leq C\exp(C^{-1}D-I)\leq\frac{1}{2}(C+DC^{-1}D);
\tag{7.5}
$$

*and*

$$
\begin{array}{rcl}
\left(\sum_{i=0}^{2k}(DC^{-1}-I\right)^i/i!)^{-1}D & \leq & D\exp(I-C^{-1}D) \\
& = & \exp(I-C^{-1}D)D \\
& \leq & \left(\sum_{i=0}^{2k-1}(DC^{-1}-I\right)^i/i!)^{-1}D,
\end{array}
\tag{7.6}
$$

*which, for $k = 1$, gives*

$$2(C^{-1}DC^{-1} + D^{-1})^{-1} \leq D\exp(I - C^{-1}D) \leq C. \tag{7.7}$$

Now set instead of the couple $(C, D)$, the couples $(A_r, G_r)$, $(G_r, H_r)$, $(A_r, H_r)$, $(M_r^{[t]}, M_r^{[s]})$ (from Theorem 2.3) or $(G_r, A_r,)$, $(H_r, A_r)$, $(H_r, G_r)$ from Theorem 2.2 and we can obtain various extensions of (7.3) and (7.4) as well as many related results.

A corresponding result was also obtained in [3]. Namely, the result obtained for real numbers in [5] also holds in the matrix case. In fact we have the following result for exponential matrix functions.

**Theorem 7.2.** *Let $C$ and $D$ be two positive definite Hermitian matrices such that $C \geq D$. Then*

$$C - D \leq C\exp(C^{-1}D) - D\exp(-D^{-1}C) \leq \frac{3}{e}(C - D). \tag{7.8}$$

*The constant is best possible and equality holds iff $C = D$.*

In fact, a matrix analogue of a result from [5] is a simple consequence of Theorem 7.2 for the arithmetic mean $A_r$ and the geometric mean $G_r$ instead of $C$ and $D$, respectively. Of course using other results we can again obtain many similar theorems for matrix means.

## References

1. M. Alić, P.S. Bullen, J. Pečarić and V. Volenec, *Gerber's and related inequalities.* Submitted for publication.

2. M. Alić, B. Mond, J. Pečarić and V. Volenec, *The arithmetic-geometric-harmonic means and related matrix inequalities.* Submitted for publication.

3. M. Alić, B. Mond, J. Pečarić and V. Volenec, *Bounds for the differences of matrix means.* To appear in SIAM J. on Matrix Analysis.

4. M. Alić, J. Pečarić and V. Volenec, *On the arithmetic mean-the geometric mean matrix inequality.* Submitted for publication.

5. H. Alzer, *A lower bound for the difference between the arithmetic and geometric means.* Nieuw Archief voor Wiskunde, **8** (1990), 195–197.

6. W.N. Amderspm, Jr. and R.J. Duffin, *Series and parallel addition of matrices.* J. Math. Anal. Appl. **26** (1969), 576–594.

7. J.K. Baksalary and S. Puntanen, *Generalized matrix versions of the Cauchy-Schwarz and Kantorovich inequalities.* Aequationes Math. **41** (1991), 103–110.

8. J. Bendat and S. Sherman, *Monotone and convex operator functions.* Trans. Amer. Math. Soc. **79** (1955), 58–71.

9. K.V. Bjagwat and R. Subramanian, *Inequalities between means of positive operators.* Math. Proc. Camb. Phil. Soc. **83** (1978), 393–401.

10. C. Davis, *Notions generalizing convexity for functions defined on spaces of matrices.* Proc. Symp. Pure Math., Vol. 7: Convexity, Amer. Math. Soc. (1963), 187–201.

11. A. Gopvagnoli and H.P. Wynn, *G-majorization with application to matrix orderings.* Lin. Alg. Appl. **67** (1985), 111–135.

12. F. Holland, *On a mixed arithmetic-mean, geometric-mean inequality.* Mathematics Competitions **5** (1992), 60–64.

13. R.A. Horn and C.R. Johnson, *Matrix Analysis.* Cambridge University Press, New York, 1985.

14. D.G. Kaffes, *An inequality for matrices.* Bull. Greek Math. Soc. **22** (1981), 143–159.

15. D.G. Kaffes, T. Mathew, M.B. Rao and K. Subramanyam, *On the matrix convexity of the Moore-Penrose inverse and some applications.* Lin. Multilin. Alg. **24** (1989), 265–271.

16. K. Kedlaya, *Proof of a mixed arithmetic-mean, geometric-mean inequality.* Amer. Math. Monthly **101** (1994), 355–357.

17. A.W. Marshall and I. Olkin, *Matrix versions of the Cauchy and Kantorovich inequalities.* Aequationes Math. **40** (1990), 89–93.

18. D.S. Mitrinović, *Analytic Inequalities.* Springer Verlag, Berlin-Heidelberg-New York, 1970.

19. D.S. Mitrinović, J.E. Pečarić and A.M. Fink, *Classical and new inequalities in analysis.* Kluwer Acad. Pub., Dordrecht-Boston-London, 1993.

20. B. Mond and J.E. Pečarić, *A simple proof of generalized inequalities of Bhagwat and Subramanian and some converse results.* Indian Journal of Mathematics **37** (1995).

21. B. Mond and J.E. Pečarić, *Remarks on Jensen's inequality for operator convex functions.* Ann. Univ. Mariae Curie-Sklodowska, Sec. A, **47** (1993), 96–103.

22. B. Mond and J.E. Pečarić, *Mixed means inequalities for positive linear operators.* Submitted for publication.

23. B. Mond and J.E. Pečarić, *Reverse forms of a convex matrix inequality.* Lin. Alg. Appl. **220** (1995), 359–364.

24. B. Mond and J.E. Pečarić, *On matrix convexity of the Moore-Penrose inverse.* International Journal of Mathematics and Mathematical Sciences. To appear.

25. B. Mond and J.E. Pečarić, *A matrix version of the Ky Fan generalization of the Kantorovich inequality.* Linear and Multilinear Algebra **36** (1994), 217–221.

26. B. Mond and J.E. Pečarić, *Matrix versions of some means inequalities.* Australian Mathematical Society Gazette **20** (1993), 117–120.

27. B. Mond and J.E. Pečarić, *A matrix version of the Ky Fan generalization of the Kantorovich inequality II.* Linear and Multilinear Algebra **38** (1995), 309–313.

28. B. Mond and J.E. Pečarić, *A mixed arithmetic-mean harmonic-mean matrix inequality.* Lin. Alg. Appl. To appear.

29. M.H. Moore, *A convex matrix function.* Amer. Math. Monthly **80** (1973), 408–409.

30. I. Olkin and J. Pratt, *A multivariate Tchebycheff inequality.* Ann. Math. Statist. **29** (1958), 226–234.

31. J. Pečarić, S. Puntanen and G.P.M. Styan, *Some further matrix extensions of the Cauchy-Schwarz and Kantorovich inequalities with some statistical applications.* Lin. Alg. Appl. To appear.

32. M. Sagae and K. Tanabe, *Upper and lower bounds for the arithmetic-geometric-harmonic means of positive definite matrices.* Linear and Multilinear Algebra **37** (1994), 279–282.

33. P. Whittle, *A multivariate generalization of Tchebychev's inequality.* Quart. J. Math. Oxford, Ser [2] **9** (1958), 232–240.

J. Pečarić, Faculty of Textile Technology, University of Zagreb, Zagreb, Croatia.

B. Mond, School of Mathematics, La Trobe University, Bundoora, Victoria 3083, Australia.

# Inequalities for Eigenvalue Problems

International Series of Numerical Mathematics, Vol. 123, © 1997 Birkhäuser Verlag Basel

# Inequalities for the first eigenvalues of the clamped plate and buckling problems

M.S. Ashbaugh*, R.D. Benguria†, and R.S. Laugesen‡

**Abstract**

In 1877 Rayleigh conjectured that the lowest frequency of vibration of a clamped plate of given area occurs when the plate is circular. This conjecture was finally proved by Nadirashvili in 1992, with important earlier contributions due to Szegő and Talenti. An analogous conjecture concerning the critical buckling load of a clamped plate was made by Pólya and Szegő around 1950. This conjecture remains open. This paper surveys the state of our knowledge of these and related problems, including their $n$-dimensional generalizations. In particular, we discuss our recent work proving Rayleigh's clamped plate conjecture for dimension 3 (and 2) and proving related, but presumably nonoptimal, inequalities for the clamped plate problem for dimension $n \geq 4$ and for the buckling problem. In the latter cases, the inequalities have the form of the conjectured lower bounds but contain an unwanted factor slightly less than 1. Our bounds for the clamped plate problem follow from detailed analysis involving Bessel functions, and compare favorably to earlier bounds of a similar form due to Talenti. Our bounds for the buckling problem follow by similar methods, and also from earlier bounds due to Payne and Krahn, as first noted by Bramble and Payne.

## 1. Introduction

In this paper we consider the eigenvalue problem

$$\Delta^2 w = \Gamma w \quad \text{in} \quad \Omega \subset \mathbb{R}^n, \tag{1.1}$$

$$w = 0 = \frac{\partial w}{\partial \nu} \quad \text{on} \quad \partial\Omega, \tag{1.2}$$

*Partially supported by National Science Foundation (USA) grants DMS-9114162, INT-9123481, and DMS-9500968.

†Partially supported by Fondecyt (Chile) projects 193-0561 and 196-0462 and Conicyt (Chile).

‡Partially supported by National Science Foundation (USA) grants DMS-9414149 and DMS-9304580.

where $\Omega$ is a bounded domain in $\mathbb{R}^n$ and $\partial/\partial\nu$ denotes the outward normal derivative with respect to $\partial\Omega$. This problem is known as the clamped plate problem ("clamped" due to the boundary conditions (1.2)); when $n = 2$ its eigenvalues $\Gamma$ are proportional to the squares of the characteristic frequencies of a vibrating clamped plate having the shape of $\Omega$. Here we shall concentrate exclusively on the lowest, or principal, eigenvalue, which we shall denote by $\Gamma_1(\Omega)$ to make its dependence on the domain $\Omega$ explicit. Indeed, our primary interest will be in understanding something of the dependence of $\Gamma_1(\Omega)$ on $\Omega$ and in particular on the area/volume of $\Omega$ (denoted $|\Omega|$).

Specifically, we consider the problem of finding lower bounds to $\Gamma_1(\Omega)$ of the form

$$\Gamma_1(\Omega) \geq c\Gamma_1(\Omega^\star), \tag{1.3}$$

where $\Omega^\star$ denotes the disk/ball in $\mathbb{R}^n$ having the same area/volume as $\Omega$ and $c$ is some constant with $0 < c \leq 1$. This problem was first formulated by Lord Rayleigh in the first edition of *The Theory of Sound* (1877) where he stated simply "...When the edges are clamped, the form of gravest pitch is doubtless the circle" [31, vol. 1, p. 382]. Here Rayleigh is asserting that for $n = 2$ (1.3) should hold with $c = 1$. Rayleigh's statement notwithstanding, this assertion requires proof, and for this proof the world had to wait 115 years, until the recent work of Nadirashvili (see [25], and also [23], [24]).

In his book Rayleigh had also set forth an analogous statement for the principal eigenvalue of the fixed membrane problem

$$-\Delta u = \lambda u \quad \text{in} \quad \Omega \subset \mathbb{R}^n, \tag{1.4}$$

$$u = 0 \quad \text{on} \quad \partial\Omega, \tag{1.5}$$

where $\Omega$ is an arbitrary bounded domain in $\mathbb{R}^n$. That is, Rayleigh asserted that

$$\lambda_1(\Omega) \geq \lambda_1(\Omega^\star) \tag{1.6}$$

should hold for all bounded domains $\Omega$ (actually Rayleigh only talked explicitly about the two-dimensional case, but surely he would have felt that the corresponding result should hold in any dimension). Rigorous proofs of (1.6) awaited the work of Faber [14] and Krahn [18], [19] in the 1920's, and it is now known as the Faber-Krahn, or sometimes the Rayleigh-Faber-Krahn, inequality (see, for example, [6], [27], [28], [30]). The key to its proof is an understanding of what symmetrization (in particular, Steiner or spherical symmetrization; see [6], [16], [30] for more information) of a function $\varphi \in W_0^{1,2}(\Omega)$ does to the Dirichlet norm, $\int_\Omega |\nabla\varphi|^2\,dx$ : specifically, symmetrization does not increase the Dirichlet norm (here $W_0^{1,2}(\Omega)$ is used to denote a particular Sobolev space in standard notation, see [2]; this notation will be used consistently throughout the paper). In particular, since $\lambda_1(\Omega)$

can be characterized variationally by

$$\lambda_1(\Omega) = \inf_{\substack{\varphi \in W_0^{1,2}(\Omega) \\ \varphi \not\equiv 0}} \frac{\int_\Omega |\nabla\varphi|^2\,dx}{\int_\Omega \varphi^2\,dx}, \tag{1.7}$$

the Rayleigh-Faber-Krahn result follows easily using spherical symmetrization as follows:

$$\lambda_1(\Omega) = \frac{\int_\Omega |\nabla u_1|^2\,dx}{\int_\Omega u_1^2\,dx} \geq \frac{\int_{\Omega^\star} |\nabla u_1^\star|^2\,dx}{\int_{\Omega^\star} u_1^{\star 2}\,dx} \geq \lambda_1(\Omega^\star). \tag{1.8}$$

The first inequality here follows from the fact that symmetric decreasing rearrangement of $u_1$ decreases its Dirichlet norm (while leaving its $L^2$ norm fixed) and the second inequality follows from the variational characterization (1.7) for $\lambda_1$ applied to $\Omega^\star$ (and the observation that $u_1^\star \in W_0^{1,2}(\Omega^\star)$).

An analogous approach to the vibrating clamped plate problem (1.1)–(1.2) is not possible since we are now confronted with a fourth order partial differential equation. This complication manifests itself in the following specific ways: (1) we have no guarantee that the first eigenfunction $w_1$ can be taken with constant sign (as we have for $u_1$ in the fixed membrane problem) since maximum principle techniques do not apply, and (2) we can no longer attempt to rearrange $w_1$ in an equimeasurable way since this would not allow us to control the integral $\int_\Omega (\Delta w_1)^2\,dx$ under rearrangement and this is certainly an integral we will need to control, as will be seen below. The first problem is definitely a problem to be reckoned with, since even for relatively simple domains it is known that $w_1$ must change signs, for example, for a square domain or any annular domain with the ratio of outer to inner radii sufficiently large (see [8], [9], [10], [11], [12], [13], and also [17]). We will be able to get around this problem, at least in part, by adopting a rearrangement technique of Talenti [35], where one uses two separate rearrangements, one for the region $\Omega_+$ on which $w_1 > 0$ and the other for the region $\Omega_-$ on which $w_1 < 0$. To discuss the second problem, we first observe that the variational characterization of $\Gamma_1(\Omega)$ is

$$\Gamma_1(\Omega) = \inf_{\substack{\varphi \in W_0^{2,2}(\Omega) \\ \varphi \not\equiv 0}} \frac{\int_\Omega (\Delta\varphi)^2\,dx}{\int_\Omega \varphi^2\,dx}, \tag{1.9}$$

which shows that it will suffice to find a rearrangement for which we can control the integrals $\int_\Omega \varphi^2\,dx$ and $\int_\Omega (\Delta\varphi)^2\,dx$ under rearrangement. A suitable procedure is to rearrange $F \equiv -\Delta w_1$ in an equimeasurable way (denoted $F^\star$) and then to gain control over $\int_{\Omega_+} w_1^2\,dx$ in terms of the solution $u$ of the inhomogeneous equation $-\Delta u = F^\star$ on $(\Omega_+)^\star$ (and an analogous problem on $(\Omega_-)^\star$). This approach goes back at least to Talenti [35], and was certainly also exploited by Nadirashvili. It is suggested by the factorization of the fourth order equation (1.1) and by the two types of integrals appearing in (1.9). In some sense this is what allows us to make progress on this fourth order problem: we have managed to find a way to handle it based on knowing how rearrangement (specifically, symmetric decreasing rearrangement on $\Omega$) affects the second order inhomogeneous elliptic problem $-\Delta u = F$ on $\Omega_+$, with Dirichlet boundary conditions imposed on $\partial\Omega_+$.

It should be noted that it is the variational characterization of $\Gamma_1(\Omega)$ in (1.9) and the use of the appropriate Sobolev space (i.e., quadratic form techniques) that allow us to get away from the classical statement of the problem (1.1)–(1.2) and formulate it for arbitrary bounded domains $\Omega$.

As far as we know, even though Rayleigh set forth his clamped plate conjecture in 1877, no progress was made on it until 1950. In that year Szegő published a paper proving (1.3) with $c = 1$ under the assumption that the domain $\Omega$ is simply connected and possesses an eigenfunction of one sign corresponding to the eigenvalue $\Gamma_1(\Omega)$ [32] (see also the correction in [34] and the further comments and discussion contained in [33] and [30]). When Szegő wrote his 1950 paper he apparently held some hope that any domain would have a first eigenfunction of one sign but this hope was soon dispelled [13], [11], with further knowledge being acquired in more recent years [8], [9], [10], [12], [17].

Additional progress on Rayleigh's conjecture was made by Mohr in 1975 [22] when he showed that if there is a bounded domain of a given area that minimizes $\Gamma_1(\Omega)$ for $\Omega \in \mathbb{R}^2$, then this domain must be the disk of the given area. However, he left open the question of existence, and thus far no further progress has been made by this approach. On the other hand, the two- and three-dimensional cases of Rayleigh's conjecture have been settled by other, more direct, means, as will be detailed below.

The line of approach which has proved successful began with Szegő's work described above, and continued with work of Talenti in 1981 in which he introduced a method of rearrangement that treats separately the regions where $w_1$ is positive and negative. The problem then gets transferred to that of understanding how to minimize the least eigenvalue of a problem on two balls ($(\Omega_+)^\star$ and $(\Omega_-)^\star$) over all possible relative sizes of the two balls. To make progress on this problem Talenti decoupled the problems on the two balls and was able to prove (1.3) with dimension-dependent constants $c = d'_n$ where $d'_2 \approx 0.9777$, $d'_3 \approx 0.7391$, $d'_4 \approx 0.6524$, $d'_5 \approx 0.6093$, $d'_6 \approx 0.5839$, ... Talenti's constants never fall below 0.5, but they do go to 0.5 in the limit as $n \to \infty$.

Finally in 1992 Nadirashvili showed how to modify Talenti's approach, keeping the problems on the two balls coupled, in a way that allowed him to prove (1.3) with $c = 1$ for the two-dimensional case. This was announced at a conference in Cortona in 1992 [24], with the full details appearing in [25] (see also the announcement [23]).

Later, in an approach following Nadirashvili but tied closely to a detailed analysis of certain Bessel function combinations that arise (which Nadirashvili had managed to avoid), two of us managed to prove (1.3) with $c = 1$ for the three-dimensional case [3] (as well as giving an alternative proof for the two-dimensional case).

Beyond this, by following the same general approach for higher dimensions another pair of us [4] obtained (1.3) with dimension-dependent constants $c = d_n$ with $d_4 \approx 0.9537$, $d_5 \approx 0.9218$, $d_6 \approx 0.9077$, $d_7 \approx 0.9018$, $d_8 \approx 0.8998$, $d_9 \approx 0.9001, \ldots$. These constants are always better than the corresponding constants $d'_n$ of Talenti, and, indeed, they go to 1 in the limit as $n \to \infty$.

The reason that our approach yields the constant 1 when $n = 2$ or 3, but only $d_n < 1$ for $n \geq 4$ is tied up with how the "parametrized two-ball problem" behaves in the respective cases. For $n = 2$ or 3, we find that the minimum first eigenvalue of this two-ball problem occurs when either one of the two balls disappears. This leaves us with the clamped plate problem for a single ball (a slightly nonobvious fact) and the constant $c = 1$ results. For $n \geq 4$, the two-ball problem has minimum first eigenvalue when the two balls are identical, and then we only obtain the (presumbably nonoptimal) constants $c = d_n$ as listed above.

In the following sections we shall state our results a bit more precisely and give some of the details of our proofs. We shall leave aside technical details concerning properties of Bessel functions and some reasonably straightforward, if not quite standard, functional analytic technicalities from the direct method of the calculus of variations. For more details we refer the reader to the full research papers [3], [4]. Many additional facts concerning the vibrations of plates are contained in the monograph of Leissa [20].

Finally, we mention that while in this introduction we have concentrated on the vibrating clamped plate, and, in particular, on Rayleigh's conjecture for its principal eigenvalue, there is an analogous conjecture for the problem of buckling of a clamped plate which was first set forth by Pólya and Szegő around 1950 (see [30]). Szegő [32] (but see also [34] for some corrections) proved this conjecture for simply connected domains $\Omega$ having a first eigenfunction of one sign, but, as for the vibrating clamped plate problem, it is now known that not all domains satisfy this hypothesis [17]. By applying the analogous procedure to this problem (the"buckling problem") that we applied to the clamped plate problem we can obtain results analogous to (1.3). Unfortunately, in this case we do not get a constant of 1 in any dimension, but only dimension-dependent constants $c_n$ that approach 1 as $n \to \infty$, and never fall below about 0.77. The precise formulation of this problem and more details about it are given in Section 4 below, and in the original research paper [4]. It turns out that the exact same constants can be obtained by combining classical results of Krahn [19] and of Payne [26]. This fact has surprisingly been overlooked in much of the literature except in a paper of Bramble and Payne [7] where the result is stated explicitly for $n = 2$. Indeed, the Pólya-Szegő conjecture for the buckling problem has been largely overlooked even by those working on Rayleigh's conjecture for the vibrating clamped plate. As far as we know, we are the first to have made the connection between the results of Bramble and Payne (and therefore also the earlier results of Krahn and Payne) and the Pólya-Szegő conjecture for the buckling problem, and to have observed that the constants $c_n$ obtained thereby (see Theorem 3 below) go to 1 as $n \to \infty$ [4].

## 2. Rayleigh's conjecture for the clamped plate for $n = 2, 3$

Nadirashvili's result together with our own yields

**Theorem 1** [25], [3]. *Let $\Omega \subset \mathbb{R}^n$, $n = 2$ or $3$, be a bounded domain. Then for the clamped plate problem (formulated classically as (1.1)–(1.2)) with first eigenvalue given by (1.9) we have*

$$\Gamma_1(\Omega) \geq \Gamma_1(\Omega^\star), \tag{2.1}$$

*where $\Omega^\star$ is the $n$-ball having the same volume as $\Omega$.*

We now proceed with an outline of the proof. Using $w_1$ as a trial function $\varphi$ in (1.9) we have

$$\begin{aligned} \Gamma_1(\Omega) &= \frac{\int_\Omega (\Delta w_1)^2\,dx}{\int_\Omega w_1^2\,dx} \\ &= \frac{\int_{\Omega_+} F^2\,dx + \int_{\Omega_-} G^2\,dx}{\int_{\Omega_+} w_1^2\,dx + \int_{\Omega_-} w_1^2\,dx} \\ &= \frac{\int_{\Omega_+^\star} (F^\star)^2\,dx + \int_{\Omega_-^\star} (G^\star)^2\,dx}{\int_{\Omega_+^\star} (w_{1,+}^\star)^2\,dx + \int_{\Omega_-^\star} (w_{1,-}^\star)^2\,dx} \\ &\geq \frac{\int_{\Omega_+^\star} (\Delta u)^2\,dx + \int_{\Omega_-^\star} (\Delta v)^2\,dx}{\int_{\Omega_+^\star} u^2 d\,x + \int_{\Omega_-^\star} v^2\,dx} \end{aligned} \tag{2.2}$$

where we have spherically rearranged $F \equiv -\Delta w_1$ and $G \equiv \Delta w_1$ on $\Omega$ in a symmetric decreasing sense, respecting signs (i.e., $F^\star$ and $G^\star$ are decreasing and equimeasurable with $-\Delta w_1$ and $\Delta w_1$, respectively, but not necessarily with $|\Delta w_1|$), and then have replaced the rearrangement of $w_1$ on $\Omega_+$, denoted $w_{1,+}^\star$, by the solution $u$ to the Dirichlet problem

$$-\Delta u = F^\star \quad \text{on} \quad \Omega_+^\star, \tag{2.3}$$

$$u = 0 \quad \text{on} \quad \partial\Omega_+^\star, \tag{2.4}$$

and similarly for $w_{1,-}^\star$ on $\Omega_-$ where $v$ is analogous to $u$. The key fact here is that $w_{1,+}^\star \leq u$ on $\Omega_+^\star$ (and analogously for $w_{1,-}^\star$ and $v$ on $\Omega_-^\star$). This fact was also used by Talenti and Nadirashivili and, at least in certain versions, goes back much farther (for further references, see [6], Theorem 2.10 and surrounding material). Note especially that while we use $F^\star$ and $G^\star$ only on the domains $\Omega_+^\star$ and $\Omega_-^\star$, respectively (see (2.2) and also problem (2.3)–(2.4) and the analogous problem for $v$ on $\Omega_-^\star$ with $G^\star$ in place of $F^\star$) we define them using a signed equimeasurable rearrangement on *all* of $\Omega$, rather than by rearrangement on $\Omega_+$ or $\Omega_-$ individually. This is important in that it guarantees the positivity of $u$ on $\Omega_+^\star$ (respectively, the positivity of $v$ on $\Omega_-^\star$); see [3],[35] for more details. However, one could also use a signed equimeasurable rearrangement of $F$ on $\Omega_+$ and a signed equimeasurable rearrangement of $G$ on $\Omega_-$, respectively.

The last member of the inequality (2.2) above can now be used to define our "parametrized two-ball variational problem":

$$J_{a,b} \equiv \inf_{u,v} \frac{\int_{|x|\le a}(\Delta u)^2\,dx + \int_{|x|\le b}(\Delta v)^2\,dx}{\int_{|x|\le a} u^2\,dx + \int_{|x|\le b} v^2\,dx} \tag{2.5}$$

where $a$ and $b$ are the radii of $\Omega_+^\star$ and $\Omega_-^\star$, respectively, and $u$ and $v$ are radial functions in $W^{2,2}(|x|<a)$ and $W^{2,2}(|x|<b)$ respectively, with $u$ and $v$ not both vanishing identically (i.e., the denominator above should not vanish) and satisfying the boundary conditions

$$u(a) = 0 = v(b), \tag{2.6}$$

$$a^{n-1}\frac{\partial u}{\partial r}(a) = b^{n-1}\frac{\partial v}{\partial r}(b). \tag{2.7}$$

The boundary condition (2.7) follows from the clamped condition $\frac{\partial w_1}{\partial \nu} = 0$ since by the divergence theorem we have

$$\begin{aligned} 0 &= \int_{\partial\Omega} \frac{\partial w_1}{\partial \nu}\,dS = \int_\Omega \Delta w_1\,dx = -\int_{\Omega_+^\star} F^\star\,dx + \int_{\Omega_-^\star} G^\star\,dx \\ &= \int_{\Omega_+^\star} \Delta u\,dx - \int_{\Omega_-^\star} \Delta v\,dx = \int_{|x|=a} \frac{\partial u}{\partial \nu}\,dS - \int_{|x|=b} \frac{\partial v}{\partial \nu}\,dS. \end{aligned}$$

This implies (2.7) since $u$ and $v$ are radial. By standard arguments it turns out that $u$ and $v$ will be classical solutions to the Euler-Lagrange equations

$$\Delta^2 u = \lambda u \quad \text{on} \quad \Omega_+^\star, \tag{2.8}$$

$$\Delta^2 v = \lambda v \quad \text{on} \quad \Omega_-^\star, \tag{2.9}$$

together with the boundary conditions (2.6), (2.7), and

$$\Delta u(a) + \Delta v(b) = 0. \tag{2.10}$$

This last boundary condition is a natural boundary condition (see, for example, [15], [38]) for the variational problem (2.5)–(2.7); that is, it comes for free from (2.5)–(2.7) (see Appendix 2 of [3]).

The whole problem is now reduced to understanding the first eigenvalue $\lambda = J_{a,b}$ of the "two-ball problem" (2.6)–(2.10) and to identifying the minimum of $J_{a,b}$ when the parameters $a$ and $b$ are allowed to vary subject only to $a^n + b^n =$ const. $= \rho^n$, $a, b \geq 0$, where $\rho$ is the radius of $\Omega^\star$. If this minimum occurs at $a = 0$ (and hence also at $b = 0$), then from (2.2) we get

$$\Gamma_1(\Omega) \geq J_{a,b} \geq J_{0,\rho} = \Gamma_1(\Omega^\star), \tag{2.11}$$

proving (2.1). If the minimum occurs anywhere else, then we only get a bound of the form (1.3) with $c < 1$.

In solving the problem (2.6)–(2.10) for $u$ and $v$ one encounters the function

$$f_\nu(k) \equiv k^{n-1}\left[\frac{J_{\nu+1}(k)}{J_\nu(k)} + \frac{I_{\nu+1}(k)}{I_\nu(k)}\right] \tag{2.12}$$

where $\nu = (n-2)/2$, and $J_\nu$, $I_\nu$, etc., denote Bessel and modified Bessel functions in standard notation [1]. In fact, the eigenvalue condition for the first eigenvalue is just

$$f_\nu(ka) + f_\nu(kb) = 0 \tag{2.13}$$

when expressed as a condition on $k \equiv \lambda^{1/4}$. (Reduction to the case of index $\nu = (n-2)/2$ here follows from a result of Lorch [21] (see also our comments in [3]).) Therefore, to understand how $\lambda$ behaves as a function of $a$ (recall that $a$ and $b$ are related implicitly via $a^n + b^n =$ const. $= \rho^n$, and note further that we can take $\rho$ to be 1 by scaling without loss of generality as we shall assume henceforth), we need to analyze

$$F_\nu(k, a) \equiv f_\nu(ka) + f_\nu(kb), \tag{2.14}$$

and this is conveniently done in terms of the graph of how $F_\nu$ behaves in the $(k, a)$-plane (see Figure 1) and certain qualitative properties of $F_\nu$. In particular, $F_\nu$ equals 0 at $k = 0$, and $F_\nu$ increases with increasing $k$ (for fixed $a \geq 0$), going to infinity at each pole reached (see figure), and starting over from $-\infty$ beyond each pole. Thus, the first eigenvalue $\lambda = k^4$ is determined by the zero $k_\nu(a)$ of $F_\nu(k, a)$ indicated by the dashed curve in the figure. This zero starts from the value $k_\nu$, the first positive zero of $f_\nu(k)$, at $a = 0$, and proceeds upward until reaching the limiting value $2^{1/n} j_\nu$ when $a = b = 2^{-1/n}$, since it is then trapped between two poles of $F_\nu$. (The symbol $j_\nu$ denotes the first positive zero of the Bessel function $J_\nu(t)$. More generally, the $k^{\text{th}}$ positive zero of $J_\nu$ will be denoted $j_{\nu,k}$.) Since $2^{1/n} j_\nu > k_\nu$ for $n = 2, 3$ ($\nu = 0, 1/2$) it is clear that $k_\nu(a) > k_\nu(0) = k_\nu$ for $a$ sufficiently near to $2^{-1/n}$. For lower values of $a$ it suffices, by the monotonicity of $F_\nu(k, a)$ in $k$ between poles, to check that $F_\nu(k_\nu, a) < 0$. It then follows from the monotonicity that $k_\nu(a) > k_\nu$ for $a > 0$ and we are done. The fact that $F_\nu(k_\nu, a) < 0$ for $a$ in the relevant range (i.e., $0 < a < (1 - (j_\nu/k_\nu)^n)^{1/n}$) and for $\nu = 0$ or $1/2$ requires technical estimates involving Bessel functions which are not particularly illuminating, so for these we refer the interested reader to [3]. We caution the reader that the inequality $k_\nu(a) > k_\nu$ does not hold when $\nu = 1, 3/2, 2, \ldots$ (i.e., for $n \geq 4$), as will be discussed in the next section.

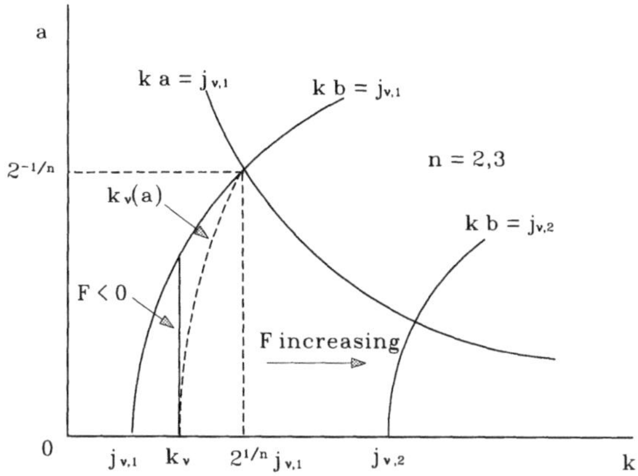


Fig 1. This sketch shows the qualitative features of $F_\nu(k,a)$ that we use in our proof for $n = 2,3$. The solid curves denote the poles of $F_\nu$, while the dashed curve denotes the function $k_\nu(a)$, defined implicitly via $F_\nu(k_\nu(a),a) \equiv 0$. Since $F_\nu$ is continuous and increasing on horizontal lines between its poles, demonstrating that $F_\nu < 0$ for $a > 0$ along the solid vertical line segment at $k = k_\nu \equiv k_\nu(0)$ proves that $k_\nu(a) > k_\nu$ for $a > 0$ and hence Rayleigh's conjecture for $n = 2$ and 3.

## 3. Results for the clamped plate problem for $n \geq 4$

In dimension 4 and greater it turns out that the approach to proving Theorem 1 as detailed in Section 2 is doomed to failure, the first indication being that the comparison between $k_\nu$ and $2^{1/n} j_\nu$ now reads $k_\nu > 2^{1/n} j_\nu$. This means that the qualitative picture of how $F_\nu$, and hence $k_\nu(a)$, behaves cannot possibly be as in Figure 1, but rather can at best be as indicated in Figure 2. Adopting the same strategy as before, but now with reference to showing that $F_\nu(2^{1/n} j_\nu, a) < 0$ for $0 \leq a < 2^{-1/n}$, two of us [4] showed that $k_\nu(a) > 2^{1/n} j_\nu$. This allowed us to obtain a result of the form (1.3), but with $c < 1$. In particular, we found

**Theorem 2** [4]. *Let $\Omega \in \mathbb{R}^n$ $n \geq 4$, be a bounded domain. Then for the first eigenvalue $\Gamma_1(\Omega)$ of the clamped plate problem as given by (1.9) we have*

$$\Gamma_1(\Omega) > d_n \Gamma_1(\Omega^\star), \tag{3.1}$$

*where $\Omega^\star$ is the n-ball having the same volume as $\Omega$ and the dimension-dependent constant $d_n$ is given by*

$$d_n = 2^{4/n} \left( \frac{j_\nu}{k_\nu} \right)^4. \tag{3.2}$$

The constants $d_n$ for low values of $n \geq 4$ are presented in Table 1, along with the value 1 for $n = 2,3$ as provided by Theorem 1 and the values of Talenti's constants $d_n'$. We have also proved [4] that $d_n \to 1$ while $d_n' \to 1/2$ as $n \to \infty$.

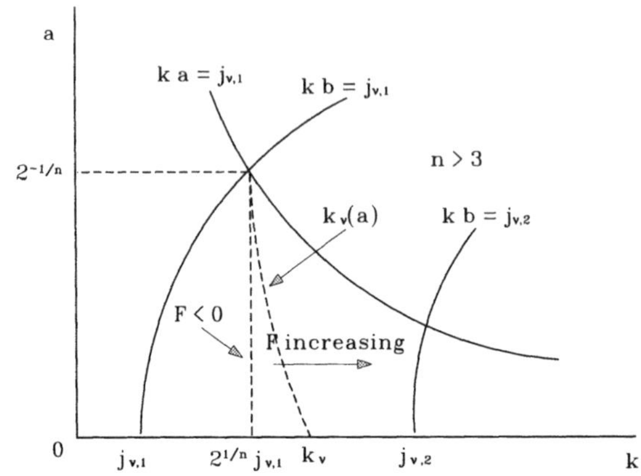


Fig 2. This sketch shows the qualitative features of $F_\nu(k,a)$ that we use in our proof for $n \geq 4$. The solid curves denote the poles of $F_\nu$, while the dashed curve denotes the function $k_\nu(a)$, defined implicitly via $F_\nu(k_\nu(a),a) \equiv 0$. Since $F_\nu$ is continuous and increasing on horizontal lines between its poles, demonstrating that $F_\nu < 0$ for $0 \leq a < 2^{-1/n}$ along the dashed vertical line segment at $k = 2^{1/n} j_{\nu,1}$ proves that $k_\nu(a) > 2^{1/n} j_{\nu,1}$ for $0 \leq a < 2^{-1/n}$ and thus proves Theorem 2, i.e., (3.1) for $n \geq 4$.

| $n$ | $d_n$ | $d'_n$ |
|---|---|---|
| 2 | 1 (Thm. 1) | 0.9777 |
| 3 | 1 (Thm. 1) | 0.7391 |
| 4 | 0.9537 | 0.6524 |
| 5 | 0.9218 | 0.6093 |
| 6 | 0.9077 | 0.5839 |
| 7 | 0.9018 | |
| 8 | 0.8998 | |
| 9 | 0.9001 | |
| $\vdots$ | $\vdots$ | $\vdots$ |
| $\to \infty$ | $\to 1$ | $\to 1/2$ |

Table 1: Our constants $d_n$ and Talenti's constants $d'_n$ for the clamped plate problem in $n$ dimensions (see Theorem 2)

**Remarks. 1.** Inequality (3.1) is strict because $J_{a,b}$ has a strict minimum when one has $a = b = 2^{-1/n}$, and this situation of having two identical balls (i.e., $a = b$) cannot be realized by any connected domain $\Omega$.

**2.** Figure 2 should make it clear why showing that $F_\nu(2^{1/n} j_\nu, a) < 0$ for $n \geq 4$ implies $k_\nu(a) > 2^{1/n} j_\nu$. The strategy is the same as in showing $k_\nu(a) > k_\nu$

by showing $F_\nu(k_\nu, a) < 0$ for $n = 2, 3$ and appropriate $a$'s for Theorem 1 (compare Figure 1).

**3.** The fact that our approach for Theorem 1 cannot prove Rayleigh's clamped plate conjecture for $n \geq 4$ is *not* an indication that the conjecture is not true, only that this strategy for obtaining a proof fails. Indeed, the surprise is more that the strategy succeeds at all in dimensions two and three, for it does not use the full strength of the hypothesis that $\frac{\partial w_1}{\partial \nu} = 0$ on $\partial\Omega$: the method above uses only that $\int_{\partial\Omega} \frac{\partial w_1}{\partial \nu}\, dS = 0$ (see the derivation of the boundary condition (2.7) given earlier). That is, in dimensions two and three, Rayleigh's conjecture is true not only for plates that are fixed and clamped on the boundary but also for plates that are fixed and "clamped on average" on the boundary. In higher dimensions (that is, for $n \geq 4$), the "clamped on average" condition only gets us the weaker inequalities of Theorem 2 (this comment also applies to the buckling problem discussed in Section 4 below in all dimensions $n \geq 2$). We still believe Rayleigh's conjecture holds for "clamped plates" in all dimensions; that is, that (1.3) holds with $c = 1$ for all dimensions $n \geq 2$.

## 4. Results for the buckling problem

Another classical eigenvalue problem, analogous to the clamped plate problem, is the buckling problem

$$\Delta^2 v = -\Lambda \Delta v \quad \text{in} \quad \Omega \subset \mathbb{R}^n, \tag{4.1}$$

$$v = 0 = \frac{\partial v}{\partial \nu} \quad \text{on} \quad \partial\Omega, \tag{4.2}$$

where $\Omega$ is a bounded domain in $\mathbb{R}^n$. When $n = 2$ this problem describes the buckling of a clamped plate in the shape of $\Omega$ under uniform compression. The first eigenvalue of this problem, $\Lambda_1(\Omega)$, is proportional to the critical buckling load (see [36], for example). Around 1950 Pólya and Szegő conjectured the isoperimetric result

$$\Lambda_1(\Omega) \geq \Lambda_1(\Omega^\star), \tag{4.3}$$

where, as before, $\Omega^\star$ denotes the disk/ball in $\mathbb{R}^n$ having the same area/volume as $\Omega$. The Pólya-Szegő conjecture (4.3) is the analog for the buckling problem of the Rayleigh conjecture for the vibrating clamped plate.

The buckling problem admits a variational formulation, just as the clamped plate problem did. In this case we have

$$\Lambda_1(\Omega) = \inf_{\substack{\varphi \in W_0^{2,2}(\Omega) \\ \nabla\varphi \not\equiv 0}} \frac{\int_\Omega (\Delta\varphi)^2\, dx}{\int_\Omega |\nabla\varphi|^2\, dx}. \tag{4.4}$$

Using this variational principle, we can use spherical symmetrization essentially as we did for the vibrating clamped plate. Just as in that problem we must introduce the domains $\Omega_\pm$ since we have no guarantee that $v_1$, the eigenfunction corresponding to $\Lambda_1(\Omega)$, can be taken to have constant sign [17]. By this approach we arrive at a lower bound to $\Lambda_1(\Omega)$ based on minimizing a new variational problem, similar to (2.5), which is again a parametrized two-ball problem where we take only radial functions as trial functions. It turns out that for all $n \geq 2$ this minimization problem is like the clamped plate problem for $n \geq 4$, that is, the minimum of the parametrized two-ball problem occurs when $a = b = 2^{-1/n}$ and not when one of the two balls vanishes, as we would need in order to prove the Pólya-Szegő conjecture (4.3). The relevant figure is analogous to Figure 2. Thus we do not prove the Pólya-Szegő conjecture for any $n \geq 2$, but rather the weaker inequalities set forth in the following theorem.

**Theorem 3** [7], [4]. *For the buckling problem for $\Omega \in \mathbb{R}^n$, $n \geq 2$,*

$$\Lambda_1(\Omega) > c_n \Lambda_1(\Omega^\star), \tag{4.5}$$

*where*

$$c_n = 2^{2/n} \left( \frac{j_\nu}{j_{\nu+1}} \right)^2 \tag{4.6}$$

*with* $\nu = \frac{n-2}{2}$.

Our constants $c_n$ for low values of $n$ are presented in Table 2. We have also observed [4] that $c_n \to 1$ as $n \to \infty$.

| $n$ | $c_n$ |
|---|---|
| 2 | 0.7877 |
| 3 | 0.7759 |
| 4 | 0.7872 |
| 5 | 0.8020 |
| 6 | 0.8163 |
| $\vdots$ | $\vdots$ |
| $\to \infty$ | $\to 1$ |

Table 2: Our constants $c_n$ for the buckling problem in $n$ dimensions (see Theorem 3)

**Remarks. 1.** The $n = 2$ case of this result was given explicitly, and the higher dimensional analogs alluded to, by Bramble and Payne in [7]. Their method of proof is different from the one we have given here and is certainly worthy of mention. By a classical result of Payne [26] (also known as Weinstein's conjecture [15], [37]),

$\Lambda_1(\Omega)$ is greater than or equal to the *second* eigenvalue $\lambda_2(\Omega)$ of the corresponding fixed membrane problem ($-\Delta u = \lambda u$ in $\Omega$, $u = 0$ on $\partial\Omega$). Combining this with the consequence of the Faber-Krahn inequality

$$\lambda_2(\Omega) > 2^{2/n}\lambda_1(\Omega^\star) \tag{4.7}$$

(which was first observed by Krahn [19] and later rediscovered by Peter Szegő as indicated by Pólya's comments in [29]), we arrive at (4.5). This argument works equally well in all dimensions, since Payne's result extends to all dimensions (see [4], [15], [27], or [28]), and, indeed, this is the proof we presented in [4]. It is remarkable that this approach leads to the same bounds as does the approach outlined above, paralleling the proof for the vibrating clamped plate for $n \geq 4$.

**2.** For completeness, we mention that for the buckling problem $f_\nu(k)$ (cf. Section 2 above) is replaced by

$$g_\nu(k) \equiv k^{2\nu+1}\frac{J_{\nu+1}(k)}{J_\nu(k)} \tag{4.8}$$

and $F_\nu(k,a)$ is replaced by $G_\nu(k,a) \equiv g_\nu(ka) + g_\nu(kb)$. The arguments are analogous, though simpler, than those for the vibrating clamped plate (see Sections 2 and 3, and [3], [4]).

**3.** The Pólya-Szegő buckling plate conjecture (4.3) was proved by Szegő [32] (see also [34]) under the additional assumption that $\Omega$ is a simply connected plane domain and has a nonnegative first eigenfunction (i.e., a nonnegative eigenfunction $v_1$ associated with its principal eigenvalue $\Lambda_1(\Omega)$).

**4.** B. Kawohl has informed us that Weinberger and Willms might be close to proving the Pólya-Szegő conjecture (i.e., (4.3)) in two dimensions, by a domain variation method. No preprint is yet available, however.

**Acknowledgements.** We would like to express our appreciation to the organizers of "General Inequalities 7", Catherine Bandle, Norrie Everitt, Laszlo Losonczi, and Wolfgang Walter, for their kind invitation to a most stimulating conference.

## References

[1] M. Abramowitz and I.A. Stegun, editors, *Handbook of Mathematical Functions.* National Bureau of Standards Applied Mathematics Series, vol. **55**, U.S. Government Printing Office, Washington, D.C., 1964.

[2] R.A. Adams, *Sobolev Spaces.* Academic Press, New York, 1975.

[3] M.S. Ashbaugh and R.D. Benguria, *On Rayleigh's conjecture for the clamped plate and its generalization to three dimensions.* Duke Math. J. **78** (1995), 1–17.

[4] M.S. Ashbaugh and R.S. Laugesen, *Fundamental tones and buckling loads of clamped plates.* Ann. Scuola Norm. Sup. Pisa Cl. Sci., Ser. IV. To appear.

[5] R. Askey, editor, *Gabor Szegő: Collected Papers*, Volume 3: 1945-1972, Birkhäuser, Boston, 1982.

[6] C. Bandle, *Isoperimetric Inequalities and Applications.* Pitman Monographs and Studies in Mathematics, vol. **7**, Pitman, Boston, 1980.

[7] J.H. Bramble and L.E. Payne, *Pointwise bounds in the first biharmonic boundary value problem.* J. Math. and Phys. **42** (1963), 278–286.

[8] C.V. Coffman, *On the structure of solutions to $\Delta^2 u = \lambda u$ which satisfy the clamped plate conditions on a right angle.* SIAM J. Math. Anal. **13** (1982), 746–757.

[9] C.V. Coffman and R.J. Duffin, *On the fundamental eigenfunctions of a clamped punctured disk.* Adv. Appl. Math. **13** (1992), 142–151.

[10] C.V. Coffman, R.J. Duffin, and D.H. Shaffer, *The fundamental mode of vibration of a clamped annular plate is not of one sign.* In: "Constructive Approaches to Mathematical Models", C.V. Coffman and G.J. Fix, editors, Academic Press, New York, 1979, pp. 267-277.

[11] R.J. Duffin, *Nodal lines of a vibrating plate.* J. Math. and Phys. **31** (1953), 294–299.

[12] R.J. Duffin, *Some problems of mathematics and science.* Bull. Amer. Math. Soc. **80** (1974), 1053–1070 [an expanded version of this article also appeared under the title *Some Problems Arising from Mathematical Model* as pp. 3-32 of "Constructive Approaches to Mathematical Models" (see reference [10] above for full bibliographic information)].

[13] R.J. Duffin and D.H. Shaffer, *On the modes of vibration of a ring-shaped plate.* Bull. Amer. Math. Soc. **58** (1952), 652.

[14] G. Faber, *Beweis, dass unter allen homogenen Membranen von gleicher Fläche und gleicher Spannung die kreisförmige den tiefsten Grundton gibt.* Sitzungberichte der mathematisch-physikalischen Klasse der Bayerischen Akademie der Wissenschaften zu München Jahrgang, 1923, pp. 169–172.

[15] S.H. Gould, *Variational Methods for Eigenvalue Problems: An Introduction to the Weinstein Method of Intermediate Problems*, second edition, revised and enlarged, University of Toronto Press (Mathematical Expositions, Number 10), Toronto, 1966 (see pp. 135–137 in particular).

[16] G.H. Hardy, J.E. Littlewood, and G. Pólya, *Inequalities*, second edition, Cambridge University Press, Cambridge, 1952.

[17] V.A. Kozlov, V.A. Kondrat'ev, and V.G. Maz'ya, *On sign variation and the absence of "strong" zeros of solutions of elliptic equations.* Izv. Akad. Nauk SSSR, Ser. Mat. **53** (1989), 328–344 (in Russian) [English translation in Math. USSR-Izv. **34** (1990), 337–353].

[18] E. Krahn, *Über eine von Rayleigh formulierte Minimaleigenschaft des Kreises.* Math. Ann. **94** (1925), 97–100.

[19] E. Krahn, *Über Minimaleigenschaften der Kugel in drei und mehr Dimensionen.* Acta Comm. Univ. Tartu (Dorpat) **A9** (1926), 1–44 [English translation in *Edgar Krahn 1894-1961: A Centenary Volume*, Ü. Lumiste and J. Peetre, editors, IOS Press, Amsterdam, 1994, pp. 139–174].

[20] A.W. Leissa, *Vibration of Plates.* National Aeronautics and Space Administration (NASA SP-160), U.S. Government Printing Office, Washington, D.C., 1969.

[21] L. Lorch, *Monotonicity of the zeros of a cross product of Bessel functions.* Methods and Appls. of Anal. **1** (1994), 75–80.

[22] E. Mohr, *Über die Rayleighsche Vermutung: Unter allen Platten von gegebener Fläche und konstanter Dichte und Elastizität hat die kreisförmige den tiefsten Grundton.* Ann. Mat. Pura Appl. (Ser. 4) **104** (1975), 85–122.

[23] N.S. Nadirashvili, *An isoperimetric inequality for the principal frequency of a clamped plate.* Dokl. Akad. Nauk **332** (1993), 436–439 (in Russian) [English translation in Phys. Dokl. **38** (1993), 419–421].

[24] N.S. Nadirashvili, *New isoperimetric inequalities in mathematical physics.* In: "Partial Differential Equations of Elliptic Type", Cortona, 1992, A. Alvino, E. Fabes, and G. Talenti, editors, Cambridge University Press (Symposia Mathematica, vol. **35**), Cambridge, 1994, pp. 197–203.

[25] N.S. Nadirashvili, *Rayleigh's conjecture on the principal frequency of the clamped plate.* Arch. Rational Mech. Anal. **129** (1995), 1–10.

[26] L.E. Payne, *Inequalities for eigenvalues of membranes and plates.* J. Rational Mech. Anal. **4** (1955), 517–529.

[27] L.E. Payne, *Isoperimetric inequalities for eigenvalues and their applications.* Autovalori e autosoluzioni, Centro Internazionale Matematico Estivo (C.I.M.E.) 2° Ciclo, Chieti, 1962, pp. 1–58.

[28] L.E. Payne, *Isoperimetric inequalities and their applications.* SIAM Review **9** (1967), 453–488.

[29] G. Pólya, *On the characteristic frequencies of a symmetric membrane.* Math. Z. **63** (1955), 331–337.

[30] G. Pólya and G. Szegő, *Isoperimetric Inequalities in Mathematical Physics.* Annals of Mathematics Studies, Number 27, Princeton University Press, Princeton, 1951.

[31] J.W.S. Rayleigh, *The Theory of Sound*, second edition revised and enlarged (in 2 volumes), Dover Publications, New York, 1945 (republication of the 1894/96 edition).

[32] G. Szegő, *On membranes and plates.* Proc. Nat. Acad. Sci. **36** (1950), 210–216 [reprinted as pp. 187–193 of [5] with comments by R. Askey on p. 194; see also [34] for corrections to Sections III and IV].

[33] G. Szegő, *On the vibrations of a clamped plate.* Atti del Quarto Congresso dell'Unione Matematica Italiana, Taormina, 25–31 Ott., 1951, vol. II, Casa Editrice Perrella, Roma, 1953, pp. 573–577 [reprinted as pp. 311–315 of [5] with comments by R. Askey on p. 316].

[34] G. Szegő, *Note to my paper "On membranes and plates".* Proc. Nat. Acad. Sci. **44** (1958), 314–316 [reprinted as pp. 481–483 of [5]].

[35] G. Talenti, *On the first eigenvalue of the clamped plate.* Ann. Mat. Pura Appl. (Ser. 4) **129** (1981), 265–280.

[36] S. Timoshenko and J.M. Gere, *Theory of Elastic Stability*, second edition, McGraw-Hill, New York, 1961.

[37] A. Weinstein, *Études des spectres des équations aux dérivées partielles de la théorie des plaques élastiques.* Mémorial des Sciences Mathématiques **88**, Paris, 1937.

[38] A. Weinstein and W. Stenger, *Methods of Intermediate Problems for Eigenvalues: Theory and Ramifications.* Academic Press, New York, 1972.

Mark S. Ashbaugh, Department of Mathematics, University of Missouri, Columbia, MO 65211, USA.

E-mail address: mark@ashbaugh.math.missouri.edu

Rafael D. Benguria, Facultad de Física, P. Universidad Católica de Chile, Casilla 306, Santiago 22, Chile.

E-mail address: rbenguri@lascar.puc.cl

Richard S. Laugesen, Department of Mathematics, University of Michigan, Ann Arbor, MI 48109, USA.

E-mail address: laugesen@umich.edu

International Series of Numerical Mathematics, Vol. 123, © 1997 Birkhäuser Verlag Basel

# On the Payne-Pólya-Weinberger conjecture on the $n$-dimensional sphere

Mark S. Ashbaugh* and Rafael D. Benguria†

**Abstract**

For a domain $\Omega$ contained in a hemisphere of the $n$-dimensional sphere $\mathcal{S}^n$ we outline the proof of the optimal result $\lambda_2/\lambda_1(\Omega) \leq \lambda_2/\lambda_1(\Omega^*)$ for the ratio of its first two Dirichlet eigenvalues where $\Omega^*$, the symmetric rearrangement of $\Omega$ in $\mathcal{S}^n$, is a geodesic ball in $\mathcal{S}^n$ having the same $n$-volume as $\Omega$. Along the way we show that $\lambda_2/\lambda_1$ for geodesic balls of geodesic radius $\theta_1$ less than or equal to $\pi/2$ is an increasing function of $\theta_1$. Our key result is that among all domains contained in a hemisphere of $\mathcal{S}^n$ and having a fixed value of $\lambda_1$ the one with the maximal value of $\lambda_2$ is the geodesic ball of the appropriate radius.

## 1. Introduction

With our proof [2], [4] of the Payne-Pólya-Weinberger conjecture [19], [20] the bound

$$\lambda_2/\lambda_1(\Omega) \leq \lambda_2/\lambda_1(\Omega^*) = j_{n/2,1}^2/j_{n/2-1,1}^2 \tag{1.1}$$

was established for the ratio of the first two eigenvalues of the Laplacian $-\Delta$ on a bounded domain $\Omega \subset \mathbb{R}^n$ with Dirichlet boundary conditions imposed on $\partial\Omega$. Here $\Omega^*$ represents the $n$-dimensional ball having the same volume as $\Omega$ (but, in fact, by scaling any ball will do) and $j_{\nu,k}$ represents the $\mathrm{k}^{th}$ positive zero of the Bessel function $J_\nu(x)$ [1]. Equality obtains in (1.1) if and only if $\Omega$ is a ball to begin with. Here we present an outline of the proof of the analog of this result for domains in a hemisphere of $\mathcal{S}^n$ ($-\Delta$ is now, of course, the Laplacian on $\mathcal{S}^n$).

As it happens it is better to view (1.1) as

$$\lambda_2(\Omega) \leq \lambda_2(B_{\lambda_1}), \tag{1.2}$$

where $B_{\lambda_1}$ represents the $n$-dimensional ball that has the value $\lambda_1(\Omega)$ as its first Dirichlet eigenvalue. This is, in fact, the way our proof proceeded ([2], [3], [5], see

*Partially supported by National Science Foundation (USA) grants DMS-9114162, INT-9123481, and DMS-9500968.

†Partially supported by Fondecyt (Chile) project 196-0462 and a Cátedra Presidencial en Ciencias (Chile).

also [12]). Of course, the choice of ball here just involves choosing an appropriate radius and this choice is always unique since the first eigenvalue of a ball is a strictly decreasing function of its radius and goes from infinity to zero as the radius goes from zero to infinity. In words, (1.2) says that among all $n$-dimensional domains having the same first eigenvalue the $n$-dimensional ball has maximal second eigenvalue. One might compare this statement with the statement of the Faber-Krahn inequality [13], [16], [17], [8]: among all $n$-dimensional domains having the same volume the $n$-dimensional ball has minimal first (Dirichlet) eigenvalue. Also of interest is the Szegő-Weinberger inequality [22], [24]: among all $n$-dimensional domains having the same volume the $n$-dimensional ball has maximal first nonzero Neumann eigenvalue. Both of these other inequalities are relevant here; the first because it figures in our proof and the second because in many ways the proof of this result is analogous to (though considerably simpler than) our proof of the Payne-Pólya-Weinberger conjecture.

What we do in this paper is transfer the strategy outlined in the last paragraph over to bounded domains in $\mathcal{S}^n$ contained in hemispheres. It turns out that, properly interpreted, everything that we have said above concerning domains in Euclidean space also holds in $\mathcal{S}^n$. Thus we prove that (1.2) holds if $\Omega$ is a domain in a hemisphere in $\mathcal{S}^n$ and $B_{\lambda_1}$ is the geodesic ball in $\mathcal{S}^n$ having $\lambda_1(\Omega)$ as its first eigenvalue. There is a Faber-Krahn result in $\mathcal{S}^n$ [21] (see also [14]), too, and $\lambda_1$ of a geodesic ball is still strictly monotone decreasing and goes to infinity at zero so a unique $B_{\lambda_1}$ which is a hemisphere or less will exist. The generalization of the Szegő-Weinberger result to domains in hemispheres of $\mathcal{S}^n$ is a recent result of ours [6] (see also prior work of Chavel [10]). Our proof here parallels the one in [6] as well as our proof of the Payne-Pólya-Weinberger conjecture [2], [3], [4] (see especially our proof in [4]) for the Euclidean case.

The biggest difference between our work on $\lambda_2/\lambda_1$ for domains in $\mathcal{S}^n$ versus those in $\mathbb{R}^n$ revolves around the difference between (1.1) and (1.2). In $\mathbb{R}^n$ (1.1) and (1.2) are equivalent since $\lambda_2/\lambda_1$ is the same for any ball, whatever its size. When one passes to $\mathcal{S}^n$ this is no longer the case. If we let $\theta_1$ denote the radius of our geodesic ball then the first and second eigenvalues of that ball, which we denote by $\lambda_1(\theta_1)$ and $\lambda_2(\theta_1)$, are in general more complicated functions of $\theta_1$. By domain monotonicity these are, of course, strictly decreasing functions but more precise knowledge of these functions (or combinations thereof) requires considerable effort. For example, for the case of $\mathcal{S}^n$, to pass from (1.2) back to the first part of (1.1) (i.e., $\lambda_2/\lambda_1(\Omega) \le \lambda_2/\lambda_1(\Omega^*)$, where $\Omega^*$ is the symmetric rearrangement of $\Omega$ on $\mathcal{S}^n$) one needs to know that $\lambda_2/\lambda_1$ for geodesic balls is an increasing function of $\theta_1$. We do this below in Section 5. In general, most of what could almost be taken for granted for the case of balls in $\mathbb{R}^n$ expands to some problem about how $\lambda_1$, $\lambda_2$ or a combination of the two behaves as a function of $\theta_1$.

For orientation we outline the elements of our proof here. Aside from item 4 (which we have just discussed) these elements were also present in our proof in the Euclidean case.

1) Rayleigh-Ritz inequality for estimating $\lambda_2$:

$$\lambda_2 - \lambda_1 \le \frac{\int_\Omega |\nabla P|^2 u_1^2 \, d\omega}{\int_\Omega P^2 u_1^2 \, d\omega}$$

if $\int_\Omega P u_1^2 \, d\omega = 0$ and $P \not\equiv 0$. Here $u_1$ denotes the normalized eigenfunction for $\lambda_1$. (In general, we denote by $\lambda_1 < \lambda_2 \le \lambda_3 \cdots$ the Dirichlet eigenvalues of the Laplacian on $\Omega$, and $u_1, u_2, \ldots$ the corresponding eigenfunctions.) This inequality applies on manifolds as in Euclidean space; one only has to view $|\nabla P|$ as a norm with respect to the metric on the manifold. Also $d\omega$ represents the intrinsic volume element for the manifold.

2) Center of mass result: A Brower fixed point theorem argument that allows us to insure that the condition $\int_\Omega P u_1^2 d\omega = 0$ is satisfied for $n$ specific choices of the function $P$. In our recent paper [6] we gave a version of this argument for $\mathcal{S}^n$ which also applies here. The original argument of this type (to our knowledge) was given by Weinberger in [24].

3) Rearrangement results for functions and domains: These results are measure theoretic in nature and easily extend to problems on manifolds. There is also an easily effected preliminary rearrangement in $\mathcal{S}^n$ (see also [6]).

4) Properties of the eigenvalues and eigenfunctions of geodesic balls in $\mathcal{S}^n$: Here we only sketch some of their properties in Section 3 below. The details depend on Legendre and associated Legendre functions though we manage to keep these functions in the background. In $\mathbb{R}^n$, of course, the special functions that occur are Bessel functions.

5) Chiti's comparison argument: This is a specialized comparison result which establishes a crossing property of the symmetric-decreasing rearrangement of the eigenfunction $u_1$ with the first eigenfunction of the geodesic ball $B_{\lambda_1}$. This result of Chiti [11],[12] for the Euclidean case is based upon a rearrangement result for partial differential equations due to Talenti [23] which in turn is based on the classical isoperimetric inequality in $\mathbb{R}^n$. All of these results generalize to spaces of constant curvature. A precise statement of these is given in Section 4.

6) Properties of certain special combinations of the eigenfunctions for $\lambda_1$ and $\lambda_2$ for geodesic balls contained in hemispheres: These functions are special to the problem of maximizing $\lambda_2/\lambda_1$ (or $\lambda_2$ subject to $\lambda_1 = \text{const.}$).

Our main results are summarized in the following theorems:

**Theorem 1.** *Let $\Omega$ be contained in a hemisphere of $\mathcal{S}^n$ and let $B_1$ denote the geodesic ball in $\mathcal{S}^n$ having the same value of $\lambda_1$ as $\Omega$ (i.e., $\lambda_1(\Omega) = \lambda_1(B_1)$). Then*

$$\lambda_2(\Omega) \le \lambda_2(B_1)$$

*with equality if and only if $\Omega$ is itself a geodesic ball in $\mathcal{S}^n$.*

The *proof* of this Theorem is outlined in Section 2.

**Theorem 2.** *The first Dirichlet eigenvalue for a geodesic ball in $S^n$ of geodesic radius $\theta_1$, $\lambda_1(\theta_1)$ is such that $\theta_1^2\lambda_1(\theta_1)$ is a decreasing function of $\theta_1$ for $0 < \theta_1 \leq \pi$.*

**Theorem 3.** *The quotient between the first two Dirichlet eigenvalues for a geodesic ball in $S^n$, of geodesic radius $\theta_1$, $\lambda_2/\lambda_1$, is an increasing function of $\theta_1$ for $0 < \theta_1 \leq \pi/2$.*

These two theorems are proved in Section 5.

Finally, we come to the PPW conjecture for domains in hemispheres of $S^n$:

**Theorem 4.** *Let $\Omega$ be contained in a hemisphere of $S^n$. Then*

$$\lambda_2/\lambda_1(\Omega) \leq \lambda_2/\lambda_1(\Omega^*)$$

*with equality if and only if $\Omega$ is itself a geodesic ball in $S^n$.*

This theorem follows from Theorems 1 and 3 above. If $B_{\lambda_1}$ is a ball having the same $\lambda_1$ as $\Omega$ we have

$$\lambda_2/\lambda_1(\Omega) \leq \lambda_2/\lambda_1(B_{\lambda_1}) \leq \lambda_2/\lambda_1(\theta_1),$$

where $\theta_1$ is the geodesic radius of $B_{\lambda_1}$. Now, Theorem 4 follows from Theorem 3 and Sperner's optimal Faber-Krahn type result [21] for domains in $S^n$, which implies that $\theta_1(\Omega^*) \geq \theta_1(B_1)$.

## 2. Proof of Theorem 1

As in our proof of the PPW conjecture for domains in $\mathbb{R}^n$, the starting point here is the use of the gap formula, which is a variational estimate for the difference between the first two eigenvalues of the Laplacian. In fact we use the inequality

$$\lambda_2 - \lambda_1 \leq \frac{\int_\Omega |\nabla P|^2 u_1^2\, d\omega}{\int_\Omega P^2 u_1^2\, d\omega}, \tag{2.1}$$

provided $\int_\Omega P u_1^2\, d\omega = 0$ and $P \not\equiv 0$. Here $\Omega$ is a domain in $S^n$ and $d\omega$ is the standard volume element in $S^n$.

The gap formula follows from the Rayleigh-Ritz inequality for $\lambda_2$ using $Pu_1$ as the trial function (hence the side condition $Pu_1 \perp u_1$) after a suitable integration by parts. To get the desired isoperimetric result out of this one must make very special choices of the function $P$: in particular, choices such that (2.1) is an equality if $\Omega$ is a geodesic ball.

A key element needed to guarantee the orthogonality of the trial functions $P_i u_1$ which we will use in the sequel with $u_1$, is a *center of mass* argument.

The center of mass result we use here is as follows.

**Theorem 5.** *Let $\Omega$ be a domain in $S^n$ which is contained in a hemisphere and let $\widetilde{G}$ be a continuous function on $[0,\pi]$ which is positive on $(0,\pi)$. Then there is a choice of Cartesian coordinates $x_1, x_2, \ldots, x_{n+1}$ for $\mathbb{R}^{n+1}$ such that*

$$\int_\Omega \widetilde{G}(\theta) x_i \, d\mu = 0 \quad \text{for } i = 1, 2, \ldots, n, \tag{2.2}$$

*where $\theta$ represents the angle from the positive $x_{n+1}$-axis and $d\mu$ represents any positive measure on $\Omega$. The $x_i$, of course, are the Cartesian coordinates restricted to $S^n$.*

**Remarks.** i) In our later applications $d\mu$ will be $u_1^2\, d\omega$, where $u_1$ is the first eigenfunction of $-\Delta$ on $\Omega$ with Dirichlet boundary conditions imposed on $\partial\Omega$ and $d\omega$ is the standard volume element for $S^n$.

ii) The proof of this theorem is given in [6]. The proof is based on the Brower fixed point theorem and derives from Weinberger's similar argument for the Euclidean case [24]. A stronger center of mass result for domains in $S^n$, with a simpler proof will be presented elsewhere [7].

Having stated the center of mass result, we go back to the gap formula and the choice of trial functions. Thinking of $S^n$ as the unit sphere in $\mathbb{R}^{n+1}$ and with the *center of mass* point for $\Omega$ fixed at the north pole one should take

$$P_i = g(\theta) \frac{x_i}{\sin\theta}, \quad i = 1, 2, \ldots, n, \tag{2.3}$$

where $\theta$ represents the angle of a point from the positive $x_{n+1}$-axis (the direction of the north pole). In $S^n$ the variable $\theta$ is the geodesic radial variable with respect to the north pole. Division by $\sin\theta = \sqrt{1 - x_{n+1}^2}$ normalizes the $n$-vector $(x_1, x_2, \ldots, x_n)$.

The choice of $g(\theta)$, as in the Euclidean case, is determined by the fact that we must have equality in (2.1) when $\Omega$ is a geodesic ball. For a geodesic ball of radius $\theta_1$, $u_1 = c_1 y_1$ where $y_1$ satisfies

$$y_1'' + (n-1)\cot\theta\, y_1' + \lambda_1 y_1 = 0 \tag{2.4}$$

with boundary conditions $y_1(0)$ finite and $y_1(\theta_1) = 0$. Also, $u_2 = c_2(x_i/\sin\theta)y_2$ (for any $i = 1, 2, \ldots, n$) where $y_2$ satisfies

$$y_2'' + (n-1)\cot\theta\, y_2' + \left(\lambda_2 - \frac{n-1}{\sin^2\theta}\right) y_2 = 0 \tag{2.5}$$

with boundary conditions $y_2(0) = y_2(\theta_1) = 0$. (Borrowing from the language of quantum mechanics, $\lambda_2(\theta_1)$ is an $\ell = 1$ eigenvalue. That is, $\lambda_2(\theta_1)$ is an eigenvalue corresponding to an eigenfunction with "total angular momentum" quantum number $\ell = 1$. In general the separation constant coming from separating off all the angles aside from $\theta$ (in spherical coordinates) will be $\ell(\ell+n-2)$ for $\ell$ a nonnegative integer. Here $\ell = 1$ leads to the term $(n-1)/\sin^2\theta$ (while for $\lambda_1(\theta_1)$ we have $\ell = 0$ leaving us free of the $\ell(\ell+n-2)/\sin^2\theta$ term; cf. equation (2.4))). The values $\lambda_1$ and $\lambda_2$ are to be taken as the least eigenvalues of these one-dimensional radial problems (2.4) and (2.5) (with boundary conditions). One can express $y_1$ and $y_2$ in terms of associated Legendre functions. In fact

$$y_1(\theta) = (\sin\theta)^{1-n/2}\, P_{\nu_1}^{n/2-1}(\cos\theta)$$

and

$$y_2(\theta) = (\sin\theta)^{1-n/2}\, P_{\nu_2}^{n/2}(\cos\theta)$$

[1], where the parameters $\nu_1$ and $\nu_2$ are related to the eigenvalues $\lambda_1$ and $\lambda_2$ respectively by

$$\nu(\nu+1) = \lambda + \frac{n(n-2)}{4}.$$

Since we want equality in (2.1) when $\Omega$ is a geodesic ball, using the form of $u_1$ and $u_2$ for a geodesic ball we see that $g(\theta)$ must be essentially the quotient of $y_2$ and $y_1$.

Let $\Omega \subset \mathcal{S}^n$ be contained in a hemisphere. Let $B_{\lambda_1}$ denote the geodesic ball in $\mathcal{S}^n$ having the same value of $\lambda_1$ as $\Omega$. By the Faber-Krahn inequality [21] (see also [14]) $\lambda_1(B_1) = \lambda_1(\Omega) \geq \lambda_1(\Omega^*)$ and by the properties of $\lambda_1$ for geodesic balls (in paticular $\lambda_1$ decreases as the radius of the ball increases, see Theorem 2), $\theta_1 \leq \pi/2$, where $\theta_1$ denotes the geodesic radius of $B_{\lambda_1}$.

We take $g(\theta) = y_2(\theta)/y_1(\theta)$ for $0 \leq \theta \leq \theta_1$, $g(\theta) = g(\theta_1)$ for $\theta_1 \leq \theta \leq \pi/2$ and we extend $g(\theta)$ to a function on $[0, \pi]$ by reflection about $\theta = \pi/2$. We then apply the *center of mass* argument (Theorem 5) to $\Omega$ and $\widetilde{G}(\theta) = g(\theta)/\sin\theta$ obtaining a choice of Cartesian coordinate axes such that

$$\int_\Omega g(\theta)\frac{x_i}{\sin\theta} u_1^2\, d\omega = 0 \quad \text{for} \quad i = 1, 2, \ldots, n.$$

Using the functions $P_i = g(\theta)x_i/\sin(\theta)$ in (2.1) we get

$$(\lambda_2 - \lambda_1)\int_\Omega g(\theta)^2 \left(\frac{x_i}{\sin\theta}\right)^2 u_1^2\, d\omega \leq \int_\Omega \left|\nabla(g(\theta)\frac{x_i}{\sin\theta}\right|^2 u_1^2\, d\omega \tag{2.6}$$

for $1 \le i \le n$ and summing them from 1 to $n$ we get

$$\lambda_2 - \lambda_1 \le \frac{\int_\Omega B(\theta) u_1^2 \, d\omega}{\int_\Omega g(\theta)^2 u_1^2 \, d\omega},$$

where

$$B(\theta) = g'(\theta)^2 + \frac{(n-1)g^2(\theta)}{\sin^2(\theta)}. \tag{2.7}$$

We observe that, in spite of the fact that $\Omega$ is contained in a hemisphere, our use of the center of mass result may imply that $\Omega$ does not lie precisely on the northern hemisphere (i.e., $\theta$ would not be in $(0, \pi)$). To fix this, we observe that since $\Omega$ is contained in a hemisphere, $-\Omega \subset S^n \backslash \Omega$. Thus, if we let $\Omega_\pm = \{\vec{x} \in \Omega \mid \pm x_{n+1} > 0\}$ we have $(-\Omega_-) \cap \Omega_+ = \emptyset$. Since $g(\theta)$ and $B(\theta)$ are both symmetric with respect to $\theta = \pi/2$ it follows that the integrals over $\Omega$ can be replaced by integrals over $\widetilde{\Omega} = \Omega_+ \cup \{-\Omega_-\}$ with no change in their values if we agree to transplant $u_1$ along with $\Omega_-$ to $-\Omega_-$. This follows since $g$ and $B$ were defined to be symmetric about $\theta = \pi/2$ and therefore they transplant into themselves. After moving the whole problem to the northern hemisphere (where $g$ is increasing and $B$ is decreasing) we can carry out all the further rearrangements in exact parallel with the Euclidean case, encountering no further difficulties.

To conclude this section we state the two chains of inequalities that go into the proof of Theorem 1. Recall that we start from

$$\lambda_2 - \lambda_1 \le \frac{\int_\Omega B(\theta) u_1^2}{\int_\Omega g(\theta)^2 u_1^2}. \tag{2.8}$$

We have

$$\int_\Omega B(\theta) u_1^2 \, d\omega = \int_{\tilde{\Omega}} B(\theta) \tilde{u}_1^2 \, d\omega \le \int_{\Omega^*} B(\theta) u_1^{*2} \, d\omega \le \int_{B_{\lambda_1}} B(\theta) z^2 \, d\omega \tag{2.9}$$

and

$$\int_\Omega g(\theta)^2 u_1^2 \, d\omega = \int_{\tilde{\Omega}} g(\theta)^2 \tilde{u}_1^2 \, d\omega \ge \int_{\Omega^*} g(\theta)^2 u_1^{*2} \, d\omega \ge \int_{B_{\lambda_1}} g(\theta)^2 z^2 \, d\omega. \tag{2.10}$$

Here $\tilde{u}_1$ represents $u_1$ as transplanted to $\tilde{\Omega}$ and $z$ is the first eigenfunction of $-\Delta$ on $B_{\lambda_1}$ with Dirichlet boundary conditions and normalized so that $\int_\Omega u_1^2 \, d\omega = \int_{B_{\lambda_1}} z^2 \, d\omega$. The functions $g$ and $B$ are likewise based on the eigenfunctions of the ball $B_{\lambda_1}$ (so that the $\theta_1$ that goes into the boundary value problems (2.4) and (2.5) that define them is the radius of the ball $B_{\lambda_1}$). In each of (2.9) and (2.10), the equality is trivial, the first inequality follows simply form rearrangements (see Section 4 for our notation) and by virtue of the monotonicity properties of $g(\theta)$ and $B(\theta)$, and the last inequality follows from our $S^n$ analog of Chiti's comparison result (see Section 4 for details) and also uses the monotonicity properties of $g$ and $B$ again. Finally, from (2.8), (2.9) and (2.10) one obtains

$$\lambda_2 - \lambda_1 \leq \frac{\int_{B_{\lambda_1}} B(\theta) z^2 \, d\omega}{\int_{B_{\lambda_1}} g(\theta)^2 z^2 \, d\omega} = \lambda_2 - \lambda_1 \quad \text{for} \quad B_{\lambda_1} \tag{2.11}$$

and hence (since $\lambda_1(B_{\lambda_1}) = \lambda_1(\Omega)$)

$$\lambda_2(\Omega) \leq \lambda_2(B_{\lambda_1}) \tag{2.12}$$

which is Theorem 1.

## 3. Monotonicity properties of $g$ and $B$

In this section we give the main ideas of the proofs concerning the properties of $g$ and $B$. As in [4], [6] we will introduce an auxiliary function $q$, which for domains in $\mathcal{S}^n$ we define as

$$q(\theta) = \sin\theta \frac{g'(\theta)}{g(\theta)}. \tag{3.1}$$

Since $g$ and $\sin\theta$ are positive, that $g$ is increasing will be a simple consequence of showing $q \geq 0$. Similarly, from the equation

$$B'(\theta) = 2\left[qq' - (\cos\theta - q)(q^2 + n - 1)\frac{1}{\sin\theta}\right]\left(\frac{g(\theta)}{\sin\theta}\right)^2 \tag{3.2}$$

it follows that $B$ decreasing will be a consequence of showing that $q' \leq 0$ and $q \leq \cos\theta$. Our strategy follows the same general method used in [4]. By Taylor expansion at $\theta = 0$ one finds that $q(0) = 1$ and that $q < \cos\theta$ on an interval just to the right of 0. Similarly, by Taylor expansion around $\theta = \theta_1$ one finds $q(\theta_1) = 0$ and that $q > 0$ on an interval just to the left of $\theta_1$. The expansions here are straightforward though the key comparisons depend on additional facts concerning eigenvalues for geodesic balls. The key inequalities that enter here are

$$\lambda_1 - \lambda_2 > \frac{n-1}{\sin^2(\theta_1)} \tag{3.3}$$

(needed to show that $q > 0$ in an interval to the left of $\theta_1$) and

$$\frac{\lambda_2 - n}{\lambda_1} > \frac{n+2}{n} \tag{3.4}$$

(needed to show that $q < \cos\theta$ in an interval to the right of 0 for $\theta_1 < \pi/2$). Here $\lambda_1$ and $\lambda_2$ are the first two Dirichlet eigenvalues for a geodesic ball of geodesic radius $0 < \theta_1 < \pi/2$. The proof of these two inequalities is carried out as in the proof of the corresponding inequalities needed in the Euclidean case [4], by appealing to the variational characterization of the eigenvalues and to a careful analysis of the special functions involved (see [7] for details).

In order to prove that $q \geq 0$, $q' \leq 0$ and $q \leq \cos\theta$, we analyze the ordinary differential equation satisfied by $q$. By differentiating (3.1) with respect to $\theta$, using our choice of $g = y_2/y_1$, and the equations (2.4) and (2.5) satisfied by $y_1$ and $y_2$, respectively, we obtain the following differential equation for $q$,

$$q' = -2pq - (n-2)q\cot\theta - \frac{q^2+1-n}{\sin\theta} - (\lambda_2 - \lambda_1)\sin\theta. \tag{3.5}$$

Here, the function $p \equiv y_1'/y_1$ obeys the Riccati equation

$$p' + p^2 + (n-1)\cot\theta\, p + \lambda_1 = 0 \tag{3.6}$$

associated to (2.4) (note that (3.5) is also a Riccati equation). The function $p$ is negative, decreasing, and concave in $(0, \theta_1)$, $p(0) = 0$, and $p \to -\infty$ as $\theta \to \theta_1$. All these properties follow from the definition of $p$ and by analyzing equation (3.6).

Once we have derived the equation for $q$, we are ready to prove the necessary facts about $q$. We start by showing $q \geq 0$ in $[0, \theta_1]$, which we prove by contradiction. Assume $q$ is negative somewhere in $[0, \theta_1]$. Since $q(0) = 1$ and $q$ is positive to the left of $\theta_1$ (and $q$ is continuous) this implies that there are two points $\alpha$, $\beta$, say, with $0 < \alpha < \beta < \theta_1$ such that $q(\alpha) = q(\beta) = 0$ and $q'(\alpha) \leq 0$, $q'(\beta) \geq 0$. At points where $q = 0$ it follows from (3.5) that

$$q' = \frac{n-1}{\sin\theta} - (\lambda_2 - \lambda_1)\sin\theta. \tag{3.7}$$

Since $\lambda_2 > \lambda_1$ and $\sin\theta$ is increasing in $\theta$ for $0 < \theta < \pi/2$, the right side of (3.7) is decreasing in $\theta$, hence it is not possible to have $\alpha < \beta$ with $q(\alpha) = q(\beta) = 0$ and $q'(\alpha) \leq 0$, $q'(\beta) \geq 0$. Therefore $q \geq 0$ in $[0, \theta_1]$.

The proof that $q \leq \cos\theta$ follows the same ideas. One defines the function $\psi = \cos\theta - q$, and derives the equation for $\psi$ from the equation for $q$. Since we want to prove that $\psi \geq 0$ we proceed by contradiction as we did before. Instead of using that $\sin\theta$ is increasing, this time we use that $p\cot\theta$ is decreasing, a fact that is not hard to prove (see [7] for details).

Finally, we show that $q(\theta)$ is decreasing in $[0, \theta_1]$. The way we will prove this here is by considering the right side of (3.5), which we will conveniently denote by $S(q, \theta)$, as a function of $q$ and $\theta$ and analyze its dependence on $\theta$ keeping $q$ fixed. In order to analyze the behavior of $S(q, \theta)$ as a function of $\theta$, we will determine the value of $\partial S(q, \theta)/\partial\theta$ at points where $S = 0$ together with the behavior of $S$ near the endpoints $\theta = 0$ and $\theta = \theta_1$.

For $\theta$ close to 0 and for fixed $0 < q < 1$ we have

$$S(q, \theta) = \frac{1}{\theta}(n-1+q)(1-q) + O(1), \tag{3.8}$$

and therefore, $S(q,\theta) \to \infty$ as $\theta \to 0$. On the other hand, as $\theta \to \theta_1$, $p \to -\infty$, and thus from (3.5), $S(q,\theta) \to \infty$ as $\theta \to \theta_1$ (for fixed $q \in (0,1)$). Let $N = (q^2 - (n-1)) < 0$ for $n \geq 2$ (and $0 < q < 1$). From (3.5), we get

$$\frac{\partial S}{\partial \theta}(q,\theta) = -2qp' + \frac{(n-2)q}{\sin^2\theta} + N\frac{\cos\theta}{\sin^2\theta} - a\cos\theta, \tag{3.9}$$

where we have set $a \equiv \lambda_2 - \lambda_1 > 0$. At values of $\theta$ for which $S(q,\theta)$ vanishes, we get from (3.5)

$$p = -\frac{n-2}{2}\cot\theta - \frac{N}{2q\sin\theta} - \frac{a\sin\theta}{2q}. \tag{3.10}$$

Using the equation satisfied by $p$, i.e., equation (3.6), and (3.10) we can also compute $p'$ at values of $\theta$ for which $S(q,\theta)$ vanishes. In fact, we get

$$p' = \frac{n(n-2)}{4}\cot^2\theta - \lambda_1 + \left(\frac{N}{\sin^2\theta} + a\right)\left(\frac{\cos\theta}{2q} - \left(\frac{N}{\sin^2\theta} + a\right)\frac{\sin^2\theta}{4q^2}\right). \tag{3.11}$$

From (3.9) and (3.11) we obtain that at those values of $\theta$ for which $S = 0$,

$$\left.\frac{\partial S}{\partial \theta}\right|_{S=0} = \frac{1}{2q\sin^2\theta}\left(N^2 - q^2(n-2)^2 + \sin^2\theta\, h(\theta)\right), \tag{3.12}$$

where $h(\theta) \equiv q^2 n(n-2) + 4q^2\lambda_1 - 4aq\cos\theta + 2aN + a^2 - a^2\cos^2\theta$. It is clear from its definition that $h(\theta)$ is concave in $\cos\theta$, and therefore it can have at most two zeros for $\theta \in (0,\pi/2)$, and so does $\sin^2\theta h(\theta)$. Moreover, the constant $|N| - q(n-2) = (1-q)(q+(n-1)) > 0$. Hence, the right side of (3.12) can vanish at most twice in the interval $(0,\pi/2)$. Given the behavior of $\partial S/\partial\theta \,|_{S=0}$ just discussed and the boundary behavior of $S$ near 0 and $\theta_1$, we conclude that either $S$ never crosses zero, or $S$ has exactly two crossings in the interval $(0,\theta_1)$. This immediatly implies that $q$ is decreasing, since if it were not we would have at least some fixed $0 < q < 1$ for which $S$ would be first negative, then positive and then negative again, which is a contradiction.

An alternative proof that $q$ is decreasing, following the lines of our previous papers [4], [6], will be presented in [7].

## 4. Chiti's comparison argument in $S^n$

In Section 2 we have used an extension of Chiti's comparison result [11], [12], given originally for domains in $\mathbb{R}^n$, to the case of domains in $S^n$. This section will be used to fix the notation on rearrangements that we use throughout the article, as well as to give the statement of the appropriate extension of Chiti's comparison theorem to $S^n$.

We let $\mathcal{S}^n(\rho)$ denote the $n$-dimensional sphere of radius $\rho$ (hence of constant sectional curvature $\kappa = 1/\rho^2$). $\mathcal{S}^n$ will always denote $\mathcal{S}^n(1)$. Define the function

$$S_\rho(t) = \rho \sin(t/\rho).$$

It is well known that, in geodesic polar coordinates, the metric on $\mathcal{S}^n(\rho)$ is

$$ds^2 = dt^2 + S_\rho(t)^2 |d\omega|^2,$$

where $t$ represents geodesic distance from a point and $|d\omega|^2$ is the canonical metric for $\mathcal{S}^{n-1}$ (one should think of $t$ as the angle $\theta$ from the north pole times $\rho$). Thus $t$ runs from 0 to $\rho\pi$.

We will always assume that $t = 0$ is the center for our spherical rearrangements (this is not a restriction since the metric is identical in geodesic polar coordinates about any point). Then for a bounded domain $\Omega \subset \mathcal{S}^n(\rho)$ $\Omega^*$, the spherical rearrangement of $\Omega$, will denote the geodesic ball about $t = 0$ having the same volume as $\Omega$, i.e., $|\Omega^*| = |\Omega|$.

For a nonnegative function $f$ defined on $\Omega$ we define two rearranged functions $f^*$ and $f^\star$. The decreasing rearrangement $f^*$ of $f$ is a function from $[0, |\Omega|]$ to $\mathbb{R}$ which is equimeasurable with $f$ and nonincreasing. We will use $s$ as the argument of $f^*$ in most instances. The symetric decreasing (or spherical decreasing) rearrangement $f^\star$ of $f$ is a function defined on $\Omega^*$ which is invariant under rotations about $t = 0$, equimeasurable with $f$ and nonincreasing with respect to $t$. $f^\star$ is a function of $x \in \Omega^*$ but because of its symmetry we will abuse notation and write $f^\star(t)$ where $t =$ geodesic distance from the center of $\Omega^*$. With this understanding we have

$$f^\star(t) = f^*(A(t)),$$

where

$$A(t) \equiv s = nC_n \int_0^t S_\rho(\tau)^{n-1}\, d\tau \tag{4.1}$$

is the $n$-volume of the geodesic ball of radius $t$ in $\mathcal{S}^n(\rho)$. Here $C_n = \pi^{n/2}/\Gamma((n/2)+1)$ denotes the volume of the unit ball in $\mathbb{R}^n$. We shall also have ocassion to use increasing rearrangements. These will be denoted $f_*$ (the increasing rearrangement of $f$) and $f_\star$ (the symmetric or spherical rearrangement of $f$) and their definitions are analogous to those of $f^*$ and $f^\star$, respectively. For further information on rearrangemnts the reader is referred to the book of Hardy, Littlewood, and Pólya [15] and the many other references given in [3], [23].

Finally, we also need the classical isoperimetric inequality extended to $\mathcal{S}^n(\rho)$ in its sharp form. We let $L(t)$ be the function giving the $(n-1)$-dimensional volume of the geodesic ball of radius $t$, i.e.,

$$L(t) = nC_n S_\rho(t)^{n-1} = A'(t). \tag{4.2}$$

Then for any domain in $\mathcal{S}^2$ one has

$$L^2 \geq 4\pi A - \left(\frac{A}{\rho}\right)^2 \tag{4.3}$$

(see, e.g., [18]). Here $A = |\Omega|$ and $L$ is the length of the boundary of $\Omega$. Equality occurs in (4.3) if and only if $\Omega$ is a geodesic ball. In $\mathcal{S}^n$, $n > 2$, the sharp classical isoperimetric inequality cannot be given in as explicit a form as (4.3). For a bounded domain $\Omega$ we define $\sigma(\partial\Omega)$ to be the $(n-1)$-dimensional volume of $\partial\Omega$ (in the sense of Minkowski content). The isoperimetric inequality on $\mathcal{S}^n(\rho)$ then reads

$$\sigma(\partial\Omega) \geq \sigma(\partial\Omega^*) \tag{4.4}$$

with equality if and only if $\Omega$ is a geodesic ball (see Burago and Zalgaller [9], p. 86, Theorem 10.2.1). In terms of the function $L(t)$ defined above (4.4) may be written

$$\sigma(\partial\Omega) \geq L(\theta(|\Omega|)) = nC_n S_\rho(\theta(|\Omega|))^{n-1}, \tag{4.5}$$

where $\theta(s)$ is the inverse function to the function $A$ defined in (4.1).

With all these ingredients we state without proof our extension of Chiti's comparison result (for the proof, see [7]).

**Theorem 6.** *Let $\Omega$ be a bounded domain in $\mathcal{S}^n(\rho)$ and let $\lambda_1$ and $u_1$ denote the first Dirichlet eigenvalue and eigenfunction of the Laplacian operator on $\Omega$. Let $B_{\lambda_1}$ be the geodesic ball of such a radius that $\lambda_1$ is also the first Dirichlet eigenvalue of the Laplacian operator on $B_{\lambda_1}$. Let $v_1$ be the first Dirichlet eigenfunction on $B_{\lambda_1}$ and fix its normalization so that $\int_\Omega u_1^2\, d\omega = \int_{B_{\lambda_1}} v_1^2\, d\omega$.*

*Then there is a value $t_1 \in (0, \theta(|B_{\lambda_1}|)$ such that*

$$v_1(t) \geq u_1^\star(t) \quad \text{for } t \in [0, t_1], \tag{4.6a}$$

*and*

$$v_1(t) \leq u_1^\star(t) \quad \text{for } t \in [t_1, \theta(|B_{\lambda_1}|)), \tag{4.6b}$$

*where $\theta(s)$ is the function defined following* (4.5) *above.*

## 5. The dependence of $\lambda_1$ and $\lambda_2$ on the radius for geodesic balls in $\mathcal{S}^n$

In this section, we will show that $\lambda_2/\lambda_1$ for a geodesic ball contained in a hemisphere is an increasing function of its radius. We will prove this in two steps. First we will show that $\theta_1^2\lambda_1(\theta_1)$ is a decreasing function of $\theta_1$ (Theorem 2) and finally that $\lambda_2(\theta_1)/\lambda_1(\theta_1)$ is an increasing function of $\theta_1$, for $\theta_1 \in (0, \pi/2]$ (Theorem 3). The PPW result for $\mathcal{S}^n$ follows from Theorem 1 above and this property of $\lambda_2/\lambda_1$ for geodesic balls, as described in the introduction.

*Proof of Theorem* 2. The first Dirichlet eigenfunction $y_1(\theta)$ of a geodesic ball satisfies the equation

$$-\frac{d^2y_1}{d\theta^2} - \frac{f'}{f}\frac{dy_1}{d\theta} = \lambda_1(\theta_1)y_1(\theta) \quad \text{for} \quad \theta \in (0, \theta_1), \tag{5.1}$$

where $f(\theta) = \sin^{n-1}(\theta)$, with boundary conditions $y_1(0)$ finite and $y_1(\theta_1) = 0$. In order to study the dependence of $\lambda_1$ on the radius $\theta_1$, it is convenient to refer the eigenvalue problem given by (5.1) to a fixed interval. Hence, we make the change of variable $\theta = x\theta_1$, so that $x \in [0, 1]$. We will denote $\dot{y}_1 = dy_1/dx$. In this new variable, equation (5.1) reads

$$-\ddot{y}_1 - \frac{f'(x\theta_1)}{f(x\theta_1)}\theta_1\dot{y}_1 = \tilde{\lambda}_1 y_1 \quad \text{for} \quad x \in (0, 1), \tag{5.2}$$

where we have denoted $\tilde{\lambda}_1 = \theta_1^2\lambda_1(\theta_1)$.

The eigenvalue $\tilde{\lambda}_1$, for the problem (5.2), is given by the Rayleigh quotient

$$\tilde{\lambda}_1 = \frac{\int_0^1 \dot{y}_1^2 f(x\theta_1)\,dx}{\int_0^1 y_1^2 f(x\theta_1)\,dx} \tag{5.3}$$

In fact, $\tilde{\lambda}_1$ is the minimum value of this Rayleigh quotient taken over all functions in $H_0^1[B(1), (f(x\theta_1)/f(x))\,d\omega]$ (here $B(1)$ is the ball of radius 1). We will write $\tilde{\lambda}_1 = N/D$ where $N$ and $D$ are respectively the numerator and the denominator of (5.3). Since $y_1(x\theta_1)$ is a minimizer of the Rayleigh quotient, we have

$$\frac{d\tilde{\lambda}_1}{d\theta_1} = \frac{\tilde{N}}{D} - \frac{N}{D}\frac{\tilde{D}}{D} = \frac{\tilde{N} - \tilde{\lambda}_1\tilde{D}}{D},$$

where

$$\tilde{N} = \int_0^1 \dot{y}_1^2 \frac{d}{d\theta_1} f(x\theta_1)\,dx$$

and

$$\tilde{D} = \int_0^1 {y_1}^2 \frac{d}{d\theta_1} f(x\theta_1)\,dx,$$

i.e.,

$$\frac{d\tilde{\lambda}_1}{d\theta_1} = \frac{1}{D}\left[\int_0^1 (\dot{y}_1^2 - \tilde{\lambda}_1 {y_1}^2) x f'(x\theta_1)\,dx\right]. \tag{5.4}$$

Using equation (5.2) we can replace $\tilde{\lambda}_1 y_1^2$ in (5.4) and obtain

$$\frac{d\tilde{\lambda}_1}{d\theta_1} = \frac{1}{D}\int_0^1 \frac{d}{dx}(f y_1 \dot{y}_1)\frac{xf'}{f}\,dx,$$

or, in terms of the original variable $\theta$,

$$\frac{d\tilde{\lambda}_1}{d\theta_1} = \frac{\theta_1}{\int_0^{\theta_1} y_1^2 f(\theta)\, d\theta} \int_0^{\theta_1} \frac{d}{d\theta}\left(f y_1 \frac{dy_1}{d\theta}\right) \frac{\theta f'}{f}\, d\theta. \tag{5.5}$$

Integrating the right side of (5.5) by parts twice we finally get

$$\frac{d\tilde{\lambda}_1}{d\theta_1} = \frac{\theta_1}{2\int_0^{\theta_1} y_1^2 f(\theta)\, d\theta} \int_0^{\theta_1} y_1^2(\theta) R(\theta)\, d\theta, \tag{5.6a}$$

where

$$R(\theta) \equiv \frac{d}{d\theta}\left(f(\theta)\frac{d}{d\theta}\left(\theta \frac{f'(\theta)}{f(\theta)}\right)\right).$$

Since $f(\theta) = \sin^{n-1}(\theta)$, we get

$$R(\theta) = \theta_1 (n-1)\sin^{n-1}\theta\big[(n-3)M(\theta) - (n-1)\big] \equiv \theta_1(n-1)\sin^{n-1}\theta s\theta, \tag{5.6b}$$

with $M(\theta) = \csc^2\theta(1-\theta\cot\theta)$. The function $M(\theta)$ is positive, increasing, and convex for $0 < \theta < \pi$. Moreover, $M(0) = 1/3$, $M(\pi/2) = 1$, $M(\pi) = \infty$ (see [7] for details). For $n=1$, $s(\theta) = -2M(\theta) < 0$, for all $0<\theta<\pi$. For $n=2$, $s(\theta) = -M(\theta) - 1 < 0$, for all $0<\theta<\pi$. For $n=3$, $s(\theta) = -2$, for all $0<\theta<\pi$. Finally for $n \geq 4$, $s(0) = -2n/3$, $s(\pi/2) = -2$ and $s''(\theta) = (n-3)M''(\theta) \geq 0$. Therefore $s(\theta) < 0$ in $[0,\pi/2]$ for $n \geq 4$. Thus $R < 0$ for all $0 < \theta < \pi/2$ and the theorem follows. □

**Remark.** The result of this theorem is still valid for $0 \leq \theta \leq \pi$. To go beyond the hemisphere, one integrates by parts only once in (5.5), to get

$$\frac{d\tilde{\lambda}}{d\theta_1} = -\theta_1 \frac{\int_0^{\theta_1} y_1 y_1' F(\theta)\, d\theta}{\int_0^{\theta_1} y_1^2 f(\theta)\, d\theta},$$

where

$$F(\theta) \equiv f(\theta)\frac{d}{d\theta}\left(\frac{\theta f'(\theta)}{f(\theta)}\right),$$

and uses the fact that $F(\theta) < 0$ for all $0 < \theta < \pi$, and that $y_1(\theta) > 0$, and $y'(\theta) < 0$ for all $0 < \theta < \theta_1$. See [7] for details.

The second Dirichlet eigenvalue for a geodesic ball in $\mathcal{S}^n$, of geodesic radius $\theta_1$, $\lambda_2(\theta_1)$ corresponds to the lowest eigenvalue of the operator

$$-\frac{d^2}{d\theta^2} - \frac{f'}{f}\frac{d}{d\theta} + \frac{n-1}{\sin^2(\theta)}$$

with zero boundary conditions at $\theta = 0$ and $\theta = \theta_1$ (i.e., $\lambda_2(\theta_1)$ corresponds to an $\ell = 1$ eigenvalue, see remarks after equation (2.5) above). Let us denote by $y_2$ the corresponding eigenfunction. Then $y_2(\theta)$ satisfies the equation

$$-y_2'' - \frac{f'}{f} y_2 + \frac{n-1}{\sin^2\theta} = \lambda_2 y_2 \quad \text{in} \quad (0, \theta_1),$$

with $y_2(0) = y_2(\theta_1) = 0$. Using the same change of variables as in the proof of Theorem 2, i.e., $\theta = x\theta_1$ with $x \in [0,1]$, this equation can be written as

$$-\ddot{y}_2 - \frac{f'(x\theta_1)}{f(x\theta_1)}\theta_1 \dot{y}_2 + \frac{n-1}{\sin^2(x\theta_1)}\theta_1^2 y_2 = \tilde{\lambda}_2 y_2 \quad \text{for} \quad x \in (0,1), \tag{5.7}$$

with zero boundary conditions. Here, $\tilde{\lambda}_2 \equiv \theta_1^2 \lambda_2(\theta_1)$. Now $\lambda_2$ is given by the Rayleigh quotient

$$\tilde{\lambda}_2(\theta_1) = \frac{\int_0^1 \dot{y}_2 f(x\theta_1)\,dx + \int_0^1 (n-1) y_2^2 \sin^{n-3}(x\theta_1)\theta_1^2\,dx}{\int_0^1 y_2^2 f(x\theta_1)\,dx}.$$

Since $y_2$ minimizes this Rayleigh quotient, we can write as we did with $\tilde{\lambda}_1$ before,

$$\frac{d\tilde{\lambda}_2}{d\theta_1} = \frac{\tilde{N}_2 - \tilde{\lambda}_2 \tilde{D}_2}{D_2},$$

where $D_2 = \int_0^1 y_2^2 f(x\theta_1)\,dx$ and

$$\begin{aligned} \tilde{N}_2 - \tilde{\lambda}_2 \tilde{D}_2 &= \int_0^1 (\dot{y}_2^2 - \tilde{\lambda}_2 y_2^2) f(x\theta_1) x\,dx \\ &\quad + \int_0^1 y_2^2 (n-1)(n-3)\sin^{n-4}(x\theta_1)\cos(x\theta_1)\theta_1^3\,dx. \end{aligned} \tag{5.8}$$

Replacing $\tilde{\lambda}_2$ in (5.8) using (5.7), and integrating by parts as in the proof of Theorem 2, we finally get

$$\frac{d\tilde{\lambda}_2}{d\theta_1} = \frac{(n-1)\theta_1}{2\int_0^{\theta_1} y_2^2 \sin^{n-1}\theta\,d\theta} \int_0^{\theta_1} y_2^2((n+1)M(\theta) - (n-1))\sin^{n-1}\theta\,d\theta. \tag{5.9}$$

After these calculations we are ready to prove Theorem 3.

*Proof of Theorem* 3. Let $r \equiv \lambda_2/\lambda_1 = \tilde{\lambda}_2/\tilde{\lambda}_1$. Then,

$$\frac{dr}{d\theta_1} = \frac{1}{\tilde{\lambda}_1}\frac{d\tilde{\lambda}_2}{d\theta_1} - \frac{\tilde{\lambda}_2}{\tilde{\lambda}_1^2}\frac{d\tilde{\lambda}_1}{d\theta_1} \geq \frac{1}{\tilde{\lambda}_1}\left(\frac{d\tilde{\lambda}_2}{d\theta_1} - \frac{d\tilde{\lambda}_1}{d\theta_1}\right), \tag{5.10}$$

where we have used that $d\tilde{\lambda}_1/d\theta_1 < 0$, by of Theorem 2, and $\tilde{\lambda}_2 > \tilde{\lambda}_1$. To finish, it will suffice to show that

$$\frac{d\tilde{\lambda}_2}{d\theta_1} > \frac{d\tilde{\lambda}_1}{d\theta_1}. \tag{5.11}$$

From equations (5.6a), (5.6b) and (5.9) this is equivalent to proving

$$(n+1)\frac{\int_0^{\theta_1} y_2^2(\theta)M(\theta)\sin^{n-1}(\theta)\,d\theta}{\int_0^{\theta_1} y_2^2(\theta)\sin^{n-1}(\theta)\,d\theta} > (n-3)\frac{\int_0^{\theta_1} y_1^2(\theta)M(\theta)\sin^{n-1}(\theta)\,d\theta}{\int_0^{\theta_1} y_1^2(\theta)\sin^{n-1}(\theta)\,d\theta}$$

which is obviously true for $n \leq 3$ since $M(\theta)$ is positive. For $n \geq 4$ it suffices to show

$$\int_0^{\theta_1} \hat{y}_2^2(\theta)M(\theta)\sin^{n-1}(\theta)\,d\theta > \int_0^{\theta_1} \hat{y}_1^2(\theta)M(\theta)\sin^{n-1}(\theta)\,d\theta,$$

where $\hat{y}_2 = y_2/\sqrt{\int_0^{\theta_1} y_2^2 \sin^{n-1}\theta\,d\theta}$ and $\hat{y}_1 = y_1/\sqrt{\int_0^{\theta_1} y_1^2 \sin^{n-1}\theta\,d\theta}$.

The fact that $g \equiv y_2/y_1$ is an increasing function of $\theta$ (see Section 3 above) implies that $\hat{y}_2$ and $\hat{y}_1$ must have exactly one crossing in $(0,\theta_1)$. Since $M(\theta)$ is positive and increasing in $[0,\pi]$ the desired inequality (5.11) follows by applying Chiti's argument (see e.g., [3], p. 607). □

**Acknowledgments.** We would like to thank the organizing committee of the General Inequalities Conference, Catherine Bandle, Norrie Everitt, Laszlo Losonczi, and Wolfgang Walter, for their kind invitation to participate in this conference, and for their warm hospitality at the Mathematisches Forschungsinstitut Oberwolfach.

## References

[1] M. Abramowitz and I.A. Stegun, eds., *Handbook of Mathematical Functions*. National Bureau of Standards, Applied Mathematical Series, vol. **55**, U.S. Government Printing Office, Washington, D.C., 1964.

[2] M.S. Ashbaugh and R.D. Benguria, *Proof of the Payne-Pólya-Weinberger conjecture*. Bull. Amer. Math. Soc. **25** (1991), 19–29.

[3] M.S. Ashbaugh and R.D. Benguria, *A sharp bound for the ratio of the first two eigenvalues of Dirichlet Laplacians and extensions*. Annals of Math. **135** (1992), 601–628.

[4] M.S. Ashbaugh and R.D. Benguria, *A second proof of the Payne-Pólya-Weinberger conjecture*. Commun. Math. Phys. **147** (1992), 181–190.

[5] M.S. Ashbaugh, and R.D. Benguria, *Isoperimetric inequalities for eigenvalue ratios.* In "Partial Differential Equations of Elliptic Type", A. Alvino, E. Fabes, G. Talenti, editors. Cambridge University Press, Cambridge, 1994, pp. 1–36.

[6] M.S. Ashbaugh and R.D. Benguria, *Sharp upper bound to the first nonzero Neumann eigenvalue for bounded domains in spaces of constant curvature.* J. London Math. Soc. (2) **52** (1995), 402–416.

[7] M.S. Ashbaugh and R.D. Benguria, *Sharp bound for the ratio of the first two eigenvalues of a domain in a hemisphere of $\mathcal{S}^n$*. Preprint, 1996.

[8] C. Bandle, *Isoperimetric Inequalities and Applications.* Pitman Monographs and Studies in Mathematics, vol. **7**, Pitman, Boston, 1980.

[9] Yu.D. Burago and V.A. Zalgaller, *Geometric Inequalities* (Grundlehren der mathematischen Wissenschaften 285). Springer, Berlin, 1988.

[10] I. Chavel, *Lowest-eigenvalue inequalities.* In Proc. Symp. Pure Math., vol. **36**, "Geometry of the Laplace Operator", R. Osserman and A. Weinstein, editors. Amer. Math. Soc., Providence, R.I., 1980, pp. 79–89.

[11] G. Chiti, *A reverse Hölder inequality for the eigenfunctions of linear second order elliptic operators.* J. Appl. Math. and Phys. (ZAMP) **33** (1982), 143–148.

[12] G. Chiti, *A Bound for the ratio of the first two eigenvalues of a membrane.* SIAM J. Math. Anal. **14** (1983), 1163–1167.

[13] G. Faber, *Beweis, dass unter allen homogenen Membranen von gleicher Fläche und gleicher Spannung die kreisförmige den tiefsten Grundton gibt.* Sitzungberichte der mathematisch-physikalischen Klasse der Bayerischen Akademie der Wissenschaften zu München Jahrgang, 1923, pp. 169–172.

[14] S. Friedland and W.K. Hayman, *Eigenvalue inequalities for the Dirichlet problem on spheres and the growth of subharmonic functions.* Comment. Math. Helvetici **51** (1976), 133–161.

[15] G.H. Hardy, J.E. Littlewood, and G. Pólya, *Inequalities.* 2nd ed., Cambridge Univ. Press, Cambridge, 1952.

[16] E. Krahn, *Über eine von Rayleigh formulierte Minimaleigenschaft des Kreises.* Math. Ann. **94** (1925), 97–100.

[17] E. Krahn, *Über Minimaleigenschaften der Kugel in drei und mehr Dimensionen.* Acta Comm. Univ. Tartu (Dorpat) **A9** (1926), 1–44. [English translation in "Edgar Krahn 1894–1961: A Centenary Volume", Ü. Lumiste and J. Peetre, editors. IOS Press, Amsterdam, 1994, pp. 139–174.]

[18] R. Osserman, *The isoperimetric inequality.* Bull. Amer. Math. Soc. **84**, (1978), 1182–1238.

[19] L.E. Payne, G. Pólya, and H.F. Weinberger, *Sur le quotient de deux fréquences propres consécutives.* Comptes Rendus Acad. Sci. Paris **241** (1955), 917–919.

[20] L.E. Payne, G. Pólya, and H.F. Weinberger, *On the ratio of consecutive eigenvalues.* J. Math. and Phys. **35** (1956), 289–298.

[21] E. Sperner, *Zur Symmetrisierung von Funktionen auf Sphären.* Math. Z. **134** (1973), 317–327.

[22] G. Szegő, *Inequalities for certain eigenvalues of a membrane of given area.* J. Rat. Mech. Anal. **3** (1954), 343–356.

[23] G. Talenti, *Elliptic equations and rearrangements.* Ann. Scuola Norm. Sup. Pisa (4) **3** (1976), 697–718.

[24] H.F. Weinberger, *An isoperimetric inequality for the n-dimensional free membrane problem.* J. Rat. Mech. Anal. **5** (1956), 633–636.

Mark S. Ashbaugh, Department of Mathematics, University of Missouri, Columbia, MO 65211, USA.

E-mail address: mark@ashbaugh.math.missouri.edu

Rafael D. Benguria, Facultad de Física, P. Universidad Católica de Chile, Casilla 306, Santiago 22, Chile.

E-mail address: rbenguri@lascar.puc.cl

International Series of Numerical Mathematics, Vol. 123, © 1997 Birkhäuser Verlag Basel

# Norm Eigenvalue bounds for some weighted Sturm-Liouville problems

R.C. Brown and D.B. Hinton

**Abstract**

The problem of obtaining upper and lower bounds for eigenvalues for regular Sturm-Liouville problems is considered in this paper. The goal is to find bounds in terms of integrals of the coefficients and avoid the use of pointwise bounds. The general weighted problem is studied.

## 1. Introduction

We consider the problem of finding good *a priori* upper and lower bounds on eigenvalues and on ratios of eigenvalues for the weighted Sturm-Liouville problem

$$L(y) := -(py')' + qy = \lambda wy\,, \quad a \le x \le b\,, \tag{1.1}$$

$$\begin{aligned}(\cos\alpha)y(a) - (\sin\alpha)(py')(a) = 0, &\quad 0 \le \alpha < \pi,\\ (\cos\beta)y(b) - (\sin\beta)(py')(b) = 0, &\quad 0 < \beta \le \pi,\end{aligned} \tag{1.2}$$

where $p, q, w$ satisfy minimal conditions in the sense of Naimark [17] for a regular problem; i.e., they are real measurable functions on the finite interval $[a, b]$ such that $p, w > 0$ on $(a, b)$ and $p^{-1}, q, w$ are Lebesgue integrable on $[a, b]$. In this case (1.1)–(1.2) have the same fundamental properties as a conventional problem with continuous coefficients. Indeed the asymptotic expressions for the eigenvalues in this general setting have been given by Atkinson and Mingarelli [2]. In particular, the eigenvalues grow asymptotically as $n^2$.

Certain cases of (1.2) are of particular importance and we use the following notation for the eigenvalues.

$$\begin{aligned}&\text{Dirichlet, i.e., } \alpha = 0, \beta = \pi: &&\lambda_0 < \lambda_1 < \cdots\\ &\text{Dirichlet with } q = 0: &&\mu_0 < \mu_1 < \cdots\end{aligned} \tag{1.3}$$

$$\text{Neumann, i.e., } \alpha = \beta = \pi/2: \quad \gamma_0 < \gamma_1 < \cdots \tag{1.4}$$

$$\text{Mixed Dirichlet-Neumann, i.e., } \alpha = 0,\ \beta = \pi/2 \text{ or } \alpha = \pi/2,\ \beta = \pi: \quad \eta_0 < \eta_1 < \cdots \tag{1.5}$$

In comparison with other work, results will be stated also with this notation.

For $q = 0$ in (1.1) we mention two known bounds. For the problem $-y'' = \lambda wy$ with $y(a) = 0 = y(b)$, Krein [12] obtained the following bounds:

$$\mu_n \geq \frac{4(n+1)^2}{(b-a)\int_a^b w\,dx}; \tag{1.6}$$

$$\text{if also } w(x) \leq H \text{ on } [a,b], \text{ then } \mu_n \leq \frac{\pi^2 H(n+1)^2}{\left(\int_a^b w\,dx\right)^2}. \tag{1.7}$$

In both (1.6) and (1.7) the bounds are best possible under the condition that $\int_a^b w(x)\,dx =$ const. in the sense that some $w$ satisfying this condition will produce equality in (1.7) or within $\varepsilon$ in (1.6). However the upper bound in (1.7) is necessary; if it is removed then $\sup \mu_n = \infty$ subject to $\int_a^b w(x)\,dx =$ const. A refinement of (1.6) is proved in [12] if also an upper bound for $w$ is assumed. These best possible bounds of Krein's result show the critical dependence of the eigenvalues on $w$.

For the problem $-(py')' = \lambda y$ with $(py')(a) = 0 = (py')(b)$, Troesch [18] obtained the bound

$$\gamma_n \leq \frac{12n^2}{(b-a)^3}\int_a^b p\,dx, \tag{1.8}$$

and he proved the bound was best possible in the sense that for each $n$, there exists a $p$ which makes (1.8) an equality. The problem of Troesch was further investigated by Bandle [3] under a norm constraint, $\int_a^b p^r\,dx$ given, and with a pointwise bound on $p$.

For $w = p = 1$ and Dirichlet boundary conditions in (1.1), the problem of optimal upper and lower bounds for $\lambda_n$ under an $\mathcal{L}_1$ norm constraint on $q$ has been investigated by several authors. In particular we refer to Egnell [6], Essen [7] Harrell [10], and Talenti [20] for a discussion of this problem. Our work is also motivated by a recent paper of Ashbaugh and Benguria [1]. They proved the following theorem:

**Theorem A.** *If in (1.1) $p, q, w$ are continuous on $[a,b]$, $h \leq p(x)w(x) \leq H$ on $[a,b]$ where $h, H$ are positive constants, and $q \geq 0$ on $[a,b]$, then for $n = 0, 1, \ldots$*

$$\frac{\lambda_n}{\lambda_0} \leq \frac{H}{h}(n+1)^2. \tag{1.9}$$

*More generally for $n > l \geq 1$,*

$$\frac{\lambda_n}{\lambda_l} \leq \frac{H}{h}\left\{\left(\frac{n+1}{l+1}\right)\right\}^2, \tag{1.10}$$

*where "$\{x\}$" denotes the least integer greater than or equal to $x$ function. Further equality in (1.9)–(1.10) holds if and only if $q = 0$ and $pw = h = H$.*

A remarkable feature of the bound in Theorem A is its independence of $q$. However, (1.9)–(1.10) are quite sensitive to the constants $H$ and $h$; they give poor results if $pw$ is large or small at even one point on $[a, b]$ and they cannot be used at all if $pw$ has a zero or becomes infinite on the interval. This disadvantage is also shared by earlier pointwise bounds on eigenvalue ratios such as those of McLaughlin [15]. Additionally the presence of $\{\cdot\}$ in (1.10) entails some loss of sharpness; for example, this estimated upper bound on $\lambda_n/\lambda_l$ may exceed $(H/h)(n+1)^2/(l+1)^2$ by $\approx 2(H/h)(n+1)/(l+1)$.

In this paper we develop upper and lower bounds for eigenvalues in terms of norms of the coefficients. In Section 2 the general equation with $q \neq 0$ is considered. In Section 3 a Prüfer transformation is used for the $q = 0$ case. Finally in Section 4 we discuss how these results may be used to obtain ratio bounds.

We also wish to point out that our methods differ from those of Ashbaugh and Benguria which are based on the ingenious employment of a comparison theorem for the Prüfer transformation. While we occasionally employ the Prüfer transformation as well, we rely mainly on variational arguments and estimates of the best constants of certain Sobolev-type inequalites. In this respect the paper is a natural continuation of [4] where such methods were employed to derive upper and lower bounds of eigenvalues, estimate eigenvalue gaps, count the number of eigenvalues in the intervals $(-\infty, \lambda)$ for Sturm-Liouville problems satisfying Dirichlet and Neumann boundary conditions, and to prove various generalised Lyapunov and de La Valée Poussin type inequalites.

The authors express their appreciation to Allan Krall for assistance with the numerical computations.

## 2. Some upper and lower bounds with $q \neq 0$

We begin with a few remarks setting notation and reviewing some preliminary facts. If $f$ is a real function function on $[a, b]$ we define

$$\{f(x)\}_+ := \max\{f(x), 0\}, \quad \{f(x)\}_- := \{f(x)\}_+ - f(x)\,.$$

We denote the absolutely continuous functions on $[a, b]$ by $AC(a, b)$ and we define

$$\mathcal{D} \equiv \mathcal{D}([a,b]) := \left\{ y \in AC(a,b) : y(a) = y(b) = 0;\ \int_a^b wy^2\,dx,\ \int_a^b {y'}^2\,dx < \infty \right\}.$$

It is well known that the change of variable

$$t := \int_a^x \frac{ds}{p(s)} \tag{2.1}$$

transforms (1.1) into a new regular Sturm-Liouville problem

$$-\ddot{y} + Q(t)y = \lambda W(t)y,\; Q(t) = p(x)q(x),\; W(t) = p(x)w(x),\; \cdot = d/dt, \tag{2.2}$$

having boundary conditions on the interval $[0, T]$, $T = \int_a^b \frac{1}{p}$, and the same eigenvalues as the untransformed problem. Consequently no generality is lost if we begin by obtaining bounds on the $n$-th eigenvalue of (1.1) when $p = 1$. For the tranformation (2.1), note that for $Q(t) := p(x)q(x)$, $W(t) := p(x)w(x)$, we have

$$\int_0^T W(t)^s dt = \int_a^b w(x)^s p(x)^{s-1}\, dx\,, \quad 1 \le s < \infty\,, \tag{2.3}$$

with a similar expression for $Q$.

One can also reduce the $w$ in (1.1) to unity by the transformation $t = \int_a^x w(s)ds$ in which case (1.1) is transformed into

$$-(P(t)\dot{y})^{\cdot} + Q(t)y = \lambda y\,, \quad P(t) := w(x)p(x)\,, \quad Q(t) := q(x)/w(x)\,,$$

on $0 \le t \le T$, $T := \int_a^b w(x)\, dx$. The relation corresponding to (2.3) is

$$\int_0^T P(t)^s dt = \int_a^b w(x)^{s+1} p(x)^s\, dx\,. \tag{2.4}$$

These integral relations can be used to transform results. Thus for $-(py')' = \lambda y$, $y(a) = 0 = y(b)$, Krein's result gives using (2.3) with $s = 1$,

$$\mu_n \ge \frac{4(n+1)^2}{T \int_0^T W(t)dt} = \frac{4(n+1)^2}{\left(\int_a^b dx/p(x)\right)(b-a)}\,, \tag{2.5}$$

and if $p(x) \le H$, then

$$\mu_n \le \frac{\pi^2 H(n+1)^2}{\left(\int_0^T W(t)dt\right)^2} = \frac{\pi^2 H(n+1)^2}{(b-a)^2}\,. \tag{2.6}$$

For $-y'' = \lambda w y$, $y'(a) = 0 = y'(b)$, Troesch's bound (1.8) with (2.4) gives

$$\gamma_n \le \frac{12n^2}{T^3} \int_0^T P(t)dt = \frac{12n^2 \int_a^b w(x)^2\, dx}{\left(\int_a^b w(x)\, dx\right)^3}\,. \tag{2.7}$$

Note that (2.5) and (2.7) give new results while (2.6) follows easily from the Rayleigh quotient expression for $\lambda_n$ and the Courant-Fischer minimization theorem [5].

For $q \equiv 0$, there is another simple transformation that can be used. If $-(py')' = \lambda y$, $(py')(a) = 0 = (py')(b)$, then $z = py'$ satisfies $-z'' = \lambda(1/p)z$,

$z(a) = z(b) = 0$, and thus $\lambda$ is an eigenvalue if $y' \neq 0$. Similarly if $-z'' = \lambda wz$, $z(a) = z(b) = 0$, then $y = z'$ satisfies $-(y'/w)' = \lambda y$, $(y'/w)(a) = (y'/w)(b) = 0$. Using these transformations together with (1.6)–(1.8) yields: for $-z'' = \lambda wz$, $z(a) = z(b) = 0$, that (note $\mu_n = \gamma_{n+1}$ since $\gamma_0 = 0$)

$$\mu_n \leq \frac{12(n+1)^2}{(b-a)^3} \int_a^b dx/w(x) \tag{2.8}$$

and for $-(py')' = \lambda y$, $(py')(a) = (py')(b) = 0$,

$$\gamma_n \geq \frac{4n^2}{(b-a)\int_a^b dx/p(x)},$$

and if also $p(x) \geq h > 0$ on $[a,b]$,

$$\gamma_n \leq \frac{\pi^2 n^2}{h\left(\int_a^b dx/p(x)\right)^2}.$$

If the eigenvalue problem (1.1) with Dirichlet boundary conditions $y(c) = y(d) = 0$ is considered on some subinterval $J := [c,d] \subset [a,b]$ we distinguish its eigenvalues from the eigenvalues $\lambda_0,\lambda_1,\ldots$ of (1.1) by the notation $\lambda_0(J),\lambda_1(J),\ldots$. Similarly the eigenvalues of (1.1) with $q = 0$ defined on some subinterval $J \subset [a,b]$ will be written $\mu_0(J), \mu_1(J), \ldots$. Finally, we will occasionally abbreviate the $L^p$ norm of a function $y$ on an interval $J$ by writing $||y||_{p,J}$, etc; and if $J$ is an interval the notation $|J|$ denotes its length.

The following result may be found in Talenti [19]. For further discussion of the properties and applications of Sobolev inequalites of this kind see Fink [8], [9] as well as the book of Mitrinović, Pecarić, and Fink [16] and the references contained therein.

**Lemma 2.1.** *Let $1 \leq p \leq \infty$ and $1 < r < \infty$. Then the best constant $K \equiv K(p,r)$ of the one dimensional Sobolev-type inequality*

$$\left(\int_a^b |y|^p\,dx\right)^{1/p} \leq K(b-a)^{1+1/p-1/r}\left(\int_a^b |y'|^r\,dx\right)^{1/r}$$

*defined on the domain $\{y \in AC(a,b) : y(a) = y(b) = 0;\ ||y'||_{[a,b],r} < \infty\}$ is*

$$K(p,r) = \frac{p(1+r'/p)^{1/r}\Gamma(1/p+1/r')}{2(1+p/r')^{1/p}\Gamma(1/p)\Gamma(1/r')}. \tag{2.9}$$

In particular we note from (2.9) that

$$K(\infty,r) = 1/2, \quad K(2,2) = 1/\pi, \quad K(1,2) = 1/\sqrt{12}.$$

**Remark 2.1.** If $\{a_i\}$, $i = 1, \ldots, n$, are the $n$ interior zeros of the eigenfunction $\phi_n$ of (1.1) corresponding to $\lambda_n$, let $J_i := (a_{i-1}, a_i)$, $a_0 := a$, $a_{n+1} = b$, and consider the sequence of Sturm-Liouville problems

$$\begin{aligned} L_{J_i}(y) := -(py')' + qy = \lambda wy\,, \quad i = 1, \ldots n+1\,, \\ y(a_{i-1}) = y(a_i) = 0\,. \end{aligned} \tag{2.10}$$

Since $\phi_n$ is an eigenfunction which contains no interior zero on any $J_i$, $\lambda_n$ must be the minimal eigenvalue $\lambda_0(J_i)$ of each problem in (2.10). By the standard variational characterization of the minimal eigenvalue we have that for $\lambda \in \mathbb{R}$,

$$\lambda_0(J_j) = \min_{y \in \mathcal{D}_i} \left\{ \int_{J_i} \left[ y'^2 + (q - \lambda w) y^2 \right] dx \right\}, \tag{2.11}$$

where now

$$\mathcal{D}_i := \left\{ y \in AC(J_i) : y(a_{i-1}) = y(a_i) = 0; \int_{J_i} wy^2\, dx = 1, \int_{J_i} (y')^2\, dx < \infty \right\}.$$

Next we divide $(a, b)$ into $n+1$ equal subintervals $\Delta_i := (p_{i-1}, p_i)$ by setting $p_i = a + i(b-a)/(n+1)$, $i = 0, \ldots, n+1$. We claim that at least one subinterval $\Delta_i$ is contained in a $J_k$, $1 \le k \le n$. Certainly this is true for $n = 1$. Assume it is true up to $n = l$ and let $n = l+1$. If $\Delta_1$ is not contained in $J_1$, $p_1 > a_1$. In general if $\Delta_i$ is not contained in $J_i$, $i = 1, \ldots, l+1$, then $p_{l+1} > a_{l+1}$ in which case $\Delta_{l+2}$ is contained in $J_{l+2}$. By a similar argument we can show that at least one subinterval $J_i$ is contained in a $\Delta_k$.

Our first bound is the analogue of Theorem 2.1 of [4] where $w = 1$. A different type of bound for (1.1) is Theorem 2.3 of [4].

**Theorem 2.1.** *For $p = 1$, $1 \le s \le \infty$, $n = 0, 1, \ldots$, and $\lambda < \lambda_n$, the following inequality holds:*

$$\begin{aligned} \lambda_n \ge \lambda \; + \; & \left( \int_a^b w^s\, dx \right)^{-\frac{1}{s}} \\ & \cdot \left\{ \frac{(n+1)^2}{K(2s', 2)(b-a)^{2-\frac{1}{s}}} \left( \int_a^b (\lambda w - q)_+^{\frac{s}{s}}\, dx \right)^{\frac{1}{s}} \right\}. \end{aligned} \tag{2.12}$$

*Proof.* If $\phi_n$ is an eigenfunction corresponding $\lambda_n$, then $\phi_n$ has exactly $n$ zeros in $(a, b)$. We label them as in Remark 2.1. Consider first $s \ne \infty$. An integration by parts gives

$$(\lambda_n - \lambda) \int_{a_{i-1}}^{a_i} w \phi_n^2\, dx = \int_{a_{i-1}}^{a_i} \left[ (\phi_n')^2 + (q - \lambda w) \phi_n^2 \right] dx\,.$$

Application of Hölder's inequality followed by application of Lemma 2.1 yields the inequalities

$$\begin{aligned}
&\int_{a_{i-1}}^{a_i} \left[(\phi_n')^2 - (\lambda w - q)_+\right] dx \\
&\quad \le (\lambda_n - \lambda) \int_{a_{i-1}}^{a_i} w\phi_n^2 \, dx \\
&\quad \le (\lambda_n - \lambda) \left( \int_{a_{i-1}}^{a_i} w^s \, dx \right)^{\frac{1}{s}} \left( \int_{a_{i-1}}^{a_i} |\phi_n|^{2s'} \, dx \right)^{\frac{1}{s'}} \\
&\quad \le (\lambda_n - \lambda) \left( \int_{a_{i-1}}^{a_i} w^s \, dx \right)^{\frac{1}{s}} K(2s',2)^2 (a_i - a_{i-1})^{1+\frac{1}{s'}} \int_{a_{i-1}}^{a_i} (\phi_n')^2 \, dx \, .
\end{aligned}$$

Also, in the same way we obtain that

$$\begin{aligned}
&\int_{a_{i-1}}^{a_i} ((\lambda w - q)_+ \phi_n^2) \, dx \\
&\le \left( \int_{a_{i-1}}^{a_i} (\lambda w - q)_+^s \, dx \right)^{\frac{1}{s}} K(2s',2)^2 (a_i - a_{i-1})^{1+\frac{1}{s'}} \int_{a_{i-1}}^{a_i} (\phi_n')^2 \, dx.
\end{aligned}$$

Hence

$$\begin{aligned}
&\int_{a_{i-1}}^{a_i} (\phi_n')^2 \, dx - \left( \int_{a_{i-1}}^{a_i} (\lambda w - q)_+^2 \, dx \right)^{\frac{1}{s}} K(2s',2)^2 (a_i - a_{i-1})^{1+\frac{1}{s'}} \int_{a_{i-1}}^{a_i} (\phi_n')^2 \, dx \\
&\quad \le (\lambda_n - \lambda) \left( \int_{a_{i-1}}^{a_i} w^s \, dx \right)^{\frac{1}{s}} K(2s',2)^2 (a_i - a_{i-1})^{1+\frac{1}{s'}} \int_{a_{i-1}}^{a_i} (\phi_n')^2 \, dx \, .
\end{aligned}$$

Upon rearrangement and division by $\int_{a_{i-1}}^{a_i} (\phi_n')^2 \, dx$, the last inequality implies that

$$\begin{aligned}
\frac{1}{(a_i - a_{i-1})^{2-\frac{1}{s}}} \quad \le \quad K(2s',2)^2 \Bigg\{ & (\lambda_n - \lambda) \left( \int_{a_{i-1}}^{a_i} w^s \, dx \right)^{\frac{1}{s}} \\
& + \left( \int_{a_{i-1}}^{a_i} (\lambda w - q)_+^s \, dx \right)^{\frac{1}{s}} \Bigg\}.
\end{aligned} \tag{2.13}$$

Now raise both sides of (2.13) to the $s$ power and sum over $i = 1, \ldots, n+1$. On the left side we use the fact that a minimum is obtained for subintervals of equal length, i.e.,

$$\sum_{i=1}^{n+1} \frac{1}{(a_i - a_{i-i})^{2s-1}} \ge (n+1) \left( \frac{n+1}{b-a} \right)^{2s-1} .$$

On the right side we use Minkowski's inequality. We then obtain

$$\frac{(n+1)^2}{(b-a)^{2-\frac{1}{s}}} \leq K(2s',2)^2 \left\{ (\lambda_n - \lambda) \left( \int_a^b w^s \, dx \right)^{\frac{1}{s}} + \left( \int_a^b (\lambda w - q)_+^s \, dx \right)^{\frac{1}{s}} \right\}$$

which is equivalent to (2.12).

For $s = \infty$, (2.13) holds, and we may replace the norms on the right on $[a_{i-1}, a_i]$ by norms on $[a,b]$. The proof then follows by summing over $i = 1, \ldots, n+1$ and repeating the previous argument for the left side.

**Corollary 2.1.** *If* $\sup\{q(x)/w(x) : a \leq x \leq b\} < \lambda_n$, *then*

$$\lambda_n \geq \left( \int_a^b w \, dx \right)^{-1} \left( \frac{4(n+1)^2}{(b-a)} + \int_a^b q \, dx \right). \tag{2.14}$$

*Proof.* We choose $\lambda < \lambda_n$ such that $q(x) \leq \lambda w(x)$ on $[a,b]$. Then $(\lambda w - q)_+ = \lambda w - q$ and taking $s = 1$ in Theorem 2.1 we obtain (2.14) after simplifying.

**Remark 2.2.** Boundary conditions other than $y(a) = y(b) = 0$ in Theorem 2.1 may be considered. For the Neumann boundary conditions $y'(a) = y'(b) = 0$ with eigenvalues $\gamma_0, \gamma_1, \ldots$ the result is that obtained by replacing $\lambda_n$ by $\gamma_n$ on the left side of (2.12) and replacing $n+1$ by $n$ on the right side of (2.12). For the boundary conditions $y(a) = y'(b) = 0$ or $y'(a) = y(b) = 0$, $n+1$ is replaced by $n + \frac{1}{2}$ and $\lambda_n$ is replaced by $\eta_n$. For $w = 1$ these details are worked out in [4]; similar arguments will apply here.

**Remark 2.3.** For $q = 0$ and $\lambda = 0$ in (2.12) we have

$$\lambda_n \geq \frac{(n+1)^2}{\left( \int_a^b w^s \, dx \right)^{\frac{1}{s}} K(2s',2)^2 (b-a)^{2-\frac{1}{s}}}. \tag{2.15}$$

For $s = 1$, this is the lower bound (1.6) of Krein.

**Remark 2.4.** For $q \equiv 0$ and $w = 1$ in (1.1) we may make the transformation (2.1). Then (1.1) becomes

$$-z'' = \lambda W(t) z, \quad 0 \leq t \leq T, \quad z(t) = y(x), \quad W(t) = p(x).$$

Using (2.3), the lower bound (2.15) reads for this case of (1.1) that

$$\lambda_n \geq \frac{(n+1)^2}{\left( \int_a^b p^{s-1} \, dx \right)^{\frac{1}{s}} K(2s',2)^2 \left( \int_a^b dx/p \right)^{2-\frac{1}{s}}}.$$

**Remark 2.5.** Let $\mu_0(\Delta_i)$ be the least eigenvalue of the Sturm-Liouville problem $-y'' = \lambda w y$ with Dirichlet boundary conditions defined on $\Delta_i$. Then as in (2.11)

$$\mu_0(\Delta_i) = \min_{y \in \tilde{\mathcal{D}}_i} \int_{\Delta_i} y'^2 \, dx .$$

This observation leads to an another lower bound on $\lambda_n$.

**Theorem 2.2.** *If $p = 1$ in (1.1) and $\lambda$ is real, then $\lambda_n$ satisfies the lower bound*

$$\lambda_n \geq \lambda + \min_i \mu_0(\Delta_i) \left\{ 1 - \frac{b-a}{4(n+1)} \max_i \int_{\Delta_i} \{\lambda w - q\}_+ \, dx \right\} \tag{2.16}$$

*whenever the number in brackets is nonnegative.*

*Proof.* According to Remark 2.1 we can choose $J_j$, $\Delta_k$ so that $J_j \subset \Delta_k$. We now consider the sequence of eigenvalue problems $\{L_{\Delta_i}\}$ like (2.10) defined on the intervals $\Delta_i$ instead of $J_i$. In analogy to $\mathcal{D}_i$ let $\tilde{\mathcal{D}}_i$, $i = 1, \ldots, n$, be the absolutely continuous functions $y$ on $\Delta_i$ such that $y(p_{i-1}) = y(p_i) = 0$ and $\int_{\Delta_i} w y^2 \, dx = 1$.

Since $\mathcal{D}_j$ can be embedded in $\tilde{\mathcal{D}}_k$ by extending its members by 0 on $\Delta_k \setminus J_j$, (2.11) implies that

$$\lambda_n \equiv \lambda_0(J_j) \geq \lambda_0(\Delta_k) .$$

Starting with this lower bound, the variational principle (2.11) applied to $\lambda_0(\Delta_k)$, an obvious estimate, Hölder's inequality, and application of Lemma 2.1 successively yields the inequalities

$$\begin{aligned}
\lambda_n - \lambda &\geq \min_{y \in \tilde{\mathcal{D}}_k} \left\{ \int_{\Delta_k} [y'^2 + (q - \lambda w) y^2] \, dx \right\} \\
&\geq \min_{y \in \tilde{\mathcal{D}}_k} \left\{ \int_{\Delta_k} y'^2 \, dx - \int_{\Delta_k} \{\lambda w - q\}_+ ||y||^2_{\infty, \Delta_k} \, dx \right\} \\
&\geq \min_{y \in \tilde{\mathcal{D}}_k} \left( \int_{\Delta_k} y'^2 \, dx \right) \left\{ 1 - \frac{b-a}{4(n+1)} \max_i \int_{\Delta_i} \{\lambda w - q\}_+ \, dx \right\} \\
&\geq \min_i \mu_0(\Delta_i) \left\{ 1 - \frac{b-a}{4(n+1)} \max_i \int_{\Delta_i} \{\lambda w - q\}_+ \, dx \right\} .
\end{aligned}$$

**Theorem 2.3.** *Let the hypotheses of Theorem 2.1 hold. Then an upper bound on $\lambda_0$ is given by*

$$\lambda_0 \leq \lambda + \mu_0 \left( 1 + \frac{b-a}{4} \int_a^b \{\lambda w - q\}_- \, dx \right) . \tag{2.17}$$

*Proof.* Since

$$\lambda_0 = \min_{y \in \mathcal{D}} \left\{ \int_a^b (y'^2 + q y^2) \, dx \right\} ,$$

we have using Hölder's inequality and Lemma 2.1 the sequence of inequalities

$$\begin{aligned}
\lambda_0 - \lambda &= \min_{y\in\mathcal{D}} \left\{ \int_a^b (y'^2 + (q - w\ lambda)y^2)\, dx \right\} \\
&\leq \min_{y\in\mathcal{D}} \left\{ \int_a^b y'^2\, dx + \left( \int_a^b \{\lambda w - q\}_-\, dx \right) \|y\|_\infty^2 \right\} \\
&\leq \min_{y\in\mathcal{D}} \left( \int_a^b y'^2\, dx \right) \left\{ 1 + (1/4)(b-a) \int_a^b \{\lambda w - q\}_-\, dx \right\} \\
&\leq \mu_0 \left\{ 1 + (1/4)(b-a) \int_a^b \{\lambda w - q\}_-\, dx \right\}.
\end{aligned}$$

**Theorem 2.4.** *Under the same hypotheses as Theorem 2.1 an upper bound on $\lambda_n$ is given by*

$$\lambda_n \leq \lambda + \max \mu_0(\Delta_i) \left( 1 + \frac{(b-a)}{4(n+1)} \max_i \int_{\Delta_i} \{\lambda w - q\}_-\, dx \right). \tag{2.18}$$

*Proof.* The argument is similar to the proof of Theorem 2.2 and is omitted.

**Corollary 2.2.** *Let $p = w = 1$. If $q$ is increasing, then*

$$\lambda_n \leq \left( \frac{\pi(n+1)}{b-a} \right)^2 + \frac{\pi^2(n+1)}{4(b-a)} \int_{b-\frac{4(b-a)}{\pi^2(n+1)}}^b q\, dx. \tag{2.19}$$

*If $q$ is decreasing, then*

$$\lambda_n \leq \left( \frac{\pi(n+1)}{b-a} \right)^2 + \frac{\pi^2(n+1)}{4(b-a)} \int_a^{a+\frac{4(b-a)}{\pi^2(n+1)}} q\, dx. \tag{2.20}$$

*Proof.* First let $n = 0$ and denote the right side of (2.17) by $\psi(\lambda)$. If $q$ is increasing, a calculation using $\mu_0(\Delta_0) = \pi^2/(b-a)^2$ shows that

$$\psi'(\lambda) = 1 - \frac{\pi^2}{4(b-a)}(b - \{q^{-1}\}(\lambda)),$$

where $\{q^{-1}\}$ denotes the inverse of $q$. Thus $\psi$ has an interior minimum when $\lambda = q(b - 4(b-a)/\pi^2)$. Substitution into (2.17) followed by simplification yields (2.19).

If $n \geq 1$, let $J_k \supset \Delta_j$. Then $\lambda_0(\Delta_j) \geq \lambda_0(J_k) = \lambda_n$ and

$$\begin{aligned}\lambda_n &\leq \left(\frac{\pi}{|\Delta_j|}\right)^2 + \frac{\pi^2}{4|\Delta_j|}\int_{p_j - \frac{4|\Delta_j|}{\pi^2}}^{p_j} q\,dx \\ &\leq \max_i \left\{\left(\frac{\pi}{|\Delta_i|}\right)^2 + \frac{\pi^2}{4|\Delta_i|}\int_{p_i - \frac{4|\Delta_i|}{\pi^2}}^{p_i} q\,dx\right\} \\ &\leq \left\{\left(\frac{\pi(n+1)}{b-a}\right)^2 + \frac{\pi^2(n+1)}{4(b-a)}\int_{b-\frac{4(b-a)}{\pi^2(n+1)}}^{b} q\,dx\right\}.\end{aligned}$$

The decreasing case is similar.

**Remark 2.6.** In the previous theorems we have in large part reduced the problem of finding upper and lower bounds of $\lambda_n$ of (1.1) to the problem of bounding the *first* eigenvalue $\mu_0(\Delta_i)$ of $-y'' = \mu w y$ subject to Dirichlet boundary conditions on some interval $\Delta_i$, $i = 1, \ldots, n+1$. This is a difficult problem to which we will return in the next section. We offer the following preliminary bounds here.

(i) According to (1.6) or Corollary 2.1

$$\mu_0(\Delta_i) > \left(\int_{\Delta_i} w\,dx\right)^{-1} \frac{4(n+1)^2}{b-a}.$$

Note that this lower bound is also an immediate consequence of the Lyapunov inequality.

(ii) If $w(x) \leq H < \infty$ on $[a,b]$, it follows from Lemma 2.1 that

$$\mu_0(\Delta_i) \geq H^{-1}\pi^2\left(\frac{n+1}{b-a}\right)^2. \tag{2.21}$$

Similarly if $w(x) \geq h > 0$ on $[a,b]$, then

$$\mu_0(\Delta_i) \leq \frac{\pi^2}{h}\left(\frac{n+1}{b-a}\right)^2. \tag{2.22}$$

(iii) Suppose $w^{-1}$ is integrable. Using $n = 0$ in (2.8) gives

$$\mu_0(\Delta_i) \leq 12\int_{\Delta_i} w^{-1}\left(\frac{n+1}{b-a}\right)^3 dx.$$

(iv) If $5w'^2 - 4ww'' \geq 0$ on $[a,b]$ Makai [16] has shown that

$$\mu_0(\Delta_i) \leq \left(\frac{\pi}{\int_{\Delta_i}\sqrt{w}\,dx}\right)^2 \tag{2.23}$$

with equality if $5w'^2 - 4ww'' = 0$. Similarly if $5w'^2 - 4ww'' < 0$ on $[a, b]$, then

$$\mu_0(\Delta_i) > \left(\frac{\pi}{\int_{\Delta_i} \sqrt{w}\, dx}\right)^2. \tag{2.24}$$

## 3. Additional bounds for the $q = 0$ case

In this section we study the problem of bounding the eigenvalues of (1.1) when $q = 0$. Define the functions $g_1$ and $g_2$ on $(0, \infty)$ by

$$g_1(\gamma) = \gamma^{-\frac{1}{2}} \int_a^b \max\left\{w, \frac{\gamma}{p}\right\} dx, \quad g_2(\gamma) = \gamma^{-\frac{1}{2}} \int_a^b \min\left\{w, \frac{\gamma}{p}\right\} dx. \tag{3.1}$$

Note that $\gamma^{-\frac{1}{2}} \int_a^b w\, dx$ and $\gamma^{\frac{1}{2}} \int_a^b p^{-1}\, dx$ are lower bounds for $g_1$ and upper bounds for $g_2$. Hence there are unique positive numbers $g_*, g^*$ defined by

$$g^* = \inf_{0<\gamma<\infty} g_1(\gamma), \quad g_* = \sup_{0<\gamma<\infty} g_2(\gamma).$$

**Theorem 3.1.** *Suppose in* (1.1)–(1.2) *that* $q = 0$ *and the boundary conditions are* (1.3), (1.4) *or* (1.5). *Denote the eigenvales by* $\xi_0 < \xi_1 < \cdots$. *Then for* $n = 0, 1, \ldots,$

$$\frac{[n\pi + \beta - \alpha]^2}{g^{*\,2}} \le \xi_n \le \frac{[n\pi + \beta - \alpha]^2}{g_*^2}. \tag{3.2}$$

*Proof.* First we make the Prüfer transformation in (1.1),

$$y = r \sin\theta, \quad (py') = (\gamma\xi_n)^{\frac{1}{2}} r \cos\theta, \tag{3.3}$$

where $y$ is the eigenfunction corresponding to $\lambda = \xi_n$. Choose $\theta(a) = \alpha$ so that $\theta(b) = \beta + n\pi$. Differentiation of (3.2) and solving for $\theta'$ using (1.1) gives

$$\theta' = \frac{(\gamma\xi_n)^{\frac{1}{2}}}{p} + \left(\frac{\xi_n}{\gamma}\right)^{\frac{1}{2}} \left(w - \frac{\gamma}{p}\right) \sin^2\theta. \tag{3.4}$$

Integrating (3.4) over $[a, b]$ yields that

$$n\pi + \beta - \alpha = \theta(b) - \theta(a) = \left(\frac{\xi_n}{\gamma}\right)^{\frac{1}{2}} \int_a^b \left[\frac{\gamma}{p} + \left(w - \frac{\gamma}{p}\right)\sin^2\theta\right] dx.$$

Thus

$$n\pi + \beta - \alpha \le \left(\frac{\xi_n}{\gamma}\right)^{\frac{1}{2}} \int_a^b \left[\frac{\gamma}{p} + \left(w - \frac{\gamma}{p}\right)_+\right] dx = \xi_n^{\frac{1}{2}} g_1(\gamma). \tag{3.5}$$

Minimizing the right-hand side of (3.5) over $\gamma$ yields the lower bound in (3.2). The upper bound in (3.2) follows from

$$\theta(b) - \theta(a) \geq \left(\frac{\xi_n}{\gamma}\right)^{\frac{1}{2}} \int_a^b \left[\frac{\gamma}{p} - \left(w - \frac{\gamma}{p}\right)_-\right] dx = \xi_n^{\frac{1}{2}} g_2(\gamma)$$

and minimization over $\gamma$.

**Example 3.1.** With $p \equiv 1$, $q \equiv 0$, and $w(x) = x^{-\frac{1}{2}}$ on $0 \leq x \leq 1$, the program SLEIGN2 gives the values in Table 1.

| $n$ | 0 | 1 | 2 | 3 | 4 | 5 |
|---|---|---|---|---|---|---|
| $\frac{\mu_n}{\pi^2(n+1)^2}$ | .649 | .608 | .593 | .586 | .581 | .578 |

Table 1

By calculus it is found that $g_*$ in (3.2) is $g_* = g_2(1.5)$ and $g^* = g_1(\sqrt{3})$. Thus $1/g_*^2 = .843$ and $1/g^{*2} = .325$. By Krein's result there is no upper bound for $\mu_n/\pi^2(n+1)^2$ among all $w$ satisfying $\int_0^1 w\,dx = \int_0^1 x^{-\frac{1}{2}}\,dx = 2$. Krein's lower bound (1.6) is $2/\pi^2 = .203$. Troesch's transformed upper bound (2.8) for $\mu_n/\pi^2(n+1)^2$ is $8/\pi^2 = .811$.

**Example 3.2.** With $p \equiv 1$, $q \equiv 0$, and $w(x) = x$ on $0 \leq x \leq 1$, the program SLEIGN2 gives the values in Table 2.

| $n$ | 0 | 1 | 2 | 3 | 4 | 5 |
|---|---|---|---|---|---|---|
| $\frac{\mu_n}{\pi^2(n+1)^2}$ | 1.92 | 2.07 | 2.13 | 2.16 | 2.18 | 2.19 |

Table 2

The bounds in (3.2) are $1/g_*^2 = 3.35$ and $1/g^{*2} = 1.30$. The optimal upper bound given by Krein for all functions $w$ on $[0,1]$ satisfying $w(x) \leq 1$ and $\int_0^1 w\,dx = \int_0^1 x\,dx = \frac{1}{2}$ is 4. Krein's lower bound (1.6) is $8/\pi^2 = .811$. Troesch's upper bound does not apply. The lower bound of Krein [12] for functions $w$ satisfying $w(x) \leq 1$ and $\int_0^1 w\,dx = \frac{1}{2}$ is 1.20.

If $p$ and $w$ satisfying $0 < h \leq p(x)w(x) \leq H$, then we have

$$g_1(h) = h^{-\frac{1}{2}} \int_a^b w\,dx, \quad g_1(H) = H^{\frac{1}{2}} \int_a^b dx/p,$$

$$g_2(h) = h^{\frac{1}{2}} \int_a^b dx/p, \quad g_2(H) = H^{-\frac{1}{2}} \int_a^b w\,dx\,.$$

Thus we have the bounds

$$g^* \le \min\left\{h^{-\frac{1}{2}}\int_a^b w\,dx,\ H^{\frac{1}{2}}\int_a^b dx/p\right\}$$

$$g_* \ge \max\left\{h^{\frac{1}{2}}\int_a^b dx/p,\ H^{-\frac{1}{2}}\int_a^b w\,dx\right\}.$$

Note that Krein's bound (1.7) follows from the bound for $g_*$.

## 4. Bounds on eigenvalue ratios

As noted in [1], Mahar and Willner [13] and Keller [11] have shown that the optimal upper bounds for $\mu_1/\mu_0$ tend to infinity as $h/H$ tends to zero with $p = 1$. However combining (1.6) and (2.8) yields for $m < n$,

$$\frac{\mu_n}{\mu_m} \le \frac{3(n+1)^2 \int_a^b dx/w \int_a^b w\,dx}{(m+1)^2(b-a)^2}.$$

In particular $\mu_1/\mu_0$ is bounded by $12\int_a^b dx/w \int_a^b w\,dx/(b-a)^2$. For $w(x) = x^{-\frac{1}{2}}$ on $[0,1]$ this bound is 16; from Example 3.1, $\mu_1/\mu_0 \approx 3.75$. For $w$ with $5(w')^2 - 4ww'' \le 0$ on $[a,b]$, then (2.8) and (2.24) yield that

$$\frac{\mu_1}{\mu_0} \le \frac{48\int_a^b dx/w\left(\int_a^b w^{\frac{1}{2}}\,dx\right)^2}{\pi^2(b-a)^2}$$

which gives $\mu_1/\mu_0 \le 4.32$ for $w(x) = x^{-\frac{1}{2}}$ on $[0,1]$.

Theorem 2.1 yields a lower bound for eigenvalues with $q$ present as does Theorem 2.2. An upper bound is given by Theorem 2.4. These theorems can be used to obtain bounds for ratios of eigenvalues. We prove a final theorem on upper bounds.

**Theorem 4.1.** *Let $\gamma > 0$ and $g_2$ be as in (3.1). Then for $n$ such that $\lambda_n > 0$,*

$$\lambda_n \le \frac{(n+1)^2\pi^2}{g_2(\gamma)^2} + \frac{2\int_a^b q_+\,dx}{\gamma^{\frac{1}{2}} g_2(\gamma)}. \tag{4.1}$$

*Proof.* Again make the Prüfer transformation (3.3) for the eigenfunction $y$ corresponding to $\lambda = \lambda_n$. Then $\theta(a) = 0$ and $\theta(b) = (n+1)\pi$. The function $\theta$ now satisfies the equation

$$\theta' = \frac{(\gamma\lambda_n)^{\frac{1}{2}}}{p} + \left(\frac{\lambda_n}{\gamma}\right)^{\frac{1}{2}}\left(w - \frac{\gamma}{p}\right)\sin^2\theta - \frac{q}{(\gamma\lambda_n)^{\frac{1}{2}}}\sin^2\theta \tag{4.2}$$

and thus

$$\theta' \geq \frac{(\gamma\lambda_n)^{\frac{1}{2}}}{p} - \left(\frac{\lambda_n}{\gamma}\right)^{\frac{1}{2}} \left(w - \frac{\gamma}{p}\right)_- - \frac{q_+}{(\gamma\lambda_n)^{\frac{1}{2}}}.$$

Integration over $[a, b]$ yields that

$$\begin{aligned}(n+1)\pi = \theta(b) - \theta(a) &\geq \left(\frac{\lambda_n}{\gamma}\right)^{\frac{1}{2}} \int_a^b \left[\frac{\gamma}{p} - \left(w - \frac{\gamma}{p}\right)\right]_- dx - \int_a^b \frac{q_+}{(\gamma\lambda_n)^{\frac{1}{2}}} dx \\ &= \lambda_n^{\frac{1}{2}} g_2(\gamma) - \int_a^b \frac{q_+}{(\gamma\lambda_n)^{\frac{1}{2}}} dx.\end{aligned} \tag{4.3}$$

We may rewrite (4.3) in the form

$$B\lambda_n^{\frac{1}{2}} \geq \lambda_n - C, \quad B := \frac{(n+1)\pi}{g_2(\gamma)}, \quad C := \int_a^b q_+ \, dx / \gamma^{\frac{1}{2}} g_2(\gamma). \tag{4.4}$$

Inequality (4.4) yields an upper bound to $\lambda_n^{\frac{1}{2}}$ which is the largest root of the equality $Bx = x^2 - C$. Thus

$$\lambda_n^{\frac{1}{2}} \leq \frac{1}{2}\left[B + \sqrt{B^2 + 4C}\right] = (B/2)\left[1 + \sqrt{1 + 4C/B^2}\right]. \tag{4.5}$$

Squaring (4.5) and using $(1+x)^{\frac{1}{2}} \leq 1 + x/2$ yields (4.1).

## References


[1] M.S. Ashbaugh and R.D. Benguria, *Eigenvalue ratios for Sturm-Liouville operators.* J. Diff. Equations **103** (1993), 205–219.

[2] F.V. Atkinson and A.B. Mingarelli, *Asymptotics of zeros and of the eigenvalues of general weighted Sturm-Liouville problems.* J. Reine Angew. Math. **375/376** (1987), 380–393.

[3] C. Bandle, *Extremal problems for eigenvalues of the Sturm-Liouville type.* In "General Inequalities 5", Birkhäuser Verlag, Basel, 1987, pp. 319–339.

[4] R.C. Brown and D.B. Hinton, *Applications of a one-dimensional Sobolev inequality to eigenvalue problems.* J. Diff. and Int. Eqs. To appear.

[5] R. Courant and D. Hilbert, *Methods of Mathematical Physics*, Vol. I. Interscience, New York, 1953.

[6] H. Egnell, *Extremal properties of the first eigenvalue of a class of elliptic eigenvalue problems.* Ann. Scuola Norm. Sup. Pisa, ser. 4 **14** (1987), 1–48.

[7] M. Essén, *On estimating eigenvalues of a second order linear differential operator.* In "General Inequalities 5", Birkhäuser Verlag, Basel, 1987, pp. 347–366.

[8] A.M. Fink, *Conjugate inequalities for functions and their derivatives.* Siam J. Math. Anal. **5** (1974), 399–411.

[9] A.M. Fink, *Differential inequalites and disconjugacy.* J. Math. Anal. Appl. **49**, 758–772.

[10] E.M. Harrell, III, *Hamiltonian operators with maximal eigenvalues.* J. Math. Physics **25** (1984), 48–51.

[11] J.B. Keller, *The mimimum ratio of two eigenvalues.* SIAM J. Appl. Math **31** (1976), 485–491.

[12] M.G. Krein, *On certain problems on the maximum and minimum of characteristic values and on the Lyapunov zones of stability.* AMS translations, ser. 2 **1** (1955), 164–187.

[13] T.J. Mahar and B.E. Willner, *An extremal eigenvalue problem.* Comm. Pure Appl. Math. **29** (1976), 517–529.

[14] E. Makai, *Über die Nullstellen von Funktionen, die Lösungen Sturm-Liouville'scher Differentialgleichungen sind.* Commentarii Mathematici Helvetici **16** (1943-4), 153–199.

[15] J.R. McLaughlin, *Upper and lower bounds on eigenvalues of second-order Sturm-Liouville systems.* J. Diff. Eqs. **19** (1975), 201–213.

[16] D.S. Mitrinović, J.E. Pecarić, and A. M. Fink, *Inequalities involving functions and their derivatives.* Mathematics and its applications, Kluwer Academic Publishers, Dordrecht–Boston–London, 1991.

[17] M.A. Naimark, *Linear differential operators*, Part I. Unger, London, 1968.

[18] B.A. Troesch, *An isoperimetric sloshing problem.* Comm. Pure Appl. Math. **18** (1965), 319–338.

[19] G. Talenti, *Best constant in Sobolev's inequality.* Annali di Mat. Pura ed appl. **110** (1976), 353–372.

[20] G. Talenti, *Estimates for eigenvalues of Sturm-Liouville problems.* In "General Inequalities 4", Birkhäuser Verlag, Basel, 1984, pp. 341–350.

R.C. Brown, Mathematics Department, University of Alabama, Tuscaloosa, AL 35487, USA.

D.B. Hinton, Mathematics Department, University of Tennessee, Knoxville, TN 37996, USA.

International Series of Numerical Mathematics, Vol. 123, 

# Discontinuous dependence of the $n$-th Sturm-Liouville eigenvalue

W.N. Everitt, M. Möller, and A. Zettl

**Abstract**

The $n$-th eigenvalue of a regular Sturm-Liouville problem is not a continuous function of the boundary conditions. On the other hand if the index $n$ is allowed to "jump", then each eigenvalue can be embedded in an eigenvalue "branch" which is not only continuous but differentiable.

Consider the Sturm-Liouville problem consisting of the equation

$$-(py')' + qy = \lambda wy \quad \text{on} \quad [a,b], \quad -\infty < a < b < \infty, \tag{1}$$

together with (separated) boundary conditions

$$A_1 y(a) + A_2(py')(a) = 0, \ (A_1, A_2) \neq (0,0), \ A_1, A_2 \in \mathbb{R}, \tag{2}$$

$$B_1 y(b) + B_2(py')(b) = 0, \ (B_1, B_2) \neq (0,0), \ B_1, B_2 \in \mathbb{R}. \tag{3}$$

Assume that

$$p, q, w : (a,b) \to \mathbb{R}, \ \frac{1}{p}, q, w \in L^1(a,b), \ p > 0, w > 0 \ \text{a.e. on} \ (a,b). \tag{4}$$

It is well known [3], [7, Theorem 13.2] that there exists an infinite but countable number of eigenvalues $\{\lambda_n : n \in \mathbb{N}_0 = \{0,1,2,\ldots\}\}$, these are all real and simple and can be indexed to satisfy

$$-\infty < \lambda_0 < \lambda_1 < \lambda_2 \ldots; \ \text{and} \ \lambda_n \to +\infty \ \text{as} \ n \to +\infty.$$

The (unique up to constant multiples) eigenfunction $u_n$ of $\lambda_n$ has exactly $n$ zeros in the open interval $(a,b)$ for each $n \in \mathbb{N}_0$.

We wish to study the variation of $\lambda_n$ with respect to $A_2$ with all other parameters fixed; and, similarly, the variation of $\lambda_n$ as a function of $B_2$ with all other parameters fixed.

**Theorem.** *Let* (1), (2), (3), (4) *hold.*

(a) *Fix* $B_1$, $B_2$ *and let* $A_1 = 1$. *Consider* $\lambda_n = \lambda_n(A_2)$ *as a function of* $A_2 \in \mathbb{R}$. *Then for each* $n \in \mathbb{N}_0$, $\lambda_n(A_2)$ *is continuous in* $A_2$ *for* $A_2 > 0$ *and* $A_2 < 0$ *but has a jump discontinuity at* $A_2 = 0$. *More precisely we have:*

(i) $\lim_{A_2 \to 0^-} \lambda_n(A_2) = \lambda_n(0)$, $n \in \mathbb{N}_0$.

(ii) $\lim_{A_2 \to 0^+} \lambda_0(A_2) = -\infty$.

(iii) $\lim_{A_2 \to 0^+} \lambda_{n+1}(A_2) = \lambda_n(0)$, $n \in \mathbb{N}_0$.

(b) *Fix* $A_1$, $A_2$ *and let* $B_1 = 1$. *Consider* $\lambda_n = \lambda_n(B_2)$ *as a function of* $B_2 \in \mathbb{R}$. *Then for each* $n \in \mathbb{N}_0$, $\lambda_n(B_2)$ *is continuous in* $B_2$ *for* $B_2 > 0$ *and* $B_2 < 0$ *but has a jump discontinuity at* $B_2 = 0$. *More precisely we have:*

(i) $\lim_{B_2 \to 0^+} \lambda_n(B_2) = \lambda_n(0)$, $n \in \mathbb{N}_0$.

(ii) $\lim_{B_2 \to 0^-} \lambda_0(B_2) = -\infty$.

(iii) $\lim_{B_2 \to 0^-} \lambda_{n+1}(B_2) = \lambda_n(0)$, $n \in \mathbb{N}_0$.

**Remark 1.** Note that $\lambda_0(A_2)$ has an infinite jump discontinuity at $A_2 = 0$, but for all $n \geq 1$, $\lambda_n(A_2)$ has a finite jump discontinuity at $A_2 = 0$; $\lambda_n(A_2)$ is left but not right continuous at 0. Similarly, $\lambda_0(B_2)$ has an infinite jump discontinuity at $B_2 = 0$, but for all $n \geq 1$, $\lambda_n(B_2)$ has a finite jump discontinuity at $B_2 = 0$; $\lambda_n(B_2)$ is right but not left continuous at 0. In all cases $\lambda_n(0)$ is embedded in a continuous branch of eigenvalues as $A_2$ or $B_2$ passes through zero but this branch is not given by a fixed index $n$, the index "jumps" from $n$ to $n+1$ as $A_2$ or $B_2$ passes through 0 from the appropriate direction.

**Remark 2.** This phenomenon of "index jumping" has important consequences in some of the algorithms and their numerical implementations used in the code SLEIGN2 [1]; these results will be the subject of a separate note [2].

**Remark 3.** Kong and Zettl [4, Theorem 4.2] have shown that each continuous eigenvalue branch is in fact differentiable everywhere including the point $A_2 = 0$ (or $B_2 = 0$) where the index jumps. This also follows from Möller and Zettl [6, Section 5].

Below we give the proof of part (a), the proof of (b) is omitted since it is entirely similar.

The boundary conditions (2), (3) have the well-known canonical representations:

$$\cos\alpha\, y(a) - \sin\alpha\, (py')(a) = 0,\ 0 \leq \alpha < \pi, \tag{5}$$

$$\sin\beta\, y(b) - \sin\beta\, (py')(b) = 0,\ 0 < \beta \leq \pi. \tag{6}$$

The relationship between $\alpha$ and $A_1$, $A_2$ is given by $\tan\alpha = -A_2/A_1$ for $A_1 \neq 0$ and $\alpha = \pi/2$ when $A_1 = 0$. In case $A_1 \neq 0$ it can be normalized to $A_1 = 1$; thus

$$\tan\alpha = -A_2. \tag{7}$$

Therefore studying $\lambda_n$ as a function of $A_2$ is equivalent to studying $\lambda_n$ as a function of $\alpha$.

Before proceeding with the proof of the Theorem we illustrate the results of part (a) with the following graphs. For this it is helpful to note that the restrictions for $\alpha$ given in (5) are arbitrary and are chosen to determine $\alpha$ uniquely. Any other interval of length $\pi$ would do just as well since $\lambda_n(\alpha)$ is a periodic function of $\alpha$ with period $\pi$.

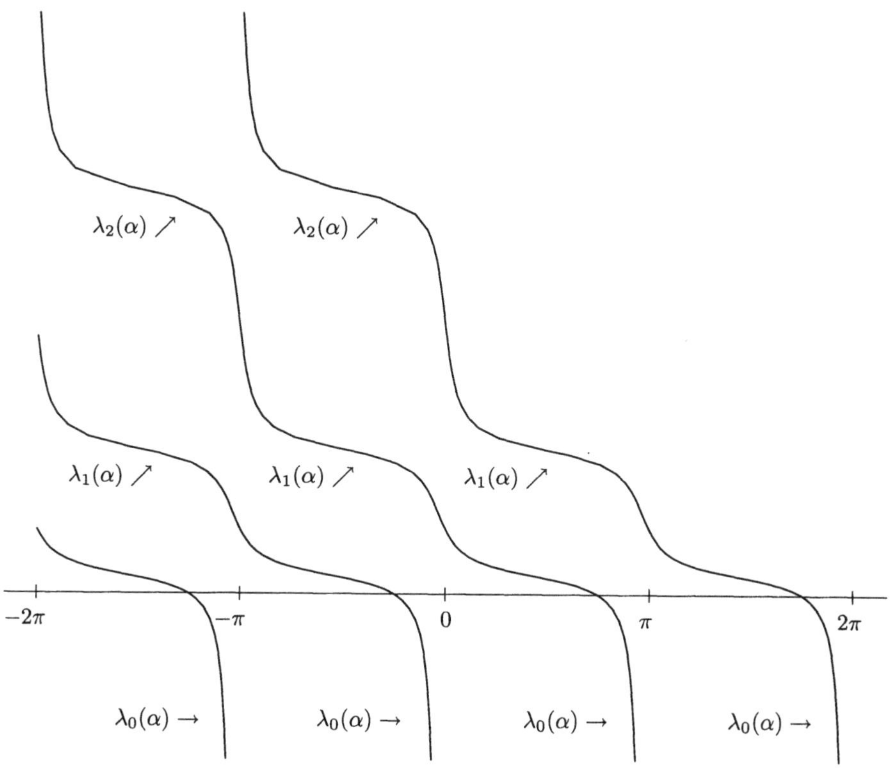


Figure 1: Graph for $\lambda_n(\alpha)$

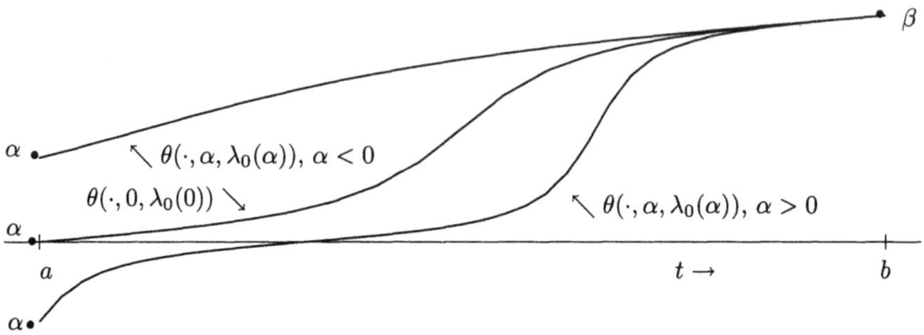


Figure 2: Graph for $\lambda_n(A_2)$

**Remark 4.** Note that for any fixed (separated) boundary condition at $b$, as $\alpha$ varies through the interval $[0, \pi)$ every real number $\lambda$ occcurs as an eigenvalue of $\alpha$.

**Remark 5.** Starting with $A_2 = 0$, as $A_2$ decreases toward $-\infty$, $\lambda_n(A_2)$ decreases continuously through $-\infty$, reappears at $+\infty$ and then continues to decrease continuously to $-\infty$ for $n = 0$ but meets up with $\lambda_{n-1}(0)$ as $A_2$ moves from $+\infty$ toward 0 for $n \geq 1$.

Our proof is based on the Prüfer transformation. It is well known [7, p. 196] that every nontrivial solution $y$ of (1) can be factored into

$$y = \rho \sin \theta \quad \text{on} \quad [a, b],$$

where $\rho(t) \neq 0$, $a \leq t \leq b$, and $\theta$ satisfies the first order nonlinear equation

$$\theta' = \frac{1}{p} \cos^2 \theta + (\lambda w - q) \sin^2 \theta \quad \text{on} \quad [a, b].$$

From the basic theory of differential equations it is well known that each initial value problem

$$\theta(a, \lambda) = \alpha, \ \alpha \in \mathbb{R}$$

has a unique solution $\theta = \theta(\cdot, \alpha, \lambda)$ which is defined on all of $[a, b]$ and that this solution is a continuous function of the parameters $\alpha \in \mathbb{R}$, $\lambda \in \mathbb{R}$ uniformly on $[a, b]$. Also it is known [3], [7, proof of Theorem 13.2] that, when $\theta$ is determined by (8), we have, for fixed $\alpha$, $0 \leq \alpha < \pi$:

$$\theta(b, \alpha, \lambda) \to +\infty \quad \text{as} \quad \lambda \to +\infty;$$
$$\theta(b, \alpha, \lambda) \to 0^+ \quad \text{as} \quad \lambda \to -\infty;$$

the eigenvalue $\lambda_n(\alpha)$ is the unique solution of the equation

$$\theta(b, \lambda) = \beta + n\pi, \ n \in \mathbb{N}_0, \tag{8}$$

and from [4, Theorem 4.2],

$$\lambda_n'(\alpha) = -u_n^2(a, \alpha) - (pu_n')^2(a, \alpha), \tag{9}$$

where $u_n$ is a normalized eigenfunction of $\lambda_n$, i. e. $\int_a^b u_n^2 w = 1$, $u_n = u_n(\cdot, \alpha)$.

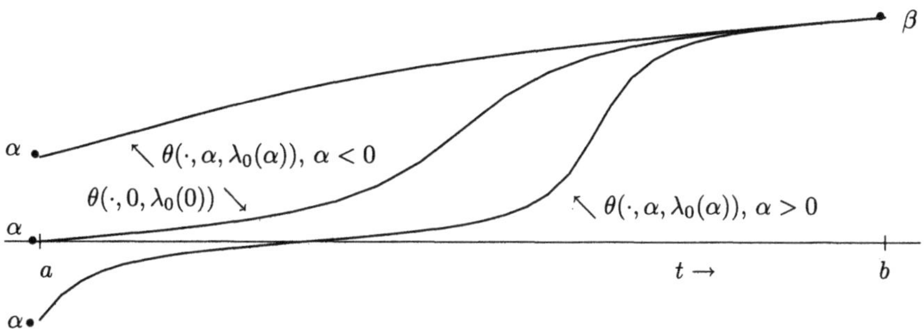


Figure 3: Graph for $\theta(\cdot, \alpha, \lambda_0(\alpha))$

From (8) and the continuous dependence of $\theta$ on $\alpha$ and $\lambda$, uniformly on $[a, b]$, it follows that $\lambda_n(\alpha)$ is continuous in $\alpha$ for $0 < \alpha < \pi$ and $\lambda_n(\alpha) \rightarrow \lambda_n(0)$ as $\alpha \rightarrow 0^+$ for each $n \in \mathbb{N}_0$. From (7), $\alpha \rightarrow 0^+$ is equivalent with $A_2 \rightarrow 0^-$ and (i) is proved.

To prove (iii) note that when $\alpha$ moves through 0 from positive to negative values $\theta$ must cross the $t$-axis exactly once, hence $\lambda_0(\alpha)$ becomes $\lambda_1(\alpha)$, $\lambda_1(\alpha)$ becomes $\lambda_2(\alpha)$, etc. Again this is equivalent, in view of (7), to $A_2$ going through zero from negative to positive values and (iii) is established.

Now for (ii) note first that (9) implies that $\lambda_0(\alpha)$ is decreasing as $\alpha \rightarrow \pi^-$. Suppose it has a finite limit, say $L$. Consider the characteristic function $\Delta(\lambda)$ – see [5, Section 3] – whose zeros are the eigenvalues. This function $\Delta$ is continuous in $\alpha$ and hence $L$ is an eigenvalue for $\alpha = \pi$, but this contradicts the fact that $\lambda_0(0) = \lambda_0(\pi)$ is the smallest eigenvalue for $\alpha = \pi$, since $L < \lambda_0(0)$. Hence $\lambda_0(\alpha) \rightarrow -\infty$ as $\alpha \rightarrow \pi^-$. Now $\alpha \rightarrow \pi^-$ is equivalent to $A_2 \rightarrow 0^+$ completing the proof of (ii) and of all of (a).

We close with some questions. For which self-adjoint coupled boundary conditions does $\lambda_n$ have a jump discontinuity? Can such discontinuities occur when the variation of $\lambda_n$ is studied with respect to $q$? $\frac{1}{p}$? $w$? $a$? $b$?

## References


[1] P. B. Bailey, W. N. Everitt and A. Zettl, "Sleign 2" a FORTRAN code for the numerical approximation of eigenvalues, eigenfunctions and continuous spectrum of Sturm-Liouville problems; available through WWW: ftp://ftp.math.niu.edu/pub/papers/Zettl/Sleign2

[2] W. N. Everitt, M. Möller and A. Zettl, "Sturm-Liouville problems, discontinuous eigenvalues and the numerical code SLEIGN2", in preparation.

[3] K. Jörgens, "Spectral theory of second-order differential operators", Lectures delivered at Aaarhus Universitet 1962/63, Matematisk Institut, Aarhus Universitet, Aaarhus 1964.

[4] Q. Kong and A. Zettl, "Eigenvalues of regular Sturm-Liouville problems", pre-print.

[5] Q. Kong, H. Wu and A. Zettl, "Dependence of eigenvalues on the problem", pre-print.

[6] M. Möller and A. Zettl, "Differentiable dependence of simple eigenvalues of operators in Banach spaces", pre-print.

[7] J. Weidmann, "Spectral theory of ordinary differential operators", Lecture Notes in Mathematics 1258, Springer-Verlag, 1987.



W.N. Everitt, School of Mathematics and Statistics, University of Birmingham, Birmingham B15 2TT, United Kingdom.

M. Möller, Department of Mathematic, University of the Witwatersrand, WITS 2050, South Africa.

A. Zettl, Department of Mathematical Sciences, Northern Illinois University, DeKalb, IL 60115, U.S.A.

# Inequalities for Differential Operators

International Series of Numerical Mathematics, Vol. 123, © 1997 Birkhäuser Verlag Basel

# Note on Wirtinger's inequality

Horst Alzer

**Abstract**

In this note we refine the following theorem due to W. Wirtinger: If $f$ has period $2\pi$ and satisfies $\int_0^{2\pi} f(x)dx = 0$, then

$$\int_0^{2\pi} f^2(x)dx \leq \int_0^{2\pi} f'^2(x)dx$$

with strict inequality unless $f(x) = a\cos(x) + b\sin(x)$, $(a, b \in \mathbb{R})$.

One of the most fascinating subjects in the theory of inequalities are integral inequalities involving a function and its derivative. Among the theorems in this field the so-called Wirtinger inequality has attracted special interest, because of its close connection to linear differential equations and differential geometry;see [1–5]. Wirtinger's inequality, which compares the integral of a square of a function with that of a square of its first derivative, appeared in 1916 in W. Blaschke's classical book "Kreis und Kugel" [3, p. 105].

**Proposition.** *Let $f$ be a real function with period $2\pi$ and $\int_0^{2\pi} f(x)dx = 0$. If $f' \in L^2$, then*

$$\int_0^{2\pi} f^2(x)dx \leq \int_0^{2\pi} f'^2(x)dx \tag{1}$$

*with strict inequality unless $f(x) = a\cos(x) + b\sin(x)$, where $a$ and $b$ are real constants.*

Three elegant proofs of (1) are given in [1]. An interesting survey on Wirtinger's and related inequalities can be found in a recently published monograph [5], which represents numerous extensions, refinements, variants, discrete analogues and applications of (1), and provides more than 200 references on this subject.

It is the aim of this paper to establish a refinement of Wirtinger's inequality which we couldn't locate in the literature. More precisely we investigate the following problem:

Let $r, s, t$ and $q$ be non-negative integers with $r > \max\{s, t\}$ and $q \geq 2$. What is the best possible value $K(r, s, t; q)$ such that the inequality

$$K(r,s,t;q)\int_0^{2\pi}\left(\frac{1}{q}\sum_{i=1}^{q}f^{(t)}(x+2\pi i/q)\right)^2dx+\int_0^{2\pi}f^{(s)}(x)^2dx\le\int_0^{2\pi}f^{(r)}(x)^2dx$$

holds for all real functions $f$ satisfying $f(0)=f(2\pi)$, $\int_0^{2\pi}f(x)dx=0$ and $f^{(r)}\in C[0,2\pi]$?

The answer will in particular provide a new sharpening of (1), if we set $r=1$, $s=t=0$ in inequality (2) given below.

**Theorem.** *Let $r,s,t$ and $q$ be integers with $r>s\ge 0$, $r>t\ge 0$ and $q\ge 2$. Further, let $f$ be a real function with period $2\pi$ and $\int_0^{2\pi}f(x)dx=0$, and let $f_q(x)$ be the arithmetic mean of $f(x),f(x+2\pi/q),\cdots,f(x+2(q-1)\pi/q)$, that is $f_q(x)=\frac{1}{q}\sum_{i=1}^{q}f(x+2\pi i/q)$. If $f^{(r)}\in C[0,2\pi]$, then*

$$\left(q^{2(r-t)}-q^{2(s-t)}\right)\int_0^{2\pi}f_q^{(t)}(x)^2dx+\int_0^{2\pi}f^{(s)}(x)^2dx\le\int_0^{2\pi}f^{(r)}(x)^2dx,\qquad(2)$$

*with equality holding if and only if $f(x)=a_1\cos(x)+a_q\cos(qx)+b_1\sin(x)+b_q\sin(qx)$, where $a_1,a_q,b_1,b_q$ are real constants.*

*Proof.* We have the Fourier series representation

$$f(x)=\frac{a_0}{2}+\sum_{k=1}^{\infty}\left(a_k\cos(kx)+b_k\sin(kx)\right).$$

Because of $\int_0^{2\pi}f(x)dx=0$ we obtain $a_0=0$. Let $n\in\{0,1,\cdots r\}$; then there exists real numbers $\alpha_n(k)$ and $\beta_n(k)$ such that

$$f^{(n)}(x)=\sum_{k=1}^{\infty}k^n\left(\alpha_n(k)\cos(kx)+\beta_n(k)\sin(kx)\right)\qquad(3)$$

with

$$\alpha_n(k)^2+\beta_n(k)^2=a_k^2+b_k^2\text{ for }k=1,2,\cdots.$$

From (3) we conclude

$$\begin{aligned}f_q^{(n)}(x)&=\sum_{k=1}^{\infty}k^n\left(\alpha_n(k)\frac{1}{q}\sum_{k=1}^{q}\cos\left(kx+2\pi ki/q\right)+\beta_n(k)\frac{1}{q}\sum_{k=1}^{q}\sin(kx+2\pi ki/q)\right)\\&=\sum_{k=1,q|k}^{\infty}k^n\left(\alpha_n(k)\cos(kx)+\beta_n(k)\sin(kx)\right).\end{aligned}$$

Using Parsevals theorem we obtain

$$\int_0^{2\pi} f_q^{(n)}(x)^2 dx = \pi \sum_{k=1,q|k}^{\infty} k^{2n}(a_k^2 + b_k^2) \tag{4}$$

and

$$\int_0^{2\pi} f^{(n)}(x)^2 dx = \pi \sum_{k=1}^{\infty} k^{2n}(a_k^2 + b_k^2). \tag{5}$$

Let $r > s \geq 0$ and $r > t \geq 0$; then we conclude from (4) and (5):

$$\begin{aligned}\int_0^{2\pi} &\{f^{(r)}(x)^2 - f^{(s)}(x)^2 - \left(q^{2(r-t)} - q^{2(s-t)}\right) f_q^{(t)}(x)^2\}dx \\ &= \pi \sum_{k=1}^{\infty} S_k(r,s,t;q)(a_k^2 + b_k^2)\end{aligned} \tag{6}$$

with

$$S_k(r,s,t;q) = \begin{cases} k^{2r} - k^{2s} - \left(q^{2(r-t)} - q^{2(s-t)}\right) k^{2t}, & \text{if } q|k, \\ k^{2r} - k^{2s}, & \text{if } q \not| k. \end{cases}$$

If $q \not| k$, then we get $S_k(r,s,t;q) \geq 0$ with equality holding if and only if $k = 1$. Let $q|k$,say $k = qm$, $(m \in \mathrm{N})$. Then we have

$$(qm)^{-2t} S_k(r,s,t;q) = (qm)^{-2(r-t)} - (qm)^{-2(s-t)} - q^{2(r-t)} + q^{2(s-t)}.$$

Since the function $g(x) = (qx)^{2(r-t)} - (qx)^{2(s-t)}$ is strictly increasing on $[1,\infty)$ we obtain $g(m) \geq g(1)$ with equality holding if and only if $m = 1$. Hence,

$$S_1(r,s,t;q) = S_q(r,s,t;q) = 0 \text{ and } S_k(r,s,t;q) > 0 \text{ for } k \in \mathbf{N} - \{1,q\}.$$

From (6) we conclude the validity of inequality (2). Moreover, the sign of equality holds in (2) if and only if $a_k = b_k = 0$ for all $k \in n - \{1,q\}$, that is, $f(x) = a_1\cos(x) + a_q\cos(qx) + b_1\sin(x) + b_q\sin(qx)$.

**Remark.** The proof of the Theorem reveals: If $s > 0$ and $t > 0$, then inequality (2) is valid even if we drop the assumption $\int_0^{2\pi} f(x)dx = 0$.

## References


[1] E.F. Beckenbach and R. Bellman, *Inequalities.* Berlin1983.

[2] P.R. Beesack, *Integral inequalities involving a function and its derivative.* Amer.Math.Monthly **78** (1971), 705–741.

[3] W. Blaschke, *Kreis und Kugel.* Leipzig, 1916.

[4] D.S. Mitrinović, *Analytic Inequalities.* New York, 1970.

[5] D.S. Mitrinović, J.E. Pecarić and A.M. Fink, *Inequalities involving functions and their integrals and derivatives.* Dordrecht, 1991.


Horst Alzer, Morsbacher Straße 10, D-51545 Waldbröl, Germany.

International Series of Numerical Mathematics, Vol. 123, © 1997 Birkhäuser Verlag Basel

# Opial-type inequalities involving higher order partial derivatives of two functions

Ravi P. Agarwal and Peter Y.H. Pang

**Abstract**

In this paper we offer very general Opial-type inequalities involving higher order partial derivatives of two functions of two independent variables. From these inequalities we then deduce extended and improved versions of several recent results.

## 1. Introduction

Opial's inequality, in its modified form, states that if $x(t)$ is absolutely continuous on $[a,T]$ with $x(a)=0$, and $\int_a^T (x'(t))^2\,dt < \infty$, then

$$\int_a^T |x(t)x'(t)|\,dt \;\le\; \frac{(T-a)}{2}\int_a^T (x'(t))^2\,dt. \tag{1}$$

This simple inequality has motivated in a large number of research papers giving its successively simpler proofs, providing various generalizations, and finding discrete analogs. (See [2-6], and [15] for an extensive bibliography consisting of 83 articles.) In the year 1981 (April 26 – May 2) during the General Inequalities 3 meeting at Oberwolfach Agarwal proved a two-independent variable analog of the inequality (1). This result can be stated as follows: If $u(t,s)\in C^{(1,1)}([a,T]\times[c,S])$, $u(a,s) = u(t,c) = 0$, then

$$\int_a^T\!\!\int_c^S |u(t,s)u_{ts}(t,s)|\,dt\,ds \;\le\; C(T-a)(S-c)\int_a^T\!\!\int_c^S |u_{ts}(t,s)|^2\,dt\,ds, \tag{2}$$

where $C=\frac{1}{2\sqrt{2}}$. The constant $C$ in (2) cannot be replaced by $\frac{1}{4}$. In fact, W. Walter and R. Redheffer gave an example which showed that $C \ge \frac{3+\sqrt{13}}{24}$. The details of this were included in the proceedings of this meeting which was published in 1983 [1]. In 1982 the inequality (2) was also proved by Yang [24], but he obtained $C=\frac{1}{2}$. Since 1982, (2) has been extended in various directions [7-14,16-23]. While the problem of obtaining the optimal constant $C$ remains unresolved, recently in [5] we have obtained extensive generalizations of (2). The particular cases of these generalizations correct as well as improve several known inequalities.

Among the generalizations of (2), there is a class of inequalities which instead of involving one function $u(t,s)$ involves two functions. The first such result is due to Pachpatte [16]. In this paper we shall present several similar inequalities involving higher order partial derivatives of two functions. These inequalities are very sharp and accommodate several recent results.

## 2. Main Results

Let the intervals $[a,T]$, $[c,S]$ and the rectangle $[a,T]\times[c,S]$ be denoted as $I_1$, $J_1$ and $R_1$, respectively. We shall assume that the functions $p(t,s)$ and $q(t,s)$ are non-negative and measurable on the rectangle $R_1$. For the function $u(t,s)$ the partial derivatives $\partial u(t,s)/\partial t$ and $\partial u(t,s)/\partial s$ are denoted as $D_1u(t,s)$ and $D_2u(t,s)$, respectively. Let $0\le i\le n-1$, $0\le j\le m-1$ be fixed, and let $u(t,s)\in C^{(n-1,m-1)}(R_1)$ be such that $D_1^{\kappa}u(a,s)=0$, $i\le\kappa\le n-1$, $s\in J_1$, $D_2^{\lambda}u(t,c)=0$, $j\le\lambda\le m-1$, $t\in I_1$, and $D_1^nD_2^{m-1}u(t,s)$ and $D_1^{n-1}D_2^mu(t,s)$ are absolutely continuous on $R_1$. Then, for $i\le\kappa\le n-1$, $j\le\lambda\le m-1$ it follows that

$$|D_1^{\kappa}D_2^{\lambda}u(t,s)| \;\le\; K_{\kappa,\lambda}\int_a^t\int_c^s h_{\kappa,\lambda}(t,s,\nu,\mu)|D_1^nD_2^mu(\nu,\mu)|\,d\nu\,d\mu; \tag{3}$$

$$|D_1^{\kappa}D_2^{m}u(t,s)| \;\le\; K_{\kappa,m-1}\int_a^t h_{\kappa,m-1}(t,s,\nu,s)|D_1^nD_2^mu(\nu,s)|\,d\nu, \tag{4}$$

and

$$|D_1^{n}D_2^{\lambda}u(t,s)| \;\le\; K_{n-1,\lambda}\int_c^s h_{n-1,\lambda}(t,s,t,\mu)|D_1^nD_2^mu(t,\mu)|\,d\mu, \tag{5}$$

where

$$h_{\kappa,\lambda}(t,s,\nu,\mu) \;=\; (t-\nu)^{n-\kappa-1}(s-\mu)^{m-\lambda-1},$$

and

$$K_{\kappa,\lambda} \;=\; \frac{1}{(n-\kappa-1)!\,(m-\lambda-1)!}.$$

Of course, to obtain (4) we only need the conditions $D_1^{\kappa}u(a,s)=0$, $i\le\kappa\le n-1$, $s\in J_1$, whereas for (5) we only require $D_2^{\lambda}u(t,c)=0$, $j\le\lambda\le m-1$, $t\in I_1$.

Hence, from Hölder's inequality with indices $r>1$ and $r/(r-1)$, for $i\le\kappa\le n-1$, $j\le\lambda\le m-1$ we have

$$\begin{aligned}
&|D_1^{\kappa}D_2^{\lambda}u(t,s)|\\
&\le K_{\kappa,\lambda}\int_a^t\int_c^s h_{\kappa,\lambda}(t,s,\nu,\mu)(p(\nu,\mu))^{-1/r}(p(\nu,\mu))^{1/r}|D_1^nD_2^mu(\nu,\mu)|\,d\nu\,d\mu\\
&\le K_{\kappa,\lambda}[P_{\kappa,\lambda}(t,s)]^{(r-1)/r}\left[\int_a^t\int_c^s p(\nu,\mu)|D_1^nD_2^mu(\nu,\mu)|^r\,d\nu\,d\mu\right]^{1/r};
\end{aligned}$$

$$|D_1^\kappa D_2^m u(t,s)| \;\le\; K_{\kappa,m-1}[P_{\kappa,m}(t,s)]^{(r-1)/r}\left[\int_a^t p(\nu,s)|D_1^n D_2^m u(\nu,s)|^r\,d\nu\right]^{1/r},$$

and

$$|D_1^n D_2^\lambda u(t,s)| \;\le\; K_{n-1,\lambda}[P_{n,\lambda}(t,s)]^{(r-1)/r}\left[\int_c^s p(t,\mu)|D_1^n D_2^m u(t,\mu)|^r\,d\mu\right]^{1/r},$$

where

$$P_{\kappa,\lambda}(t,s) \;=\; \int_a^t\int_c^s (h_{\kappa,\lambda}(t,s,\nu,\mu))^{r/(r-1)}(p(\nu,\mu))^{-1/(r-1)}\,d\nu\,d\mu;$$

$$P_{\kappa,m}(t,s) \;=\; \int_a^t (h_{\kappa,m-1}(t,s,\nu,s))^{r/(r-1)}(p(\nu,s))^{-1/(r-1)}\,d\nu,$$

and

$$P_{n,\lambda}(t,s) \;=\; \int_c^s (h_{n-1,\lambda}(t,s,t,\mu))^{r/(r-1)}(p(t,\mu))^{-1/(r-1)}\,d\mu.$$

We set

$$z(t,s) \;=\; \int_a^t\int_c^s p(\nu,\mu)|D_1^n D_2^m u(\nu,\mu)|^r\,d\nu\,d\mu$$

so that

$$D_1 z(t,s) \;=\; \int_c^s p(t,\mu)|D_1^n D_2^m u(t,\mu)|^r\,d\mu;$$

$$D_2 z(t,s) \;=\; \int_a^t p(\nu,s)|D_1^n D_2^m u(\nu,s)|^r\,d\nu;$$

$$D_1 D_2 z(t,s) \;=\; p(t,s)|D_1^n D_2^m u(t,s)|^r,$$

and

$$z(a,s) \;=\; z(t,c) \;=\; D_1 z(t,c) \;=\; D_2 z(a,s) \;=\; 0.$$

Let $i\le k\le n,\ j\le \ell\le m,\ k+\ell<n+m$, (but fixed), and $r_{i,j},\ r_{k,\ell}\ge 0,\ r_{n,m}>0$. Further, in the above considerations we let $\kappa=i,\ \lambda=j,\ u=u_1=u_1(t,s),\ z=z_1=z_1(t,s)$ and $\kappa=k,\ \lambda=\ell,\ u=u_2=u_2(t,s),\ z=z_2=z_2(t,s)$. Then, for the function

$$U_{k,\ell} \;=\; U_{k,\ell}(t,s) \;=\; q|D_1^i D_2^j u_1|^{r_{i,j}}|D_1^k D_2^\ell u_2|^{r_{k,\ell}}|D_1^n D_2^m u_1|^{r_{n,m}}$$

it follows that

$$U_{k,\ell} \;\le\; Q_{k,\ell}\left[z_1^{r_{i,j}} z_2^{r_{k,\ell}}(D_1 D_2 z_1)^{r_{n,m}}\right]^{1/r},$$

where

$$Q_{k,\ell}=Q_{k,\ell}(t,s) \;=\; qK_{i,j}^{r_{i,j}}K_{k,\ell}^{r_{k,\ell}}\left[P_{i,j}^{r_{i,j}}P_{k,\ell}^{r_{k,\ell}}\right]^{(r-1)/r}p^{-r_{n,m}/r}$$

provided $i \le k \le n-1,\ j \le \ell \le m-1$;

$$U_{k,m} \ \le\ Q_{k,m}\left[z_1^{r_{i,j}}(D_2z_2)^{r_{k,m}}(D_1D_2z_1)^{r_{n,m}}\right]^{1/r},$$

where

$$Q_{k,m} = Q_{k,m}(t,s) \ =\ qK_{i,j}^{r_{i,j}}K_{k,m-1}^{r_{k,m}}\left[P_{i,j}^{r_{i,j}}P_{k,m}^{r_{k,m}}\right]^{(r-1)/r}p^{-r_{n,m}/r}$$

provided $i \le k \le n-1$, and

$$U_{n,\ell} \ \le\ Q_{n,\ell}\left[z_1^{r_{i,j}}(D_1z_2)^{r_{n,\ell}}(D_1D_2z_1)^{r_{n,m}}\right]^{1/r},$$

where

$$Q_{n,\ell} = Q_{n,\ell}(t,s) = qK_{i,j}^{r_{i,j}}K_{n-1,\ell}^{r_{n,\ell}}\left[P_{i,j}^{r_{i,j}}P_{n,\ell}^{r_{n,\ell}}\right]^{(r-1)/r}p^{-r_{n,m}/r}$$

provided $j \le \ell \le m-1$.

Now, integrating the above inequalities over $R_1$ and applying Hölder's inequality with indices $r/r_{n,m} > 1$ and $r/(r-r_{n,m})$, we obtain for $i \le k \le n-1,\ j \le \ell \le m-1$ that

$$\int_a^T\!\!\int_c^S U_{k,\ell}\,dt\,ds \ \le\ L_0^{k,\ell}\left[\int_a^T\!\!\int_c^S \left[z_1^{r_{i,j}}z_2^{r_{k,\ell}}\right]^{1/r_{n,m}}D_1D_2z_1\,dt\,ds\right]^{r_{n,m}/r}; \qquad (6)$$

$$\int_a^T\!\!\int_c^S U_{k,m}\,dt\,ds \ \le\ L_0^{k,m}\left[\int_a^T\!\!\int_c^S \left[z_1^{r_{i,j}}(D_2z_2)^{r_{k,m}}\right]^{1/r_{n,m}}D_1D_2z_1\,dt\,ds\right]^{r_{n,m}/r}, \qquad (7)$$

and

$$\int_a^T\!\!\int_c^S U_{n,\ell}\,dt\,ds \ \le\ L_0^{n,\ell}\left[\int_a^T\!\!\int_c^S \left[z_1^{r_{i,j}}(D_1z_2)^{r_{n,\ell}}\right]^{1/r_{n,m}}D_1D_2z_1\,dt\,ds\right]^{r_{n,m}/r}, \qquad (8)$$

where for $i \le k \le n,\ j \le \ell \le m,\ k+\ell < m+n$,

$$L_0^{k,\ell} \ =\ L_0(p,q,r_{i,j},r_{k,\ell},r_{n,m},r) \ =\ \left(\int_a^T\!\!\int_c^S (Q_{k,\ell}(t,s))^{r/(r-r_{n,m})}\,dt\,ds\right)^{(r-r_{n,m})/r}.$$

Next, for $i \le k \le n,\ j \le \ell \le m,\ k+\ell < m+n$, we define the function

$$V_{k,\ell} = V_{k,\ell}(t,s) \ =\ q|D_1^iD_2^ju_2|^{r_{i,j}}|D_1^kD_2^\ell u_1|^{r_{k,\ell}}|D_1^nD_2^mu_2|^{r_{n,m}}.$$

Then, as above for $i \le k \le n-1,\ j \le \ell \le m-1$ it follows that

$$\int_a^T\!\!\int_c^S V_{k,\ell}\,dt\,ds \ \le\ L_0^{k,\ell}\left[\int_a^T\!\!\int_c^S \left[z_2^{r_{i,j}}z_1^{r_{k,\ell}}\right]^{1/r_{n,m}}D_1D_2z_2\,dt\,ds\right]^{r_{n,m}/r}; \qquad (9)$$

$$\int_a^T\int_c^S V_{k,m}\,dt\,ds \;\le\; L_0^{k,m}\left[\int_a^T\int_c^S \left[z_2^{r_{i,j}}(D_2z_1)^{r_{k,m}}\right]^{\frac{1}{r_{n,m}}} D_1D_2z_2\,dt\,ds\right]^{\frac{r_{n,m}}{r}}, \tag{10}$$

and

$$\int_a^T\int_c^S V_{n,\ell}\,dt\,ds \;\le\; L_0^{n,\ell}\left[\int_a^T\int_c^S \left[z_2^{r_{i,j}}(D_1z_1)^{r_{n,\ell}}\right]^{\frac{1}{r_{n,m}}} D_1D_2z_2\,dt\,ds\right]^{\frac{r_{n,m}}{r}}. \tag{11}$$

From the above inequalities we shall obtain a number of interesting results. For this, we need the following elementary inequalities

$$2^{\rho-1}(x^\rho+y^\rho) \;\le\; (x+y)^\rho \;\le\; x^\rho+y^\rho,\; x,\; y\ge 0,\; 0\le\rho\le 1 \tag{12}$$

$$x^\rho+y^\rho \;\le\; (x+y)^\rho \;\le\; 2^{\rho-1}(x^\rho+y^\rho),\; x,\; y\ge 0,\; \rho\ge 1. \tag{13}$$

When $r_{k,\ell}=0$, an addition of (6) and (9) gives

$$S_1 \;=\; \int_a^T\int_c^S q\left[|D_1^iD_2^ju_1|^{r_{i,j}}|D_1^nD_2^mu_1|^{r_{n,m}} + |D_1^iD_2^ju_2|^{r_{i,j}}|D_1^nD_2^mu_2|^{r_{n,m}}\right]dt\,ds$$

$$\le L_1\left(\left[\int_a^T\int_c^S z_1^{r_{i,j}/r_{n,m}}D_1D_2z_1\,dt\,ds\right]^{r_{n,m}/r}+\left[\int_a^T\int_c^S z_2^{r_{i,j}/r_{n,m}}D_1D_2z_2\,dt\,ds\right]^{r_{n,m}/r}\right),$$

where $L_1=L_0(p,q,r_{i,j},0,r_{n,m},r)$. However, from the fact $D_2z_\xi\ge 0,\;\xi=1,2$, we have

$$\begin{aligned}\int_a^T\int_c^S z_\xi^{r_{i,j}/r_{n,m}}D_1D_2z_\xi\,dt\,ds \;&\le\; \int_a^T (z_\xi(t,S))^{r_{i,j}/r_{n,m}}D_1z_\xi(t,S)\,dt\\ &=\; \frac{r_{n,m}}{r_{i,j}+r_{n,m}}(z_\xi(T,S))^{(r_{i,j}+r_{n,m})/r_{n,m}}\end{aligned} \tag{14}$$

and hence in view of (12) and (13) it follows that

$$S_1 \;\le\; L_2A_1(p,u_1,u_2,r_{i,j},r_{n,m},r), \tag{15}$$

where

$$A_1(p,u_1,u_2,r_{i,j},r_{n,m},r)=\left[\int_a^T\int_c^S p[|D_1^nD_2^mu_1|^r+|D_1^nD_2^mu_2|^r]dtds\right]^{\frac{(r_{i,j}+r_{n,m})}{r}},$$

$$L_2 \;=\; L_2(p,q,r_{i,j},r_{n,m},r) \;=\; \theta_1\left(\frac{r_{n,m}}{r_{i,j}+r_{n,m}}\right)^{r_{n,m}/r} K_{i,j}^{r_{i,j}} \;\times \tag{16}$$

$$\left[\int_a^T\int_c^S \left(q^rp^{-r_{n,m}}\right)^{1/(r-r_{n,m})}P_{i,j}^{r_{i,j}(r-1)/(r-r_{n,m})}\,dt\,ds\right]^{(r-r_{n,m})/r},$$

and

$$\theta_1 = \theta_1(r_{i,j}, r_{n,m}, r) = \begin{cases} 2^{1-((r_{i,j}+r_{n,m})/r)}, & r_{i,j}+r_{n,m} \le r \\ 1, & r_{i,j}+r_{n,m} \ge r. \end{cases}$$

Similarly, when $r_{k,m} = 0$, addition of (7) and (10), or when $r_{n,\ell} = 0$ addition of (8) and (11) gives the same inequality (15).

We summarize the above case in the following:

**Theorem 1.** *Assume that*

(i) $p(t,s)$, $q(t,s)$ *are non-negative and measurable functions on* $R_1$, *and* $0 \le i \le n-1$, $0 \le j \le m-1$, *but fixed,*

(ii) $r_{i,j} \ge 0$, $r_{n,m} > 0$, $r > 1$, $r > r_{n,m}$ *are given numbers,*

(iii) *for* $\xi = 1,2$ *functions* $u_\xi(t,s) \in C^{(n-1,m-1)}(R_1)$ *are such that* $D_1^\kappa u_\xi(a,s) = 0$, $i \le \kappa \le n-1$, $s \in J_1$, $D_2^\lambda u_\xi(t,c) = 0$, $j \le \lambda \le m-1$, $t \in I_1$, *and* $D_1^n D_2^{m-1} u_\xi(t,s)$, $D_1^{n-1} D_2^m u_\xi(t,s)$ *are absolutely continuous on* $R_1$,

(iv) *the function* $P_{i,j}(t,s)$ *is defined on* $R_1$ *and* $L_2 < \infty$,

(v) *both the integrals in the right-hand side of* (16) *are finite.*

*Then, the inequality* (15) *holds.*

When $r_{i,j} = 0$, an addition of (6) and (9) gives

$$S_2 = \int_a^T \int_c^S q\left[|D_1^k D_2^\ell u_2|^{r_{k,\ell}} |D_1^n D_2^m u_1|^{r_{n,m}} + |D_1^k D_2^\ell u_1|^{r_{k,\ell}} |D_1^n D_2^m u_2|^{r_{n,m}}\right] dt\, ds$$

$$\le L_3 \left( \left[ \int_a^T \int_c^S z_2^{r_{k,\ell}/r_{n,m}} D_1 D_2 z_1 \, dt\, ds \right]^{\frac{r_{n,m}}{r}} + \left[ \int_a^T \int_c^S z_1^{r_{k,\ell}/r_{n,m}} D_1 D_2 z_2 \, dt\, ds \right]^{\frac{r_{n,m}}{r}} \right),$$

where $L_3 = L_0(p,q,0,r_{k,\ell},r_{n,m},r)$. However, since $z_\xi(t,s) \le z_\xi(t,S)$, $\xi = 1,2$ in view of the inequality (12) it follows that

$$S_2 \le 2^{(r-r_{n,m})/r} L_3 I_\alpha^{r_{n,m}/r}, \tag{17}$$

where $I_\alpha$ on defining $z_\xi(t,S) = y_\xi(t)$, $\xi = 1,2$ and $\alpha = r_{k,\ell}/r_{n,m}$ appears as

$$I_\alpha = \int_a^T \{(y_2(t))^\alpha y_1'(t) + (y_1(t))^\alpha y_2'(t)\}\ dt.$$

Let $\theta_2 = \theta_2(r_{k,\ell}, r_{n,m}) = \theta_2(\alpha) = \begin{cases} 1, & \alpha \ge 1 \\ 2^{1-\alpha}, & \alpha \le 1. \end{cases}$ Then, from (12) and (13) we have

$$\begin{aligned} I_\alpha &= \int_a^T [(y_1(t))^\alpha + (y_2(t))^\alpha] (y_1'(t) + y_2'(t)) \, dt \\ &\qquad - \int_a^T [(y_1(t))^\alpha y_1'(t) + (y_2(t))^\alpha y_2'(t)] \, dt \\ &\le \theta_2 \int_a^T (y_1(t) + y_2(t))^\alpha (y_1(t) + y_2(t))' \, dt \\ &\qquad - \frac{1}{1+\alpha} \left[(y_1(T))^{1+\alpha} + (y_2(T))^{1+\alpha}\right] \\ &= \frac{1}{1+\alpha} \left[\theta_2 (y_1(T) + y_2(T))^{1+\alpha} - (y_1(T))^{1+\alpha} - (y_2(T))^{1+\alpha}\right] \\ &\le \frac{(\theta_2 2^\alpha - 1)}{1+\alpha} \left[(y_1(T))^{1+\alpha} + (y_2(T))^{1+\alpha}\right] \qquad (18) \\ &\le \frac{\theta_3}{1+\alpha} [y_1(T) + y_2(T)]^{1+\alpha}, \qquad (19) \end{aligned}$$

where

$$\theta_3 = \theta_3(r_{k,\ell}, r_{n,m}) = \theta_3(\alpha) = \begin{cases} 2^\alpha - 1, & \alpha \ge 1 \\ 1, & \alpha \le 1. \end{cases}$$

On combining (17) and (19), we obtain the inequality

$$S_2 \le L_4 A_1(p, u_1, u_2, r_{k,\ell}, r_{n,m}, r), \qquad (20)$$

where

$$L_4 = L_4(p, q, r_{k,\ell}, r_{n,m}, r) = 2^{(r-r_{n,m})/r} \left(\frac{r_{n,m}\theta_3(r_{k,\ell}, r_{n,m})}{r_{k,\ell} + r_{n,m}}\right)^{r_{n,m}/r} K_{k,\ell}^{r_{k,\ell}} \times$$

$$\left[\int_a^T \int_c^S \left(q^r p^{-r_{n,m}}\right)^{1/(r-r_{n,m})} P_{k,\ell}^{r_{k,\ell}(r-1)/(r-r_{n,m})} \, dt \, ds\right]^{(r-r_{n,m})/r}.$$

We present the above case in the following:

**Theorem 2.** *Let in Theorem* 1 *integers* $i, j$ *and the number* $r_{i,j}$ *be replaced by* $k, \ell$ *and* $r_{k,\ell}$, *respectively. Then, the inequality* (20) *holds.*

When $r_{i,j} = 0$, an addition of (7) and (10) gives

$$S_3 = \int_a^T \int_c^S q\left[|D_1^k D_2^m u_2|^{r_{k,m}} |D_1^n D_2^m u_1|^{r_{n,m}} + |D_1^k D_2^m u_1|^{r_{k,m}} |D_1^n D_2^m u_2|^{r_{n,m}}\right] dtds$$

$$\leq L_5 \left( \left[ \int_a^T \int_c^S (D_2 z_2)^{\frac{r_{k,m}}{r_{n,m}}} D_1 D_2 z_1 dtds \right]^{\frac{r_{n,m}}{r}} + \left[ \int_a^T \int_c^S (D_2 z_1)^{\frac{r_{k,m}}{r_{n,m}}} D_1 D_2 z_2 dtds \right]^{\frac{r_{n,m}}{r}} \right),$$

where $L_5 = L_0(p, q, 0, r_{k,m}, r_{n,m}, r)$. Thus, for fixed $s$, if we define $D_2 z_\xi(t,s) = y_\xi(t)$, $\xi = 1, 2$ and $\alpha_1 = r_{k,m}/r_{n,m}$, then as above it follows that

$$S_3 \leq 2^{(r - r_{n,m})/r} L_5 \left[ \int_c^S I_{\alpha_1} ds \right]^{r_{n,m}/r} \leq L_6 A_2(p, u_1, u_2, r_{k,m}, r_{n,m}, r), \quad (21)$$

where

$$A_2(p, u_1, u_2, r_{k,m}, r_{n,m}, r)$$

$$= \left[ \int_c^S \left\{ \int_a^T p\left[\, |D_1^n D_2^m u_1|^r + |D_1^n D_2^m u_2|^r \right] dt \right\}^{(r_{k,m} + r_{n,m})/r_{n,m}} ds \right]^{r_{n,m}/r}, \quad (22)$$

and

$$L_6 = 2^{(r - r_{n,m})/r} \left( \frac{r_{n,m} \theta_3(r_{k,m}, r_{n,m})}{r_{k,m} + r_{n,m}} \right)^{r_{n,m}/r} K_{k,m-1}^{r_{k,m}} \times$$

$$\left[ \int_a^T \int_c^S (q^r p^{-r_{n,m}})^{1/(r - r_{n,m})} P_{k,m}^{r_{k,m}(r-1)/(r - r_{n,m})} dt\, ds \right]^{(r - r_{n,m})/r}.$$

We state the above case in the following:

**Theorem 3.** *Assume that*

(i) $p(t,s)$, $q(t,s)$ *are non-negative and measurable functions on* $R_1$, *and* $0 \leq k \leq n-1$, *but fixed,*

(ii) $r_{k,m} \geq 0$, $r_{n,m} > 0$, $r > 1$, $r > r_{n,m}$ *are given numbers,*

(iii) *for* $\xi = 1, 2$ *functions* $u_\xi(t,s) \in C^{(n-1,m-1)}(R_1)$ *are such that* $D_1^\kappa u_\xi(a,s) = 0$, $k \leq \kappa \leq n-1$, $s \in J_1$, *and* $D_1^n D_2^{m-1} u_\xi(t,s)$, $D_1^{n-1} D_2^m u_\xi(t,s)$ *are absolutely continuous on* $R_1$,

(iv) *the function* $P_{k,m}(t,s)$ *is defined on* $R_1$ *and* $L_6 < \infty$,

(v) *both the integrals in the right-hand side of* (22) *are finite.*

*Then, the inequality* (21) *holds.*

Following as above, once again when $r_{i,j} = 0$, an addition of (8) and (11) gives the inequality

$$\begin{aligned} S_4 &= \int_a^T \int_c^S q\left[|D_1^n D_2^\ell u_2|^{r_{n,\ell}} |D_1^n D_2^m u_1|^{r_{n,m}} + |D_1^n D_2^\ell u_1|^{r_{n,\ell}} |D_1^n D_2^m u_2|^{r_{n,m}}\right] dt\, ds \\ &\le L_7 A_3(p, u_1, u_2, r_{n,\ell}, r_{n,m}, r), \end{aligned} \tag{23}$$

where

$$\begin{aligned} &A_3(p, u_1, u_2, r_{n,\ell}, r_{n,m}, r) \\ &= \left[\int_a^T \left\{\int_c^S p[\,|D_1^n D_2^m u_1|^r + |D_1^n D_2^m u_2|^r]\, ds\right\}^{\frac{(r_{n,\ell}+r_{n,m})}{r_{n,m}}} dt\right]^{\frac{r_{n,m}}{r}}, \end{aligned} \tag{24}$$

and

$$\begin{aligned} L_7 &= 2^{(r-r_{n,m})/r} \left(\frac{r_{n,m}\theta_3(r_{n,\ell}, r_{n,m})}{r_{n,\ell} + r_{n,m}}\right)^{r_{n,m}/r} K_{n-1,\ell}^{r_{n,\ell}} \times \\ &\left[\int_a^T \int_c^S \left(q^r p^{-r_{n,m}}\right)^{1/(r-r_{n,m})} P_{n,\ell}^{r_{n,\ell}(r-1)/(r-r_{n,m})}\, dt\, ds\right]^{(r-r_{n,m})/r}. \end{aligned}$$

Thus, we have the following:

**Theorem 4.** *Assume that*

(i) *$p(t,s)$, $q(t,s)$ are non-negative and measurable functions on $R_1$, and $0 \le \ell \le m-1$, but fixed,*

(ii) *$r_{n,\ell} \ge 0$, $r_{n,m} > 0$, $r > 1$, $r > r_{n,m}$ are given numbers,*

(iii) *for $\xi = 1,2$ functions $u_\xi(t,s) \in C^{(n-1,m-1)}(R_1)$ are such that $D_2^\lambda u_\xi(t,c) = 0$, $\ell \le \lambda \le m-1$, $t \in I_1$, and $D_1^n D_2^{m-1} u_\xi(t,s)$, $D_1^{n-1} D_2^m u_\xi(t,s)$ are absolutely continuous on $R_1$,*

(iv) *the function $P_{n,\ell}(t,s)$ is defined on $R_1$ and $L_7 < \infty$,*

(v) *both the integrals in the right-hand side of (24) are finite.*

*Then, the inequality (23) holds.*

Theorems 1 and 2 can be unified. For this, we define

$$\sigma_1 = r_{i,j} + r_{k,\ell},\ \sigma_2 = r_{i,j} + r_{k,\ell} + r_{n,m},$$

use the arithmetic-geometric means inequality in the right side of (6), follow as in (14), and then use (12), to obtain

$$\begin{aligned}
&\int_a^T\int_c^S U_{k,\ell}(t,s)\,dt\,ds \\
&\le L_0^{k,\ell}\left[\int_a^T\int_c^S \left(\frac{r_{i,j}}{\sigma_1}z_1^{\sigma_1/r_{n,m}} + \frac{r_{k,\ell}}{\sigma_1}z_2^{\sigma_1/r_{n,m}}\right) D_1D_2z_1\,dt\,ds\right]^{r_{n,m}/r} \\
&\le L_0^{k,\ell}\left[\left(\frac{r_{i,j}r_{n,m}}{\sigma_1\sigma_2}\right)^{r_{n,m}/r}(z_1(T,S))^{\sigma_2/r}\right. \\
&\qquad \left. + \left(\frac{r_{k,\ell}}{\sigma_1}\right)^{r_{n,m}/r}\left(\int_a^T (z_2(t,S))^{\sigma_1/r_{n,m}} D_1z_1(t,S)\,dt\right)^{r_{n,m}/r}\right],
\end{aligned} \tag{25}$$

and similarly, from (9) we find

$$\begin{aligned}
&\int_a^T\int_c^S V_{k,\ell}(t,s)\,dt\,ds \\
&\le L_0^{k,\ell}\left[\left(\frac{r_{i,j}r_{n,m}}{\sigma_1\sigma_2}\right)^{r_{n,m}/r}(z_2(T,S))^{\sigma_2/r}\right. \\
&\qquad \left. + \left(\frac{r_{k,\ell}}{\sigma_1}\right)^{r_{n,m}/r}\left(\int_a^T (z_1(t,S))^{\sigma_1/r_{n,m}} D_1z_2(t,S)\,dt\right)^{r_{n,m}/r}\right].
\end{aligned} \tag{26}$$

Now an addition of (25) and (26) in view of (12) gives

$$\begin{aligned}
S_5 =& \int_a^T\int_c^S [\,U_{k,\ell}(t,s) + V_{k,\ell}(t,s)]\,dt\,ds \\
\le& L_0^{k,\ell}\left[\left(\frac{r_{i,j}r_{n,m}}{\sigma_1\sigma_2}\right)^{\frac{r_{n,m}}{r}}\left((z_1(T,S))^{\frac{\sigma_2}{r}} + (z_2(T,S))^{\frac{\sigma_2}{r}}\right) + 2^{\frac{(r-r_{n,m})}{r}}\left(\frac{r_{k,\ell}}{\sigma_1}\right)^{\frac{r_{n,m}}{r}}\right. \\
&\times\left.\left(\int_a^T\left\{(z_2(t,S))^{\frac{\sigma_1}{r_{n,m}}}D_1z_1(t,S) + (z_1(t,S))^{\frac{\sigma_1}{r_{n,m}}}D_1z_2(t,S)\right\}dt\right)^{\frac{r_{n,m}}{r}}\right].
\end{aligned} \tag{27}$$

Now following as above for the cases $r_{i,j} = 0$ and $r_{k,\ell} = 0$, we get

$$S_5 \le L_8A_1(p, u_1, u_2, r_{i,j} + r_{k,\ell}, r_{n,m}, r), \tag{28}$$

where $L_8 = L_0^{k,\ell} M_0(r)$,

$$M_0(r) \;=\; \left(\frac{r_{n,m}}{\sigma_1\sigma_2}\right)^{r_{n,m}/r} \left[\, r_{i,j}^{r_{n,m}/r}\theta_4 + 2^{(r-r_{n,m})/r}(r_{k,\ell}\theta_5)^{r_{n,m}/r} \,\right],$$

and $\theta_4 = \theta_1(\sigma_1, r_{n,m}, r)$, $\theta_5 \;=\; \theta_3(\sigma_1, r_{n,m})$.

Thus, a result which unifies Theorems 1 and 2 is the following:

**Theorem 5.** *Let $0 \le i \le k \le n-1$, $0 \le j \le \ell \le m-1$ and $r_{i,j}$, $r_{k,\ell} \ge 0$, $r_{n,m} > 0$, $r > 1$, $r > r_{n,m}$ be given numbers. Further, let $p(t,s)$, $q(t,s)$, $u_1(t,s)$, $u_2(t,s)$ be as in Theorem 1, functions $P_{i,j}(t,s)$, $P_{k,\ell}(t,s)$ be defined on $R_1$ and $L_8 < \infty$. Then, the inequality (28) holds.*

It is clear that we can obtain results similar to that of Theorem 5 which unify Theorems 1 and 3, and Theorems 1 and 4.

**Remark 1.** From the arithmetic-geometric means inequality we note that

$$S_5 \le \frac{r_{i,j}}{\sigma_1}\int_a^T\int_c^S q\Big[|D_1^iD_2^ju_1|^{\sigma_1}|D_1^nD_2^mu_1|^{r_{n,m}} + |D_1^iD_2^ju_2|^{\sigma_1}|D_1^nD_2^mu_2|^{r_{n,m}}\Big]dtds$$

$$+\frac{r_{k,\ell}}{\sigma_1}\int_a^T\int_c^S q\big[|D_1^kD_2^\ell u_2|^{\sigma_1}|D_1^nD_2^mu_1|^{r_{n,m}} + |D_1^kD_2^\ell u_1|^{\sigma_1}|D_1^nD_2^mu_2|^{r_{n,m}}\big]dtds.$$

Thus, from Theorems 1 and 2 it follows that in the inequality (28) the constant $L_8$ can be replaced by $L_9$, where

$$L_9 \;=\; \frac{r_{i,j}}{r_{i,j}+r_{k,\ell}}L_2(p,q,r_{i,j}+r_{k,\ell},r_{n,m},r) + \frac{r_{k,\ell}}{r_{i,j}+r_{k,\ell}}L_4(p,q,r_{i,j}+r_{k,\ell},r_{n,m},r).$$

From the above considerations (cf. (6) with $r_{i,j} = \zeta \ge 0$, $r_{k,\ell} = \zeta+\eta$, $r_{n,m} = \eta > 0$, and a similar inequality), and (12), we have

$$\begin{aligned} S_6 \;&=\; \int_a^T\int_c^S q|D_1^iD_2^ju_1D_1^kD_2^\ell u_2|^\zeta \left[\, |D_1^kD_2^\ell u_2 D_1^nD_2^m u_1|^\eta + |D_1^kD_2^\ell u_1 D_1^nD_2^m u_2|^\eta \,\right] dt\,ds \\ &\le\; 2^{(r-\eta)/r}L_{10}\left[\int_a^T\int_c^S (z_1z_2)^{\zeta/\eta}(z_2D_1D_2z_1 + z_1D_1D_2z_2)\,dt\,ds\right]^{\eta/r} \\ &\le\; 2^{(r-\eta)/r}L_{10}\left[\int_a^T (y_1(t)y_2(t))^{\zeta/\eta}(y_2(t)y_1'(t) + y_1(t)y_2'(t))\,dt\right]^{\eta/r}, \end{aligned}$$

where $L_{10} = L_0(p,q,\zeta,\zeta+\eta,\eta,r)$, and once again $z_\xi(t,S) = y_\xi(t)$, $\xi = 1,2$.

Thus, it follows that

$$\begin{aligned} S_6 &\le 2^{(r-\eta)/r} L_{10} \left[ \frac{\eta}{\zeta+\eta} \left( z_1(T,S) z_2(T,S) \right)^{(\zeta+\eta)/\eta} \right]^{\eta/r} \\ &\le 2^{(r-\eta)/r} L_{10} \left( \frac{\eta}{\zeta+\eta} \right)^{\eta/r} \qquad (29) \\ &\quad \cdot \left( \frac{1}{2} \right)^{2(\zeta+\eta)/r} [\, z_1(T,S) + z_2(T,S) \,]^{2(\zeta+\eta)/r}. \end{aligned}$$

Hence, the following inequality holds

$$S_6 \le L_{11} A_1(p, u_1, u_2, 2\zeta+\eta, \eta, r), \qquad (30)$$

where

$$L_{11} = 2^{(r-2\zeta-3\eta)/r} \left( \frac{\eta}{\zeta+\eta} \right)^{\eta/r} L_0(p,q,\zeta,\zeta+\eta,\eta,r).$$

Thus, we have established the following:

**Theorem 6.** *Let* $0 \le i \le k \le n-1,\ 0 \le j \le \ell \le m-1$ *and* $\zeta \ge 0,\ \eta > 0,\ r > 1,\ r > \eta$ *be given numbers. Further, let* $p(t,s),\ q(t,s),\ u_1(t,s),\ u_2(t,s)$ *be as in Theorem* 1, *functions* $P_{i,j}(t,s),\ P_{k,\ell}(t,s)$ *be defined on* $R_1$ *and* $L_{11} < \infty$. *Then, the inequality* (30) *holds.*

**Remark 2.** For $i \le \kappa \le n-1,\ j \le \lambda \le m-1$ it is clear that

$$\lim_{r\to 1} (P_{\kappa,\lambda}(t,s))^{r-1} = \bar{P}_{\kappa,\lambda}(t,s) = \text{ess.sup}_{\nu\in[a,t],\mu\in[c,s]} \left[ h_{\kappa,\lambda}(t,s,\nu,\mu)(p(\nu,\mu))^{-1} \right].$$

Thus, for the case $r = 1$ inequalities (28) and (30) hold by replacing $L_8$ and $L_{11}$, respectively, by

$$\bar{L}_8 = M_0(1) K_{i,j}^{r_{i,j}} K_{k,\ell}^{r_{k,\ell}} \left[ \int_a^T \int_c^S \left( q p^{-r_{n,m}} \bar{P}_{i,j}^{r_{i,j}} \bar{P}_{k,\ell}^{r_{k,\ell}} \right)^{1/(1-r_{n,m})} dt\, ds \right]^{1-r_{n,m}},$$

and

$$\bar{L}_{11} = 2^{1-(2\zeta+3\eta)} K_{i,j}^{\zeta} K_{k,\ell}^{\zeta+\eta} \left( \frac{\eta}{\zeta+\eta} \right)^{\eta} \left[ \int_a^T \int_c^S \left( q p^{-\eta} \bar{P}_{i,j}^{\zeta} \bar{P}_{k,\ell}^{\zeta+\eta} \right)^{1/(1-\eta)} dtds \right]^{1-\eta}.$$

**Remark 3.** As in Remark 2 we note that for $r = r_{n,m}$ the inequality (28) holds with $L_8$ replaced by

$$L_8^* = M_0(r_{n,m}) K_{i,j}^{r_{i,j}} K_{k,\ell}^{r_{k,\ell}}$$
$$\times \ \mathrm{ess.\,sup}_{t\in[a,T],s\in[c,S]} \left[ qp^{-1} P_{i,j}^{r_{i,j}(r_{n,m}-1)/r_{n,m}} P_{k,\ell}^{r_{k,\ell}(r_{n,m}-1)/r_{n,m}} \right].$$

Similarly, for $r = \eta$ the inequality (30) holds with $L_{11}$ replaced by

$$L_{11}^* = 2^{-2(\zeta+\eta)/\eta} K_{i,j}^{\zeta} K_{k,\ell}^{\zeta+\eta} \left(\frac{\eta}{\zeta+\eta}\right)$$
$$\times \ \mathrm{ess.\,sup}_{t\in[a,T],s\in[c,S]} \left[ qp^{-1} P_{i,j}^{\zeta(\eta-1)/\eta} P_{k,\ell}^{(\zeta+\eta)(\eta-1)/\eta} \right].$$

**Remark 4.** For the case $p(t,s) = q(t,s) = 1$ the constants $L_i$, $\bar{L}_i$, $L_i^*$, $i = 8, 11$, which we rename $W_i$, $\bar{W}_i$, $W_i^*$, $i = 8, 11$ can be computed. For this, we define

$$J_{\kappa,\lambda} = J_{\kappa,\lambda}(r) = \frac{r-1}{r(n-\kappa)-1} \times \frac{r-1}{r(m-\lambda)-1};\ i \le \kappa \le n-1,\ j \le \lambda \le m-1$$

$$N = N(r_{i,j}, r_{k,\ell}, r_{n,m}, r) = \frac{r - r_{n,m}}{r\left[r_{i,j}(n-i) + r_{k,\ell}(n-k) + 1\right] - \sigma_2},$$

and

$$Z = Z(r_{i,j}, r_{k,\ell}, r_{n,m}, r) = \frac{r - r_{n,m}}{r\left[r_{i,j}(m-j) + r_{k,\ell}(m-\ell) + 1\right] - \sigma_2},$$

to obtain

$$W_8 = M_0(r) \left(K_{i,j} J_{i,j}^{(r-1)/r}\right)^{r_{i,j}} \left(K_{k,\ell} J_{k,\ell}^{(r-1)/r}\right)^{r_{k,\ell}} [NZ]^{(r-r_{n,m})/r} \times$$
$$(T-a)^{[r_{i,j}(n-i)+r_{k,\ell}(n-k)+1]-\sigma_2/r} (S-c)^{[r_{i,j}(m-j)+r_{k,\ell}(m-\ell)+1]-\sigma_2/r};$$

$$W_{11} = 2^{(r-2\zeta-3\eta)/r} \left(\frac{\eta}{\zeta+\eta}\right)^{\eta/r} \left(K_{i,j} J_{i,j}^{(r-1)/r}\right)^{\zeta} \left(K_{k,\ell} J_{k,\ell}^{(r-1)/r}\right)^{\zeta+\eta} \times$$
$$[N(\zeta, \zeta+\eta, \eta, r) Z(\zeta, \zeta+\eta, \eta, r)]^{(r-\eta)/r} \times$$
$$(T-a)^{[\zeta(n-i)+(\zeta+\eta)(n-k)+1]-2(\zeta+\eta)/r} (S-c)^{[\zeta(m-j)+(\zeta+\eta)(m-\ell)+1]-2(\zeta+\eta)/r};$$

$$\bar{W}_8 = M_0(1) K_{i,j}^{r_{i,j}} K_{k,\ell}^{r_{k,\ell}} \left[N(r_{i,j}, r_{k,\ell}, r_{n,m}, 1) Z(r_{i,j}, r_{k,\ell}, r_{n,m}, 1)\right]^{1-r_{n,m}} \times$$
$$(T-a)^{r_{i,j}(n-i)+r_{k,\ell}(n-k)+1-\sigma_2} (S-c)^{r_{i,j}(m-j)+r_{k,\ell}(m-\ell)+1-\sigma_2};$$

$$\bar{W}_{11} = 2^{1-(2\zeta+3\eta)} \left(\frac{\eta}{\zeta+\eta}\right)^{\eta} K_{i,j}^{\zeta} K_{k,\ell}^{\zeta+\eta} \left[N(\zeta, \zeta+\eta, \eta, 1) Z(\zeta, \zeta+\eta, \eta, 1)\right]^{1-\eta} \times$$
$$(T-a)^{\zeta(n-i)+(\zeta+\eta)(n-k)+1-2(\zeta+\eta)} (S-c)^{\zeta(m-j)+(\zeta+\eta)(m-\ell)+1-2(\zeta+\eta)};$$

$$W_8^* = M_0(r_{n,m}) \left(K_{i,j} J_{i,j}^{(r_{n,m}-1)/r_{n,m}}(r_{n,m})\right)^{r_{i,j}} \left(K_{k,\ell} J_{k,\ell}^{(r_{n,m}-1)/r_{n,m}}(r_{n,m})\right)^{r_{k,\ell}} \times$$

$$(T-a)^{[r_{i,j}(n-i)+r_{k,\ell}(n-k)]-\sigma_1/r_{n,m}}(S-c)^{[r_{i,j}(m-j)+r_{k,\ell}(m-\ell)]-\sigma_1/r_{n,m}},$$

and

$$\begin{aligned} W_{11}^* &= 2^{-2(\zeta+\eta)/\eta}\left(\frac{\eta}{\zeta+\eta}\right)\left(K_{i,j}J_{i,j}^{(\eta-1)/\eta}(\eta)\right)^{\zeta}\left(K_{k,\ell}J_{k,\ell}^{(\eta-1)/\eta}(\eta)\right)^{\zeta+\eta}\times \\ &\quad (T-a)^{[\zeta(n-i)+(\zeta+\eta)(n-k)-1]-2\zeta/\eta}(S-c)^{[\zeta(m-j)+(\zeta+\eta)(m-\ell)-1]-2\zeta/\eta}. \end{aligned}$$

**Remark 5.** For $u_1 = u(t,s),\ u_2 = 0$, or $u_1 = 0,\ u_2 = u(t,s)$ from the proof of the inequality (15) (but not from (15)), we find

$$\int_a^T\int_c^S q|D_1^iD_2^ju|^{r_{i,j}}|D_1^nD_2^mu|^{r_{n,m}}dtds \le L_{12}\left[\int_a^T\int_c^S p|D_1^nD_2^mu|^r dtds\right]^{\frac{(r_{i,j}+r_{n,m})}{r}}, \tag{31}$$

where

$$L_{12} = \begin{cases} L_8 \text{ provided } r_{i,j}\ge 0,\ r_{n,m}>0,\ r>1,\ r>r_{n,m} \\ \bar{L}_8 \text{ provided } r_{i,j}\ge 0,\ r_{n,m}>0,\ r=1,\ r>r_{n,m} \\ L_8^* \text{ provided } r_{i,j}\ge 0,\ r_{n,m}>0,\ r>1,\ r=r_{n,m} \end{cases}$$

with $\theta_4 = 1,\ r_{k,\ell} = 0$. Recently, in [5] this inequality has been extensively generalized. For $p(t,s) = q(t,s) = n = m = 1,\ r = r_{0,0}+r_{1,1}$, the inequality (31) reduces to

$$\int_a^T\int_c^S |u|^{r_{0,0}}|D_1D_2u|^{r_{1,1}}\,dt\,ds \le L_{13}\int_a^T\int_c^S |D_1D_2u|^{r_{0,0}+r_{1,1}}\,dt\,ds, \tag{32}$$

where

$$L_{13} = \frac{r_{1,1}}{r_{0,0}+r_{1,1}}\left(\frac{1}{r_{1,1}(r_{0,0}+r_{1,1})}\right)^{r_{0,0}/(r_{0,0}+r_{1,1})}(T-a)^{r_{0,0}}(S-c)^{r_{0,0}},$$

$$r_{0,0}\ge 0,\ r_{1,1}>0,\ r_{0,0}+r_{1,1}\ge 1.$$

In particular, for $r_{0,0} = r_{1,1} = 1$, the inequality (32) reduces to (2) with $C = \frac{1}{2\sqrt{2}}$. For $r_{0,0}\ge 0,\ r_{1,1}\ge 1$, the inequality (32) has also been obtained in [13], however, with a larger constant $L_{13} = \frac{r_{1,1}}{r_{0,0}+r_{1,1}}(T-a)^{r_{0,0}}(S-c)^{r_{0,0}}$.

**Remark 6.** Let $p(t,s) = q(t,s)$, where $q(t,s)$ is a non-negative, non-increasing (with respect to each variable), and measurable function on $R_1$. Then, for $r = r_{i,j}+r_{k,\ell}+r_{n,m}$ from the inequality (28) it follows rather easily that

$$S_5 \le W_8A_1(q,u_1,u_2,r_{i,j}+r_{k,\ell},r_{n,m},r_{i,j}+r_{k,\ell}+r_{n,m}). \tag{33}$$

Similarly, for $r = 2(\zeta+\eta)$ the inequality (30) leads to

$$S_6 \le W_{11} A_1(q, u_1, u_2, 2\zeta+\eta, \eta, 2(\zeta+\eta)). \tag{34}$$

For $n = m = 1$ the inequality (34) reduces to

$$\begin{aligned} S_7(\zeta,\eta,T,S) &= \int_a^T \int_c^S q|u_1u_2|^\zeta \left[\, |u_2 D_1D_2u_1|^\eta + |u_1 D_1 D_2 u_2|^\eta \,\right] dt\, ds \\ &\le L_{14} A_4(q, T, S, \zeta, \eta), \end{aligned} \tag{35}$$

where

$$A_4(q,T,S,\zeta,\eta) = \int_a^T \int_c^S q \left[\, |D_1D_2u_1|^{2(\zeta+\eta)} + |D_1D_2u_2|^{2(\zeta+\eta)} \,\right] dt\, ds,$$

and

$$L_{14} = \frac{\eta}{2(\zeta+\eta)} \left( \frac{1}{2\eta(\zeta+\eta)} \right)^{(2\zeta+\eta)/2(\zeta+\eta)} (T-a)^{2\zeta+\eta}(S-c)^{2\zeta+\eta}$$

$$\zeta \ge 0,\ \eta > 0,\ 2(\zeta+\eta) \ge 1.$$

Inequality (35) with the constant $L_{14} = \frac{\eta}{2(\zeta+\eta)}(T-a)^{2\zeta+\eta}(S-c)^{2\zeta+\eta}$ was first proved by Pachpatte [16] for the case $q(t,s) = 1,\ \zeta \ge 0,\ \eta = 1$, and then by Chen and Lin [7] and Cheung [8] for the case $\zeta \ge 0,\ \eta \ge 1$. Thus, our inequality (35) not only extends the range of the parameters $\zeta$ and $\eta$, but also improves some earlier results.

**Remark 7.** Once again let $p(t,s) = q(t,s)$, where $q(t,s)$ is measurable and $0 < \gamma_1 \le q(t,s) \le \gamma_2$ on $R_1$. Then, as in Remark 6 inequalities (33) and (34) hold with the right-hand sides multiplied by $(\gamma_2/\gamma_1)^{(r_{i,j}+r_{k,\ell})/(r_{i,j}+r_{k,\ell}+r_{n,m})}$, and $(\gamma_2/\gamma_1)^{(2\zeta+\eta)/2(\zeta+\eta)}$, respectively. Thus, in particular the inequality (35) holds with the right-hand side multiplied by $(\gamma_2/\gamma_1)^{(2\zeta+\eta)/2(\zeta+\eta)}$. This resulting inequality was proved by Cheung [8] for the case $\zeta \ge 0,\ \eta \ge 1$ with the right-hand side multiplied by $(2\eta(\zeta+\eta))^{(2\zeta+\eta)/2(\zeta+\eta)}$ $(\ge 1)$.

**Remark 8.** The inequality (27) for $r = r_{n,m}$ in view of (18) with $\alpha = (r_{i,j} + r_{k,\ell})/r_{n,m}$ takes the form

$$S_5 \le L_{15} \left[\, (z_1(T,S))^{\sigma_2/r_{n,m}} + (z_2(T,S))^{\sigma_2/r_{n,m}} \,\right], \tag{36}$$

where $L_{15}$ is the same as $L_8^*$ with $\theta_4 = 1$. Now let $p(t,s) = (q(t,s))^{r_{n,m}/\sigma_2}$, so that for $\xi = 1, 2$

$$z_\xi(T,S) = \int_a^T \int_c^S (g(t,s))^{-r_{n,m}/\sigma_2} (g(t,s)q(t,s))^{r_{n,m}/\sigma_2} |D_1^n D_2^m u_\xi(t,s)|^{r_{n,m}}\, dt\, ds,$$

where $g(t,s)$ is a positive and measurable function on $R_1$. Thus, if $r_{i,j}$ or $r_{k,\ell} > 0$, Hölder's inequality with indices $\sigma_2/r_{n,m}$ and $\sigma_2/\sigma_1$ gives

$$\begin{aligned}(z_\xi(T,S))^{\sigma_2/r_{n,m}} \;\le\; & \left(\int_a^T\int_c^S (g(t,s))^{-r_{n,m}/\sigma_1}\,dt\,ds\right)^{\sigma_1/r_{n,m}} \\ & \times \int_a^T\int_c^S g(t,s)q(t,s)|D_1^n D_2^m z_\xi(t,s)|^{\sigma_2}\,dt\,ds.\end{aligned} \tag{37}$$

Further, if $\sigma_1 \ge r_{n,m}$ from Hölder's inequality with indices $\sigma_1/r_{n,m}$ and $\sigma_1/(\sigma_1 - r_{n,m})$, we have

$$\left(\int_a^T\int_c^S g^{\frac{-r_{n,m}}{\sigma_1}}\,dt\,ds\right)^{\frac{\sigma_1}{r_{n,m}}} \le [(T-a)(S-c)]^{\frac{(\sigma_1-r_{n,m})}{r_{n,m}}}\int_a^T\int_c^S g^{-1}\,dt\,ds. \tag{38}$$

Thus, if $r = r_{n,m} > 1$, $r_{i,j}$ or $r_{k,\ell} > 0$ and $\sigma_1 \ge r_{n,m}$, a combination of (36)–(38) gives

$$S_5 \;\le\; L_{16}\left(\int_a^T\int_c^S (g(t,s))^{-1}\,dt\,ds\right)A_1(gq,u_1,u_2,\sigma_1,r_{n,m},\sigma_2), \tag{39}$$

where

$$\begin{aligned}L_{16} \;=\; & M_1(r_{n,m})[(T-a)(S-c)]^{(\sigma_1-r_{n,m})/r_{n,m}}K_{i,j}^{r_{i,j}}K_{k,\ell}^{r_{k,\ell}} \\ & \times\ \text{ess.}\sup_{t\in[a,T],s\in[c,S]}\left[\,q^{\sigma_1/\sigma_2}P_{i,j}^{*\,r_{i,j}(r_{n,m}-1)/r_{n,m}}P_{k,\ell}^{*\,r_{k,\ell}(r_{n,m}-1)/r_{n,m}}\,\right],\end{aligned}$$

and

$$M_1(r_{n,m}) \;=\; \frac{r_{n,m}[r_{i,j}+(2^{\sigma_1/r_{n,m}}-1)r_{k,\ell}]}{\sigma_1\sigma_2},$$

$$P_{\kappa,\lambda}^*(t,s) \;=\; \int_a^t\int_c^s (h_{\kappa,\lambda}(t,s,\nu,\mu))^{r_{n,m}/(r_{n,m}-1)}(q(\nu,\mu))^{-(r_{n,m}/\sigma_2(r_{n,m}-1))}\,d\nu\,d\mu.$$

If $r_{n,m} = 1$ then $L_{16}$ renamed as $\bar{L}_{16}$ becomes

$$\begin{aligned}\bar{L}_{16} \;=\; & M_1(1)[(T-a)(S-c)]^{\sigma_1-1}K_{i,j}^{r_{i,j}}K_{k,\ell}^{r_{k,\ell}} \\ & \times\ \text{ess.}\sup_{t\in[a,T],s\in[c,S]}\left[\,q^{\sigma_1/(\sigma_1+1)}Q_{i,j}^{r_{i,j}}Q_{k,\ell}^{r_{k,\ell}}\,\right],\end{aligned}$$

where

$$Q_{\kappa,\lambda}(t,s) \;=\; \text{ess.}\sup_{\nu\in[a,t],\mu\in[c,s]}\left[\,h_{\kappa,\lambda}(t,s,\nu,\mu)(q(\nu,\mu))^{-1/(\sigma_1+1)}\,\right].$$

If in addition $q(t,s)$ is non-increasing on $R_1$, then the inequality (39) holds with $L_{16}$ ($\bar{L}_{16}$ when $r_{n,m} = 1$) replaced by $W_{16}$ ($\bar{W}_{16}$), where

$$W_{16} = M_1(r_{n,m})\left(K_{i,j}J_{i,j}^{(r_{n,m}-1)/r_{n,m}}(r_{n,m})\right)^{r_{i,j}}\left(K_{k,\ell}J_{k,\ell}^{(r_{n,m}-1)/r_{n,m}}(r_{n,m})\right)^{r_{k,\ell}}$$
$$\times (T-a)^{r_{i,j}(n-i)+r_{k,\ell}(n-k)-1}(S-c)^{r_{i,j}(m-j)+r_{k,\ell}(m-\ell)-1},$$

and

$$\bar{W}_{16} = M_1(1)K_{i,j}^{r_{i,j}}K_{k,\ell}^{r_{k,\ell}}(T-a)^{r_{i,j}(n-i)+r_{k,\ell}(n-k)-1}(S-c)^{r_{i,j}(m-j)+r_{k,\ell}(m-\ell)-1}. \tag{40}$$

Finally, we note that if instead of $q(t,s)$ being non-increasing, $0 < \gamma_1 \le q(t,s) \le \gamma_2$ on $R_1$, then the inequality (39) holds with $L_{16}$ ($\bar{L}_{16}$ when $r_{n,m} = 1$) replaced by $(\gamma_2/\gamma_1)^{\sigma_1/\sigma_2} W_{16}$ $\left((\gamma_2/\gamma_1)^{\sigma_1/(\sigma_1+1)}\bar{W}_{16}\right)$.

In particular, when $r_{i,j} = 0$, $r_{k,\ell} = r_{n,m} = n = m = 1$ and $q(t,s)$ is non-increasing on $R_1$, the inequality (39) in view of (40) reduces to

$$S_7(0,1,T,S) \le \frac{1}{2}\left(\int_a^T\int_c^S g^{-1}\,dt\,ds\right)A_4(gq,T,S,0,1). \tag{41}$$

**Remark 9.** For $r = \eta > 1$, $\zeta \ge 0$ the inequality (29) leads to

$$S_6 \le \frac{\eta}{2(\zeta+\eta)}L_0(p,q,\zeta,\zeta+\eta,\eta,\eta)\left[\,(z_1(T,S))^{2(\zeta+\eta)/\eta} + (z_2(T,S))^{2(\zeta+\eta)/\eta}\,\right].$$

Thus, on following as in Remark 8 we obtain the inequality

$$S_6 \le L_{17}\left(\int_a^T\int_c^S g^{-1}\,dt\,ds\right)A_1(gq,u_1,u_2,2\zeta+\eta,\eta,2(\xi+\eta)), \tag{42}$$

where $L_{17}$ is the same as $L_{16}$ ($W_{16}$ provided $q(t,s)$ is non-increasing on $R_1$) with $r_{i,j} = \zeta$, $r_{k,\ell} = \zeta+\eta$, $r_{n,m} = \eta$ and $M_1(r_{n,m}) = \eta/(2(\zeta+\eta))$. Further, when $\eta = 1$ in (42) the constant $L_{17}$ has to be replaced by $\bar{L}_{17}$, where $\bar{L}_{17}$ is the same as $\bar{L}_{16}$ ($\bar{W}_{16}$ provided $q(t,s)$ is non-increasing on $R_1$) with $r_{i,j} = \zeta$, $r_{k,\ell} = \zeta+1$ and $M_1(1) = 1/(2(\mu+1))$.

In particular, when $n = m = 1$, $\zeta \ge 0$, $\eta \ge 1$ and $q(t,s)$ is non-increasing on $R_1$, the inequality (42) gives the following generalization of (41)

$$S_7(\zeta,\eta,T,S)$$
$$\le \frac{\eta}{2(\zeta+\eta)}[(T-a)(S-c)]^{2\zeta+\eta-1}\left(\int_a^T\int_c^S g^{-1}\,dt\,ds\right)A_4(gq,T,S,\zeta,\eta). \tag{43}$$

This inequality has been proved in the year 1986 by Lin and Yang [14], and later in the year 1991 by Cheung [10]. It is clear that for $g(t,s) = 1$ the inequality (35) is sharper than (43). Finally, as in Remark 8 we note that if instead of $q(t,s)$ being non-increasing, $0 < \gamma_1 \le q(t,s) \le \gamma_2$ on $R_1$, then the inequality (43) holds with the right-hand side multiplied by $(\gamma_2/\gamma_1)^{(2\zeta+\eta)/2(\zeta+\eta)}$. This resulting inequality has appeared in [10].

**Remark 10.** Let the interval $[T,b]$ and the rectangle $[T,b] \times [c,S]$ be denoted as $I_2$ and $R_2$, respectively. Further, let $p(t,s)$ and $q(t,s)$ be non-negative and measurable on the rectangle $R_2$, and let $u(t,s) \in C^{(n-1,m-1)}(R_2)$ be such that $D_1^{\kappa}u(b,s) = 0,\ i \le \kappa \le n-1,\ s \in J_1,\ D_2^{\lambda}u(t,c) = 0,\ j \le \lambda \le m-1,\ t \in I_2$, and $D_1^n D_2^{m-1}u(t,s)$ and $D_1^{n-1}D_2^m u(t,s)$ are absolutely continuous on $R_2$. Then, for $u = u_1(t,s),\ u = u_2(t,s)$ our above discussion up to Remark 9 remains valid provided we replace $[a,T]$, $\int_a^t$, $\int_a^T$, $(t-\nu)$ and $T-a$ by $[T,b]$, $\int_t^b$, $\int_T^b$, $(\nu - t)$ and $b-T$. Further, in Remarks 6, 8 and 9 the non-increasing behavior of $q(t,s)$ in both the variables $t$ and $s$ has to be replaced by the non-decreasing with respect to $t$ and non-increasing with respect to $s$. In particular, if in the inequalities (33) and (34) we denote $W_8 = W_8(T-a, S-c)$ and $W_{11} = W_{11}(T-a, S-c)$, then with such a replacement these inequalities become

$$\int_T^b\int_c^S [\, U_{k,\ell}(t,s) + V_{k,\ell}(t,s)\,]\, dt\, ds \le W_8(b-T, S-c)\left[\int_T^b\int_c^S q\,[\,|D_1^n D_2^m u_1|^{\sigma_2} + |D_1^n D_2^m u_2|^{\sigma_2}\,]\, dt\, ds\right], \tag{44}$$

and

$$\int_T^b\int_c^S q|D_1^i D_2^j u_1 D_1^k D_2^\ell u_2|^\zeta\, [\,|D_1^k D_2^\ell u_2 D_1^n D_2^m u_1|^\eta + |D_1^k D_2^\ell u_1 D_1^n D_2^m u_2|^\eta\,]\, dt\, ds$$
$$\le W_{11}(b-T, S-c)\left[\int_T^b\int_c^S q\,[\,|D_1^n D_2^m u_1|^{2(\zeta+\eta)} + |D_1^n D_2^m u_2|^{2(\zeta+\eta)}\,]\, dt\, ds\right]. \tag{45}$$

**Remark 11.** Let the interval $[S,d]$ and the rectangle $[a,T] \times [S,d]$ be denoted as $J_2$ and $R_3$, respectively. Further, let $p(t,s)$ and $q(t,s)$ be non-negative and measurable on the rectangle $R_3$, and let $u(t,s) \in C^{(n-1,m-1)}(R_3)$ be such that $D_1^{\kappa}u(a,s) = 0,\ i \le \kappa \le n-1,\ s \in J_2,\ D_2^{\lambda}u(t,d) = 0,\ j \le \lambda \le m-1,\ t \in I_1$, and $D_1^n D_2^{m-1}u(t,s)$ and $D_1^{n-1}D_2^m u(t,s)$ are absolutely continuous on $R_3$. Then, for $u = u_1(t,s),\ u = u_2(t,s)$ our above discussion up to Remark 9 remains valid provided we replace $[c,S]$, $\int_c^s$, $\int_c^S$, $(s-\mu)$ and $S-c$ by $[S,d]$, $\int_s^d$, $\int_S^d$, $(\mu - s)$ and $d-S$. Further, in Remarks 6, 8 and 9 the non-increasing behavior of $q(t,s)$ in both the variables $t$ and $s$ has to be replaced by the non-increasing with respect to $t$ and non-decreasing with respect to $s$. With such a replacement, the inequalities (33) and (34) will be denoted as $(33)'$ and $(34)'$.

**Remark 12.** As in Remark 11 our discussion with appropriate changes remains valid in the rectangle $R_4 = [T, b] \times [S, d]$. In this case the inequalities (33) and (34) we shall denote as $(33)^*$ and $(34)^*$.

Inequalities (33), (44), $(33)'$ and $(33)^*$ can be added to obtain the following interesting result.

**Theorem 7.** *Assume that*

(i) $q(t,s)$ *is non-negative, monotone and measurable function on* $R = [a,b] \times [c,d]$,

(ii) $0 \le i \le k \le n-1$, $0 \le j \le \ell \le m-1$, *but fixed, and* $r_{i,j}$, $r_{k,\ell} \ge 0$, $r_{n,m} > 0$, $\sigma_2 = r_{i,j} + r_{k,\ell} + r_{n,m} \ge 1$ *are given numbers,*

(iii) *for* $\xi = 1, 2$ *functions* $u_\xi(t,s) \in C^{(n-1,m-1)}(R)$ *are such that* $D_1^\kappa u_\xi(a,s) = D_1^\kappa u_\xi(b,s) = 0$, $i \le \kappa \le n-1$, $s \in J = [c,d]$, $D_2^\lambda u_\xi(t,c) = D_2^\lambda u_\xi(t,d) = 0$, $j \le \lambda \le m-1$, $t \in I = [a,b]$, *and* $D_1^n D_2^{m-1} u_\xi(t,s)$, $D_1^{n-1} D_2^m u_\xi(t,s)$ *are absolutely continuous on* $R$, *also* $\int_a^b \int_c^d q(t,s) |D_1^n D_2^m u_\xi(t,s)|^{\sigma_2}\, dt\, ds < \infty$.

*Then, the following inequality holds*

$$\begin{aligned} &\int_a^b \int_c^d [\, U_{k,\ell}(t,s) + V_{k,\ell}(t,s)\, ]\, dt\, ds \\ &\le\ W_8(b-a, d-c) \left[ \int_a^b \int_c^d q\, [\, |D_1^n D_2^m u_1|^{\sigma_2} + |D_1^n D_2^m u_2|^{\sigma_2}\, ]\, dt\, ds \right]. \end{aligned} \tag{46}$$

*Further, if* $0 < \gamma_1 \le q(t,s) \le \gamma_2$ *on* $R$, *then in* (46) *the constant* $W_8(b-a, d-c)$ *can be replaced by* $(\gamma_2/\gamma_1)^{\sigma_1/\sigma_2}\, W_8\left(\frac{1}{2}(b-a), \frac{1}{2}(d-c)\right)$.

Similarly, on adding the inequalities (34), (45), $(34)'$ and $(34)^*$ we get the following result.

**Theorem 8.** *Let the conditions (i) and (iii) of Theorem* 7 *with* $r_{i,j} = \zeta \ge 0$, $r_{k,\ell} = \zeta + \eta$, $r_{n,m} = \eta > 0$ *be satisfied. Then, the following inequality holds*

$$\begin{aligned} &\int_a^b \int_c^d q |D_1^i D_2^j u_1 D_1^k D_2^\ell u_2|^\zeta \left[\, |D_1^k D_2^\ell u_2 D_1^n D_2^m u_1|^\eta + |D_1^k D_2^\ell u_1 D_1^n D_2^m u_2|^\eta \,\right] dt\, ds \\ &\le\ W_{11}(b-a, d-c) \left[ \int_a^b \int_c^d q \left[ |D_1^n D_2^m u_1|^{2(\zeta+\eta)} + |D_1^n D_2^m u_2|^{2(\zeta+\eta)} \right] dt\, ds \right]. \end{aligned} \tag{47}$$

*Further, if* $0 < \gamma_1 \le q(t,s) \le \gamma_2$ *on* $R$, *then in* (47) *the constant* $W_{11}(b-a, d-c)$ *can be replaced by* $(\gamma_2/\gamma_1)^{(2\zeta+\eta)/2(\zeta+\eta)}\, W_{11}\left(\frac{1}{2}(b-a), \frac{1}{2}(d-c)\right)$.

**Remark 13.** When $0 < \gamma_1 \le q(t,s) \le \gamma_2$ on $R$ and $n = m = 1$ the inequality (47) reduces to

$$S_7(\zeta, \eta, b, d) \;\le\; L_{18} A_4(q, b, d, \zeta, \eta), \tag{48}$$

where

$$\begin{aligned} L_{18} \;=\; & \frac{\eta}{2(\zeta+\eta)} \left(\frac{1}{2\eta(\zeta+\eta)}\right)^{(2\zeta+\eta)/2(\zeta+\eta)} \left(\frac{\gamma_2}{\gamma_1}\right)^{(2\zeta+\eta)/2(\zeta+\eta)} \\ & \times \left[\frac{(b-a)(d-c)}{4}\right]^{2\zeta+\eta}, \qquad \zeta \ge 0,\ \eta > 0,\ 2(\zeta+\eta) \ge 1. \end{aligned}$$

Inequality (48) with $L_{18} = \frac{\eta}{2(\zeta+\eta)} (\gamma_2/\gamma_1)^{(2\zeta+\eta)/2(\zeta+\eta)} [(b-a)(d-c)/4]^{2\zeta+\eta}$ was first proved by Pachpatte [16] for the case $\gamma_1 = \gamma_2 = 1,\ \zeta \ge 0,\ \eta = 1$, and then by Chen and Lin [7] and Cheung [8] for the case $\zeta \ge 0,\ \eta \ge 1$. Thus, our inequality (48) improves these results.

**Remark 14.** Results similar to Theorems 7 and 8 for the inequalities (39) and (42) can be stated rather easily. In particular, when $0 < \gamma_1 \le q(t,s) \le \gamma_2$ on $R$, from (43) the following inequality follows

$$S_7(\zeta, \eta, b, d) \;\le\; L_{19} A_4(gq, b, d, \zeta, \eta), \tag{49}$$

where

$$L_{19} = \frac{E\eta}{2(\zeta+\eta)} \left(\frac{\gamma_2}{\gamma_1}\right)^{(2\zeta+\eta)/2(\zeta+\eta)} \left[\frac{(b-a)(d-c)}{4}\right]^{2\zeta+\eta},$$
$$\zeta \ge 0,\ \eta > 0,\ 2(\zeta+\eta) \ge 1,$$

and

$$\begin{aligned} E \;=\; \max\Bigg\{ & \int_a^{(a+b)/2}\int_c^{(c+d)/2} (g(t,s))^{-1}\,dt\,ds,\ \int_a^{(a+b)/2}\int_{(c+d)/2}^{d} (g(t,s))^{-1}\,dt\,ds, \\ & \int_{(a+b)/2}^{b}\int_c^{(c+d)/2} (g(t,s))^{-1}\,dt\,ds,\ \int_{(a+b)/2}^{b}\int_{(c+d)/2}^{d} (g(t,s))^{-1}\,dt\,ds \Bigg\}. \end{aligned}$$

For $\zeta \ge 0,\ \eta \ge 1$ inequality (49) has been proved by Lin and Yang [14], and Cheung [10].

**Remark 15.** Extensions of the above inequalities involving three or more functions are immediate. For example, an addition of (48) and the inequalities similar to (48) for the pair of functions $u_2 = u_2(t,s)$, $u_3 = u_3(t,s)$ and $u_3 = u_3(t,s)$, $u_1 = u_1(t,s)$ leads to the inequality

$$\begin{aligned}\int_a^b \int_c^d q\,\{&|u_1u_2|^\zeta\,[\,|u_2D_1D_2u_1|^\eta + |u_1D_1D_2u_2|^\eta\,]\\ &+|u_2u_3|^\zeta\,[\,|u_3D_1D_2u_2|^\eta + |u_2D_1D_2u_3|^\eta\,]\\ &+|u_3u_1|^\zeta\,[\,|u_1D_1D_2u_3|^\eta + |u_3D_1D_2u_1|^\eta\,]\}\,dt\,ds\\ \le 2L_{18}\int_a^b\int_c^d q&\left[|D_1D_2u_1|^{2(\zeta+\eta)} + |D_1D_2u_2|^{2(\zeta+\eta)} + |D_1D_2u_3|^{2(\zeta+\eta)}\right]dt\,ds. \quad (50)\end{aligned}$$

This inequality for the case $\gamma_1 = \gamma_2 = 1$, $\zeta \ge 0$, $\eta = 1$, with the right-hand side multiplied by the constant $(2(\zeta+1))^{(2\zeta+1)/2(\zeta+1)}$ $(\ge 1)$ has been proved by Pachpatte [16].

## References

1. R.P. Agarwal, *An integro-differential inequality.* General Inequalities III, eds. E.F. Beckenbach and W. Walter, Birkhäuser, Basel, 1983, 501–503.

2. R.P. Agarwal, *Opial's and Wirtinger's type discrete inequalities in two independent variables.* Applicable Analysis **43** (1992), 47–62.

3. R.P. Agarwal and P.Y.H. Pang, *Opial-type inequalities involving higher order derivatives.* J. Math. Anal. Appl. **189** (1995), 85–103.

4. R.P.Agarwal and P.Y.H.Pang, *Remarks on the generalizations of Opial's inequality.* J. Math. Anal. Appl. **190** (1995), 559–577.

5. R.P. Agarwal and P.Y.H. Pang, *Sharp Opial-type inequalities in two variables.* Applicable Analysis (to appear).

6. R.P. Agarwal and Qin Sheng, *Sharp integral inequalities in n independent variables.* Nonlinear Analysis **26** (1996), 179–210.

7. M.S. Chen and C.T. Lin, *A note on Opial's inequality.* Chinese J. Math. **14** (1986), 53–58.

8. W.S. Cheung, *On Opial-type inequalities in two variables.* Aequationes Math. **38** (1989), 236–244.

9. W.S. Cheung, *Some generalized Opial-type inequalities.* J. Math. Anal. Appl. **162** (1991), 317–321.

10. W.S. Cheung, *Generalizations of Opial-type inequalities in two variables.* Tamkang J. Math. **22** (1991), 43–50.

11. W.S. Cheung, *Opial-type inequalities with $m$ functions in $n$ variables.* Mathematika **39** (1992), 319–326.

12. C.T. Lin, *A note on an integrodifferential inequality.* Tamkang J. Math. **23** (1985), 349–354.

13. C.T. Lin and G.S. Yang, *A generalized Opial's inequality in two variables.* Tamkang J. Math. **15** (1984), 115–122.

14. C.T. Lin and G.S. Yang, *On some new Opial-type inequalities in two variables.* Tamkang J. Math. **17** (1986), 31–36.

15. D.S. Mitrinović, J.E. Pecarić and A.M. Fink, *Inequalities Involving Functions and their Integrals and Derivatives.* Kluwer, Dordrecht, 1991.

16. B.G. Pachpatte, *On Opial type inequalities in two independent variables.* Proc. Royal Soc. Edinburgh **A100** (1985), 263–270.

17. B.G. Pachpatte, *On two inequalities similar to Opial's inequality in two independent variables.* Period. Math. Hung. **18** (1987), 137–141.

18. B.G. Pachpatte, *On Yang type integral inequalities.* Tamkang J. Math. **18** (1987), 89–96.

19. B.G. Pachpatte, *On multidimensional Opial-type inequalities.* J. Math. Anal. Appl. **126** (1987), 85–89.

20. B.G. Pachpatte, *On some new integral inequalities in two independent variables.* J. Math. Anal. Appl. **129** (1988), 375–382.

21. B.G. Pachpatte, *On a generalized Opial type inequality in two variables.* Anal. sti. Univ. "Al. I. Cuza" din Iasi **35** (1989), 231–235.

22. B.G. Pachpatte, *On certain two dimensional integral inequalities.* Chinese J. Math. **17** (1989), 273–279.

23. B.G. Pachpatte, *Opial type inequality in several variables.* Tamkang J. Math. **22** (1991), 7–11.

24. G.S. Yang, *Inequality of Opial-type in two variables.* Tamkang J. Math. **13** (1982), 255–259.

25. R.P. Agarwal and P.Y.H. Pang, *Opial Inequalities with Applications in Differential and Difference Equations.* Kluwer, Dordrecht, 1995.


Ravi P. Agarwal and Peter Y.H. Pang, Department of Mathematics, National University of Singapore, Kent Ridge, Singapore 0511.

International Series of Numerical Mathematics, Vol. 123, © 1997 Birkhäuser Verlag Basel

# The HELP type integral inequalities for $2n^{th}$ order differential operators

B.M. Brown and N.G.J. Dias

**Abstract**

In 1971 W.N. Everitt published his seminal work on the family of HELP inequalities. This paper, written in the twenty-fifth year of the HELP paper, is respectfully dedicated to him.

## 1. Introduction

Everitt's HELP inequality is an inequality between a function and it's first two derivatives. More specifically, suppose that there are real valued functions $p, q$ such that $p > 0, 1/p, q \in L^1_{loc}[0, \infty)$. The inequality

$$\left(\int_0^\infty p|y'|^2\,dx + q|y|^2\,dx\right) < K\int_0^\infty |y|^2\,dx \int_0^\infty |-(py')' + qy|^2\,dx \tag{1}$$

is said to be valid with finite best constant $K$ for all functions $y$ in $L^2[0, \infty)$ which are such that the right-hand side of (1) is defined and finite. Otherwise it is said to be invalid and we write $K = +\infty$. Although (1) is an extension of the amous inequality

$$\int_0^\infty |y'|^2\,dx < 4\int_0^\infty |y|^2\,dx \int_0^\infty |y''|^2\,dx$$

due to Hardy and Littlewood it must also be regarded as the pro-genitor of its type.

A feature of Everitt's treatment of (1) is the connection he establishes between the existence and where appropriate the value of the best constant $K$, and the spectral theory of the differential expression

$$-(py')' + qy. \tag{2}$$

In particular he showed that the existence of a finite $K$ and its value, is determined by the behaviour of the Titchmarsh-Weyl $m$ function. This function [4] carries with it the spectral properties of (2).

Although the treatment of the HELP inequality in [8] is within the context of the variational calculus an alternative way of looking at the inequality is provided by abstract function space theory. The inequality (1) may be viewed as an inequality between the $L^2[0,\infty)$ norms of the vector $y$, the self-adjoint operator generated by (2) in $L^2[0,\infty)$ and the natural Dirichlet integral associated with this differential expression. This is the stand-point taken by Evans and Zettl [7], Evans and Everitt [6] and by Bennewitz [1] who view (1) as an example of an inequality between pairs of self-adjoint operators. However Bennewitz's treatment of the inequality, whilst adopting a similar approach to that in [7], [6], reduces the problem of determining a valid inequality to one on members of the deficiency spaces of a certain self-adjoint operator. This does not lend itself easily to a constructive method for obtaining examples.

There have been several extensions of Everitt's work on (1). Russell in [12] and [13] showed that a similar inequality was possible for the fourth-order symmetric operator

$$(py'')'' + (qy')' + ry$$

and the criterion for its existence could be expressed in terms of the elements of the spectral $M$ matrix associated with the problem. This extended some work of Bradley and Everitt on an inequality based upon the fourth-order Fourier operator.

The purpose of this paper is to report on recent work on the $2n^{th}$ order HELP type inequality

$$\left(\int_a^{\rightarrow b}\left(\sum_{r=0}^{n} p_r|f^{(r)}|^2\right)\right)^2 \le K\int_a^b w|f|^2 \int_a^b w|w^{-1}M_{2n}[f]|^2, \tag{3}$$

which is required to hold for all $f \in \Delta$ which make the right-hand side of the integral inequality (3) defined and finite. The coefficient functions $p_r$ $(r = 0, 1, \ldots, n)$ and $w$, the $2n^{th}$ order quasi-differential expression $M_{2n}$ and the domain $\Delta$ are defined in Section 2. As in [6] the existence and value of the best constant will depend upon the behaviour of the Titchmarsh-Weyl spectral matrix associated with the operator.

Section 2 is devoted to introducing the background concepts that we shall need. Section 3 contains the main theorem which establishes the inequality, while Section 4 discuses the inequality when the $M$ matrix has prescribed behaviour at $\lambda = 0$. Section 5 introduces a numerical treatment of the inequality, Section 6 is devoted to an example.

## 2. Background results

We are concerned with an inequality which involves the symmetric $2n^{th}$ order differential expression

$$M_{2n}[f] = \sum_{r=0}^{n}(-1)^r(p_r f^{(r)})^{(r)} \tag{4}$$

defined on the interval $[a,b)$ where the coefficients $p_r$, $0 \leq r \leq n$ are real valued functions of $x \in [a,b)$ and satisfy

$$p_n > 0 \text{ on } [a,b) \text{ and } p_n^{-1}, p_{n-1}, \ldots, p_1, p_0 \in L_{loc}[a,b). \tag{5}$$

The expression (4) will be regarded as a quasi-differential expression defined on

$$D = \{f : [a,b) \to \mathbb{C} \mid f^{[j-1]} \in AC_{loc}[a,b), j = 1,2,\ldots,2n\} \tag{6}$$

where $f^{[j]}$ denotes the $j^{th}$ quasi-derivative of $f$ with respect to $x$. For a definition and discussion of quasi-derivatives see [11].

We shall use (4) to define operators in the weighted Hilbert space $L_w^2(a,b)$ having inner-product and norm

$$(f,g)_w = \int_a^b f\bar{g}w\,dx \text{ and } \|f\|_w = (f,f)_w^{\frac{1}{2}} \tag{7}$$

and the assumptions made on the coefficients ensure that $M$ is regular at $a$ and singular at $b$. Then as in [11] we may define maximal and minimal symmetric operators in $L_w^2(a,b)$ which we denote by $T$ and $T_0$ respectively. The domain of $T$ is denoted by $\Delta$ while that of $T_0$ is denoted by $\Delta_0$. It follows from [11] that the adjoint of $T_0$, $T_0^*$ satisfies $T_0^* = T$ and the von Neumann formula

$$\Delta = \Delta_0 \dot{+} N(\lambda) \dot{+} N(\bar{\lambda}) \tag{8}$$

holds. $N(\lambda)$ and $N(\bar{\lambda})$ are the deficiency spaces of $T_0$ which are also the $L_w^2(a,b)$ solution spaces of $w^{-1}M[f] = \lambda f$ and $w^{-1}M[f] = \bar{\lambda} f$ respectively. The dimensions of the deficiency spaces, the deficiency indices $n_{\pm}$, are independent of the strictly complex number $\lambda$, and satisfy

$$n \leq n_+ = n_- \leq 2n.$$

When $n_+ = n$ we say that the expression (4) is *limit-n* at $b$. This is the situation that we shall now assume prevails.

An important concept that we shall need is that of *strong limit-n*. We say that the symmetric linear ordinary quasi-differential expression $M$ is *strong limit-n* at $b$ if

$$\lim_{x \to b-} \left[ \sum_{r=1}^{n} f^{[r-1]}(x)\bar{g}^{[2n-r]}(x) \right] = 0,$$

for all $f, g \in \Delta$. Conditions under which (4) is *strong limit-n* at $b$ have been reported in a number of papers which include Everitt et al. (1973), Everitt et al. (1974a), Everitt (1975 and 1976), Hinton (1974), Kalf (1974), Race (1985 and 1990), Brown & Evans (1973), Everitt et al. (1974b). It follows immediately that *strong limit-n* implies *limit-n*, but not conversely. An integration by parts, together with the definitions of the *strong limit-n* shows that

$$\int_a^{\to b} \left( \sum_{r=1}^{n} p_r f^{(r)} \bar{g}^{(r)} \right) = - \left[ \sum_{r=1}^{n} f^{[2n-r]} \bar{g}^{[r-1]} \right] (a) + \int_a^b w(w^{-1} M[f]) \bar{g}, \qquad (9)$$

for all $f, g \in \Delta$, and hence the Dirichlet integral

$$\int_a^{\to b} \left( \sum_{r=1}^{n} p_r f^{(r)} \bar{g}^{(r)} \right) dx, \qquad (10)$$

is finite for all $f, g \in \Delta$. If for all $f \in \Delta$,

$$|p_r|^{\frac{1}{2}} f^{(r)} \in L^2(a,b), \quad (r = 0, 1, \ldots, n),$$

then we say that $M$ is *Dirichlet* at $b$, and in which case we may write the Dirichlet integral as

$$\int_a^b \left( \sum_{r=1}^{n} p_r f^{(r)} \bar{g}^{(r)} \right) dx.$$

In Everitt's treatment of (1) he makes the connection between the existence of the inequality, value of the best constant and the spectral behaviour of the operator as characterised by the Titchmarsh-Weyl $m$ function. A feature of the work that we now report on uses a similar characterisation, where now the $m$ function is replaces be a $n \times n$ matrix valued function. We review this generalisation in the yet wider context of linear Hamiltonian systems which include the $2n^{th}$ order symmetric expression (4). Since it is shown that the $2n^{th}$ order differential expression (4) may be written as a linear Hamiltonian system, this will cover the present discussion.

The spectral theory of linear Hamiltonian systems has been developed by Hinton and Shaw in a series of papers which include [9],[10]. Their work covers the system

$$Jf' = (\lambda A + B) f, \quad a \le x < b, \qquad (11)$$

where

(a)

$$J = \begin{pmatrix} 0 & -I_n \\ I_n & 0 \end{pmatrix},$$

(b) $A = (a_{ij})$ is the non-negative definite real symmetric matrix of order $2n \times 2n$ defined by

$$a_{ij} = \begin{cases} w(x), & i = j = 1; \\ 0, & \text{otherwise,} \end{cases}$$

and

(c) $B = (b_{ij})$ is the real symmetric matrix of order $2n \times 2n$ defined by

$$b_{ij} = \begin{cases} -p_{i-1}, & i = j \text{ and } i = 1, 2, \ldots, n \\ p_n^{-1}, & i = j = 2n \\ 1, & j = n + i - 1 \text{ and } i = 2, \ldots, n \\ 1, & i = n + j - 1 \text{ and } j = 2, \ldots, n \\ 0, & \text{otherwise} \end{cases}$$

which is the Hamiltonian formulation of the $2n^{th}$ order differential equation

$$My = \lambda wy. \tag{12}$$

Among other results they show that in the *limit-n* case, there is an $n \times n$ unique symmetric matrix which satisfies $M(\lambda)^* = M^T(\bar{\lambda})$, of rank $n$ which is a complex valued function defined on the whole of $\mathbb{C}$ and analytic away from the real line. Further if $N$ is the $2n \times 2n$ matrix

$$N = \begin{pmatrix} 0 & -\alpha^T \\ \alpha & 0 \end{pmatrix} \tag{13}$$

and $\alpha$ the $n \times n$ matrix with $i, j$ entry

$$\alpha_{ij} = \begin{cases} (-1)^{i+1}, & j = n + 1 - i \\ 0, & \text{otherwise,} \end{cases}$$

$\Psi(x, \lambda)$ is the $2n \times n$ matrix whose columns are the $n$ linearly independent solutions of (12) which belong to $L^2_w[0, \infty)$, then

$$\Psi(a, \lambda) = N \begin{pmatrix} I_n \\ M(\lambda) \end{pmatrix}. \tag{14}$$

## 3. The main theorems

In order to characterise the validity of the integral inequality (3), in terms of the Titchmarsh-Weyl matrix $M(\lambda)$, we need the following notation and definitions. Let:

(i) $\lambda = \rho e^{i\theta}$ ($\rho \in (0, \infty)$ and $\theta \in [0, 2\pi)$),

(ii) $L_\pm(\theta)$ : $(\theta \in (0, \frac{\pi}{2}])$ denotes the line segments defined by

$$L_+(\theta) := \{\rho e^{i\theta} |\ \rho \in (0, \infty)\}$$

and

$$L_-(\theta) := \{\ \rho e^{i(\theta+\pi)}\ |\ \rho \in (0, \infty)\},$$

(iii) the arguments $\theta_\pm \in [0, \frac{\pi}{2}]$ defined by

$$\theta_\pm := \inf \left\{ \theta \in \left(0, \frac{\pi}{2}\right] \mid \quad \text{for all} \ \ \phi \in \left[\theta, \frac{\pi}{2}\right] \right.$$
$$\left. \text{and} \ \ \mp \mathrm{Im}\,[\lambda^2 M_\pm(\lambda)] \geq 0 \quad (\lambda \in L_\pm(\phi)) \right\},$$

(iv) $\theta_0 := \max(\theta_+, \theta_-)$,

(v) the real number sets $E_\pm \subset (0, \infty)$ be defined by

$$E_\pm := \{\ \rho \in (0, \infty)\ |\ \lambda \in L_\pm(\theta_0) \ \text{ and } \ \mathrm{Im}\,[\lambda^2 M_\pm(\lambda)] = 0\ \},$$

(vi) $Y_{r\pm}$ : $[a, b) \times (0, \infty) \to \mathbb{R}$ be defined by

$$Y_{r\pm}(x, \rho) = \mathrm{Im}\,[\lambda \psi_{r\pm}(x, \lambda)],$$

where $x \in [a, b)$ and $\lambda \in L_\pm(\theta_0)$.

With this notation we present a theorem which establishes the existence of the inequality (3), and where appropriate classify the cases of equality and also determine the value of the best constant.

**Theorem 3.1** *Let the quasi-differential expression* $M_{2n}$ *be* strong limit-$n$ *at* $b$. *Then the following results hold:*

(a) $\theta_0 \in (0, \frac{\pi}{2}]$, *i. e.* $\theta_0 \neq 0$ *and at least one of* $\theta_+ > 0, \theta_- > 0$ *must hold;*

(b) *the inequality* (3), *i. e.*

$$\left( \int_a^{\to b} \left( \sum_{r=1}^{n} p_r |f^{(r)}|^2 \right) \right)^2 \leq K \int_a^b w|f|^2 \int_a^b w|w^{-1} M_{2n}[f]|^2 \ (f \in \Delta)$$

*where* $\Delta \subset L^2_w(a, b)$ *is the linear manifold defined by* (3), *is valid if and only if* $\theta_0 \in (0, \frac{\pi}{2})$, *i. e.*

$$\theta_0 \neq \frac{\pi}{2}; \tag{15}$$

(c) *if* (15) *is satisfied then the best possible constant* $K$ *in* (3) *is given by*

$$K = (\cos\theta_0)^{-2}; \tag{16}$$

(d) *if* (15) *is satisfied and* $K$ *is given by* (16), *then all elements* $f \in \Delta$ *that give equality in* (3) *are determined by the following three mutually exclusive conditions:*

(i) $f = 0$,

(ii) *there exists* $f \in \Delta$ *with* $f \neq 0, M[f] = 0$ *and* $\left[\sum_{r=1}^{n} f^{[2n-r]} f^{[r-1]}\right](a) = 0$, *in which case both sides of* (3) *are equal to zero,*

(iii) $E_+ \cup E_- \neq \phi$ *and the element* $f \in \Delta$ *defined by* $f(x) = \sum_{r=1}^{n} A_r Y_{r\pm}(x,\rho)$, $(x \in [a,b),\ \lambda \in L_\pm(\theta_0))$ *with* $A_r \in \mathbb{R}\setminus\{0\}$ $(r = 1,\ldots,n)$ *in which case both sides of* (3) *are equal but not zero.*

The proof of this theorem is to be found in [5].

As in the case of the classical HELP inequality, we may say a little more on the validity of (3) provided we impose stronger conditions on the coefficients of the differential expression. In particular we show that when $w = 1$, the value $\theta_0 = \frac{\pi}{2}$ can be achieved only as $|\lambda| \to 0$. In Theorem 3.2 we first show that as $|\lambda| \to \infty$, $\theta_0 \neq \frac{\pi}{2}$, while in Theorem 3.3 we show that $\theta_0 \neq \frac{\pi}{2}$ for any $|\lambda| > r_0 > 0$.

**Theorem 3.2** *Let the conditions of Theorem* 3.1 *hold, and in addition let* $w = 1$. *Then there exists an* $R > 0$ *such that for all* $r > R$ *and for some* $0 < \theta < \frac{\pi}{2}$

$$\mp \mathrm{Im}\, \lambda^2 M(\lambda) > 0$$

*Proof.* In view of the technical details we give the proof for the $4^{th}$ order $(n = 2)$ inequality. A proof on similar lines may be given for other values of $n$.

It is known [3] that under the conditions of the theorem

$$M(\lambda) = M_0(\lambda)[I_n + o(1)] \tag{17}$$

as $|\lambda| = \rho \to \infty$, where $o(1)$ denotes a matrix term which tends to 0 as $\rho \to \infty$, and

$$M_0(\lambda) = \begin{pmatrix} (1+i)\lambda^{\frac{-1}{4}} & i\lambda^{\frac{-1}{2}} \\ i\lambda^{\frac{-1}{2}} & (i-1)\lambda^{\frac{-3}{4}} \end{pmatrix}.$$

Thus with $\lambda = \rho\exp(i\theta) \in L_+$ and

$$A(\rho,\theta) := \begin{pmatrix} -\sqrt{2}\rho^{\frac{7}{4}} \sin\left(\dfrac{7\theta+\pi}{4}\right) & -\rho^{\frac{3}{2}} \cos\left(\dfrac{3\theta}{2}\right) \\ -\rho^{\frac{3}{2}} \cos\left(\dfrac{3\theta}{2}\right) & \sqrt{2}\rho^{\frac{5}{4}} \sin\left(\dfrac{5\theta-\pi}{4}\right) \end{pmatrix}$$

$$-\,\mathrm{Im}\,[\lambda^2 M_+(\lambda)] = A(\rho,\theta)[I_n + o(1)] \tag{18}$$

and for all $\theta \in [3\pi/7, \pi/2]$, both $-\sin(7\theta+\pi/4)$ and $\sin(7\theta+\pi/4)$ are non-negative. Further $\det A(\rho, 3\pi/7) < 0$ and $\det A(\rho, \pi/2) > 0$. Thus for all $\rho > R$ for some $R > 0$

$$\mp \mathrm{Im}\, \lambda^2 M(\lambda) > 0.$$

A similar proof holds for $\lambda \in L_-(\theta)$ giving a lower bound for $\theta_-$ which is less than the lower bound for $\theta_+$.

We remark that this theorem provides a lower bound for the best constant $K$ in the inequality. The numerical value of this lower bound for $\theta_0$ is 1.457921 which gives a lower bound on the value of the constant $K$ of 78.821.

We now show that if the conditions of Theorem 3.2 hold then the validity of the inequality is dependant only upon the behaviour of $M(\lambda)$ in an interval of the origin of the spectral plane $\mathbb{C}$.

**Theorem 3.3** *Let the conditions of the Theorem* 3.2 *hold and* $M(\lambda)$ *be the Titchmarsh-Weyl* $M$*-matrix associated with the differential equation* $M_4[f] = \lambda w f$ *on the interval* $[a, b)$. *If there exists* $\delta_\pm > 0$ *and* $\theta_{S\pm} \in (0, \frac{\pi}{2})$ *such that*

$$\mp \mathrm{Im}\, [\lambda^2 M_\pm(\lambda)] \geq 0 \ \textit{ for all } \ \rho \in (0, \delta_\pm] \ \textit{ and } \ \theta \in [\theta_{S\pm}, \frac{\pi}{2}],$$

*then there is a valid integral inequality.*

*Proof.* Let $\lambda \in L_\pm(\theta)$. Since the conditions of the Theorem 3.2 hold, then there exists $R > 0$ and $\theta_{L\pm} \in (0, \frac{\pi}{2})$ such that

$$\mp \mathrm{Im}\, [\lambda^2 M_\pm(\lambda)] \geq 0 \ \text{ for all } \ \rho \geq R \ \text{ and } \ \theta \in \left[\theta_{L\pm}, \frac{\pi}{2}\right]. \tag{19}$$

By the hypothesis there exists $\delta_\pm > 0$ and $\theta_{S\pm} \in (0, \frac{\pi}{2})$ such that

$$\mp \mathrm{Im}\, [\lambda^2 M_\pm(\lambda)] \geq 0 \ \text{ for all } \ \rho \in (0, \delta_\pm] \ \text{ and } \ \theta \in \left[\theta_{S\pm}, \frac{\pi}{2}\right]. \tag{20}$$

Since $\mp\mathrm{Im}\, [\lambda^2 M_\pm(\lambda)] > 0$ for $\lambda \in L_\pm(\frac{\pi}{2})$ and $\mp\mathrm{Im}\, [\lambda^2 M_\pm(\lambda)]$ is continuous in $\theta$, on the compact interval $[\delta_\pm, R]$, by continuity arguments, we can choose $\theta_{M\pm} \in (0, \frac{\pi}{2})$ such that

$$\mp \mathrm{Im}\, [\lambda^2 M_\pm(\lambda)] \geq 0 \ \text{ for all } \ \rho \in [\delta_\pm, R] \ \text{ and } \ \theta \in \left[\theta_{M\pm}, \frac{\pi}{2}\right]. \tag{21}$$

If now we choose $\theta_{C\pm} = \max(\theta_{L\pm}, \theta_{S\pm}, \theta_{M\pm})$, then $\theta_{C\pm} \in (0, \frac{\pi}{2})$ and from (19), (20) and (21) we get

$$\mp \mathrm{Im}\, [\lambda^2 M_\pm(\lambda)] \geq 0 \ \text{ for all } \ \rho > 0 \ \text{ and } \ \theta \in \left[\theta_{C\pm}, \frac{\pi}{2}\right]. \tag{22}$$

We take $\theta_C = \max(\theta_{C+}, \theta_{C-})$ and the result follows.

In the next section, we investigate the properties of the $M(\lambda)$ matrix about the origin, which lead to the determination of a valid integral inequality.

## 4. Behaviour of the $M(\lambda)$ matrix about the origin

We first define the concepts that we shall need. We say that the $n \times n$ matrix $M(\lambda)$ has a zero at $\lambda = \lambda_0$ if all its entries have a zero at $\lambda = \lambda_0$; while we say that it has a simple pole at $\lambda = \lambda_0$ if at least one of its entries has a simple pole at $\lambda = \lambda_0$. In this later case it has the Laurent series expansion

$$M(\lambda) = \sigma_{-1}(\lambda - \lambda_0)^{-1} + \sigma_0 + \sigma_1(\lambda - \lambda_0) + \sigma_2(\lambda - \lambda_0)^2 + \dots,$$

in a neighbourhood of $\lambda_0$, where $\sigma_k$'s are $n \times n$ complex constant matrices.

It is shown [9], that the diagonal entries of $M(\lambda)$ belong to the Pick-Nevanlinna class; i.e.,

$$\text{Im}\,[\lambda].\text{Im}\,[m_{rr}(\lambda)] > 0 \;\; r = 1, \dots, n.$$

Thus the diagonal entries can have at most simple poles at any isolated singularity, and this in turn implies that all the off-diagonal entries of the matrix have at most simple poles [9],[10]. Thus, all the entries of the matrix $M(\lambda)$ have, at most simple poles, at any isolated singularity. Since $m_{rr}(\lambda, \alpha)$ $(r = 1, \dots, n)$ is a Nevanlinna function $-\dfrac{1}{m_{rr}(\lambda, \alpha)}$ $(r = 1, \dots, n)$ is also a Nevanlinna function and therefore has at most simple poles at any isolated singularity. From this we may deduce that $M(\lambda)$ has at most simple zeros.

We discuss the validity of the inequality when:

1. $M$ is regular at 0 and $M(0) \neq 0$;
2. $M(\lambda)$ has a simple pole at the origin.

**Theorem 4.1** *Let the conditions of Theorem* 3.1 *hold. Suppose that the matrix* $M(\lambda)$ *is regular at* 0 *with* $M(0) \neq 0$. *If* rank $M(0) = n$ *then the inequality* (3) *is not valid.*

*Proof.* Since $M(\lambda)$ is regular and non-zero at 0

$$M(\lambda) = M(0) + \sigma_1\lambda + \sigma_2\lambda^2 + \dots, \tag{23}$$

where $\sigma_1, \sigma_2, \dots$ are $n \times n$ constant complex matrices. It is shown in [9] that $M^*(0) = M(0)$, and so $M(0)$ is a non-zero $n \times n$ real-symmetric matrix. Since $M(0)$ also has rank $n$ then either $M(0) > 0$ or $M(0) < 0$ are possible.

We consider the case $M(0) > 0$. Let $\lambda = \rho e^{i\theta} \in L_+(\theta)$. Then from (23) we get

$$\lambda^2 M(\lambda) = M(0)\rho^2 e^{2i\theta} + \sigma_1 \rho^3 e^{3i\theta} + \sigma_2 \rho^4 e^{4i\theta} + \dots . \tag{24}$$

and so

$$-\operatorname{Im}[\lambda^2 M(\lambda)] = \rho^2[-M(0)\sin 2\theta + O(\rho)] \tag{25}$$

as $\rho \to 0$. Writing $A(\theta) = -M(0)\sin 2\theta$, then $A(\theta) < 0$ for all $\theta \in (0, \frac{\pi}{2})$, we can choose a non-zero vector $\underline{x} \in \mathbb{R}^n$ such that $\underline{x}^T A(\theta)\underline{x} < 0$ for all $\theta \in (0, \frac{\pi}{2})$.

Since $A(\theta)$ is independent of $\rho$, we can choose $\delta > 0$ such that

$$\left|\underline{x}^T O(\rho)\underline{x}\right| < -\frac{1}{2}\underline{x}^T A(\theta)\underline{x},$$

for all $\rho \in (0, \delta)$ and $\theta \in (0, \frac{\pi}{2})$. Hence from (25) we have

$$\underline{x}^T(-\operatorname{Im}[\lambda^2 M(\lambda)])\underline{x} < \rho^2\left[\underline{x}^T A(\theta)\underline{x} - \frac{1}{2}\underline{x}^T A(\theta)\underline{x}\right] = \frac{1}{2}\underline{x}^T A(\theta)\underline{x} < 0,$$

for all $\rho \in (0, \delta)$ and $\theta \in (0, \frac{\pi}{2})$. So, there exists $\delta > 0$ such that $-\operatorname{Im}[\lambda^2 M(\lambda)] < 0$ for all $\rho \in (0, \delta)$ and $\theta \in (0, \frac{\pi}{2})$, i.e., this implies that there is no $\delta_+ > 0$ and $\theta_{S+} \in (0, \frac{\pi}{2})$ such that $-\operatorname{Im}[\lambda^2 M(\lambda)] \geq 0$ for all $\rho \in (0, \delta_+)$ and $\theta \in [\theta_{S+}, \frac{\pi}{2}]$. A similar argument holds good if $M(0) < 0$ and also if $\lambda \in L_-(\theta)$. Hence, in either case there are no valid integral inequalities and $K = +\infty$. This completes the proof of the theorem.

We remark that if rank $(M(0)) < n$ then the integral inequality (3) may or may not be valid.

**Theorem 4.2** *Let the conditions of the Theorem* 3.1 *hold. Suppose that* $M(\lambda)$ *has a simple pole at the origin and has the Laurent series expansion*

$$M(\lambda) = \sigma_{-1}\lambda^{-1} + \sigma_0 + \sigma_1\lambda + \sigma_2\lambda^2 + \dots, \tag{26}$$

*in a neighbourhood of the origin, where* $\sigma_k$*'s are* $n \times n$ *complex constant. Then provided* rank $(\sigma_{-1}) = n$, *the integral inequality* (3) *is valid.*

*Proof.* It is shown in [9] that $\sigma_k^* = \sigma_k$ for $k = -1, 0, 1, \dots$. If we choose $\lambda = i\nu$, then from (26) we get

$$M(\lambda) = \sigma_{-1}\frac{1}{i\nu} + \sigma_0 + \sigma_1(i\nu) + \sigma_2(i\nu)^2 + \dots . \tag{27}$$

Then from (27) we have

$$M(\lambda) - M^*(\lambda) = 2\sigma_{-1}\frac{1}{i\nu} + 2\sigma_1(i\nu) + \dots, \tag{28}$$

then

$$\nu \operatorname{Im}[M(\lambda)] = -\sigma_{-1} + O(\nu^2) \tag{29}$$

as $\nu \to 0$. From Theorem 3.3, we have that for all $\lambda \in \mathbb{C}_+ \cup \mathbb{C}_-$,

$$\nu \operatorname{Im}[M(\lambda)] > 0. \tag{30}$$

Hence from (29) and (30), we get $\sigma_{-1} \leq 0$. Since rank $(\sigma_{-1}) = n$ and $\sigma_{-1} \leq 0$, we get $\sigma_{-1} < 0$. From (26) we get

$$\lambda^2 M(\lambda) = \sigma_{-1}\lambda + \sigma_0\lambda^2 + \sigma_1\lambda^3 + \ldots. \tag{31}$$

Let $\lambda = \rho e^{i\theta} \in L_+(\theta)$. Then from (31) we have

$$\lambda^2 M(\lambda) = \sigma_{-1}\rho e^{i\theta} + \sigma_0\rho^2 e^{2i\theta} + \sigma_1\rho^3 e^{3i\theta} + \ldots. \tag{32}$$

which gives

$$-\operatorname{Im}[\lambda^2 M(\lambda)] = \rho[-\sigma_{-1}\sin\theta + O(\rho)] \tag{33}$$

as $\rho \to 0$. The rest of the proof follows in a manner similar to that of Theorem 3.2. Hence, the integral inequality is valid with best constant $K \in (0, \infty)$.

We note that if rank $(\sigma_{-1}) < n$ then the integral inequality (3) may or may not be valid. This is considered further in the example in Section 6.

## 5. Numerical method

In view of the sparsity of examples of $M$ matrices that are known in closed form, and the considerable analytic difficulties in resolving the validity of the inequality, we have developed a numerical approach to gleam information about the validity of the inequality and value of the best constant. We now summarise this.

We need a definition of $\theta_\pm$ which is more amendable to computation than that given in Section 3. The following Lemma 5.1 is the matrix analogue of the Corollary 7.1 in Evans & Everitt (1982, Section 7), and it leads to an alternative definition for $\theta_\pm$, which is suitable for developing a numerical algorithm to estimate $\theta_0$.

**Lemma 5.1** *Suppose that for some $\theta_1 \in (0, \frac{\pi}{2})$, $\mp\operatorname{Im}[\lambda^2 M_\pm(\lambda)] \geq 0$ $(\lambda \in L_\pm(\theta_1))$. Then for any $\theta \in (\theta_1, \frac{\pi}{2}]$, $\mp\operatorname{Im}[\lambda^2 M_\pm(\lambda)] > 0$, $(\lambda \in L_\pm(\theta))$.*

*Proof.* This lemma follows from Theorem 5.1.

From the above Lemma 5.1, it follows that,

$$\theta_\pm = \sup\left\{\theta \in \left(0, \frac{\pi}{2}\right] \mid \operatorname{Im}[\lambda^2 M(\lambda)] = 0 \ \text{ for some } \ \lambda \in \ L_\pm(\theta)\right\} \tag{34}$$

which is the basis of our numerical algorithm to estimate $\theta_\pm$ and hence $\theta_0$. In order to estimate $\theta_\pm$, we examine the set $\{\rho,\theta\}$ for which $\mp\text{Im}\,[\lambda^2 M(\lambda)] = 0$, when $\lambda \in L_\pm(\theta)$. The value of $\theta_\pm$ will then be the maximum $\theta$ for which $\text{Im}[\lambda^2 M(\lambda)] = 0$ for all $\rho > 0$ when $\lambda \in L_\pm(\theta)$.

Note that from Lemma 5.1 above, for fixed $\rho$, if Im $[\lambda^2 M(\lambda)] = 0$ for some $\theta \in (0, \frac{\pi}{2})$, then the real symmetric matrix $-\text{Im}\,[\lambda^2 M(\lambda)]$ changes sign as $\arg(\lambda)$ increases through this value of $\theta$. Now, if we choose the largest $\theta \in (0, \frac{\pi}{2})$ such that $\text{Im}\,[\lambda^2 M(\lambda)] = 0$, then the real symmetric matrix $\pm\text{Im}\,[\lambda^2 M(\lambda)]$ changes sign from negative to positive as $\arg(\lambda)$ increases through this largest value of $\theta$. For fixed $\rho$, as $\arg(\lambda)$ increases through this largest value of $\theta$, the sign change of the real symmetric matrix $\pm\text{Im}\,[\lambda^2 M(\lambda)]$ is equivalent to the change of sign of the least eigenvalue $\chi_\pm(\rho,\theta)$ of the same matrix. Hence, our method of estimation of $\theta_\pm$ consists of investigating the value of $\theta$ such that the least eigenvalue $\chi_\pm(\rho,\theta) = 0$, for each $\rho$ in a finite set of different values of $\rho \in (0, \rho_{max})$, where $\rho_{max}$ is a suitably chosen positive real number which depends on the particular problem. For each $\rho$, a bisection search is performed to find $\theta \in (0, \frac{\pi}{2})$ until the least eigenvalue $\chi_\pm(\rho,\theta) = 0$. At each iteration of the bisection search, the real symmetric matrix $\mp\text{Im}\,[\lambda^2 M(\lambda)]$ and the least eigenvalue $\chi_\pm(\rho,\theta)$ can be estimated for fixed $\rho$ and $\theta$. Thus, the three main parts in our numerical method are:

1. Calculation of the real symmetric matrix Im $[\lambda^2 M(\lambda)]$, for fixed $\rho$ and $\theta$. In the example that we discuss in Section 7 this is achieved by first obtaining the closed form expression for $M$ in terms of the gamma function.
2. Estimation of the value of the least eigenvalue $\chi_\pm(\rho,\theta)$ of the real symmetric matrix $\mp\text{Im}\,[\lambda^2 M(\lambda)]$. This is performed with a simple code that uses the explicit formulae for the eigenfunction.
3. Estimation of the value of $\theta \in (0, \frac{\pi}{2})$ such that $\chi_\pm(\rho,\theta) = 0$, for fixed $\rho$. A simple bisection algorithm has been written to perform this task.

## 6. Example

In this section we discuss an example of the inequality generated by the $4^{th}$ order differential expression

$$M_{4,\tau}[f] = (M_4 - \tau)[f] = f^{(4)} - (2x^2 f^{(1)})^{(1)} + (x^4 - 2 - \tau)f. \tag{35}$$

The candidate inequality is

$$\left(\int_0^\infty (|f^{(2)}|^2 + 2x^2|f^{(1)}|^2 + (x^4 - 2 - \tau)|f|^2)\right)^2 \le K_\tau \int_0^\infty |f|^2 \int_0^\infty |M_{4,\tau}[f]|^2 \tag{36}$$

where the expression (35) is generated as the formal square of the second order expression

$$-y'' + x^2 y. \tag{37}$$

It follows from [11] that the spectrum of the operator (35) with Neumann initial conditions consists only of isolated eigenvalues and that the $2 \times 2$ $M$ matrix associated with (35) is meromorphic with poles at the eigenvalues of (35).

Following the construction given in [5], we have

$$\begin{aligned} m_{11,\tau}(\lambda) &= \frac{2}{\sqrt{\lambda+\tau}[m(-\sqrt{\lambda+\tau}, \frac{\pi}{2}) - m(\sqrt{\lambda+\tau}, \frac{\pi}{2})]}, \\ m_{12,\tau}(\lambda) &= m_{21,\tau}(\lambda) \\ &= \frac{m(-\sqrt{\lambda+\tau}, \frac{\pi}{2}) + m(\sqrt{\lambda+\tau}, \frac{\pi}{2})}{\sqrt{\lambda+\tau}[m(-\sqrt{\lambda+\tau}, \frac{\pi}{2}) - m(\sqrt{\lambda+\tau}, \frac{\pi}{2})]}, \\ m_{22,\tau}(\lambda) &= \frac{2m(\sqrt{\lambda+\tau}, \frac{\pi}{2})m(-\sqrt{\lambda+\tau}, \frac{\pi}{2})}{\sqrt{\lambda+\tau}[m(-\sqrt{\lambda+\tau}, \frac{\pi}{2}) - m(\sqrt{\lambda+\tau}, \frac{\pi}{2})]}, \end{aligned} \tag{38}$$

where

$$m = \frac{\Gamma(1/4 - \lambda/4)}{2\Gamma(3/4 - \lambda/4)} \tag{39}$$

is the Titchmarsh-Weyl Neumann $m$ function associated with 37), (see [2] for details).

Thus it follows that $M$ has simple poles at $\tau_n = (4n)^2, 0 \leq n < \infty$ with residues matrix

$$\sigma_{-1,0} = \begin{pmatrix} -\frac{8}{\pi}\frac{\Gamma(\frac{3}{4})}{\Gamma(\frac{1}{4})} & -\frac{4}{\pi} \\ -\frac{4}{\pi} & -\frac{2}{\pi}\frac{\Gamma(\frac{1}{4})}{\Gamma(\frac{3}{4})} \end{pmatrix}, \quad \text{and} \quad \sigma_{-1,n} = \begin{pmatrix} -\frac{16}{\pi}\frac{\Gamma(n+\frac{3}{4})}{\Gamma(n+\frac{1}{4})} & -\frac{8}{\pi} \\ -\frac{8}{\pi} & -\frac{4}{\pi}\frac{\Gamma(n+\frac{1}{4})}{\Gamma(n+\frac{3}{4})} \end{pmatrix}$$

when $n \in \mathbb{N}$. A simple calculation shows that $M(\tau) \neq 0$ for $\tau \neq (4n)^2$, $0 \leq n < \infty$ and that the residue matrix is of rank 1 for all $\tau_n$. Thus in view of Theorem 4.1 it follows that unless $\tau = \tau_n$ the inequality will not be valid. However in contrast to the analogous situation for the classical HELP inequality when $\tau$ is an eigenvalue of the Dirichlet or Neumann problems, we are not able to deduce the existence of a valid inequality. We must therefore examine

$$\mp \operatorname{Im} \lambda^2 M(\lambda) \tag{40}$$

in both the first and third quadrants of $\mathbb{C}$.

In view of the analytic difficulties involved in this procedure we have resorted to the numerical means that we outlined in Section 6, to gleam information on the likely existence of a valid inequality.

Our computation indicates that there will not be an inequality when $\tau = 0$ but there are valid inequalities for all other values on $\tau_n$ with best constant $K_{\tau_n}$ satisfying

$$78.821 < K_{\tau_n} < 78.823 \quad n = 1,2,3,4$$
$$78.823 < K_{\tau_n} < \infty \quad n > 5$$

with $K_{\tau_n}$, $n > 5$ a monotone function of $n$.

## References


[1] C. Bennewitz, *A general version of the Hardy-Littewood-Polya-Everitt (HELP) inequality.* Proc. Roy. Soc. Edin. A **97** (1984), 9–20.

[2] C. Bennewitz and W.N. Everitt, *Some remarks on the Titchmarsh-Weyl m-coefficient.* In: Proceedings of the Pleijel conference, Uppsala, 1980.

[3] C. Bennewitz and A.D. Wood, *Spectral asymptotics for higher order ordinary differential equations.* Proc. Lond. Math. Soc. **23** (1994), 43–55.

[4] J. Chaudhuri and W.N. Everitt, *On the spectrum of ordinary second order differential operators.* Proc. Roy. Soc. Edin. A **68** (1968), 95–119.

[5] N.G.J. Dias, *On an integral inequality associated with a $2n^{th}$ order quasi-differential expression.* PhD thesis, Cardiff: University of Wales, 1994.

[6] W.D. Evans and W.N. Everitt, *A return to the Hardy-Littlewood inequality.* Proc. Roy. Soc. Lond. A **380** (1982), 447–486.

[7] W.D. Evans and A. Zettl, *Norm inequalities involving derivatives.* Proc. Roy. Soc. Lond. A **82** (1978), 51–70.

[8] W.N. Everitt, *On an extension to an integro-differential inequality of Hardy, Littlewood and Pólya.* Proc. Roy. Soc. Edin., A **69** (1989), 295–333.

[9] D.B. Hinton and J.K. Shaw, *On Titchmarsh-Weyl $m(\lambda)$-functions for Hamiltonian systems.* J. Diff. Equations **40** (1981), 316–342.

[10] D.B. Hinton and J.K. Shaw, *On the spectrum of a singular Hamiltonian system.* Quaestiones Mathematicae **5** (1982), 29–81.

[11] M.A. Naimark, *Linear differential operators II.* Ungar, New York, 1968.

[12] A. Russell, *Integral inequalities and spectral theory of certain ordinary differential operators.* PhD thesis, University of Dundee, Scotland, U.K., 1975.

[13] A. Russell, *On a fourth-order singular integral inequality.* Proc. Roy. Soc. Edin. A **80** (1978), 249–260.



B.M. Brown, Department of Computer Science, Cardiff, University of Wales, Cardiff, CF2 3XF, U.K.

N.G.J. Dias, Department of Mathematics, University of Kelaniya, Kelaniya, Sri Lanka

International Series of Numerical Mathematics, Vol. 123, © 1997 Birkhäuser Verlag Basel

# An estimate related to the Gagliardo-Nirenberg inequality

Moshe Marcus

## 1. Introduction

Let $\Omega$ be a domain of unit volume in $\mathbb{R}^n$ satisfying the cone condition. If $u \in L_\infty(\Omega)$ and $\|u\|_{L_\infty} \neq 0$, denote

$$O(u) := \|u\|_{L_\infty} / \|u\|_{L_1}. \tag{1.1}$$

If $u \geq 0$, the quantity $O(u)$ is a measure of the 'amplitude' of $u$, i.e. of its deviation from a constant.

For $u \in W_{m,q}(\Omega) \cap L_p(\Omega)$ let $I_{m,q;p}(u)$ denote the energy of $u$ defined as follows,

$$I_{m,q;p}(u) := \int_\Omega \Big( \sum_{|\alpha|=m} |D^\alpha u|^q + |u|^p \Big)\, dx. \tag{1.2}$$

We shall assume that

$$1 \leq q < p \quad \text{and} \quad mq > n. \tag{1.3}$$

Thus, $W_{m,q}(\Omega) \hookrightarrow C_b(\Omega)$ where $C_b$ is the space of continuous bounded functions. If $u \in W_{m,q}(\Omega)$,

$$\overline{u} := \int_\Omega |u| \leq (I_{m,q;p}(u))^{1/p}. \tag{1.4}$$

Accordingly the minimum energy in the class of functions

$$\mathcal{U}_a := \{u \in W_{m,q}(\Omega) : \overline{u} = a\}$$

is attained by the constant function $u$ with value $a$.

In this note we shall derive an estimate of the energy depending on the amplitude. More precisely, for $u \in W_{m,q}(\Omega)$ we shall obtain a sharp estimate for the rate of growth of $I_{m,q;p}(u) - \overline{u}^p$ in terms of $\|u\|_{L_\infty}$, which is valid for functions whose amplitude is 'sufficiently large'. An estimate of this type was used in [1] (in the one dimensional case, with $m = q = 2$) in order to obtain a-priori bounds for the minimizers of a variational problem. It will also be shown that this estimate implies a special case of the Gagliardo-Nirenberg inequality and that the two inequalities are not equivalent.

**Acknowledgment.** I wish to thank professors H. Brezis and V. Maz'ja for helpful discussions concerning the relation between the energy estimate and the Gagliardo-Nirenberg inequality.

## 2. Energy estimate

In this section we establish the following estimate.

**Theorem 2.1** *Let $\Omega$ be a domain of unit volume satisfying the cone condition. Assume (1.3) and denote $\beta = (p-q)/mq$. Then there exist positive constants $c, k$ such that,*

$$I_{m,q;p}(u) - \overline{u}^p \geq c\|u\|_{L_\infty}^{p-n\beta} \tag{2.1}$$

*for all $u \in W_{m,q}(\Omega)$ satisfying $O(u) > k$.*

For the proof of the theorem we need the following lemma.

**Lemma 2.2** *Under the assumptions of the theorem, there exist positive constants $c', k'$ such that,*

$$\|u\|_{m,q} := \int_\Omega \sum_{|\alpha|\leq m} |D^\alpha u|^q \, dx \leq c'|u|_{m,q} := c' \int_\Omega \sum_{|\alpha|=m} |D^\alpha u|^q \, dx, \tag{2.2}$$

*for all $u \in W_{m,q}(\Omega)$ satisfying $O(u) > k'$.*

**Proof of lemma.** Let $\sigma$ be a positive number and suppose that the assertion of the lemma does not hold if $k' = \sigma$. Then there exists a sequence $\{u_j\}$ in $W_{m,q}(\Omega)$ such that,

$$\|u_j\|_{m,q} > j|u_j|_{m,q}, \tag{2.3}$$

$$\|u_j\|_{m-1,q} = 1 \text{ and } O(u_j) > \sigma. \tag{2.4}$$

The sequence $\{u_j\}$ is bounded in $W_{m,q}(\Omega)$. Indeed if there exists a subsequence $\{u_{j_i}\}$ such that $\|u_{j_i}\|_{m,q} \to \infty$ then, by (2.4),

$$\|u_{j_i}\|_{m,q}/|u_{j_i}|_{m,q} \to 1$$

which contradicts (2.3). This fact and (2.3) imply that $|u_j|_{m,q} \to 0$ and that there exists a subsequence $\{u_{j_i}\}$ which converges in $W_{m,q}(\Omega)$ to an element $U_\sigma$ such that

$$|U_\sigma|_{m,q} = 0, \quad \|U_\sigma\|_{m-1,q} = 1 \text{ and } O(U_\sigma) \geq \sigma. \tag{2.5}$$

This implies that $U_\sigma$ is a polynomial of order at most $m-1$.

Now let $\mathcal{P}_m$ denote the set of polynomials of order $< m$, restricted to $\Omega$. Since this is a finite dimensional space, all norms on $\mathcal{P}_m$ are equivalent. Assume that the assertion of the lemma fails so that the above argument can be applied to every positive $\sigma$. Then, by (2.5), the set $\{U_\sigma : \sigma > 0\}$ is bounded with respect to the norm $\|\cdot\|_{m,q}$. Therefore it is bounded with respect to the $L_\infty$ norm. Hence, by the last inequality in (2.5), $\|U_\sigma\|_{L_1} \to 0$ as $\sigma \to \infty$. In view of the equivalence of norms this contradicts the second statement in (2.5).

**Proof of theorem.** Let $u \in W_{m,q}(\Omega)$ and denote $\bar{b} := \sup_\Omega |u|$. Let $K_0$ be an open cone associated with the domain $\Omega$ as in the cone condition. In the first part of the proof we shall assume that $\bar{b}$ is sufficiently large so that,

$$|K_0| \geq (\bar{b}/2)^{-n\beta}. \tag{2.6}$$

Here we use the notation $|A|$ :=meas$(A)$, for measurable sets in $\mathbb{R}^n$.

Let $b$ be a number in $(\bar{b}/2, \bar{b})$. Since $u$ is continuous, there exists $x^0 \in \Omega$ such that $|u(x^0)| \geq b$. Since $\Omega$ satisfies the cone condition, there exists a cone $Q_0$ with vertex $x^0$ such that $Q_0 \subset \Omega$ and $Q_0$ is congruent to $K_0$. Without loss of generality we shall assume that $x^0$ is the origin. Denote

$$\rho := b^{-\beta}|Q_0|^{-1/n}, \quad \text{and} \quad Q := \rho Q_0. \tag{2.7}$$

In view of (2.6) $\rho \leq 1$ so that $Q \subseteq Q_0$. We note that,

$$|Q| = b^{-n\beta} \quad \text{and} \quad \|u\|_{L_\infty(Q)} =: b_0 \geq b. \tag{2.8}$$

Let

$$T := b^\beta, \quad Q' := TQ \quad \text{and} \quad v(\xi) := b_0^{-1} u(\xi/T) \quad \forall \xi \in Q'. \tag{2.9}$$

Then, by (2.8),

$$|Q'| = 1, \quad v \in W_{m,q}(Q') \quad \text{and} \quad \|v\|_{L_\infty(Q')} = 1.$$

Note that $Q' = |K_0|^{-1/n} K_0$ so that $Q'$ does not depend on $b$. By Sobolev's inequality there exists a constant $c_0$, depending only on $K_0, m, q$ such that,

$$\int_{Q'} \Big( \sum_{|\alpha|=m} |D_\xi^\alpha v|^q + |v|^q \Big) \, d\xi \geq c_0. \tag{2.10}$$

Since $p > q$ this implies,

$$\int_{Q'} \Big( \sum_{|\alpha|=m} |D_\xi^\alpha v|^q + |v|^p \Big)\, d\xi \ \geq c_1, \tag{2.11}$$

where $c_1 = \min(c_0/2, (c_0/2)^{p/q})$. By the definition of $b_0$ and $T$,

$$b_0^q T^{mq} \geq b^{q(1+m\beta)} = b^p.$$

Therefore, by (2.9) and (2.11),

$$\int_Q \Big( \sum_{|\alpha|=m} |D^\alpha u|^q + |u|^p \Big)\, dx \ \geq c_1 b^{p-n\beta}. \tag{2.12}$$

We note that, in view of (1.3), $p - n\beta > q$.

If $u$ satisfies the inequality,

$$\overline{u}^p \leq (c_1/2) b^{p-n\beta},$$

then (2.12) implies (2.1). Therefore we may assume that,

$$\overline{u}^p > (c_1/2) b^{p-n\beta}.$$

Consequently, if

$$b^{n(p-1)\beta} \geq (\bar{b}/2)^{n(p-1)\beta} \geq 2^{p+1}/c_1, \tag{2.13}$$

then,

$$\overline{u} > b^{1-n\beta}. \tag{2.14}$$

Let $E := \Omega \setminus Q$. Assuming that $\bar{b}$ satisfies condition (2.13), inequality (2.14) and (2.8) imply,

$$\int_E |u| = \int_\Omega |u| - \int_Q |u| \geq \overline{u} - b^{1-n\beta} =: \theta > 0. \tag{2.15}$$

Therefore, by Jensen's inequality,

$$|E|^{-1} \int_E |u|^p \geq \Big(|E|^{-1} \int_E |u|\Big)^p \geq (\theta/|E|)^p.$$

Since $|E| < 1$, it follows that

$$\int_E |u|^p \geq \theta^p. \tag{2.16}$$

From this inequality and (2.12) we conclude that,

$$\begin{aligned} \int_\Omega \Big( \sum_{|\alpha|=m} |D^\alpha u|^q + |u|^p \Big)\, dx \quad &\geq \quad c_1 b^{p-n\beta} + \int_E |u|^p \\ &\geq \quad c_1 b^{p-n\beta} + \theta^p. \end{aligned} \tag{2.17}$$

Recall that by assumption, $\bar{b} \geq k\bar{u}$. Hence, by (2.15),

$$\begin{aligned} \bar{u}^p - \theta^p &\leq p\bar{u}^{p-1}(\bar{u} - \theta) \\ &\leq p(\bar{b}/k)^{p-1}b^{1-n\beta} \\ &\leq p(2/k)^{p-1}b^{p-n\beta}. \end{aligned} \tag{2.18}$$

This inequality and (2.17) imply that,

$$\int_\Omega \Big( \sum_{|\alpha|=m} |D^\alpha u|^q + |u|^p \Big)\, dx \geq \tag{2.19}$$
$$c_1 b^{p-n\beta} + \bar{u}^p - p(2/k)^{p-1}b^{p-n\beta}.$$

Hence, if $k$ is sufficiently large so that

$$p(2/k)^{p-1} \leq c_1/2, \tag{2.20}$$

inequality (2.1) holds. Thus the theorem is proved in the case that $\bar{b}$ is sufficiently large as required in (2.6) and (2.13). Both of these conditions are satisfied if $\bar{b}$ is larger than a certain number $B$ which depends only on $K_0, m, q$.

To complete the proof we turn to the case,

$$\bar{b} = \sup_\Omega |u| \leq B. \tag{2.21}$$

By Sobolev's inequality and Lemma 2.2, if $O(u) \geq k'$, then there exists a positive constant $c_2$ such that,

$$\int_\Omega \Big( \sum_{|\alpha|=m} |D^\alpha u|^q \Big)\, dx \geq c_2 \bar{b}^q. \tag{2.22}$$

Since $p - n\beta > q$ this inequality and (2.21) imply,

$$\int_\Omega \Big( \sum_{|\alpha|=m} |D^\alpha u|^q \Big) \geq c_3 \bar{b}^{p-n\beta}, \tag{2.23}$$

with $c_3 = c_2 B^{q-p+n\beta}$. Finally, (2.1) follows from (2.23) and Jensen's inequality. This completes the proof.

## 3. A related inequality

Inequality (2.1) is non-homogeneous. However, as we show below, it can be easily reduced to homogeneous form.

**Proposition 3.3** *Let* $\Omega, m, p, q$ *be as in theorem* 2.1. *Let*

$$\sigma := nq/(mpq - n(p-q)) = n[m(p - n\beta)]^{-1}$$

*and for* $u \in L_p(\Omega)$ *denote*

$$|u|_p^* := (\|u\|_{L_p}^p - \|u\|_{L_1}^p)^{1/p}.$$

*Then there exist positive constants* $c_1, k_1$ *such that,*

$$\|u\|_{L_\infty} \leq c_1 |u|_{m,q}^\sigma (|u|_p^*)^{1-\sigma} \tag{3.1}$$

*for all* $u \in W_{m,q}(\Omega)$ *satisfying* $O(u) > k_1$.

**Proof.** Let $u$ be a function in $W_{m,q}(\Omega)$ such that $O(u) > k$ with $k$ as in theorem 2.1. Let $\lambda$ be a positive number and apply (2.1) to the function $\lambda u$. Then,

$$\lambda^{p-q}(|u|_p^*)^p - c\lambda^{\beta-q}\bar{b}^\beta + |u|_{m,q}^q \geq 0, \tag{3.2}$$

where $\bar{b} = \|u\|_{L_\infty}$ as before. For $\lambda \in (0, \infty)$, the l.h. side of (3.2) obtains its minimum at

$$\lambda_0 = \alpha \bar{b}^{\beta/(p-\beta)} (|u|_p^*)^{-p/(p-\beta)}$$

where,

$$\alpha = [c(\beta - q)/(p - q)]^{1/(p-\beta)}.$$

Substituting $\lambda = \lambda_0$ in (3.2) we obtain,

$$\alpha^{\beta-q}(c - \alpha^{p-\beta})\bar{b}^{\beta(p-q)/(p-\beta)} \leq |u|_{m,q}^q (|u|_p^*)^{p(\beta-q)/(p-\beta)}. \tag{3.3}$$

From the definition of $\alpha$ it is clear that $c > \alpha^{p-\beta}$. Thus (3.3) implies (3.1).

**Corollary 3.4** *Let* $\Omega, m, p, q$ *be as in theorem* 2.1. *For* $u \in L_p(\Omega)$ *denote,*

$$[u]_p^* := \inf\{|u - a|_p^* \, : \, a \leq \|u\|_{L_1(\Omega)}\}.$$

*Then there exist positive constants* $c, k$ *such that,*

$$\|u\|_{L_\infty} \leq c|u|_{m,q}^\sigma ([u]_p^*)^{1-\sigma} \tag{3.4}$$

*for all* $u \in W_{m,q}(\Omega)$ *satisfying* $O(u) > k$.

**Proof.** Let $k_1$ be as in the previous proposition and set $k = 1 + 2k_1$. Suppose that $u \in W_{m,q}(\Omega)$ and $O(u) > k$. Then, for $a \leq \|u\|_{L_1(\Omega)}$,

$$\|u - a\|_{L_\infty(\Omega)} \geq (1 - \frac{1}{k})\|u\|_{L_\infty(\Omega)} \text{ and } \|u - a\|_{L_1(\Omega)} \leq 2\|u\|_{L_1(\Omega)}.$$

Hence $O(u - a) > k_1$ and consequently, by (3.1),

$$\|u\|_{L_\infty(\Omega)} \leq \frac{k}{k-1} c_1 |u|_{m,q}^\sigma (|u - a|_p^*)^{1-\sigma}. \tag{3.5}$$

Thus (3.4) holds with $c = \frac{k}{k-1}c_1$.

Inequality (3.1) is similar to the Gagliardo-Nirenberg inequality for the corresponding parameters [2, p.69-70]. Indeed for $u \in W_{m,q}(\Omega)$ with $m, p, q$ as above the Gagliardo-Nirenberg inequality states,

$$\|u\|_{L_\infty} \leq c\|u\|_{m,q}^{\sigma}\|u\|_{L_p}^{1-\sigma}. \tag{3.6}$$

In order to compare the two inequalities we shall rewrite (3.6) in a form similar to (3.4). First we observe that, in view of lemma 2.2, if $O(u)$ is sufficiently large, inequality (3.6) yields,

$$\|u\|_{L_\infty} \leq c'|u|_{m,q}^{\sigma}\|u\|_{L_p}^{1-\sigma}.$$

Applying this inequality to $u - a$ with $a \leq \|u\|_{L_1(\Omega)}$ we obtain,

$$\|u\|_{L_\infty} \leq c''|u|_{m,q}^{\sigma}(\langle u\rangle_p^*)^{1-\sigma}, \tag{3.7}$$

where

$$\langle u\rangle_p^* := \inf\{\|u-a\|_p \ : \ a \leq \|u\|_{L_1(\Omega)}\}.$$

**Proposition 3.5** *Inequality* (3.4) *implies* (3.7) *but the converse does not hold, irrespective of the magnitude of the 'amplitude'. Thus the two inequalities are not equivalent.*

**Proof.** The first assertion is obvious; therefore we turn to the proof of the second assertion. Consider the following sequence of functions on the interval [0,1]:

$$u_j(t) = \begin{cases} j & \text{if } t \in [0, \gamma_j) \\ 1 & \text{if } t \in [\gamma_j, (1-\gamma_j)/2) \\ -1 & \text{if } t \in [(1-\gamma_j)/2, 1], \end{cases} \tag{3.8}$$

where $\gamma_j = j^{-p-1}, \ j \in \mathbb{N}$. Let,

$$\alpha_j := \int_0^1 |u_j| = 1 + (j-1)\gamma_j.$$

In what follows $o(1)$ stands for a quantity which tends to zero as $j \to \infty$. A straightforward calculation yields (for $|a| \leq \alpha_j = 1 + o(1)$),

$$\begin{aligned} \int_0^1 |u_j - a| &= 1 + o(1), \\ \int_0^1 |u_j - a|^p &= [(1+|a|)^p + (1-|a|)^p]/2 + o(1). \end{aligned} \tag{3.9}$$

Hence,

$$\langle u_j\rangle_p^* = 1 + o(1), \text{ and } [u_j]_p^* = o(1). \tag{3.10}$$

In summation we obtain,

$$O(u_j) \to \infty, \quad \langle u_j \rangle_p^* \to 1 \text{ and } [u_j]_p^* \to 0 \text{ as } j \to \infty. \tag{3.11}$$

By applying an appropriate smoothing operator to each of the functions $u_j$ we obtain a sequence of smooth functions in [0,1], say $\{\tilde{u}_j\}$, which still satisfies (3.11). Finally if $\Omega$ is the unit cube $(0,1)^n$ we define functions $U_j$ in $\Omega$ by $U_j(x) = \tilde{u}_j(x_1)$. Clearly, the sequence $\{U_j\}$ is contained in $W_{m,q}(\Omega)$ and it satisfies (3.11). This completes the proof.

**Remark.** It is possible to derive energy estimates similar to (2.1) in which $L_\infty$ is replaced by $L_r$ with $1 \le q < p < r$, $mq < n$ and $r = \frac{nq}{n-mq}$. However these estimates do not provide new information. In fact, in this case, they follow directly from the corresponding Sobolev inequalities.

**References**

[1] Moshe Marcus, *Uniform estimates for a variational problem with small parameters*. Arch. Rat. Mech. Anal. **124** (1993), 67-98.

[2] Vladimir G.Maz'ja, *Sobolev spaces*. Springer-Verlag, Berlin, Heidelberg, 1985.

Moshe Marcus, Department of Mathematics, Technion, 32000 Haifa, Israel.

International Series of Numerical Mathematics, Vol. 123, © 1997 Birkhäuser Verlag Basel

# Sobolev inequalities in 2-dimensional hyperbolic space

Francesco Mugelli and Giorgio Talenti

**Abstract**

A sharp form of some Sobolev-type inequalities in 2-dimensional hyperbolic space $\mathbb{H}^2$ is discussed. Here $\mathbb{H}^2$ is modeled on the upper Euclidean half-plane endowed with the Poincaré-Bergman metric. The proof rests upon rearrangements of functions, the isoperimetric theorem in $\mathbb{H}^2$, and inequalities in the calculus of variations.

## 1. Main result

**Theorem 1** *Let $\mathbb{R}^2_+ = \{(x,y) : -\infty < x < \infty, 0 < y < \infty\}$, the upper Euclidean half-plane, and let $u$ be a real-valued function defined in $\mathbb{R}^2_+$. Suppose $u$ is sufficiently smooth (e.g. Lipschitz-continuous) and decays well (e.g. the support of $u$ is bounded and is bounded away from the $x$-axis). The following inequalities hold and are sharp.*

(i)
$$\left\{\int_{\mathbb{R}^2_+} |u| \frac{dxdy}{y^2}\right\}^2 + 4\pi \int_{\mathbb{R}^2_+} |u|^2 \frac{dydy}{y^2} \leq \left\{\int_{\mathbb{R}^2_+} y\sqrt{u_x^2+u_y^2}\,\frac{dxdy}{y^2}\right\}^2. \qquad (1.2)$$

(ii) *If $1 < p < 2$ and $q = 2p/(2-p)$ then*

$$p^{-2}\left\{\int_{\mathbb{R}^2_+} |u|^p \frac{dxdy}{y^2}\right\}^{2/p} + \frac{(4\pi/q)^2(q/2-1)^{2/q}}{sin(2\pi/q)}\left\{\int_{\mathbb{R}^2_+} |u|^q \frac{dxdy}{y^2}\right\}^{2/q}$$
$$\leq \left\{\int_{\mathbb{R}^2_+} y^p(u_x^2+u_y^2)^{p/2}\frac{dxdy}{y^2}\right\}^{2/p}. \qquad (1.2)$$

(iii) *If $p > 2$ then*

$$\sup|u| \le (4\pi)^{-1/p} \left\{ \frac{\Gamma\left(\frac{p-2}{2(p-1)}\right)\Gamma\left(\frac{1}{p-1}\right)}{\Gamma\left(\frac{p}{2(p-1)}\right)} \right\}^{1-1/p}$$

$$\times \left\{ \int_{\mathbb{R}^2_+} y^p (u_x^2 + u_y^2)^{p/2} \frac{dxdy}{y^2} \right\}^{1/p}. \tag{1.3}$$

(iv) *If $1 \le p < \infty$ then*

$$\left\{ \int_{\mathbb{R}^2_+} |u|^q \frac{dxdy}{y^2} \right\}^{2/q} \le p^p \int_{\mathbb{R}^2_+} y^p (u_x^2 + u_y^2)^{p/2} \frac{dxdy}{y^2}. \tag{1.4}$$

Theorem 1 can be interpreted as follows. The quadratic form

$$y^{-2}\left[(dx)^2 + (dy)^2\right], \tag{1.5a}$$

sometimes called *Poincaré-Bergman metric*, makes $\mathbb{R}^2_+$ a Riemannian manifold that models the *2-dimensional hyperbolic space*, is denoted by $\mathbb{H}^2$ throughout, and has the properties listed below. (Relevant references are [Bi, Chapter 14], [Bl, Sections 74 and 75], [SG, Section 9.5] and [Sie, Section 2.2], for example.)

Any differentiable mapping from $\mathbb{H}^2$ into $\mathbb{H}^2$ that leaves the Riemannian metric of $\mathbb{H}^2$ *invariant* is given either by a rational function of order 1 of the complex variable $x + iy$ — a Möbius transformation — having real coefficients, or by the conjugate of such a function. The Riemannian *angle* between two tangent vectors to $\mathbb{H}^2$ coincides with the Euclidean angle; the Riemannian *length* of a tangent vector to $\mathbb{H}^2$ at a point $(x, y)$ equals $y\times$(the Euclidean length). The *geodesics* of $\mathbb{H}^2$ are the half-lines and the half-circles orthogonal to the $x$-axis; the Riemannian *distance* between two points $(x_1, y_1)$ and $(x_2, y_2)$ equals

$$\log\left[\frac{\sqrt{(x_2-x_1)^2+(y_2+y_1)^2}+\sqrt{(x_2-x_1)^2+(y_2-y_1)^2}}{\sqrt{(x_2-x_1)^2+(y_2+y_1)^2}-\sqrt{(x_2-x_1)^2+(y_2-y_1)^2}}\right]. \tag{1.5b}$$

The Riemannian *2-dimensional measure* on $\mathbb{H}^2$, $\mathcal{M}$, is given by

$$d\mathcal{M} = y^{-2}dxdy. \tag{1.5c}$$

The *Laplace-Beltrami operator* on $\mathbb{H}^2, \Delta$, obeys $\Delta = y^2\left((\partial/\partial x)^2 + (\partial/\partial y)^2\right)$; the *curvature* of $\mathbb{H}^2$ is identically $-1$.

Inequalities (1.1)–(1.4) are closely related to the Riemannian structure of $\mathbb{H}^2$. In fact, if we think of $u$ as a scalar field on $\mathbb{H}^2$ we have

$$\left\{\int_{\mathbb{R}^2_+} |u|^q \, \frac{dxdy}{y^2}\right\}^{1/q} = \left\{\int_{\mathbb{H}^2} |u|^q \, d\mathcal{M}\right\}^{1/q} \tag{1.6a}$$

*the norm of $u$ in Lebesgue space $L^q(\mathbb{H}^2)$.* On the other hand, the covariant derivative of $u$, $\nabla u$, is a tangent vector field to $\mathbb{H}^2$ whose components are $u_x$ and $u_y$, and whose Riemannian length, $|\nabla u|$, equals $y\sqrt{u_x^2+u_y^2}$. Therefore

$$\left\{\int_{\mathbb{R}^2_+} y^p(u_x^2+u_y^2)^{p/2} \, \frac{dxdy}{y^2}\right\}^{1/p} = \left\{\int_{\mathbb{H}^2} |\nabla u|^p \, d\mathcal{M}\right\}^{1/p}, \tag{1.6b}$$

*the norm of $\nabla u$ in $L^p(\mathbb{H}^2) \times L^p(\mathbb{H}^2)$.*

Thus, statements (i), (ii) and (iii) from Theorem 1 amount to *Sobolev inequalities* in hyperbolic space $\mathbb{H}^2$; statement (iv) is an *inequality à la Poincaré* in $\mathbb{H}^2$. Inequalities of this sort are well-established: the point is that the constants displayed in Theorem 1 are *the best possible.* (Standard references on Sobolev spaces and inequalities are [Ad], [Ma], [Zi]. An overview of sharp Sobolev inequalities in Euclidean $n$-dimensional spaces can be found in [Ta2, Section 2]. Sharp forms of some Sobolev inequalities on spheres appeared in [Ta1] and [Ci]. Sobolev spaces and inequalities on Riemannian manifolds are discussed in [Au1], [Au2] and [Au3].) Remarks on Theorem 1 follow.

(i) The proof of Theorem 1 we offer in the present paper rests on the *isoperimetric theorem* for hyperbolic space $\mathbb{H}^2$, though an alternative approach briefly mentioned in Section 2 may work as well. We would like to stress that, conversely, the isoperimetric theorem for $\mathbb{H}^2$ can be *derived* from inequality (1.1) — details of such a derivation are much as in [Ta2, Theorem 2A] or [Zi, Section 2.7], for example. In other words, statement (i) from Theorem 1 is *equivalent* to the isoperimetric theorem for $\mathbb{H}^2$.

(ii) The leading constant involved in inequality (1.2) coincides with (an appropriate power of) the *Sobolev constant* for the *Euclidean* 2-dimensional space — consistently with results by Aubin, [Au1] and [Au2].

(iii) Inequality (1.4) is peculiar to hyperbolic space $\mathbb{H}^2$. It implies also that

$$\int_{\mathbb{H}^2} u^2 d\mathcal{M} \leq 4 \int_{\mathbb{H}^2} u(-\Delta u) d\mathcal{M} \tag{1.7}$$

for every test function $u$ — an inequality already observed by McKean [Mk] showing a cognate peculiarity of $\mathbb{H}^2$ : *the spectrum of $\Delta : L^2(\mathbb{H}^2) \mapsto L^2(\mathbb{H}^2)$ lies below* $-1/4$. As one may infer from proofs, the peculiarities in question are concerned with the *negative curvature* of $\mathbb{H}^2$.

The present work was prompted by a paper of L.E. Fraenkel [Fr], who — aiming at an existence theory for a partial differential equation in fluid mechanics — supplied a proof that (in our notations) if $q \geq 2$ then a constant $A$ exists such that

$$\left\{\int_{\mathbb{R}^2_+} |\varphi|^q y^{-q/2-2} dxdy\right\}^{1/q} \leq A\left\{\int_{\mathbb{R}^2_+} (\varphi_x^2+\varphi_y^2)\,\frac{dxdy}{y}\right\}^{1/2} \tag{1.8a}$$

for every real-valued compactly supported smooth function $\varphi$ defined in $\mathbb{R}^2_+$.

Fraenkel had an eye to the smallest constant $A$ that renders the above inequality true, and was able to show that such a constant equals $2^{6/5}\cdot 15^{-1/2}\cdot\pi^{-1/5}$ in the special case where $q=\frac{10}{3}$.

The change of variables $\varphi(x,y)=\sqrt{y}\,u(x,y)$ and an integration by parts turn the inequality above into

$$\left\{\int_{\mathbb{H}^2} |u|^q\,d\mathcal{M}\right\}^{1/q} \leq A\left\{\int_{\mathbb{H}^2} |\nabla u|^2\,d\mathcal{M}+\frac{3}{4}\int_{\mathbb{H}^2}|u|^2\,d\mathcal{M}\right\}^{1/2}. \tag{1.8b}$$

Thus, Fraenkel's inequality can be conveniently regarded as a *borderline case* — i.e. a case where the leading exponent equals the dimension — of Sobolev inequalities in hyperbolic space $\mathbb{H}^2$. Such a case can be approached using the methods of the present paper though it is not included in Theorem 1, and will be the subject of a future paper.

## 2. Rearrangements

Let $u$ be any real-valued function defined on hyperbolic space $\mathbb{H}^2$ such that the Riemannian measure of the level set $\{(x,y)\in\mathbb{H}^2 : |u(x,y)|>t\}$ is finite for every positive $t$. The following definitions mimic those introduced by Hardy and Littlewood and elaborated by several authors (see [HLP, Chapter 10], [PS, Chapters 1 and 2], [Ban, Chapter 2], [Kw, Chapter 2], [Bae], [Ta2], and the references quoted therein).

The *distribution function* of $u$, $\mu$, is the Riemannian measure of the level sets of $u$, i.e. is defined by

$$\mu(t)=\int_{\{(x,y)\in\mathbb{H}^2:|u(x,y)|>t\}} d\mathcal{M} \tag{2.1}$$

for every nonnegative $t$. The *decreasing rearrangement* of $u, u^*$, is defined by

$$u^*(s)=\inf\{t\geq 0:\mu(t)\leq s\} \tag{2.2}$$

for every nonnegative $s$. The *symmetric rearrangement* of $u$, $u^\star$, is defined by

$$u^\star(x,y)=u^*\left(\frac{\pi}{y}\left(x^2+(y-1)^2\right)\right) \tag{2.3}$$

for every $(x,y)$ from $\mathbb{H}^2$.

Clearly, $\mu$ is *nonnegative*, *decreasing* and *right-continuous*, and $u^*$ coincides with the distribution function of $\mu$. It is easy to show that

$$\{s \geq 0 : u^*(s) > t\} = [0, \mu(t)[ \tag{2.4}$$

for every nonnegative $t$ — i.e. *the set where $u^*$ exceeds $t$ is an interval on the real line whose end-points are 0 and the value of $\mu$ at $t$.*

Recall that the *geodesic disk* in $\mathbb{H}^2$ with center at $(a, b)$ and radius $r$ coincides with the Euclidean disk whose center is $(a, b\cosh r)$ and whose radius is $b \sinh r$; and that the Riemannian area and the Riemannian perimeter of a geodesic disk in $\mathbb{H}^2$ with radius $r$ are $4\pi[\sinh(r/2)]^2$ and $2\pi \sinh r$, respectively. In other words, if $a, b$ and $s$ are real parameters, and $b$ and $s$ are positive, then the inequality $(x-a)^2+(y-b)^2 < (s/\pi)by$ defines the geodesic disk in $\mathbb{H}^2$ such that: center$=(a, b)$, measure$=s$, perimeter$=\sqrt{s^2 + 4\pi s}$, and radius$=\log\left(1 + \frac{1}{2\pi}s + \frac{1}{2\pi}\sqrt{s^2 + 4\pi s}\right)$. Thus, equation (2.3) implies that the value of $u^\star$ at any point $(x, y)$ depends only upon the Riemannian distance between $(x, y)$ and $(0, 1)$, and decreases as such a distance increases. Equations (2.3) and (2.4) give precisely that

$$\left\{(x, y) \in \mathbb{H}^2 : |u^\star(x, y)| > t\right\} = \textit{the open geodesic disk whose center is } (0,1) \textit{ and whose Riemannian measure equals } \mu(t) \tag{2.5}$$

for every nonnegative $t$.

Equations (2.4) and (2.5) tell us that $u, u^*, u^\star$ are *equidistributed*, i.e. both the distribution function of $u^*$ and the distribution function of $u^\star$ coincide with $\mu$, the distribution function of $u$. It follows that

$$\operatorname{ess\,sup}|u| = u^*(0) = u^\star(0, 0), \tag{2.6a}$$

and

$$\int_{\mathbb{H}^2} \Phi\left(|u(x, y)|\right) d\mathcal{M} = \int_0^\infty \Phi\left(u^*(s)\right) ds = \int_{\mathbb{H}^2} \Phi\left(|u^\star(x, y)|\right) d\mathcal{M} \tag{2.6b}$$

if $\Phi$ is a nonnegative increasing function defined in $[0, \infty[$ such that $\Phi(0) = 0$.

As shown in Section 4, a proof of Theorem 1 rests upon equations (2.6a) and (2.6b), and the following theorem.

**Theorem 2** *Let $u$ be a real-valued function defined on hyperbolic space $\mathbb{H}^2$. Assume $u$ is Lipschitz-continuous and the Riemannian measure of $\left\{(x, y) \in \mathbb{H}^2 : |u(x, y)| > t\right\}$ is finite for every positive $t$. Let $\Phi$ be any Young function — i.e. assume $\Phi$ maps $[0, \infty[$ into $[0, \infty[$, $\Phi(0) = 0$, $\Phi$ is increasing and convex. Then*

$$\int_{\mathbb{H}^2} \Phi\left(|\nabla u|\right) d\mathcal{M} \geq \int_0^\infty \Phi\left(-\sqrt{4\pi s + s^2}\, \frac{du^*}{ds}(s)\right) ds, \tag{2.7a}$$

*and*

$$\text{the right-hand side of (2.7a)} = \int_{\mathbb{H}^2} \Phi\left(|\nabla u^\star|\right) d\mathcal{M}. \tag{2.7b}$$

Theorem 2 amounts to saying that Dirichlet-type integrals decrease under the symmetric rearrangement. (A counterpart of this, where $\mathbb{H}^2$ is replaced by an Euclidean space, is a popular tool, exhaustively discussed in [BrZi], for example.) Interestingly, Theorem 2 can be derived from rearrangement inequalities that have their roots in *merely combinatorial* arguments. A proof along these lines is due to A. Baernstein and W. Beckner, and is outlined in [Bae, Sections 3 and 4]. Here we insist on a more geometric approach, and provide a proof that is based on the isoperimetric theorem for $\mathbb{H}^2$ and may help to understand how the negative curvature of $\mathbb{H}^2$ comes into play.

The *isoperimetric theorem* for hyperbolic space $\mathbb{H}^2$ is as follows. Let $E$ be a sufficiently smooth subset of $\mathbb{H}^2$, and let

$$P = \int_{\partial E} \frac{1}{y}\sqrt{(dx)^2 + (dy)^2} \quad \text{and} \quad A = \int_E \frac{dxdy}{y^2}$$

— the Riemannian perimeter and the Riemannian area of $E$, respectively. If $A$ is finite, then

$$P \geq \sqrt{4\pi A + A^2}; \tag{2.8}$$

moreover, $P$ equals $\sqrt{4\pi A + A^2}$ if, and only if, $E$ is a disk. One early proof of this theorem is due to E. Schmidt [Sch], another can be found in [BuZa, Section 10].

The curvature is a clue to the isoperimetric theorem for $\mathbb{H}^2$. Indeed, as observed in [Oss, Section 4], the isoperimetric inequality on a 2-dimensional manifold *of constant curvature* — either a sphere, or the Euclidean plane, or the hyperbolic plane — reads

$$(\text{perimeter}) \geq \left[4\pi \times (\text{area}) - (\text{curvature}) \times (\text{area})^2\right]^{1/2}. \tag{2.9}$$

The proofs given below in this section and in the subsequent ones indicate that: (i) The isoperimetric theorem for $\mathbb{H}^2$ is precisely what brings the weight

$$\sqrt{C_1 s + C_2 s^2},$$

$$C_1 = 4\pi \quad \text{and} \quad C_2 = -(\text{curvature}),$$

into the inequalities of Theorem 2. (ii) This same weight, which behaves like $\sqrt{s}$ when $s$ is small and like $s$ when $s$ grows large, is precisely what causes the inequalities of Theorem 1 to hold in the form stated.

*Proof of Theorem* 2. Clearly there is no loss in generality if we assume $u \geq 0$. Federer's coarea formula implies that

$$\begin{aligned} &\int_{u^{-1}(]u^*(s+h),u^*(s)])} y\sqrt{(u_x)^2 + (u_y)^2}\,\frac{dxdy}{y^2} \\ &= \int_{u^*(s+h)}^{u^*(s)} dt \int_{u^{-1}(\{t\})} \frac{1}{y}\sqrt{(dx)^2 + (dy)^2} \end{aligned} \tag{2.10}$$

if $s \geq 0$ and $h \geq 0$. (An accessible version of Federer's coarea formula can be found in [Zi, Section 2.7], for example.) The isoperimetric theorem for hyperbolic space $\mathbb{H}^2$ implies that

$$\int_{u^{-1}(\{t\})} \frac{1}{y}\sqrt{(dx)^2+(dy)^2} \geq \sqrt{4\pi\mu(t)+[\mu(t)]^2} \tag{2.11}$$

for every nonnegative $t$. Therefore, if $s \geq 0$ and $h \geq 0$ we have

$$\int_{u^{-1}(]u^*(s+h),u^*(s)])} |\nabla u| d\mathcal{M} \geq \sqrt{4\pi s+s^2}\,[u^*(s)-u^*(s+h)]\,. \tag{2.12}$$

Inequality (2.12) is crucial: the whole Theorem 2 will be derived from it.

Let us commence by showing that $u^*$ is locally Lipschitz-continuous — a property implicit in the statement of Theorem 2. Basic properties of both distribution function and decreasing rearrangement — equations (2.1), (2.2) and (2.4) — imply that

$$\int_{u^{-1}(]u^*(s+h),u^*(s)[)} d\mathcal{M} \leq h \tag{2.13}$$

if $s \geq 0$ and $h \geq 0$. Moreover,

$$\int_{u^{-1}(\{t\})} |\nabla u| d\mathcal{M} = 0 \tag{2.14}$$

for every nonnegative $t$, because either $u^{-1}(\{t\})$ has Riemannian measure zero or $|\nabla u|$ vanishes almost everywhere on $u^{-1}(\{t\})$. Inequalities (2.12) and (2.13), and equation (2.14) imply that

$$\sqrt{4\pi s+s^2}\,[u^*(s)-u^*(s+h)] \leq h \cdot \operatorname{ess\,sup} |\nabla u| \tag{2.15}$$

if $s \geq 0$ and $h \geq 0$. The above mentioned property follows. Inequality (2.12) gives immediately that

$$\frac{d}{ds}\int_{\{(x,y)\in\mathbb{H}^2:u(x,y)>u^*(s)\}} |\nabla u| d\mathcal{M} \geq \sqrt{4\pi s+s^2}\left[-\frac{du^*}{ds}(s)\right] \tag{2.16}$$

for almost every positive $s$.

Now we prove that

$$\frac{d}{ds}\int_{\{(x,y)\in\mathbb{H}^2:u(x,y)>u^*(s)\}} \Phi\,(|\nabla u|) d\mathcal{M} \geq \Phi\left(\sqrt{4\pi s+s^2}\left[-\frac{du^*}{ds}(s)\right]\right) \tag{2.17}$$

for almost every positive $s$. There are exactly three alternatives: (i) $s$ belongs to some exceptional set having one-dimensional Lebesgue measure zero; (ii) $du^*/ds$ vanishes at $s$; (iii) a neighborhood of $s$ exists where $u^*$ decreases strictly.

If either (i) or (ii) holds, there is nothing to prove. If (iii) is in force, a simple argument shows that

$$\int_{u^{-1}(]u^*(s+h),u^*(s)])} d\mathcal{M} = h \tag{2.18}$$

if $h$ is positive and small enough. Then Jensen's inequality for convex functions gives

$$\frac{1}{h}\int_{u^{-1}(]u^*(s+h),u^*(s)])} \Phi\left(|\nabla u|\right)d\mathcal{M} \geq \Phi\left(\frac{1}{h}\int_{u^{-1}(]u^*(s+h),u^*(s)])} |\nabla u| d\mathcal{M}\right), \tag{2.19}$$

consequently we have

$$\begin{aligned} &\frac{d}{ds}\int_{\{(x,y)\in\mathbb{H}^2:u(x,y)>u^*(s)\}} \Phi(|\nabla u|)d\mathcal{M} \\ &\geq \Phi\left(\frac{d}{ds}\int_{\{(x,y)\in\mathbb{H}^2:u(x,y)>u^*(s)\}} |\nabla u| d\mathcal{M}\right). \end{aligned} \tag{2.20}$$

Inequalities (2.16) and (2.20) yield (2.17).

The proof of inequality (2.7a) is now at hand. Indeed, (2.7a) follows from inequality (2.17) and the obvious inequality

$$\int_{\mathbb{H}^2} \Phi\left(|\nabla u|\right)d\mathcal{M} \geq \int_0^\infty ds\frac{d}{ds}\int_{\{(x,y)\in\mathbb{H}^2:u(x,y)>u^*(s)\}} \Phi\left(|\nabla u|\right)d\mathcal{M}. \tag{2.21}$$

The remaining part of the proof runs this way. Define $s$ by

$$s(x,y) = \frac{\pi}{y}\left[x^2 + (y-1)^2\right], \tag{2.22}$$

and observe that: (i) $s(x,y) \geq 0$ if $y > 0$; (ii) $s$ obeys the partial differential equation

$$y^2\left(s_x^2 + s_y^2\right) = 4\pi s + s^2; \tag{2.23}$$

(iii) for every positive $t$, the level line where $s(x,y) = t$ is a geodesic circle of $\mathbb{H}^2$ whose Riemannian length equals $\sqrt{4\pi t + t^2}$. Derive from (i), (ii) and (iii) and Federer's coarea formula that

$$\int_{\mathbb{H}^2} f(s(x,y))d\mathcal{M} = \int_0^\infty f(t)dt \tag{2.24}$$

if $f$ is defined in $[0,\infty[$ and decays fast enough near 0 and $\infty$. Deduce from the definition of $u^\star$ — equation (2.3) — and from equation (2.23) that

$$|\nabla u^\star| = \sqrt{4\pi s + s^2}\left[-\frac{du^*}{ds}(s)\right]. \tag{2.25}$$

Conclude the proof by observing that equations (2.24) and (2.25) give (2.7b).

## 3. Lemmas

**Lemma 3.1** (i) *The following inequality*

$$\frac{\int_0^\infty s|du(s)|}{\int_0^\infty |u(s)|ds} \geq 1 \tag{3.1}$$

*holds for every non-zero real-valued function u such that: u has bounded variation, the integral* $\int_0^\infty s|du(s)|$ *is finite, and* $u(\infty) = 0$. *The right-hand side of* (3.1) *is the minimum value of the left-hand side, and any positive decreasing function is a minimizer.*

(ii) *If* $1 < p < \infty$, *then*

$$\frac{\int_0^\infty |su'(s)|^p ds}{\int_0^\infty |u(s)|^p ds} > p^{-p} \tag{3.2}$$

*for every non-zero real-valued function u such that: u is absolutely continuous, the integral* $\int_0^\infty |su'(s)|^p ds$ *is finite, and* $u(\infty) = 0$. *The right-hand side of (3.2) is the greatest lower bound — not attained — of the left-hand side; a minimizing sequence is given by*

$$u_k(s) = s^{-1/p+1/k} e^{-s} \qquad (k = 1, 2, 3, \dots).$$

**Lemma 3.2** (i) *The following inequality*

$$\frac{\left\{\int_0^\infty \sqrt{s}|du(s)|\right\}^2}{\int_0^\infty [u(s)]^2 ds} \geq 1 \tag{3.3}$$

*holds for every non-zero real-valued function u such that: u has bounded variation, the integral* $\int_0^\infty \sqrt{s}|du(s)|$ *is finite, and* $u(\infty) = 0$. *The right-hand side of* (3.3) *is the minimum value of the left-hand side; the characteristic function of the interval* $[0, 1]$, *and any rescaled version of it, are minimizers.*

(ii) *Let* $1 < p < 2$ *and* $q = 2p/(2 - p)$. *Then*

$$\frac{\left\{\int_0^\infty s^{p/2}|u'(s)|^p ds\right\}^{2/p}}{\left\{\int_0^\infty |u(s)|^q ds\right\}^{2/q}} \geq \frac{4\pi(q/2-1)^{2/q}}{q^2 sin(2\pi/q)} \tag{3.4}$$

*for every non-zero real-valued function u such that: u is absolutely continuous, the integral* $\int_0^\infty s^{p/2}|u'(s)|^p ds$ *is finite, and* $u(\infty) = 0$. *The right-hand side is the minimum value of the left-hand side; a minimizer is given by*

$$u(s) = \left[1 + s^{q/(q-2)}\right]^{-2/q},$$

*and any other minimizer is a rescaled version of this.*

**Lemma 3.3** (i) *Suppose $A$ and $B$ are nonnegative constants. The following inequality*

$$\frac{\left\{\int_0^\infty \sqrt{s^2+4\pi s}\,|du(s)|\right\}^2}{A\left\{\int_0^\infty |u(s)|\,ds\right\}^2 + B\int_0^\infty |u(s)|^2\,ds} \geq 1 \tag{3.5a}$$

*holds for every non-zero $u$ if, and only if,*

$$A \leq 1 \qquad \text{and} \qquad B \leq 4\pi. \tag{3.5b}$$

*Here $u$ is any real-valued function such that: $u$ has bounded variation, the integrals $\int_0^\infty s|du(s)|$ and $\int_0^\infty \sqrt{s}\,|du(s)|$ are finite, and $u(\infty)=0$.*

(ii) *Let $1 \leq p < 2$ and $q = 2p/(2-p)$, and suppose $A$ and $B$ are nonnegative constants. The following inequality*

$$\frac{\left\{\int_0^\infty (s^2+4\pi s)^{p/2}\,|u'(s)|^p\,ds\right\}^{2/p}}{A\left\{\int_0^\infty |u(s)|^p\,ds\right\}^{2/p} + B\left\{\int_0^\infty |u(s)|^q\,ds\right\}^{2/q}} \geq 1 \tag{3.6a}$$

*holds for every non-zero $u$ if, and only if,*

$$A \leq p^{-2} \qquad \text{and} \qquad B \leq \frac{(4\pi/q)^2(q/2-1)^{2/q}}{sin(2\pi/q)}. \tag{3.6b}$$

*Here $u$ is any real-valued function such that: $u$ is absolutely continuous, the integrals $\int_0^\infty |su'(s)|^p ds$ and $\int_0^\infty s^{p/2}|u'(s)|^p ds$ are finite, and $u(\infty)=0$.*

**Lemma 3.4** *Let $2 < p < \infty$, and denote $p/(p-1)$ by $p'$. Then*

$$\frac{\sup|u|}{\left\{\int_0^\infty (s^2+4\pi s)^{p/2}\,|u'(s)|^p\,ds\right\}^{2/p}} \leq (4\pi)^{-1/p}\left\{\frac{\Gamma(1-p'/2)\Gamma(p'-1)}{\Gamma(p'/2)}\right\}^{1/p'} \tag{3.7}$$

*for every non-zero real-valued function $u$ such that: $u$ is absolutely continuous, the integrals $\int_0^\infty |su'(s)|^p\,ds$ and $\int_0^\infty s^{p/2}\,|u'(s)|^p\,ds$ are finite, and $u(\infty)=0$. The right-hand side of* (3.7) *is exactly the maximum value of the left-hand side.*

*Proof of Lemma* 3.1. A proof of statement (i) can be easily figured out and is omitted here. Statement (ii) follows from [HLP, Theorem 328] — a variant of Hardy's inequality.

*Proof of Lemma* 3.2. (i) Replacing $u$ by $]0,\infty[\ni s \mapsto \int_s^\infty |du(t)|$ leaves the set of competing functions invariant and decreases the left-hand side of (3.3). Thus, there is no loss of generality if the extra assumption is used that $u$ is nonnegative-valued and decreasing.

We have

$$\int_0^s \frac{u(t)}{\sqrt{t}}dt \geq u(s)\int_0^s \frac{1}{\sqrt{t}}dt$$

because of the monotonicity of $u$, and consequently

$$u(s) \leq \frac{1}{2\sqrt{s}}\int_0^s \frac{u(t)}{\sqrt{t}}dt \quad \text{and} \quad [u(s)]^2 \leq \frac{d}{ds}\left\{\int_0^s \frac{u(t)}{2\sqrt{t}}dt\right\}^2$$

for every positive $s$. Therefore

$$\int_0^\infty [u(s)]^2 ds \leq \left\{\int_0^\infty \frac{u(t)}{2\sqrt{t}}dt\right\}^2.$$

On the other hand, the equation $u(s) = \int_s^\infty [-du(t)]$ gives

$$\int_0^\infty \frac{u(t)}{2\sqrt{t}}dt = \int_0^\infty \sqrt{s}[-du(s)].$$

We have shown

$$\int_0^\infty [u(s)]^2 ds \leq \left\{\int_0^\infty \sqrt{s}[-du(s)]\right\}^2.$$

Inequality (3.3) is demonstrated. The remaining part of statement (i) follows from a straightforward inspection.

(ii) A theorem by Bliss [Bs] says that if $1 < p < q$ then

$$\frac{\left\{\int_0^\infty |w'(t)|^p\,dt\right\}^{1/p}}{\left\{\int_0^\infty |w(t)|^q\,t^{-1-q(1-1/p)}dt\right\}^{1/q}} \geq \left(q - \frac{q}{p}\right)^{1/q}\left\{\frac{\Gamma\left(\dfrac{pq}{q-p}\right)}{\Gamma\left(\dfrac{q}{q-p}\right)\Gamma\left(p\dfrac{q-1}{q-p}\right)}\right\}^{1/p-1/q}$$

for every non-zero real-valued function $w$ such that $w$ is absolutely continuous, the integral $\int_0^\infty |w'(t)|^p dt$ is finite, and $w(0) = 0$. The right-hand side of the last inequality is the *minimum value* of the left-hand side; a *minimizer* is given by

$$w(t) = t\left(t^{q/p-1} + 1\right)^{-p/(q-p)},$$

and any other minimizer is a rescaled version of this. (Bliss' theorem has proved instrumental in investigating Sobolev-type inequalities; it relies upon typical methods of the classical calculus of variations, and on the circumstance that appropriate solutions to the relevant Euler equation — a differential equation of the Emden-Fowler type — are available in a close form. Incidentally, statement 270 in [HLP] is flawed by a misprint.)

Letting $q = 2p/(2-p)$ and applying Bliss' theorem to test functions of this form

$$]0,\infty[\ni t \mapsto u\left(t^{1-q/2}\right)$$

results in statement (ii).

*Proof of Lemma* 3.3. Let us focus on statement (ii) — the proof of (i) is similar.

Suppose $1 < p < 2$ and $q = 2p/(2-p)$, and that $A$ and $B$ obey (3.6b). If $u$ is any competing function, we have

$$\left\{\int_0^\infty \left(s^2 + 4\pi s\right)^{p/2} |u'(s)|^p \, ds\right\}^{p/2}$$
$$\geq t^{1-p/2} \int_0^\infty |su'(s)|^p \, ds + (1-t)^{1-p/2} (4\pi)^{p/2} \int_0^\infty s^{p/2} |u'(s)|^p \, ds$$

for every $t$ such that $0 \leq t \leq 1$, consequently

$$\left\{\int_0^\infty \left(s^2 + 4\pi s\right)^{p/2} |u'(s)|^p \, ds\right\}^{2/p}$$
$$\geq \left\{\int_0^\infty |su'(s)|^p \, ds\right\}^{2/p} + 4\pi \left\{\int_0^\infty s^{p/2} |u'(s)|^p \, ds\right\}^{2/p}.$$

We used the formula

$$(a+b)^k = \max\left\{t^{1-k} a^k + (1-t)^{1-k} b^k : 0 \leq t \leq 1\right\},$$

where $0 < k < 1$ and $a$ and $b$ are positive. Therefore Lemmas 3.1 and 3.2 yield

$$\left\{\int_0^\infty \left(s^2 + 4\pi s\right)^{p/2} |u'(s)|^p \, ds\right\}^{2/p}$$
$$\geq p^{-2} \left\{\int_0^\infty |u(s)|^p \, ds\right\}^{2/p} + \frac{(4\pi/q)^2 (q/2-1)^{2/q}}{\sin(2\pi/q)} \left\{\int_0^\infty |u(s)|^q \, ds\right\}^{2/q}.$$

Inequality (3.6a) follows.

Suppose $A$ and $B$ are positive constants and that inequality (3.6a) holds for every test function $u$. If $\lambda$ is any positive constant and $u$ is rescaled — i.e. replaced by $]0,\infty[\ni s \mapsto u(\lambda s)$ — inequality (3.6a) becomes

$$\frac{\left\{\int_0^\infty \left(s^2 + 4\pi\lambda s\right)^{p/2} |u'(s)|^p \, ds\right\}^{2/p}}{A\left\{\int_0^\infty |u(s)|^p \, ds\right\}^{2/p} + B\lambda \left\{\int_0^\infty |u(s)|^q \, ds\right\}^{2/q}} \geq 1.$$

Letting $\lambda \longrightarrow 0$ gives

$$\frac{\int_0^\infty |su'(s)|^p ds}{\int_0^\infty |u(s)|^p ds} \geq A,$$

letting $\lambda \longrightarrow \infty$ gives

$$\frac{\left\{\int_0^\infty s^{p/2}|u'(s)|^p ds\right\}^{2/p}}{\left\{\int_0^\infty |u(s)|^q ds\right\}^{2/q}} \geq B.$$

Inequalities (3.6b) follow, via Lemmas 3.1 and 3.2. The proof is complete.

*Proof of Lemma* 3.4. Clearly

$$\sup |u| \leq \int_0^\infty |u'(s)|\, ds,$$

and

$$\int_0^\infty |u'(s)|\, ds \leq \left\{\int_0^\infty \left(s^2+4\pi s\right)^{p/2} |u'(s)|^p\, ds\right\}^{1/p} \left\{\int_0^\infty \left(s^2+4\pi s\right)^{-p'/2} ds\right\}^{1/p'}$$

by Hölder inequality. Hence

$$\frac{\sup |u|}{\left\{\int_0^\infty \left(s^2+4\pi s\right)^{p/2} |u'(s)|^p\, ds\right\}^{1/p}} \leq \left\{\int_0^\infty \left(s^2+4\pi s\right)^{-p'/2} ds\right\}^{1/p'}.$$

Equality holds in these inequalities if $u$ obeys

$$u(\infty) = 0, \qquad u'(s) = -\left(s^2+4\pi s\right)^{-p'/2}.$$

The Lemma follows.

## 4. Proof of Theorem 1

Let $u$ be any real-valued Lipschitz-continuous compactly supported function defined in hyperbolic space $\mathbb{H}^2$. The theory outlined in Section 2 tells us that the rearrangement of $u$, called $u^\star$ there, is Lipschitz-continuous and supported by a geodesic disk of finite radius. Moreover, $u^\star$ obeys

$$\int_{\mathbb{H}^2} |\nabla u^\star|^p\, d\mathcal{M} \leq \int_{\mathbb{H}^2} |\nabla u|^p\, d\mathcal{M} \tag{4.1}$$

for every $p$ larger than (or equal to) 1, and

$$\int_{\mathbb{H}^2} (u^\star)^q\, d\mathcal{M} = \int_{\mathbb{H}^2} |u|^q\, d\mathcal{M}. \tag{4.2}$$

Hence the set of competing functions and the left-hand sides of inequalities (1.1)–(1.4) are *invariant* under the mapping $u \mapsto u^\star$, and the right-hand sides of the same inequalities *decrease* under this mapping. In other words, the competing functions that really count in the present context are *circular waves* $u$ obeying

$$u(x,y) = u^*\left(\frac{\pi}{y}\left(x^2+(y-1)^2\right)\right), \tag{4.3}$$

where $u^*$ is defined in $[0,\infty[$, is decreasing and locally Lipschitz-continuous, and vanishes in a neighborhood of infinity.

Equation (4.3) gives

$$\int_{\mathbb{R}^2_+} y^p\left(u_x^2+u_y^2\right)^{p/2}\frac{dxdy}{y^2} = \int_0^\infty \left(s^2+4\pi s\right)^{p/2}\left[-\frac{du^*}{ds}(s)\right]^p ds \tag{4.4}$$

and

$$\int_{\mathbb{R}^2_+} |u|^q \frac{dxdy}{y^2} = \int_0^\infty [u^*(s)]^q ds. \tag{4.5}$$

Equations (4.4) and (4.5) turn inequalities (1.1)–(1.4) into the following set:

(i)

$$\left\{\int_0^\infty u^*(s)ds\right\}^2 + 4\pi\int_0^\infty [u^*(s)]^2 ds \le \left\{\int_0^\infty \sqrt{s^2+4\pi s}\left[-\frac{du^*}{ds}(s)\right]ds\right\}^2; \tag{4.6}$$

(ii)

$$p^{-2}\left\{\int_0^\infty [u^*(s)]^p ds\right\}^{2/p} + \frac{(4\pi/q)^2(q/2-1)^{2/q}}{\sin(2\pi/q)}\left\{\int_0^\infty [u^*(s)]^q ds\right\}^{2/q} \le \left\{\int_0^\infty \left(s^2+4\pi s\right)^{p/2}\left[-\frac{du^*}{ds}(s)\right]^p ds\right\}^{2/p}, \tag{4.7}$$

where $1<p<2$ and $q=2p/(2-p)$;

(iii)

$$u^*(0) \le (4\pi)^{-1/p}\left\{\frac{\Gamma\left(\frac{p-2}{2(p-1)}\right)\Gamma\left(\frac{1}{p-1}\right)}{\Gamma\left(\frac{p}{2(p-1)}\right)}\right\}^{1-1/p} \times\left\{\int_0^\infty \left(s^2+4\pi s\right)^{p/2}\left[-\frac{du^*}{ds}(s)\right]^p ds\right\}^{2/p}, \tag{4.8}$$

where $p>2$;

(iv)

$$\int_0^\infty [u^*(s)]^p ds \le p^p\int_0^\infty \left(s^2+4\pi s\right)^{p/2}\left[-\frac{du^*}{ds}(s)\right]^p ds, \tag{4.9}$$

where $1\le p<\infty$. Inequalities (4.6), (4.7) and (4.9) follow from Lemma 3.3; inequality (4.8) follows from Lemma 3.4. The same lemmas — together with appropriate density arguments — show that the inequalities in question are sharp.

The proof of Theorem 1 is complete.

## References

[Ad] R.A. Adams, *Sobolev Spaces.* Academic Press, 1975.

[Au1] T. Aubin, *Problèmes isopérimétriques et espaces de Sobolev.* J. Diff. Geometry **11** (1976), 573–598.

[Au2] T. Aubin, *Espaces de Sobolev sur les variétés Riemanniennes.* Bull. Sci. Math. **100** (1976), 149–173.

[Au3] T. Aubin, *Nonlinear Analysis on Manifolds: Monge-Ampère Equations.* Grundlehren der mathematischen Wissenshaften, vol.252, Springer-Verlag, 1982.

[Bae] A. Baernstein II, *A unified approach to symmetrization.* Pages 47–91 in: Partial Differential Equations of Elliptic Type (A. Alvino & E. Fabes & G. Talenti, eds.), Symposia Mathematica 35, Cambridge Univ. Press, 1994.

[Ban] C. Bandle, *Isoperimetric Inequalities and Applications.* Monographs and Studies in Math. **7**, Pitman, 1980.

[Bi] L. Bianchi, *Lezioni di Geometria Differenziale*, vol. 1, Zanichelli, 1927.

[Bl] W. Blaschke, *Vorlesungen über Differentialgeometrie*, vol. 1, Grundlehren der mathematischen Wissenshaften, vol. 1, Springer-Verlag, 1930.

[Bs] G.A. Bliss, *An integral inequality.* J. London Math. Soc. **5** (1930), 40–46.

[BrZi] J. Brothers and W. Ziemer, *Minimal rearrangements of Sobolev functions.* J. Reine Angew. Math. **384** (1988), 153–179.

[BuZa] Yu.D. Burago and V.A. Zalgaller, *Geometric inequalities.* Springer-Verlag, 1988.

[Ci] A. Cianchi, *A sharp form of Poincaré type inequalities on balls and spheres.* J. Appl. Math. Phys. (ZAMP) **40**, (1989), 558–569.

[Fr] L.E. Fraenkel, *On steady vortex rings with swirl and a Sobolev inequality.* Pages 13–26 in: Progress in partial differential equations: calculus of variations, applications (C. Bandle, J. Bemelmans, M. Chipot, M. Grüter and J. Saint Jean Paulin, eds.), Pitman Research Notes in Math. vol. 266, 1992.

[HLP] G.H. Hardy, J.E. Littlewood and G. Pólya, *Inequalities.* Cambridge Univ. Press (first edition 1934, second edition 1952).

[Kw] B. Kawohl, *Rearrangements and convexity of level sets in PDE.* Lecture Notes in Math. 1150, Springer-Verlag, 1985.

[Ma] V.G. Maz'ja, *Sobolev Spaces.* Springer-Verlag, 1985.

[Mk] H.P. McKean, *An upper bound to the spectrum of $\Delta$ on a manifold of negative curvature.* J. Differential Geometry **4**, (1970), 359–366.

[Oss] R. Osserman, *The isoperimetric inequality*. Bull. Amer. Math. Soc. **84** (1978), 1182–1238.

[PS] G. Pólya and G. Szegö, *Isoperimetric inequalities in mathematical physics*. Princeton Univ. Press, 1951.

[SG] G. Sansone and J. Gerretsen, *Lectures on the theory of functions of a complex variable*, vol. 2. Wotters-Noordhoff, 1970.

[Sch] E. Schmidt, *Über die isoperimetrische Aufgabe im $n$-dimensionalen Raum konstanter negativer Krümmung I: Die isoperimetrischen Ungleichungen in hyperbolischen Ebenen und für Rotationskörper im $n$-dimensionalen hyperbolischen Raum*. Math. Z. **46** (1940), 204–230.

[Sie] C.L. Siegel, *Topics in Complex Functions Theory*, vol. 2: Automorphic Functions and Abelian Integrals. Interscience Tracts in Pure Appl. Math. 25, Wiley-Interscience, 1971.

[Ta1] G. Talenti, *Some inequalities of Sobolev type on two-dimensional spheres*. Pages 401-408 in: General Inequalities 5 (W. Walter, ed.), Int. Series Numer. Math. 80, Birkhäuser-Verlag, 1987.

[Ta2] G. Talenti, *Inequalities in rearrangement invariant function spaces*. Pages 177-231 in: Nonlinear Analysis, Function Spaces and Applications 5 (M. Krbec, A. Kufner, B. Opic and J. Rakosnik, eds.), Prometheus Publ., Prague, 1994.

[Zi] W.P. Ziemer, *Weakly Differentiable Functions: Sobolev Spaces and Functions of Bounded Variation*. Graduate Texts in Math. 120, Springer-Verlag, 1989.

Francesco Mugelli, Dipartimento di Matematica dell'Università, viale Morgagni 67A, I-50134 Firenze, Italy.

Giorgio Talenti, Dipartimento di Matematica dell'Università, viale Morgagni 67A, I-50134 Firenze, Italy.

# Convexity

International Series of Numerical Mathematics, Vol. 123, © 1997 Birkhäuser Verlag Basel

# On the separation with $n$-additive functions

Roman Badora

**Abstract**

We consider the problem of the separation of a pair of $n$-subadditive and $n$-superadditive functions defined on a product of amenable semigroups with values in a complete vector lattice by a $n$-additive mapping.

## 1. Introduction

In this paper we are concerned with the problem of the stability of the Hyers-Ulam type for $n$-additive mappings defined on a product of amenable semigroups with values in a complete vector lattice. The result of this type, for Abelian groups, were first obtained by M.A. Albert and J.A. Baker [1]. Next, we are going to study the problem of the separation of a pair of $n$-subadditive and $n$-superadditive functions defined on a product of amenable semigroups with values in a complete vector lattice by an $n$-additive mapping.

For $n = 1$, results concerning the problem of the separation of subadditive and superadditive functionals were obtained by R. Kaufman [7], P. Kranz [9], G. Rodé [12], H. König [8], A. Chaljub-Simon and P. Volkmann [3], Z. Gajda and Z. Kominek [4] which present some generalizations of the classical Hahn-Banach theorem.

The case of the selection in which $n = 2$ and functionals are positively homogeneous was solved by Z. Páles [10], [11].

Z. Gajda and Z. Kominek in [4] proved some version of the separation theorem (Theorem 3) of Kranz's type for functions defined on an amenable semigroup. Amenability is a condition essentially weaker than commutativity (every Abelian semigroup, every finite semigroup and every solved group is amenable). Moreover, notice that without any additional hypotheses on the semigroup the answer to the separation question may be negative (an example is given in [4]).

Let $(S, \cdot)$ be a semigroup. For any $y \in S$ we define the *left* and *right translation* of a function $f$ defined on $S$ as follows:

$$ {}_yf(x) := f(yx), \quad f_y(x) := f(xy), \quad x \in S. \tag{1} $$

Now we recall the notion of invariant mean and amenability. If $B(S,\mathbb{R})$ denotes the space of all real bounded functions on a semigroup $S$, then a real linear functional $M$ on $B(S,\mathbb{R})$ is called a *left (right) invariant mean* if and only if it is translation invariant, i.e.

$$M({}_yf) = M(f) \ \ (M(f_y) = M(f)), \ y \in S, \ f \in B(S,\mathbb{R}),$$

and normalized:

$$\inf\{f(x) : x \in S\} \le M(f) \le \sup\{f(x) : x \in S\}, \ f \in B(S,\mathbb{R}).$$

If a left (right) invariant mean exists, then $S$ is called *left (right) amenable* (see e.g. F. P. Greenleaf [6]).

In [2] (see also Z. Gajda [5]) we have shown that the concept of an invariant mean can be extended to some function spaces larger than the space of real functions. Precisely, we have proved the following

**Theorem 1** *Let $(S,\cdot)$ be a left (right) amenable semigroup and let $Y$ be a complete vector lattice. Let $B(S,Y)$ be a space of all bounded functions defined on $S$ with values in $Y$. Then there exists a linear operator $M : B(S,Y) \to Y$ such that:*

$$M({}_yf) = M(f) \ \ (M(f_y) = M(f)), \ y \in S, \ f \in B(S,Y) \tag{2}$$

*and*

$$\inf\{f(x) : x \in S\} \le M(f) \le \sup\{f(x) : x \in S\}, \ f \in B(S,Y). \tag{3}$$

By a *complete vector lattice* we understand any ordered vector space $Y$ which is a lattice with the following property: any bounded subset of $Y$ admits a greatest lower bound and a least upper bound.

The complete vector lattice $Y$ has the following property (see Z. Gajda [5], Lemma 4.1):

$$\inf\{\frac{1}{m}c \ : \ m \in \mathbb{N}\} = 0, \ \ c \in Y, \ c \ge 0. \tag{4}$$

In the present paper $n$ is a positive integer, $(S_i,+)$ stands for an amenable semigroup, for $i \in \{1,2,\dots,n\}$, and $Y$ is a complete vector lattice.

For $F : S_1 \times \ldots \times S_n \to Y$ by $C^kF[x_1,\dots,x_{k-1},(x_k,y_k),x_{k+1},\dots,x_n]$ we denote the $k$-th Cauchy difference of $F$ defined as follows

$$\begin{aligned} &C^kF[x_1,\dots,x_{k-1},(x_k,y_k),x_{k+1},\dots,x_n] \\ &:= \ F(x_1,\dots,x_{k-1},x_k+y_k,x_{k+1},\dots,x_n) \\ &\quad -F(x_1,\dots,x_{k-1},x_k,x_{k+1},\dots,x_n) - F(x_1,\dots,x_{k-1},y_k,x_{k+1},\dots,x_n), \end{aligned}$$

for $x_i \in S_i$, $i \in \{1,2,\dots,n\} \setminus \{k\}$, $x_k, y_k \in S_k$ and $k \in \{1,2,\dots,n\}$.

We recall that a function $F : S_1 \times \dots S_n \to Y$ is said to be *n-subadditive* [*n-superadditive*] iff

$$C^k F[x_1, \dots, x_{k-1}, (x_k, y_k), x_{k+1}, \dots, x_n] \leq 0 \tag{5}$$

$$[\, C^k F[x_1, \dots, x_{k-1}, (x_k, y_k), x_{k+1}, \dots, x_n] \geq 0\, ], \tag{6}$$

for all $x_i \in S_i$, $i \in \{1, 2, \dots, n\} \setminus \{k\}$, $x_k, y_k \in S_k$ and $k \in \{1, 2, \dots, n\}$. A map $F : S_1 \times \dots S_n \to Y$ is *n-additive* iff

$$C^k F[x_1, \dots, x_{k-1}, (x_k, y_k), x_{k+1}, \dots, x_n] = 0, \tag{7}$$

for all $x_i \in S_i$, $i \in \{1, 2, \dots, n\} \setminus \{k\}$, $x_k, y_k \in S_k$ and $k \in \{1, 2, \dots, n\}$.

We shall prove all results in the left invariant version. The proofs of their right invariant analogues require only minor changes, therefore, may be omitted.

## 2. Hyers-Ulam stability

We start with the following version of the Albert-Baker theorem on the Hyers-Ulam stability for $n$-additive mappings.

**Theorem 2** *Let $\varepsilon_k \in Y$, $\varepsilon_k \geq 0$, for $k \in \{1, \dots, n\}$ and let $F : S_1 \times \dots \times S_n \to Y$ satisfies, for $k = 1, 2, \dots, n$, the following condition:*

$$-\varepsilon_k \leq C^k F[x_1, \dots, x_{k-1}, (x_k, y_k), x_{k+1}, \dots, x_n] \leq \varepsilon_k, \tag{8}$$

*for all $x_i \in S_i$, $i \in \{1, \dots, n\} \setminus \{k\}$ and $x_k, y_k \in S_k$.*

*Then there exists exactly one $n$-additive function $A : S_1 \times \dots \times S_n \to Y$ such that*

$$-\varepsilon \leq A(x_1, \dots, x_n) - F(x_1, \dots, x_n) \leq \varepsilon, \;\; (x_1, \dots, x_n) \in S_1 \times \dots \times S_n,$$

*where $\varepsilon = \inf \{\varepsilon_k \; : \; k \in \{1, \dots, n\}\}$.*

*Proof.* Condition (8) for $k = 1$ means that for any fixed element $(x_1, \dots, x_n)$ of the product $S_1 \times \dots \times S_n$ the map

$$S_1 \ni y_1 \to F(x_1 + y_1, x_2, \dots, x_n) - F(y_1, x_2, \dots, x_n) \in Y$$

is bounded, i.e. belongs to the space $B(S_1, Y)$. Let $M_1$ be a generalized left invariant mean on $B(S_1, Y)$. We define the map $A_1 : S_1 \times \dots \times S_n \to Y$ by means of the formula:

$$A_1(x_1, \dots, x_n) := M_1(F(x_1 + \cdot, x_2, \dots, x_n) - F(\cdot, x_2, \dots, x_n)),$$

for $(x_1, \ldots, x_n) \in S_1 \times \ldots \times S_n$. Using the left invariance and the linearity of $M_1$, we have

$$\begin{aligned} & A_1(x_1, \ldots, x_n) + A_1(y_1, \ldots, x_n) \\ & \quad = M_1(F(x_1 + \cdot, x_2, \ldots, x_n) - F(\cdot, x_2, \ldots, x_n)) \\ & \qquad + M_1(F(y_1 + \cdot, x_2, \ldots, x_n) - F(\cdot, x_2, \ldots, x_n)) \\ & \quad = M_1(F(x_1 + y_1 + \cdot, x_2, \ldots, x_n) - F(y_1 + \cdot, x_2, \ldots, x_n)) \\ & \qquad + M_1(F(y_1 + \cdot, x_2, \ldots, x_n) - F(\cdot, x_2, \ldots, x_n)) \\ & \quad = M_1(F(x_1 + y_1 + \cdot, x_2, \ldots, x_n) - F(\cdot, x_2, \ldots, x_n)) \\ & \quad = A_1(x_1 + y_1, x_2, \ldots, x_n), \end{aligned}$$

for all $x_1, y_1 \in S_1$ and $x_2 \in S_2, \ldots, x_n \in S_n$; so $A_1$ is additive with respect to the first variable. Moreover, by (3),

$$-\varepsilon_1 \leq A_1(x_1, \ldots, x_n) - F(x_1, \ldots, x_n) \leq \varepsilon_1, \quad (x_1, \ldots, x_n) \in S_1 \times \ldots \times S_n.$$

Now, let we observe that

$$\begin{aligned} & C^2 A_1[x_1, (x_2, y_2), x_3, \ldots, x_n] \\ & \quad = C^2(A_1 - F)[x_1, (x_2, y_2), x_3, \ldots, x_n] + C^2 F[x_1, (x_2, y_2), x_3, \ldots, x_n], \end{aligned}$$

for all $x_1 \in S_1$, $x_2, y_2 \in S_2$, $x_3 \in S_3, \ldots, x_n \in S_n$. Therefore, the last inequality and (8) with $k = 2$ imply that

$$-3\varepsilon_1 - \varepsilon_2 \leq C^2 A_1[x_1, (x_2, y_2), x_3, \ldots, x_n] \leq 3\varepsilon_1 + \varepsilon_2,$$

for $x_1 \in S_1$, $x_2, y_2 \in S_2$, $x_3 \in S_3, \ldots, x_n \in S_n$.

Again, for fixed $(x_1, \ldots, x_n) \in S_1 \times \ldots \times S_n$ the map

$$S_2 \ni y_2 \to A_1(x_1, x_2 + y_2, x_3, \ldots, x_n) - A_1(x_1, y_2, x_3, \ldots, x_n) \in Y$$

belongs to the space $B(S_2, Y)$ and we define a function $A_2 : S_1 \times \ldots \times S_n \to Y$ as follows:

$$A_2(x_1, \ldots, x_n) := M_2(A_1(x_1, x_2 + \cdot, x_3, \ldots, x_n) - A_1(x_1, \cdot, x_3, \ldots, x_n)),$$

for $(x_1, \ldots, x_n) \in S_1 \times \ldots \times S_n$, where $M_2$ is the generalized left invariant mean on the space $B(S_2, Y)$. The same calculation based on the left invariance and the linearity of $M_2$ shows that $A_2$ is additive as a function of the second variable and

$$-4\varepsilon_1 - \varepsilon_2 \leq A_2(x_1, \ldots, x_n) - F(x_1, \ldots, x_n) \leq 4\varepsilon_1 + \varepsilon_2, \quad (x_1, \ldots, x_n) \in S_1 \times \ldots \times S_n.$$

Moreover, $A_2$ is additive with respect to the first variable. Indeed, for any fixed $x_1, y_1 \in S_1$ and $x_2 \in S_2, \ldots, x_n \in S_n$, functions

$$\begin{aligned} & A_1(x_1, x_2 + \cdot, x_3, \ldots, x_n) - A_1(x_1, \cdot, x_3, \ldots, x_n), \\ & A_1(y_1, x_2 + \cdot, x_3, \ldots, x_n) - A_1(y_1, \cdot, x_3, \ldots, x_n) \end{aligned}$$

are in $B(S_2, Y)$ and, by the additivity of $A_1$ and the linearity of $M_2$, we have

$$\begin{aligned} A_2(x_1+y_1, x_2, \dots, x_n) & \\ &= M_2(A_1(x_1+y_1, x_2+\cdot, x_3, \dots, x_n) - A_1(x_1+y_1, \cdot, x_3, \dots, x_n)) \\ &= M_2(A_1(x_1, x_2+\cdot, x_3, \dots, x_n) - A_1(x_1, \cdot, x_3, \dots, x_n)) \\ &\quad + M_2(A_1(y_1, x_2+\cdot, x_3, \dots, x_n) - A_1(y_1, \cdot, x_3, \dots, x_n)) \\ &= A_2(x_1, x_2, \dots, x_n) + A_2(y_1, x_2, \dots, x_n). \end{aligned}$$

Now, for $x_1 \in S_1$, $x_2 \in S_2$, $x_3, y_3 \in S_3$, $x_4, \dots, x_n \in S_n$, we get

$$\begin{aligned} C^3 A_2&[x_1, x_2, (x_3, y_3), x_4, \dots, x_n] \\ &= C^2(A_2 - F)[x_1, x_2, (x_3, y_3), x_4, \dots, x_n] + C^2 F[x_1, x_2, (x_3, y_3), x_4, \dots, x_n], \end{aligned}$$

whence,

$$-3(4\varepsilon_1 + \varepsilon_2) - \varepsilon_3 \le C^3 A_2[x_1, x_2, (x_3, y_3), x_4, \dots, x_n] \le 3(4\varepsilon_1 + \varepsilon_2) + \varepsilon_3,$$

for $x_1 \in S_1$, $x_2 \in S_2$, $x_3, y_3 \in S_3$, $x_4, \dots, x_n \in S_n$.

After $n$ steps we obtain a constant $\delta \ge 0$ in $Y$ and an $n$-additive function $A : S_1 \times \ldots \times S_n \to Y$ such that

$$-\delta \le A(x_1, \dots, x_n) - F(x_1, \dots, x_n) \le \delta, \ \ (x_1, \dots, x_n) \in S_1 \times \ldots \times S_n.$$

To prove that $\delta = \varepsilon = \inf \{\varepsilon_k \ : \ k \in \{1, \dots, n\}\}$ let we assume that there exists $(x_1, \dots, x_n) \in S_1 \times \ldots \times S_n$ such that

$$\delta \ge A(x_1, \dots, x_n) - F(x_1, \dots, x_n) > \varepsilon$$

and let $\gamma := A(x_1, \dots, x_n) - F(x_1, \dots, x_n) - \varepsilon > 0$. Then

$$\begin{aligned} \delta \ge A(2x_1&, \dots, x_n) - F(2x_1, \dots, x_n) \\ &= 2A(x_1, \dots, x_n) - 2F(x_1, \dots, x_n) - F(2x_1, \dots, x_n) + 2F(x_1, \dots, x_n) \\ &\ge 2\varepsilon + 2\gamma - \varepsilon_1 \ge 2\varepsilon + 2\gamma - 2\varepsilon_1 \end{aligned}$$

and, by induction, we have

$$\delta \ge A(2^m x_1, \dots, x_n) - F(2^m x_1, \dots, x_n) \ge 2^m \varepsilon + 2^m \gamma - 2^m \varepsilon_1, \ \ m \in \mathbb{N}.$$

It follows that

$$\frac{1}{2^m} \delta \ge \varepsilon + \gamma - \varepsilon_1, \ \ m \in \mathbb{N}.$$

Combining this inequality with (4) we conclude that $\varepsilon_1 \ge \varepsilon + \gamma$. In a similar fashion we show that

$$\varepsilon_k \ge \varepsilon + \gamma, \ \ k \in \{1, \dots, n\}.$$

From the definition of $\varepsilon$ we derive that $\gamma \leq 0$, which shows that

$$A(x_1, \dots, x_n) - F(x_1, \dots, x_n) \leq \varepsilon$$

for all $(x_1, \dots, x_n) \in S_1 \times \dots \times S_n$. The proof of the second inequality is the same.

Now we want to prove the uniqueness of $A$. Assume that there exists another $n$-additive mapping $B$ such that

$$-\gamma \leq B(x_1, \dots, x_n) - F(x_1, \dots, x_n) \leq \gamma, \quad (x_1, \dots, x_n) \in S_1 \times \dots \times S_n,$$

for some $\gamma \geq 0$. Then

$$-\varepsilon - \gamma \leq A(x_1, \dots, x_n) - B(x_1, \dots, x_n) \leq \varepsilon + \gamma, \quad (x_1, \dots, x_n) \in S_1 \times \dots \times S_n.$$

Therefore,

$$\begin{aligned} \frac{1}{m}(-\varepsilon - \gamma) &\leq A(x_1, \dots, x_n) - B(x_1, \dots, x_n) \\ &= \frac{1}{m}A(mx_1, \dots, x_n) - \frac{1}{m}B(mx_1, \dots, x_n) \leq \frac{1}{m}(\varepsilon + \gamma), \end{aligned}$$

for all $(x_1, \dots, x_n) \in S_1 \times \dots \times S_n$ and $m \in \mathbb{N}$. Thus, by (4),

$$A(x_1, \dots, x_n) = B(x_1, \dots, x_n), \quad (x_1, \dots, x_n) \in S_1 \times \dots \times S_n,$$

and the proof is finished. □

## 3. Separation theorems

Now we shall consider the question whether for any pair of functions $P,\ Q : S_1 \times \dots \times S_n \to Y$ satysfying condition (5), (6) respectively (i.e. $P$ is $n$-subadditive, $Q$ is $n$-superadditive) and such that

$$Q(x_1, \dots, x_n) \leq P(x_1, \dots, x_n), \quad (x_1, \dots, x_n) \in S_1 \times \dots \times S_n, \tag{9}$$

one can find a $n$-additive function $A : S_1 \times \dots \times S_n \to Y$ which separates $P$ and $Q$, i.e.,

$$Q(x_1, \dots, x_n) \leq A(x_1, \dots, x_n) \leq P(x_1, \dots, x_n), \tag{10}$$

for all $(x_1, \dots, x_n) \in S_1 \times \dots \times S_n$. Using Theorem 2 we are able to prove one possible solution to this problem.

**Theorem 3** *If $P,\ Q : S_1 \times \dots \times S_n \to Y$ fulfil* (5), (6), (9) *and the following condition:*

$$\sup \{P(x_1, \dots, x_n) - Q(x_1, \dots, x_n) \ : \ x_i \in S_i,\ i \in \{1, \dots, n\}\} = K \in Y, \tag{11}$$

*then there exists exactly one $n$-additive function $A : S_1 \times \dots \times S_n \to Y$ satisfying* (10).

*Proof.* Notice that, by (11) and (9), we have

$$0 \leq P(x_1, \dots, x_n) - Q(x_1, \dots, x_n) \leq K; \tag{12}$$

$$-K \leq Q(x_1, \dots, x_n) - P(x_1, \dots, x_n) \leq 0 \tag{13}$$

and

$$Q(x_1, \dots, x_n) \leq P(x_1, \dots, x_n) \leq Q(x_1, \dots, x_n) + K, \tag{14}$$

for all $(x_1, \dots, x_n) \in S_1 \times \dots \times S_n$. Hence, for arbitrarily chosen $k \in \{1, \dots, n\}$ and $x_k, y_k \in S_k$, $x_i \in S_i$, $i \in \{1, \dots, n\} \setminus \{k\}$ the following ineqalities hold:

$$\begin{aligned} 0 \geq C^k P[x_1, &\dots, x_{k-1}, (x_k, y_k), x_{k+1}, \dots, x_n] \\ = \ & C^k Q[x_1, \dots, x_{k-1}, (x_k, y_k), x_{k+1}, \dots, x_n] \\ & + C^k (P - Q)[x_1, \dots, x_{k-1}, (x_k, y_k), x_{k+1}, \dots, x_n] \\ \geq \ & 0 + 0 - K - K = -2K. \end{aligned}$$

This shows that the Cauchy differences of $P$ are bounded (by $-2K$ and $2K$). By Theorem 2 there exists an $n$-additive function $A : S_1 \times \dots \times S_n \to Y$ such that

$$-2K \leq A(x_1, \dots, x_n) - P(x_1, \dots, x_n) \leq 2K, \quad (x_1, \dots, x_n) \in S_1 \times \dots \times S_n.$$

Hence, for each $(x_1, \dots, x_n) \in S_1 \times \dots \times S_n$, we have

$$P(x_1, \dots, x_n) - 2K \leq A(x_1, \dots, x_n) \leq P(x_1, \dots, x_n) + 2K$$

and consequently (using (9))

$$Q(x_1, \dots, x_n) - 2K \leq A(x_1, \dots, x_n) \leq P(x_1, \dots, x_n) + 2K.$$

We shall prove that, actually, $A$ lies between $P$ and $Q$, that is, the constant $2K$ may be dropped in the last inequality. For, choose arbitrarily a positive integer $m$ and $(x_1, \dots, x_n) \in S_1 \times \dots \times S_n$. Then

$$\begin{aligned} mQ(x_1, \dots, x_n) - 2K &\leq Q(mx_1, x_2, \dots, x_n) - 2K \\ \leq \ & A(mx_1, x_2, \dots, x_n) = mA(x_1, \dots, x_n) \\ \leq \ & P(mx_1, x_2, \dots, x_n) + 2K \leq mP(x_1, \dots, x_n) + 2K, \end{aligned}$$

whence,

$$Q(x_1, \dots, x_n) - \frac{1}{m} 2K \leq A(x_1, \dots, x_n) \leq P(x_1, \dots, x_n) + \frac{1}{m} 2K.$$

Using (4), we infer that

$$Q(x_1, \dots, x_n) \leq A(x_1, \dots, x_n) \leq P(x_1, \dots, x_n), \quad (x_1, \dots, x_n) \in S_1 \times \dots \times S_n,$$

and thus $A$ is the required $n$-additive mapping separating $P$ and $Q$.

To prove its uniqueness, suppose that $B$ is another $n$-additive mapping with the same property. Then for any $m \in \mathbb{N}$ and $(x_1, \ldots, x_n) \in S_1 \times \ldots \times S_n$ we get

$$\begin{aligned} & A(x_1, \ldots, x_n) - B(x_1, \ldots, x_n) \\ = \ & \frac{1}{m}(A(mx_1, x_2, \ldots, x_n) - B(mx_1, x_2 \ldots, x_n)) \\ \leq \ & \frac{1}{m}(P(mx_1, x_2, \ldots, x_n) - Q(mx_1, x_2 \ldots, x_n)) \\ \leq \ & \frac{1}{m}K. \end{aligned}$$

Applying (4) once more and taking into account the exchangeable role of $A$ and $B$ we derive the identity $A = B$. □

Hypothesis (11) is a typical assumption in the separation theory but it is fairly restrictive. It can be replaced by some condition of the type assumed by Z. Gajda and Z. Kominek in [4], Theorem 3.

**Theorem 4** *Let $P$, $Q : S_1 \times \ldots \times S_n \to Y$ fulfil* (5), (6) *and* (9). *Assume that there exists a $j \in \{1, \ldots, n\}$ such that*

$$\begin{aligned} \sup \{ & P(x_1, \ldots, x_{j-1}, x_j + y_j, x_{j+1}, \ldots, x_n) \\ & -P(x_1, \ldots, x_{j-1}, y_j + x_j, x_{j+1}, \ldots, x_n) \ : \ x_j, y_j \in S_j\} \\ = \ & K(x_1, \ldots, x_{j-1}, x_{j+1}, \ldots, x_n) \in Y, \end{aligned} \tag{15}$$

*for $x_i \in S_i$, $i \in \{1, \ldots, n\} \setminus \{j\}$ and, for $k \in \{1, \ldots, n\} \setminus \{j\}$,*

$$\begin{aligned} \sup \{ & P(x_1, \ldots, x_{k-1}, x_k, x_{k+1}, \ldots, x_n) \\ & -Q(x_1, \ldots, x_{k-1}, x_k, x_{k+1}, \ldots, x_n) \ : \ x_k \in S_k\} \\ = \ & K_k(x_1, \ldots, x_{k-1}, x_{k+1}, \ldots, x_n) \in Y. \end{aligned} \tag{16}$$

*for $x_i \in S_i$, $i \in \{1, \ldots, n\} \setminus \{k\}$.*

*Then there exists an $n$-additive function $A : S_1 \times \ldots \times S_n \to Y$ satisfying* (10).

*Proof.* We can assume that $j = 1$. Then, by (9) and (6), we have

$$\begin{aligned} & P(x_1 + y_1, x_2, \ldots, x_n) - Q(y_1, x_2, \ldots, x_n) \\ \geq \ & Q(x_1 + y_1, x_2, \ldots, x_n) - Q(y_1, x_2, \ldots, x_n) \\ \geq \ & Q(x_1, x_2, \ldots, x_n), \end{aligned}$$

for $x_1, y_1 \in S_1$, $x_2 \in S_2, \ldots, x_n \in S_n$. Therefore, a map $\phi : S_1 \times \ldots \times S_n \to Y$ is well defined by

$$\begin{aligned} & \phi(x_1, x_2, \ldots, x_n) \\ & := \inf \{P(x_1 + y_1, x_2, \ldots, x_n) - Q(y_1, x_2, \ldots, x_n) \ : \ y_1 \in S_1\}, \end{aligned}$$

for $(x_1, x_2, \dots, x_n) \in S_1 \times \dots \times S_n$. By (5) we have

$$\begin{aligned}
&\phi(x_1 + y_1, x_2, \dots, x_n) \\
&\quad = \inf \{P(x_1 + y_1 + z_1, x_2, \dots, x_n) - Q(z_1, x_2, \dots, x_n) \ : \ z_1 \in S_1\} \\
&\quad \leq \inf \{P(x_1, x_2, \dots, x_n) + P(y_1 + z_1, x_2, \dots, x_n) \\
&\qquad -Q(z_1, x_2, \dots, x_n) \ : \ z_1 \in S_1\} \\
&\quad = P(x_1, x_2, \dots, x_n) + \phi(y_1, x_2, \dots, x_n)
\end{aligned}$$

and on the other hand, by (15) and (6), we have

$$\begin{aligned}
&\phi(x_1 + y_1, x_2, \dots, x_n) \\
&\quad = \inf \{P(x_1 + y_1 + z_1, x_2, \dots, x_n) - Q(z_1, x_2, \dots, x_n) \ : \ z_1 \in S_1\} \\
&\quad \geq \inf \{P(y_1 + z_1 + x_1, x_2, \dots, x_n) - K(x_2, \dots, x_n) \\
&\qquad -Q(z_1 + x_1, x_2, \dots, x_n) + Q(x_1, x_2, \dots, x_n) \ : \ z_1 \in S_1\} \\
&\quad = -K(x_2, \dots, x_n) + Q(x_1, x_2, \dots, x_n) \\
&\qquad + \inf \{P(y_1 + w_1, x_2, \dots, x_n) + Q(w_1, x_2, \dots, x_n) \ : \ w_1 \in S_1\} \\
&\quad = -K(x_2, \dots, x_n) + Q(x_1, x_2, \dots, x_n) + \phi(y_1, x_2, \dots, x_n).
\end{aligned}$$

As a result, for each fixed $(x_1, \dots, x_n) \in S_1 \times \dots \times S_n$, the mapping:

$$S_1 \ni y_1 \to \phi(x_1 + y_1, x_2, \dots, x_n) - \phi(y_1, x_2, \dots, x_n) \in Y$$

is bounded from below by $-K(x_2, \dots, x_n) + Q(x_1, x_2, \dots, x_n)$ and from above by $P(x_1, x_2, \dots, x_n)$. Now, as in the proof of Theorem 2, we define a mapping $A_1 : S_1 \times \dots \times S_n \to Y$ by the formula

$$A_1(x_1, x_2, \dots, x_n) := M_1(\phi(x_1 + \cdot, x_2, \dots, x_n) - \phi(\cdot, x_2, \dots, x_n))$$

for $(x_1, \dots, x_n) \in S_1 \times \dots \times S_n$, where $M_1$ is a generalized left invariant mean on the space $B(S_1, Y)$. Then $A_1$ is additive with respect to the first variable and

$$-K(x_2, \dots, x_n) + Q(x_1, x_2, \dots, x_n) \leq A(x_1, x_2, \dots, x_n) \leq P(x_1, x_2, \dots, x_n),$$

for $(x_1, \dots, x_n) \in S_1 \times \dots \times S_n$. In a similar fashion as in Theorem 3 we may eliminate the constant $K(x_2, \dots, x_n)$. Then

$$Q(x_1, \dots, x_n) \leq A_1(x_1, \dots, x_n) \leq P(x_1, \dots, x_n), \tag{17}$$

for all $(x_1, \dots, x_n) \in S_1 \times \dots \times S_n$.

Making use of (17), (6) and (16) we easily derive

$$\begin{aligned}
&A_1(x_1, x_2 + y_2, x_3, \dots, x_n) - A_1(x_1, y_2, x_3, \dots, x_n) \\
&\quad \geq Q(x_1, x_2 + y_2, x_3, \dots, x_n) - P(x_1, y_2, x_3, \dots, x_n) \\
&\quad = Q(x_1, x_2 + y_2, x_3, \dots, x_n) - Q(x_1, y_2, x_3, \dots, x_n) \\
&\qquad +Q(x_1, y_2, x_3, \dots, x_n) - P(x_1, y_2, x_3, \dots, x_n) \\
&\quad \geq Q(x_1, x_2, x_3, \dots, x_n) - K_2(x_1, x_3, \dots, x_n)
\end{aligned}$$

and, by (17), (5) and (16), we get

$$\begin{aligned} & A_1(x_1, x_2 + y_2, x_3, \ldots, x_n) - A_1(x_1, y_2, x_3, \ldots, x_n) \\ & \quad \le \; P(x_1, x_2 + y_2, x_3, \ldots, x_n) - Q(x_1, y_2, x_3, \ldots, x_n) \\ & \quad \le \; P(x_1, x_2 + y_2, x_3, \ldots, x_n) + P(x_1, y_2, x_3, \ldots, x_n) - Q(x_1, y_2, x_3, \ldots, x_n) \\ & \quad \le \; P(x_1, x_2, x_3, \ldots, x_n) + K_2(x_1, x_3, \ldots, x_n), \end{aligned}$$

for $x_1 \in S_1$, $x_2, y_2 \in S_2$, $x_3 \in S_3, \ldots, x_n \in S_n$. The last two inequalities imply that for any $(x_1, \ldots, x_n) \in S_1 \times \ldots \times S_n$ the function

$$S_2 \ni y_2 \to A_1(x_1, x_2 + y_2, x_3, \ldots, x_n) - A_1(x_1, y_2, x_3, \ldots, x_n) \in Y$$

is bounded, i.e. belongs to the space $B(S_2, Y)$. Let $M_2$ be a generalized invariant mean on $B(S_2, Y)$. Then the map $A_2 : S_1 \times \ldots \times S_n \to Y$ defined as follows

$$A_2(x_1, x_2, x_3, \ldots, x_n) := M_2(A_1(x_1, x_2 + \cdot, x_3, \ldots, x_n) - A_1(x_1, \cdot, x_3, \ldots, x_n))$$

for $(x_1, \ldots, x_n) \in S_1 \times \ldots \times S_n$ is additive with respect to the first and the second variable and

$$Q(x_1, \ldots, x_n) \le A_2(x_1, \ldots, x_n) \le P(x_1, \ldots, x_n), \quad (x_1, \ldots, x_n) \in S_1 \times \ldots \times S_n.$$

Consequently, continuing our procedure, we obtain some $n$-additive function $A : S_1 \times \ldots \times S_n \to Y$ such that

$$Q(x_1, \ldots, x_n) \le A(x_1, \ldots, x_n) \le P(x_1, \ldots, x_n), \quad (x_1, \ldots, x_n) \in S_1 \times \ldots \times S_n.$$

Thus the proof is finished. □

Theorem 4 remains valid if we replace condition (15) by the same assumption on $Q$. Then Theorem 4 in its original form allows us to separate $-Q$ and $-P$ by an $n$-additive function $B$. It is evident that $A := -B$ is the required $n$-additive mapping. Moreover, let us observe that condition (15) is fulfilled for Abelian semigroups.

For $n = 1$ Theorem 4 reduces to the result of by Z. Gajda and Z. Kominek in [4] (Theorem 3).

Finally, we note that if the pair $P$ and $Q$ can be represented in the form:

$$P = P_1 + \ldots + P_l, \quad Q = Q_1 + \ldots + Q_l,$$

where, for $i \in \{1, \ldots, l\}$, the pairs $P_i$, $Q_i$ fulfil the assumptions of Theorem 4, then $P$ and $Q$ are separated by some $n$-additive function. Indeed, for $i \in \{1, \ldots, l\}$, there exists an $n$-additive function $A_i : S_1 \times \ldots \times S_n \to Y$ satisfying

$$Q_i(x_1, \ldots, x_n) \le A_i(x_1, \ldots, x_n) \le P_i(x_1, \ldots, x_n), \quad (x_1, \ldots, x_n) \in S_1 \times \ldots \times S_n.$$

Then $A := A_1 + \ldots + A_l$ is the required $n$-additive mapping between $P$ and $Q$.

For example, let $a : S_1 \times S_2 \to Y$ be a 2-additive function and let, for $i = 1, 2$, $p_i : S_i \to Y$ are subadditive, $q_i : S_i \to Y$ are superadditive fulfilling

$$\sup \{p_i(x_i + y_i) - p_i(y_i + x_i) \ : \ x_i, y_i \in S_i\} \in Y$$

and

$$q_i(x_i) \leq p_i(x_i), \quad x_i \in S_i.$$

Then the pairs $(\frac{1}{2}a + p_1, \frac{1}{2}a + q_1)$ and $(\frac{1}{2}a + p_2, \frac{1}{2}a + q_2)$ satisfy the assumptions of Theorem 4 and, by our remark, we infer that the mappings $P$, $Q : S_1 \times S_2 \to Y$ defined as follows

$$P(x_1, x_2) := a(x_1, x_2) + p_1(x_1) + p_2(x_2), \quad x_1 \in S_1, \ x_2 \in S_2;$$

$$Q(x_1, x_2) := a(x_1, x_2) + q_1(x_1) + q_2(x_2), \quad x_1 \in S_1, \ x_2 \in S_2$$

are separated by a 2-additive function.

## References

[1] M.A. Albert and J.A. Baker, *Functions with bounded n-th differences.* Ann. Polon. Math. **43** (1983), 93–103.

[2] R. Badora, *On some generalized invariant means and their application to the stability of the Hyers-Ulam type.* Ann. Polon. Math. **58** (1993), 147–159.

[3] A. Chaljub-Simon, P. Volkmann, *Bemerkungen zu einem Satz von Rodé,* Arch. Math. **57** (1991), 180–188.

[4] Z. Gajda and Z. Kominek, *On separation theorems for subadditive and superadditive functionals.* Studia Math. **100** (1) (1991), 25–38.

[5] Z. Gajda, *Invariant means and representation of semigroups in the theory of functional equations.* Prace Naukowe Uniwersytetu Śląskiego, Katowice, 1992.

[6] F.P. Greenleaf, *Invariant means on topological groups and their applications.* Van Nostrand Mathematical Studies 16, New York – Toronto – London – Melbourne, 1969.

[7] R. Kaufman, *Interpolation of additive functionals.* Studia Math. **27** (1966), 269–272.

[8] P. König *On the abstract Hahn-Banach theorem due to Rodé.* Aequationes Math. **34** (1987), 89–95.

[9] P. Kranz, *Additive functionals on abelian semigroups*. Prace Mat. (Comment. Math.) **16** (1972), 239–246.

[10] Z. Páles, *Linear selections for set-valued functions and extension of bilinear forms*. Arch. Math. **62** (1994), 427–432.

[11] Z. Páles, *Separation with symmetric bilinear forms and symmetric selections of set-valued functions*, Publ. Math. Debrecen **46** /3-4 (1995), 321–331.

[12] P. Rodé, *Eine abstrakte Version des Satzes von Hahn-Banach*. Arch. Math. (Basel) **31** (1978), 474–481.

Roman Badora, Institute of Mathematics, Silesian University, Bankowa 14, PL 40-007 Katowice, Poland.

International Series of Numerical Mathematics, Vol. 123, © 1997 Birkhäuser Verlag Basel

# Convexity of power functions with respect to symmetric homogeneous means

Janusz Matkowski and Jürg Rätz

**Abstract**

It is well known that the power function $x \mapsto x^p$ on $\mathbb{R}_+^*$ is strictly Jensen-convex if $p^2 - p > 0$, strictly Jensen-concave if $p^2 - p < 0$, Jensen-convex and Jensen-concave if $p^2 - p = 0$. These Jensen type properties are based upon the arithmetic mean $A : \mathbb{R}_+^* \times \mathbb{R}_+^* \to \mathbb{R}_+^*$. It is the purpose of this paper to investigate the convexity/concavity classification of the power functions for symmetric homogeneous means on $\mathbb{R}_+^*$ other than $A$. In Section 3, a convexity/concavity criterion is presented, and in Section 4 this is applied to the families of Stolarsky means and Gini means (both containing $A$) as well as to weighted geometric means.

## 1. Introduction

Throughout the paper, $\mathbb{R}$, $\mathbb{R}_+$, $\mathbb{R}_-$, $\mathbb{R}_+^*$, $\mathbb{R}_-^*$, $\mathbb{R}^*$ denote the sets of real, nonnegative real, nonpositive real, positive real, negative real, non-zero real numbers, respectively. For $p \in \mathbb{R}$ we define

$$\varphi_p : \mathbb{R}_+^* \to \mathbb{R}_+^*, \qquad \varphi_p(x) := x^p \qquad (x \in \mathbb{R}_+^*). \tag{1}$$

*log* denotes the natural logarithm.

**Definition 1.**

a) The function $M : \mathbb{R}_+^* \times \mathbb{R}_+^* \to \mathbb{R}_+^*$ is called a *mean* on $\mathbb{R}_+^*$ if

$$\begin{aligned} &\min\{x, y\} \leq M(x, y) \leq \max\{x, y\}, \quad \text{i.e.,} \\ &M(x, y) \in \operatorname{conv}\{x, y\}, \quad \text{for all} \quad x, y \in \mathbb{R}_+^*. \end{aligned} \tag{2}$$

b) A mean $M$ on $\mathbb{R}_+^*$ is said to be *homogeneous* (more precisely: *positively homogeneous*) if

$$M(ax, ay) = aM(x, y) \qquad \text{for all } a, x, y \in \mathbb{R}_+^*, \tag{3}$$

*symmetric* if

$$M(x, y) = M(y, x) \text{ for all } x, y \in \mathbb{R}_+^*. \tag{4}$$

The set of all symmetric homogeneous means on $\mathbb{R}_+^*$ is denoted by $\mathcal{M}(\mathbb{R}_+^*)$. Note that the degree of homogeneity of a mean has to be one.

**Definition 2.** If $M$ is a mean on $\mathbb{R}_+^*$, a function $\varphi : \mathbb{R}_+^* \to \mathbb{R}_+^*$ is called

a) *$M$-convex* on $\mathbb{R}_+^*$ if

$$\varphi(M(x, y)) \leq M(\varphi(x), \varphi(y)) \text{ for all } x, y \in \mathbb{R}_+^*, \tag{5}$$

b) *strictly $M$-convex* on $\mathbb{R}_+^*$ if

$$\varphi(M(x, y)) < M(\varphi(x), \varphi(y)) \text{ for all } x, y \in \mathbb{R}_+^*, \ x \neq y, \tag{6}$$

c) *$M$-concave* on $\mathbb{R}_+^*$ if

$$\varphi(M(x, y)) \geq M(\varphi(x), \varphi(y)) \text{ for all } x, y \in \mathbb{R}_+^*, \tag{7}$$

d) *strictly $M$-concave* on $\mathbb{R}_+^*$ if

$$\varphi(M(x, y)) > M(\varphi(x), \varphi(y)) \text{ for all } x, y \in \mathbb{R}_+^*, \ x \neq y, \tag{8}$$

e) *$M$-affine* on $\mathbb{R}_+^*$ if

$$\varphi(M(x, y)) = M(\varphi(x), \varphi(y)) \text{ for all } x, y \in \mathbb{R}_+^*. \tag{9}$$

**Remark 3.** It is clear that the convexity/concavity behavior of $\varphi_p$ is strongly related to the comparison (also called dominance) problem of the means involved: strict $M$-convexity of $\varphi_p$ for $p \in \mathbb{R}^*$ on $\mathbb{R}_+^*$, e.g., means that

$$M(x, y) < (M(x^p, y^p))^{\frac{1}{p}} \qquad (x, y \in \mathbb{R}_+^*;\ x \neq y)$$

which expresses a comparison result for two means being "conjugate by $\varphi_p$". So our convexity/concavity problem is a special case of the general comparison problem. The proofs here are easier and more direct for our purposes and, accordingly, the structure of the conditions is simpler that those for the general problem (cf., e.g., [9] , [14], [13] in connection with some cases in Theorems 13 and 16). The main reason for our direct access, however, is that we intend to obtain strict inequalities also in the limiting cases (e.g., for $E_{r,r}$ and $E_{r,0}$ in Theorem 13), and this does require an extra effort anyway.

**Remark 4.** For a different point of view leading to the functional inequalities in Definition 2, cf., e.g. [16] , p. 38–39 or [18], p. 103–104.

## 2. Preliminaries

**Lemma 5.** *Let $M$ be a mean on $\mathbb{R}_+^*$.*

1°. *If $\varphi: \mathbb{R}_+^* \to \mathbb{R}_+^*$ and $\psi: \mathbb{R}_+^* \to \mathbb{R}_+^*$ are*

$$M\text{-}\genfrac{}{}{0pt}{}{\textit{convex}}{\textit{concave}} \quad \textit{on } \mathbb{R}_+^*$$

*and $\psi$ is increasing, then $\psi \circ \varphi$ is $M\text{-}\genfrac{}{}{0pt}{}{\textit{convex}}{\textit{concave}}$ on $\mathbb{R}_+^*$. (There is no analogue for $\psi$ decreasing.)*

2°. *Let $\varphi: \mathbb{R}_+^* \to \mathbb{R}_+^*$ be bijective and (strictly) $M\text{-}\genfrac{}{}{0pt}{}{\textit{convex}}{\textit{concave}}$ on $\mathbb{R}_+^*$.*
*If $\varphi$ is increasing, then $\varphi^{-1}$ is (strictly) $M\text{-}\genfrac{}{}{0pt}{}{\textit{concave}}{\textit{convex}}$ on $\mathbb{R}_+^*$.*
*If $\varphi$ is decreasing, then $\varphi^{-1}$ is (strictly) $M\text{-}\genfrac{}{}{0pt}{}{\textit{convex}}{\textit{concave}}$ on $\mathbb{R}_+^*$.*

The immediate proof is omitted.

The next statement prepares the convexity/concavity classification of the power functions $\varphi_p$ ($p \in \mathbb{R}$) and focusses the interest to $p > 1$ and $p = -1$.

**Theorem 6.** *Let $M$ be a mean on $\mathbb{R}_+^*$.*

1°. *For every $p > 1$, the function $\varphi_p$ is (strictly) $M\text{-}\genfrac{}{}{0pt}{}{\textit{convex}}{\textit{concave}}$ on $\mathbb{R}_+^*$ if, and only if, the function $\varphi_{\frac{1}{p}}$ is (strictly) $M\text{-}\genfrac{}{}{0pt}{}{\textit{concave}}{\textit{convex}}$ on $\mathbb{R}_+^*$.*

2°. *If all $\varphi_p$ $(p > 1)$ are (strictly) $M\text{-}\genfrac{}{}{0pt}{}{\textit{convex}}{\textit{concave}}$ on $\mathbb{R}_+^*$ and if $\varphi_{-1}$ is $M\text{-}\genfrac{}{}{0pt}{}{\textit{convex}}{\textit{concave}}$ on $\mathbb{R}_+^*$, then all $\varphi_q$ $(q < 0)$ are (strictly) $M\text{-}\genfrac{}{}{0pt}{}{\textit{convex}}{\textit{concave}}$ on $\mathbb{R}_+^*$.*

3°. *If all $\varphi_p$ $(p > 1)$ are $M\text{-}\genfrac{}{}{0pt}{}{\textit{convex}}{\textit{concave}}$ on $\mathbb{R}_+^*$ and if $\varphi_{-1}$ is (strictly) $M\text{-}\genfrac{}{}{0pt}{}{\textit{convex}}{\textit{concave}}$ on $\mathbb{R}_+^*$, then all $\varphi_q$ $(q < 0)$ are (strictly) $M\text{-}\genfrac{}{}{0pt}{}{\textit{convex}}{\textit{concave}}$ on $\mathbb{R}_+^*$.*

4°. *$\varphi_0$ and $\varphi_1$ are $M$-affine on $\mathbb{R}_+^*$.*

*Proof.* 1°. is obtained by the substitution $u = x^p, v = y^p$.

2°. Assume that all $\varphi_p$ $(p > 1)$ are strictly $M$-convex and $\varphi_{-1}$ is $M$-convex on $\mathbb{R}_+^*$. Suppose first $q < -1$. Then $-q > 1$, and by the assumption we have

$$(M(x,y))^{-q} < M(x^{-q}, y^{-q}) \qquad (x, y \in \mathbb{R}_+^*, x \neq y).$$

Substituting $x$, $y$ by $x^{-1}$, $y^{-1}$ we get

$$(M(x^{-1}, y^{-1}))^{-q} < M(x^q, y^q) \qquad (x, y \in \mathbb{R}_+^*,\ x \neq y). \tag{10}$$

$M$-convexity of $\varphi_{-1}$ on $\mathbb{R}_+^*$ yields

$$(M(x,y))^{-1} \leq M(x^{-1}, y^{-1}) \qquad (\forall\, x, y \in \mathbb{R}_+^*,\ x \neq y).$$

Since $\varphi_{-q}$ is strictly increasing, it follows that

$$(M(x,y))^q \leq (M(x^{-1},y^{-1}))^{-q} \qquad (\forall\, x,y \in \mathbb{R}_+^*,\ x \neq y)$$

and together with (10)

$$(M(x,y))^q < M(x^q,y^q) \qquad (\forall\, x,y \in \mathbb{R}_+^*,\ x \neq y), \tag{11}$$

i.e., that $\varphi_q$ is strictly $M$-convex on $\mathbb{R}_+^*$.

Now suppose that $q \in ]-1,0[$. Then $-\frac{1}{q} > 1$, and by the assumption, $\varphi_{-\frac{1}{q}}$ is strictly $M$-convex on $\mathbb{R}_+^*$, and since it is increasing, $\varphi_{-q}$ is strictly $M$-concave on $\mathbb{R}_+^*$ by Lemma 5.2°, i.e.,

$$(M(x,y))^{-q} > M(x^{-q},y^{-q}) \qquad (x,y \in \mathbb{R}_+^*,\ x \neq y). \tag{12}$$

$M$-convexity of $\varphi_{-1}$ on $\mathbb{R}_+^*$ implies

$$(M(x^{-q},y^{-q}))^{-1} \leq M(x^q,y^q) \qquad (x,y \in \mathbb{R}_+^*,\ x \neq y),$$

and together with (12) we obtain again (11).

The non-strict convexity case as well as the two concavity cases can be proved along the same lines.

3°. The proof follows again the same lines, but here the possible strictness of $\varphi_q$ results from that of $\varphi_{-1}$.

4°. is obvious, and the proof of Theorem 6 is complete. □

**Lemma 7.** *If $M$ is a mean on $\mathbb{R}_+^*$, $p \in \mathbb{R}$, $r \in \mathbb{R}^*$, and ${}_rM(x,y) := (M(x^r,y^r))^{\frac{1}{r}}$ for all $x,y \in \mathbb{R}_+^*$, then we have:*

1°. *${}_rM$ is a mean on $\mathbb{R}_+^*$, the* $r$-conjugate of $M$.

2°. *If $M \in \mathcal{M}(\mathbb{R}_+^*)$, then ${}_rM \in \mathcal{M}(\mathbb{R}_+^*)$.*

3°. *If $r \in \mathbb{R}^*$, then $\varphi_p$ is ${}_rM$-affine on $\mathbb{R}_+^* \iff \varphi_p$ is $M$-affine on $\mathbb{R}_+^*$.*

4°. *If $r \in \mathbb{R}_+^*$, then $\varphi_p$ is ${}_rM$-$\genfrac{}{}{0pt}{}{convex}{concave}$ on $\mathbb{R}_+^* \iff \varphi_p$ is $M$-$\genfrac{}{}{0pt}{}{convex}{concave}$ on $\mathbb{R}_+^*$.*

5°. *If $r \in \mathbb{R}_-^*$, then $\varphi_p$ is ${}_rM$-$\genfrac{}{}{0pt}{}{convex}{concave}$ on $\mathbb{R}_+^* \iff \varphi_p$ is $M$-$\genfrac{}{}{0pt}{}{concave}{convex}$ on $\mathbb{R}_+^*$.*

The obvious proof is omitted.

## 3. An M-convexity/-concavity criterion

After having dealt with arbitrary means on $\mathbb{R}_+^*$ in Section 2, we now turn to the question as to how we can recognize whether all power functions $\varphi_p (p > 1)$ have the same convexity/concavity behavior with respect to a symmetric homogeneous mean on $\mathbb{R}_+^*$.

**Theorem 8.** *If $M \in \mathcal{M}(\mathbb{R}_+^*)$, then all $\varphi_p$ $(p > 1)$ are (strictly) M- $\begin{smallmatrix}\text{convex}\\ \text{concave}\end{smallmatrix}$ on $\mathbb{R}_+^*$ if, and only if the function*

$$f_M : \mathbb{R}_+^* \to \mathbb{R}_+^*, \qquad f_M(x) := (M(e^x, 1))^{\frac{1}{x}} \qquad (x \in \mathbb{R}_+^*) \tag{13}$$

*is (strictly) $\begin{smallmatrix}\text{increasing}\\ \text{decreasing}\end{smallmatrix}$ on $\mathbb{R}_+^*$.*

*Proof.* We write the argument for the strictly convex case; immediate modifications for the three remaining cases are available.

Using symmetry of $M$, putting $t = \frac{x}{y}$, using homogeneity of $M$, dividing by $y^p$, putting $e^x = t$, and exponentiating with $\frac{1}{px}$ $(> 0)$ generate the following sequence of mutually equivalent statements:

$$\begin{array}{lcll}
(M(x,y))^p & < & M(x^p, y^p) & (x, y \in \mathbb{R}_+^*;\ x \neq y; p > 1) \\
(M(x,y))^p & < & M(x^p, y^p) & (x, y \in \mathbb{R}_+^*;\ x > y; p > 1) \\
(M(ty,y))^p & < & M(t^p y^p, y^p) & (t > 1;\ y \in \mathbb{R}_+^*;\ p > 1) \\
y^p (M(t,1))^p & < & y^p M(t^p, 1) & (t > 1;\ y \in \mathbb{R}_+^*;\ p > 1) \\
(M(t,1))^p & < & M(t^p, 1) & (t > 1,\ p > 1) \\
(M(e^x,1))^p & < & M(e^{px}, 1) & (x \in \mathbb{R}_+^*,\ p > 1) \\
(M(e^x,1))^{\frac{1}{x}} & < & (M(e^{px}, 1))^{\frac{1}{px}} & (x \in \mathbb{R}_+^*,\ p > 1) \\
f_M(x) & < & f_M(px) & (x \in \mathbb{R}_+^*,\ p > 1).
\end{array}$$

□

## 4. Applications

We first deal with the quasi-arithmetic means in $\mathcal{M}(\mathbb{R}_+^*)$, which are precisely the geometric mean $G =: M_0$ and the *power means* $M_r$ $(r \in \mathbb{R}^*)$ given by

$$M_r(x,y) := \left(\frac{x^r + y^r}{2}\right)^{\frac{1}{r}} \quad \text{for all } x, y \in \mathbb{R}_+^*$$

([6], p. 68, Theorem 84), the latter being the $r$-conjugates ${}_rA$ of the arithmetic mean $A$ (cf. Lemma 7.1°). The result stated at the beginning of the Abstract, Lemma 7, and the fact that all $\varphi_p$ $(p \in \mathbb{R})$ are $M_0$-affine on $\mathbb{R}_+^*$ lead to

$$p, r \in \mathbb{R} \Rightarrow \left[\varphi_p \text{ is } \begin{array}{c}\text{strictly } M_r\text{-convex}\\ M_r\text{-affine}\\ \text{strictly } M_r\text{-concave}\end{array} \text{ on } \mathbb{R}_+^* \Leftrightarrow r(p^2 - p) \begin{array}{c} > \\ = \\ < \end{array} 0\right]. \tag{14}$$

Later we shall treat two distinct 2-parameter families $(E_{r,s})_{r,s\in\mathbb{R}}$ and $(M_{r,s})_{r,s\in\mathbb{R}}$ of elements of $\mathcal{M}(\mathbb{R}_+^*)$ which contain the means $M_r$ $(r\in\mathbb{R})$; see Remark 17.

We begin by two specific members of the former family, the *logarithmic mean* $L$ and the *identric mean* $I$; $L$ has numerous applications in physics (cf., e.g., [3] p. 376-377; [7]; [12], sect. 8.7, 8.9; [15], sect. 6.2), and it is not a quasi-arithmetic mean ([10], Theorems 2 and 4; [7], Theorem 1).

**Theorem 9.** *Let* $L:\mathbb{R}_+^*\times\mathbb{R}_+^*\to\mathbb{R}_+^*$, *be defined by*

$$L(x,x):=x,\ L(x,y):=\frac{x-y}{\log x-\log y}\qquad (x,y\in\mathbb{R}_+^*,\ x\neq y).$$

*Then we have*

1°. $L\in\mathcal{M}(\mathbb{R}_+^*)$,

$$2^\circ.\quad p\in\mathbb{R}\Rightarrow\left[\varphi_p \textit{ is } \begin{array}{c}\textit{strictly } L\textit{-convex}\\ L\textit{-affine}\\ \textit{strictly } L\textit{-concave}\end{array} \textit{ on } \mathbb{R}_+^*\iff p^2-p\ \begin{array}{c}>\\=\\<\end{array}\ 0\right].$$

*Proof.* 1°. is well known; the intermediacy property (2) easily follows from the Mean Value Theorem.

2°. (i) For $t>1$ we have $z:=\frac{1}{2}\log t>0$, so, by looking at the Maclaurin series, $z<\sinh z$, i.e., $2z<e^z-e^{-z}$, therefore $\log t<t^{\frac12}-t^{-\frac12}(t>1)$ (cf. also [11], p.272, 3.6.15, where a different argument is used). If follows that

$$(\log t)^2<t-2+\frac{1}{t}\qquad (t>1).\tag{15}$$

Take $x,y\in\mathbb{R}_+^*$, $x>y$ and put $t=\frac{x}{y}$. Then (15) yields $(\log x-\log y)^2<\frac{x}{y}-2+\frac{y}{x}=(x-y)(\frac{1}{y}-\frac{1}{x})$, which can be written in the form $(L(x,y))^{-1}<L(\frac{1}{x},\frac{1}{y})$ $(x>y>0)$, and symmetry of $L$ implies that

$$\varphi_{-1}\ \text{is strictly}\ L\text{-convex on}\ \mathbb{R}_+^*.\tag{16}$$

(ii) $f_L(x)\overset{(13)}{=}(L(e^x,1))^{\frac{1}{x}}=(\frac{e^x-1}{x})^{\frac{1}{x}}$ $(x\in\mathbb{R}_+^*)$. We show that

$$f_L\ \text{is strictly increasing on}\ \mathbb{R}_+^*.\tag{17}$$

To this end we define $f_1,f_2:\mathbb{R}_+\to\mathbb{R}_+$ by

$$f_1(x):=\frac{xe^x}{e^x-1}-1,\ f_2(x):=\log\frac{e^x-1}{x}\ (x\in\mathbb{R}_+^*),\ f_1(0):=0,\ f_2(0):=0.\tag{18}$$

$f_1, f_2$ are continuous on $\mathbb{R}_+$ and differentiable on $\mathbb{R}_+^*$. From $\frac{e^x-1}{x} > e^{\frac{x}{2}}$ $(x \in \mathbb{R}_+^*)$ we obtain $\frac{1}{x} > \frac{xe^x}{(e^x-1)^2}$ $(x \in \mathbb{R}_+^*)$, therefore

$$f_1'(x) = \frac{e^x(e^x-1) - xe^x}{(e^x-1)^2} > \frac{e^x}{e^x-1} - \frac{1}{x} = f_2'(x) \quad (x \in \mathbb{R}_+^*). \tag{19}$$

Applying the Mean Value Theorem to $f_1 - f_2$ and using (18) and (19) yields

$$f_1(x) > f_2(x) \qquad (x \in \mathbb{R}_+^*). \tag{20}$$

Finally $f_L'(x) = f_L(x) \cdot \frac{1}{x^2} \cdot (f_1(x) - f_2(x)) > 0$ $(x \in \mathbb{R}_+^*)$, by (20), and (17) is proved.

(iii) By (17) and Theorem 8, all $\varphi_p (p > 1)$ are strictly $L$-convex on $\mathbb{R}_+^*$. By Theorem 6.1°, all $\varphi_p$ $(0 < p < 1)$ are strictly $L$-concave on $\mathbb{R}_+^*$, and by (16) and Theorem 6.2°, all $\varphi_p$ $(p < 0)$ are strictly $L$-convex on $\mathbb{R}_+^*$. Theorem 6.4° completes the "$\Longleftarrow$" part of 2°.

(iv) It is sufficient to prove the "$\Longleftarrow$" statements because on both sides of "$\Longleftrightarrow$" we have threefold alternatives, the one on the right-hand side being complete, i.e., a trichotomy. □

**Theorem 10.** *Let* $I : \mathbb{R}_+^* \times \mathbb{R}_+^* \to \mathbb{R}_+^*$, *be defined by*

$$I(x,x) := x, \; I(x,y) := \frac{1}{e}\left(\frac{y^y}{x^x}\right)^{\frac{1}{(y-x)}} \qquad (x, y \in \mathbb{R}_+^*, \; x \neq y).$$

*Then we have*

1°. $I \in \mathcal{M}(\mathbb{R}_+^*)$,

2°. $p \in \mathbb{R} \Rightarrow \left[\varphi_p \text{ is } \begin{array}{c} \textit{strictly } I\textit{-convex} \\ I\textit{-affine} \\ \textit{strictly } I\textit{-concave} \end{array} \text{ on } \mathbb{R}_+^* \Longleftrightarrow p^2 - p \begin{array}{c} > \\ = \\ < \end{array} 0\right]$.

*Proof.* 1° is well known.

2°. (i) We start from $\frac{2t}{2+t} < \log(1+t)$ $(t \in \mathbb{R}_+^*)$ ([11], p. 273, 3.6.18) and get $2 < (1 + \frac{2}{t})\log(1+t)$ $(t \in \mathbb{R}_+^*)$. For $x, y \in \mathbb{R}_+^*$, $x < y$, $t := \frac{y-x}{x}$, we obtain by a simple calculation $0 < \log I(x,y) + \log I(x^{-1}, y^{-1})$, i.e., $1 < I(x,y) \cdot I(x^{-1}, y^{-1})$ $(x, y \in \mathbb{R}_+^*, \; x < y)$. By symmetry of $I$, this is valid for all $x \neq y$, so

$$\varphi_{-1} \text{ is strictly } I\text{-convex on } \mathbb{R}_+^*. \tag{21}$$

(ii) $f_I(x) \overset{(13)}{=} (I(e^x, 1))^{\frac{1}{x}} = e^{\frac{e^x}{(e^x-1)} - \frac{1}{x}}$ $(x \in \mathbb{R}_+^*)$.

For $x \in \mathbb{R}_+^*$ we have $\cosh x > 1 + \frac{x^2}{2}$, $e^x + e^{-x} > 2 + x^2$, $e^x - 2 + e^{-x} > x^2$, $\frac{(e^x-1)^2}{e^x} > x^2$, $0 < -\frac{e^x}{(e^x-1)^2} + \frac{1}{x^2}$ and finally $\left(\frac{e^x}{e^x-1} - \frac{1}{x}\right)' > 0$, which guarantees that

$$f_I \text{ is strictly increasing on } \mathbb{R}_+^*. \tag{22}$$

(iii) The proof is completed as for Theorem 9 with (21), (22) instead of (16), (17). □

**Definition 11.** The *Stolarsky means* ([17], p. 88)

$$E_{r,s} : \mathbb{R}_+^* \times \mathbb{R}_+^* \longrightarrow \mathbb{R}_+^* \qquad (r, s \in \mathbb{R})$$

are defined by

$$E_{r,s}(x,x) := x \qquad (x \in \mathbb{R}_+^*)$$

and for $x, y \in \mathbb{R}_+^*$, $x \neq y$, by

$$E_{r,s}(x,y) := \left(\frac{r}{s} \cdot \frac{y^s - x^s}{y^r - x^r}\right)^{\frac{1}{s-r}} \quad \text{if} \quad rs(s-r) \neq 0 \tag{23}$$

$$E_{r,0}(x,y) := E_{0,r}(x,y) := \left(\frac{1}{r} \cdot \frac{y^r - x^r}{\log y - \log x}\right)^{\frac{1}{r}} \quad \text{if} \quad r \neq 0 \tag{24}$$

$$E_{r,r}(x,y) := e^{-\frac{1}{r}} \cdot \left(\frac{y^{y^r}}{x^{x^r}}\right)^{\frac{1}{y^r - x^r}} \quad \text{if} \quad r = s \neq 0 \tag{25}$$

$$E_{0,0}(x,y) := G(x,y). \tag{26}$$

**Remark 12.** a) All $E_{r,s}$ $(r, s \in \mathbb{R})$ are means on $\mathbb{R}_+^*$: they are so-called mean-value means, arising from the Cauchy Mean Value Theorem ([1], p. 345); furthermore $E_{r,s} \in \mathcal{M}(\mathbb{R}_+^*)$.

b) For numerous further properties of $E_{r,s}$ cf.[8], p. 86. Some of them will be needed here, namely

$$E_{r,2r} = M_r \qquad \text{for all } r \in \mathbb{R}^*, \tag{27}$$

$$E_{r,-r} = G \qquad \text{for all } r \in \mathbb{R}, \tag{28}$$

$$E_{r,kr} = {}_rE_{1,k} \qquad \text{for all } r \in \mathbb{R}^*,\ k \in \mathbb{R}, \tag{29}$$

$$E_{1,0} = L, \qquad \text{the logarithmic mean}, \tag{30}$$

$$E_{1,1} = I, \qquad \text{the identric mean}. \tag{31}$$

**Theorem 13.** *For $p, r, s \in \mathbb{R}$ we have*

$$\varphi_p \textit{ is } \begin{array}{c} \textit{strictly } E_{r,s}\textit{-convex} \\ E_{r,s}\textit{-affine} \\ \text{strictly } E_{r,s}\textit{-concave} \end{array} \textit{ on } \mathbb{R}_+^* \iff (r+s)(p^2-p) \begin{array}{c} > \\ = \\ < \end{array} 0.$$

*Proof.* It is sufficient to prove the "$\Longleftarrow$" statements (see (iv)) in the proof of Theorem 9.

The case $(r+s)(p^2-p)=0$ is settled by Theorem 6.4° ($p^2-p=0$) and (28) together with the remark before (14) ($r+s=0$).

According to the structure of Definition 11 and taking into account that $r=-s$ is already settled, we treat the following three cases separately:

Case 1: $r,s\in\mathbb{R}^*,\ r\neq s, r\neq -s$ (cf.(23)),

Case 2: $r\in\mathbb{R}^*,\ s=0$ (cf.(24)),

Case 3: $r=s\in\mathbb{R}^*$ (cf.(25)).

*Case* 1. (i) By (23) and (13)

$$f_{E_{r,s}}(t)=\left(\frac{r}{s}\cdot\frac{e^{ts}-1}{e^{tr}-1}\right)^{\frac{1}{t(s-r)}}\qquad(t\in\mathbb{R}_+^*).$$

Therefore

$$(\log f_{E_{r,s}}(t))'=\frac{1}{(s-r)t^2}\cdot(g_1(t)-g_2(t))\qquad(t\in\mathbb{R}_+^*),\tag{32}$$

where $g_1,g_2:\mathbb{R}_+\to\mathbb{R}$, $g_1(t):=\frac{tse^{ts}}{e^{ts}-1}-\frac{tre^{tr}}{e^{tr}-1}$ $(t\in\mathbb{R}_+^*)$, $g_1(0):=0$, $g_2(t):=\log\left(\frac{r}{s}\cdot\frac{e^{ts}-1}{e^{tr}-1}\right)$ $(t\in\mathbb{R}_+^*)$, $g_2(0):=0$.

$g_1$ and $g_2$ are continuous on $\mathbb{R}_+$ and differentiable on $\mathbb{R}_+^*$. Furthermore

$$(g_1(t)-g_2(t))'=t\cdot g_2''(t)\quad(t\in\mathbb{R}_+^*),\tag{33}$$

$$g_2''(t)=r^2\cdot\frac{e^{tr}}{(e^{tr}-1)^2}-s^2\cdot\frac{e^{ts}}{(e^{ts}-1)^2}\quad(t\in\mathbb{R}_+^*).\tag{34}$$

Because of (32), (33), $\operatorname{sgn} g_2''(t)$ is of high importance. We define

$$h_t:\mathbb{R}^*\to\mathbb{R},\quad h_t(v):=\frac{v^2e^{tv}}{(e^{tv}-1)^2}\quad(v\in\mathbb{R}^*;\ t\in\mathbb{R}_+^*\ \text{ fixed})$$

and get $t^2\cdot h_t(v)=\left(\frac{\frac{t}{2}v}{\sinh\frac{t}{2}v}\right)^2$ $(v\in\mathbb{R}^*;\,t\in\mathbb{R}_+^*)$.

The Maclaurin series (or strict convexity on $\mathbb{R}_+^*$) of *sinh* shows that $u\mapsto\frac{\sinh u}{u}$ is strictly increasing on $\mathbb{R}_+^*$, so for all $t\in\mathbb{R}_+^*$, the function $h_t$ is strictly decreasing on $\mathbb{R}_+^*$, moreover it is even.

Now by (34),(33)

$$0\ \begin{matrix}<\\>\end{matrix}\ g_2''(t)\qquad(t\in\mathbb{R}_+^*)\qquad\begin{pmatrix}0<|r|<|s|\\0<|s|<|r|\end{pmatrix},$$

$$0\ \begin{matrix}<\\>\end{matrix}\ (g_1(t)-g_2(t))'\qquad(t\in\mathbb{R}_+^*)\qquad\begin{pmatrix}0<|r|<|s|\\0<|s|<|r|\end{pmatrix}.$$

The Mean Value Theorem and $g_1(0) = 0 = g_2(0)$ imply

$$0 \begin{matrix} < \\ > \end{matrix} g_1(t) - g_2(t) \qquad (t \in \mathbf{R}_+^*) \qquad \begin{pmatrix} 0 < |r| < |s| \\ 0 < |s| < |r| \end{pmatrix}, \qquad \text{i.e.,}$$

$$0 < \frac{1}{s^2 - r^2}(g_1(t) - g_2(t)) \qquad (t \in \mathbb{R}_+^*), \qquad \text{i.e., by (32)}$$

$$0 < \frac{1}{r+s}(\log f_{E_{r,s}}(t))' \qquad (t \in \mathbb{R}_+^*),$$

therefore by Theorem 8

$$\text{all } \varphi_p (p > 1) \text{ are strictly } \begin{matrix} E_{r,s}\text{-convex} \\ E_{r,s}\text{-concave} \end{matrix} \text{ on } \mathbb{R}_+^* \quad (r + s \begin{matrix} > \\ < \end{matrix} 0). \tag{35}$$

(ii) For every $t \in \mathbb{R}_+^*$ we define

$$\tilde{h}_t(v) := \left( \frac{\sinh \frac{t}{2} v}{\frac{t}{2} \cdot v} \right)^2 \qquad (v \in \mathbb{R}^*).$$

Then $\tilde{h}_t$ is strictly increasing on $\mathbb{R}_+^*$ and even, so

$$\tilde{h}_t(s) \begin{matrix} > \\ < \end{matrix} \tilde{h}_t(r) \qquad (t \in \mathbb{R}_+^*) \qquad \begin{pmatrix} 0 < |r| < |s| \\ 0 < |s| < |r| \end{pmatrix}.$$

By putting $z := e^t$ we obtain

$$r^2(2 - z^s - z^{-s}) \begin{matrix} < \\ > \end{matrix} s^2(2 - z^r - z^{-r}) \qquad (z > 1) \qquad \begin{pmatrix} 0 < |r| < |s| \\ 0 < |s| < |r| \end{pmatrix}. \tag{36}$$

Let be $x, y \in \mathbb{R}_+^*$ arbitrary, $x \neq y$. Then $z := \max\{\frac{x}{y}, \frac{y}{x}\} > 1$, and (36) yields

$$r^2(y^{-s} - x^{-s})(y^s - x^s) \begin{matrix} < \\ > \end{matrix} s^2(y^{-r} - x^{-r})(y^r - x^r) \qquad \begin{pmatrix} 0 < |r| < |s| \\ 0 < |s| < |r| \end{pmatrix}.$$

Dividing by the (negative) right-hand side gives

$$\frac{r}{s} \cdot \frac{y^{-s} - x^{-s}}{y^{-r} - x^{-r}} \cdot \frac{r}{s} \cdot \frac{y^s - x^s}{y^r - x^r} \begin{matrix} > \\ < \end{matrix} 1 \qquad \begin{pmatrix} 0 < |r| < |s| \\ 0 < |s| < |r| \end{pmatrix}.$$

Raising to the power with exponent $\frac{1}{s-r} \begin{matrix} > \\ < \end{matrix} 0 \begin{pmatrix} r < s \\ s < r \end{pmatrix}$ and using (23) leads to

$$E_{r,s}(x^{-1}, y^{-1}) \cdot E_{r,s}(x, y) \begin{matrix} > \\ < \end{matrix} 1 \qquad \begin{pmatrix} r + s > 0 \\ r + s < 0 \end{pmatrix}, \qquad \text{i.e. ,}$$

$$\varphi_{-1} \text{ is strictly } \begin{matrix} E_{r,s}\text{-convex} \\ E_{r,s}\text{-concave} \end{matrix} \text{ on } \mathbb{R}_+^* \qquad (r + s \begin{matrix} > \\ < \end{matrix} 0). \tag{37}$$

(iii) In Case 1, the assertion follows from (35), (37), Theorem 6.1° and 2°.

*Case* 2. Here $r+s=r$. $E_{r,0} \overset{(29)}{=} {}_rE_{1,0} \overset{(30)}{=} {}_rL$. By Theorem 9.2° and Lemma 7.3°, 4°, 5°, the assertion follows.

*Case* 3. Here $r+s=2r$. $E_{r,r} \overset{(29)}{=} {}_rE_{1,1} \overset{(31)}{=} {}_rI$. By Theorem 10.2° and Lemma 7.3°, 4°, 5°, the assertion follows. □

**Definition 14.** The *Gini means* $M_{r,s} : \mathbb{R}_+^* \times \mathbb{R}_+^* \to \mathbb{R}_+^*$ $(r, s \in \mathbb{R})$ are defined by

$$M_{r,s}(x,x) := x \qquad (x \in \mathbb{R}_+^*)$$

and for $x, y \in \mathbb{R}_+^*$, $x \neq y$, by

$$M_{r,s}(x,y) = \begin{cases} \left(\frac{x^s+y^s}{x^r+y^r}\right)^{\frac{1}{(s-r)}} & \text{if } r \neq s \\ e^{\frac{x^r \log x + y^r \log y}{x^r+y^r}} & \text{if } r = s \end{cases}$$

([4]; J. Aczél informed the authors about the earlier reference [5]). There are applications in statistics. For further references cf. [2].

**Remark 15.** It is well-known that $M_{r,s} \in \mathcal{M}(\mathbb{R}_+^*)$ for all $r, s \in \mathbb{R}$. Furthermore

$$M_{r,0} = M_r \qquad \text{for all } r \in \mathbb{R}^*, \tag{38}$$

$$M_{r,-r} = G \qquad \text{for all } r \in \mathbb{R}. \tag{39}$$

**Theorem 16.** *For $p, r, s \in \mathbb{R}$ we have*

$$\varphi_p \textit{ is } \begin{matrix} \textit{strictly } M_{r,s}\textit{-convex} \\ M_{r,s}\textit{-affine} \\ \textit{strictly } M_{r,s}\textit{-concave} \end{matrix} \textit{ on } \mathbb{R}_+^* \iff (r+s)(p^2-p) \begin{matrix} > \\ = \\ < \end{matrix} 0.$$

*Proof.* As for Theorem 13, proving "$\Longleftarrow$" is enough, and the case $(r+s)(p^2-p)=0$ can again be isolated (cf. (39)), so $r \neq -s$.

*Case* 1: $r \neq s$.

(i) From Definition 14 and (13) we obtain

$$(\log f_{M_{r,s}}(t))' = \frac{1}{(s-r)t^2} \cdot (j_t(s) - j_t(r)) \qquad (t \in \mathbb{R}_+^*) \tag{40}$$

where $j_t : \mathbb{R} \to \mathbb{R}$,

$$j_t(v) := \frac{tv \cdot e^{tv}}{e^{tv}+1} - \log(e^{tv}+1) \qquad (v \in \mathbb{R};\ t \in \mathbb{R}_+^* \text{ fixed}).$$

Simple computations show that for every $t \in \mathbb{R}_+^*$, the function $j_t$ is even and $(\frac{d}{dv} j_t)(v) = \frac{t^2 v e^{tv}}{(e^{tv}+1)^2}$ $(v \in \mathbb{R})$, i.e. $j_t$ is strictly increasing on $\mathbb{R}_+^*$. Hence

$$j_t(s) - j_t(r) \begin{smallmatrix} > \\ < \end{smallmatrix} 0 \quad (t \in \mathbb{R}_+^*) \quad \begin{pmatrix} |r| < |s| \\ |s| < |r| \end{pmatrix}$$

(notice that the case $r = -s$ is excluded), i.e. $\frac{1}{s^2 - r^2} \cdot (j_t(s) - j_t(r)) > 0$ $(t \in \mathbb{R}_+^*)$, i.e., by (40), $\frac{1}{r+s} \cdot (\log f_{M_{r,s}}(t))' > 0$ $(t \in \mathbb{R}_+^*)$, and finally by Theorem 8

$$\text{all } \varphi_p (p > 1) \text{ are strictly } \begin{cases} M_{r,s}\text{-convex} \\ M_{r,s}\text{-concave} \end{cases} \text{on } \mathbb{R}_+^* \ \left(r + s \begin{smallmatrix} > \\ < \end{smallmatrix} 0\right). \tag{41}$$

(ii) For $t \in \mathbb{R}_+^*$, $\cosh tr \begin{smallmatrix} < \\ > \end{smallmatrix} \cosh ts$ $\begin{pmatrix} |r| < |s| \\ |s| < |r| \end{pmatrix}$ and putting $z := e^t$, we have

$$z^r + z^{-r} \begin{smallmatrix} < \\ > \end{smallmatrix} z^s + z^{-s} \qquad (z > 1) \qquad \begin{pmatrix} |r| < |s| \\ |s| < |r| \end{pmatrix}. \tag{42}$$

Let be $x, y \in \mathbb{R}_+^*$, $x \neq y$. Then $z := \max\left\{\frac{x}{y}, \frac{y}{x}\right\} > 1$, and (42) implies

$$(x^r + y^r)(x^{-r} + y^{-r}) \begin{smallmatrix} < \\ > \end{smallmatrix} (x^s + y^s)(x^{-s} + y^{-s}) \quad \begin{pmatrix} |r| < |s| \\ |s| < |r| \end{pmatrix},$$

i.e.,

$$1 \begin{smallmatrix} < \\ > \end{smallmatrix} \frac{x^s + y^s}{x^r + y^r} \cdot \frac{x^{-s} + y^{-s}}{x^{-r} + y^{-r}} \quad \begin{pmatrix} |r| < |s| \\ |s| < |r| \end{pmatrix}.$$

Raising to the power with exponent $\frac{1}{s-r} \begin{smallmatrix} > \\ < \end{smallmatrix} 0 \begin{pmatrix} r < s \\ s < r \end{pmatrix}$ and using Definition 14 leads to

$$1 \begin{smallmatrix} < \\ > \end{smallmatrix} M_{r,s}(x, y) \cdot M_{r,s}(x^{-1}, y^{-1}) \quad \left(r + s \begin{smallmatrix} > \\ < \end{smallmatrix} 0\right),$$

i.e.,

$$\varphi_{-1} \text{ is strictly } \begin{matrix} M_{r,s}\text{-convex} \\ M_{r,s}\text{-concave} \end{matrix} \text{ on } \mathbb{R}_+^* \quad \left(r + s \begin{smallmatrix} > \\ < \end{smallmatrix} 0\right). \tag{43}$$

(iii) In Case 1, the assertion follows from (41), (43), Theorem 6. 1° and 2°.

*Case 2:* $r = s$.

(i) From Definition (14) and (13) $f_{M_{r,r}}(t) = e^{\frac{e^{tr}}{(e^{tr}+1)}}$ $(t \in \mathbb{R}_+^*)$. This is strictly $\begin{matrix} \text{increasing} \\ \text{decreasing} \end{matrix}$ on $\mathbb{R}_+^*$ $\left(r \begin{smallmatrix} > \\ < \end{smallmatrix} 0\right)$, and since $r + s = 2r$, (41) holds again.

(ii) Let be $x, y \in \mathbb{R}_+^*$, $x \neq y$. Then

$$M_{r,r}(x,y) \cdot M_{r,r}(x^{-1},y^{-1}) \underset{<}{\overset{>}{}} 1 \Longleftrightarrow \left(\frac{x}{y}\right)^{x^r - y^r} \underset{<}{\overset{>}{}} 1. \tag{44}$$

On the other hand, $\left(\frac{x}{y}\right)^{x^r - x^r} \underset{<}{\overset{>}{}} 1 \quad \left(r \underset{<}{\overset{>}{}} 0\right)$, i.e., by (44)

$$M_{r,r}(x,y) \cdot M_{r,r}(x^{-1},y^{-1}) \underset{<}{\overset{>}{}} 1 \quad \left(r \underset{<}{\overset{>}{}} 0\right).$$

Since $r + s = 2r$, (43) holds again, and the proof ends as in Case 1. □

**Remark 17.** By (27), (28), (38), (39)

$$E_{r,2r} = M_r = M_{r,0} \qquad \text{for all } r \in \mathbb{R},$$

and now (14) appears as a very special case of both Theorems 13 and 16.

Now we turn to a family of means which contains the limiting cases *min* and *max* of the family $(M_c)_{c \in \mathbb{R}}$ of power means and of other 1-parameter families of means (cf., e.g., [1], p. 346, for the generalized logarithmic means). The convexity/concavity situation of the power functions will be completely different from that of Theorems 13 and 16 (cf. Theorem 20). We begin by some characterization results.

**Lemma 18.** *Let $M$ be a homogeneous mean on $\mathbb{R}_+^*$ and*

$$g_M : \mathbb{R}^* \to \mathbb{R}_+^*, \; g_M(t) := (M(e^t, 1))^{\frac{1}{t}} \quad (t \in \mathbb{R}^*). \tag{45}$$

*Then we have:*

1°. *For every $c \in [0,1]$*

$$g_M(t) = e^c \text{ for all } t \in \begin{matrix} \mathbb{R}_+^* \\ \mathbb{R}_-^* \\ \mathbb{R}^* \end{matrix} \Leftrightarrow M(x,y) = x^c y^{1-c} \text{ for all } x,y \in \mathbb{R}_+^*, \quad \begin{matrix} x \geq y \\ x \leq y \\ \text{arbitrary} \end{matrix}.$$

2°. *If $M \in \mathcal{M}(\mathbb{R}_+^*)$, then for every $c \in [0,1]$*

$$g_M(t) = e^c \text{ for all } t \in \begin{matrix} \mathbb{R}_+^* \\ \mathbb{R}_-^* \end{matrix} \Leftrightarrow M(x,y) = \left\{ \begin{matrix} (\max\{x,y\})^c \cdot (\min\{x,y\})^{1-c} \\ (\min\{x,y\})^c \cdot (\max\{x,y\})^{1-c} \end{matrix} \right.$$

*for all $x, y \in \mathbb{R}_+^*$.*

3°. *If* $M \in \mathcal{M}(\mathbb{R}_+^*)$, *then*

$$g_M \text{ is constant on } \mathbb{R}^* \iff M = G.$$

*Proof.* By (45), we have for every $t \in \mathbb{R}^*$

$$g_M(t) = e^c \iff M(e^t, 1) = e^{ct}, \tag{46}$$

so, by (2), necessarily $c \in [0,1]$. Conversely, for $c \in [0,1]$, the expressions for $M(x,y)$ in 1°, 2°, 3° do provide homogeneous means on $\mathbb{R}_+^*$, namely weighted geometric means; the ones in 1° need not be symmetric.

1°. $\Longleftarrow$ immediately follows from (46).

$\Longrightarrow$: From the assumption and (46) and from $M(e^0, 1) = e^{c0}$ we obtain

$$M(e^t, 1) = e^{ct} \text{ for all } t \in \begin{matrix} \mathbb{R}_+ \\ \mathbb{R}_- \\ \mathbb{R} \end{matrix} . \tag{47}$$

Now let be $x, y \in \mathbb{R}_+^*$, $\begin{matrix} x \geq y \\ x \leq y \\ arbitrary \end{matrix}$ .

For $t := \log \frac{x}{y}$, the homogeneity of $M$ and (47) yield

$$\frac{1}{y} M(x,y) = M\left(\frac{x}{y}, 1\right) = \left(\frac{x}{y}\right)^c \text{ for all } x, y \in \mathbb{R}_+^*, \quad \begin{matrix} x \geq y \\ x \leq y \\ arbitrary \end{matrix}$$

2°. $N_c(x,y) := (\max\{x,y\})^c \cdot (\min\{x,y\})^{1-c}$, for all $x, y \in \mathbb{R}_+^*$. Then

$$x^c y^{1-c} = \begin{cases} N_c(x,y) & x \geq y \\ N_{1-c}(x,y) & x \leq y \end{cases},$$

and by 1° and symmetry of $M, N_c, N_{1-c}$

$$g_M(t) = e^c \text{ for all } t \in \begin{matrix} \mathbb{R}_+^* \\ \mathbb{R}_-^* \end{matrix} \iff M(x,y) = \begin{cases} N_c(x,y) \\ N_{1-c}(x,y) \end{cases} \text{ for all } x, y \in \mathbb{R}_+^*, \quad \begin{matrix} x \geq y \\ x \leq y \end{matrix}$$

$$\iff M(x,y) = \begin{cases} N_c(x,y) \\ N_{1-c}(x,y) \end{cases} \text{ for all } x, y \in \mathbb{R}_+^*.$$

3°. $\Longleftarrow$ is immediate from 1°.

$\Longrightarrow$: By 1° (lowest case) there exists $c \in [0,1]$ such that $M(x,y) = x^c y^{1-c}$ for all $\mathbb{R}_+^*$. Symmetry of $M$ implies $c = 1 - c$, i.e., $c = \frac{1}{2}$, i.e., $M = G$. □

**Corollary 19.** *For $M \in \mathcal{M}(\mathbb{R}_+^*)$, the function $\varphi_{-1}$ is $M$-affine if and only if $M = G$.*

*Proof.* $\varphi_{-1}$ is certainly $G-affine$. Conversely, by hypothesis $(M(x,y))^{-1} = M(x^{-1},y^{-1})$ $(x,y \in \mathbb{R}_+^*)$, so $(M(x,x^{-1}))^{-1} = M(x^{-1},x) = M(x,x^{-1})$. Since $M(x,x^{-1}) \in \mathbb{R}_+^*$, we get $M(x,x^{-1}) = 1$ $(x \in \mathbb{R}_+^*)$. If $t \in \mathbb{R}^*$, $g_M(t) \overset{(45)}{=} (M(e^t,1))^{\frac{1}{t}} = (e^{\frac{t}{2}} \cdot M(e^{\frac{t}{2}},e^{-\frac{t}{2}}))^{\frac{1}{t}} = (e^{\frac{t}{2}} \cdot 1)^{\frac{1}{t}} = e^{\frac{1}{2}}$, and by Lemma 18.3°, $M = G$. □

**Theorem 20.** *If $c \in [0,1]$ and $N_c(x,y) := (\max\{x,y\})^c \cdot (\min\{x,y\})^{1-c}$ $(x,y \in \mathbb{R}_+^*)$, then*

1°. $p \in \mathbb{R}_+ \Longrightarrow \varphi_p$ *is $N_c$-affine on $\mathbb{R}_+^*$.*

$$2^\circ.\ p \in \mathbb{R}_-^* \Longrightarrow \left[\varphi_p \text{ is } \begin{array}{c} \textit{strictly } N_c\textit{-convex} \\ N_c\textit{-affine} \\ \textit{strictly } N_c\textit{-concave} \end{array} \text{ on } \mathbb{R}_+^* \Longleftrightarrow c \begin{array}{c} > \\ = \\ < \end{array} \frac{1}{2}\right].$$

*Proof.* 1°. By Lemma 18.2°, $g_{N_c}$ is constant on $\mathbb{R}_+^*$. Since $N_c \in \mathcal{M}(\mathbb{R}_+^*)$, all $\varphi_p$ $(p > 1)$ are $N_c$-affine on $\mathbb{R}_+^*$ by Theorem 8, and by Theorem 6.1° and 4° so are all $\varphi_p$ $(0 < p < 1)$ as well as $\varphi_0$ and $\varphi_1$. (There is also a simple autonomous proof for 1°).

2°. Let be $x, y \in \mathbb{R}_+^*$ arbitrary, $x \neq y$. The fact $\max\{x^{-1},y^{-1}\}=(\min\{x,y\})^{-1}$ easily leads to

$$(N_c(x,y))^{-1} \begin{array}{c} < \\ = \\ > \end{array} N_c(x^{-1},y^{-1}) \Leftrightarrow (\min\{x,y\}^{2c-1} \begin{array}{c} < \\ = \\ > \end{array} (\max\{x,y\})^{2c-1} \Leftrightarrow c \begin{array}{c} > \\ = \\ < \end{array} \frac{1}{2}$$

which says that the assertion holds for $p = -1$. The general assertion now follows from 1° and Theorem 6.3°. □

**Remark 21.** Theorem 20 shows that in Theorem 6.2° and 3°, the hypothesis on $\varphi_{-1}$ is essential: All $\varphi_p$ $(p > 1)$ are $\begin{array}{c} N_1\text{-concave} \\ N_0\text{-convex} \end{array}$ on $\mathbb{R}_+^*$, but no $\varphi_q (q < 0)$ is $\begin{array}{c} N_1\text{-concave} \\ N_0\text{-convex} \end{array}$ on $\mathbb{R}_+^*$.

Also, strict convexity/concavity of all $\varphi_q (q < 0)$ does not imply that of all $\varphi_p$ $(p > 1)$.

## 5. Miscellaneous remarks and comments

**Remark 22.** For $M \in \mathcal{M}(\mathbb{R}_+^*)$, the function $\varphi_{-1}$ is $M$-convex on $\mathbb{R}_+^*$ if, and only if,

$$\sqrt{t} \leq M(t,1) \quad \text{for all} \quad t > 1$$

(see Corollary 19 for a related result). In fact: $(M(x,y))^{-1} \leq M(x^{-1}, y^{-1})$ $(x, y \in \mathbb{R}_+^*)$ is equivalent to $(M(t,1))^{-1} \leq M(t^{-1}, 1)$ $(t > 1)$ (see the procedure in the proof of Theorem 8), and this latter is equivalent to $1 \leq M(t,1) \cdot M(t^{-1}, 1)$ $(t > 1)$, $t \leq M(t,1) \cdot M(1,t)$ $(t > 1)$ and finally to $t \leq (M(t,1))^2$ $(t > 1)$.

**Remark 23.** The fact $G(x,y) < L(x,y) < A(x,y)$ $(x, y \in \mathbb{R}_+^*,\ x \neq y)$ is well-known (cf., e.g.[1], p. 348). In the light of Remark 3, this can be interpreted as the strict $L$-convexity of $\varphi_{-1}$ and $\varphi_2$ on $\mathbb{R}_+^*$, the latter being guaranteed by Theorem 9.2°. In fact: For $x, y \in \mathbb{R}_+^*, x \neq y$, $(G(x,y))^2 < (L(x,y))^2$ is equivalent to $(L(x,y))^{-1} < L(x^{-1},\ y^{-1})$, and $(L(x,y))^2 < A(x,y) \cdot L(x,y)$ is equivalent to $(L(x,y))^2 < L(x^2, y^2)$.

**Remark 24.** All means $M \in \mathcal{M}(\mathbb{R}_+^*)$ occurring in Section 4 are continuous in each variable (notice that the contraharmonic mean $M_{1,2}$, in the sense of Definition 14, is not monotonic in $x$ and in $y$; cf. [2], bottom of p. 604). It then turned out that every $\varphi_p$ $(p \in \mathbb{R})$ is convex on $\mathbb{R}_+^*$ and/or concave on $\mathbb{R}_+^*$ (see Theorems 9, 10, 13, 16, 20). This is not so in general (a more general discussion will follow in a forthcoming paper).

**Example 25.** For $M(x,y) := \min\{x,y\} \cdot \left[\frac{\max\{x,y\}}{\min\{x,y\}}\right]$ $(x, y \in \mathbb{R}_+^*)$, we have $M \in \mathcal{M}$ $(\mathbb{R}_+^*)$ and $f_M(t) = [e^t]^{\frac{1}{t}}$ $(t \in \mathbb{R}_+^*)$, so $f_M$ is not monotonic on $\mathbb{R}_+^*$. Accordingly, by Theroem 8, $\varphi_{\frac{3}{2}}$ is neither $M$-convex nor $M$-concave on $\mathbb{R}_+^*$ : $M(1^{\frac{2}{3}}, (2^{\frac{2}{3}})^{\frac{3}{2}}) = 2 > 1 = (M(1, 2^{\frac{2}{3}}))^{\frac{3}{2}}$, while $M(1^{\frac{3}{2}}, 2^{\frac{3}{2}}) = 2 < 2^{\frac{3}{2}} = (M(1,2))^{\frac{3}{2}}$. Furthermore, $\varphi_{-1}$ has the same behavior: $M(2,3) \cdot M(2^{-1}, 3^{-1}) = \frac{2}{3} < 1$, but $M(1,2) \cdot M(1^{-1}, 2^{-1}) = 2 > 1$.

## References

[1] P.S. Bullen, D.S. Mitrinović and P.M. Vasić, *Means and their inequalities*. D. Reidel Publ. Comp. Dordrecht 1988.

[2] D. Farnsworth and R. Orr, *Gini means*. Amer. Math. Monthly **93** (1986), 603–607.

[3] S.J. Gdula, *Podstawy techniki cieplnej dla automatyków*. Politechnika Śląska, Skrypty Uczelniane Nr. 543, Gliwice 1974.

[4] C. Gini, *Di una formula comprensiva delle medie.* Metron **13**, no. 2 (1938), 3–22.

[5] K. Goldziher, *Über die Verwendung von Mittelwertprozessen in der Bevölkerungsstatistik und in der Zinsrechnung.* Skandinavisk Aktuarietidskrift **1920**, 72–96. (Jahrb. Fortschr. Math. **47** (1919–1920), 496-497).

[6] G.H. Hardy, J.E. Littlewood and G. Pólya, *Inequalities.* Cambridge Univ. Press 1934.

[7] P. Kahlig and J. Matkowski, *On functional equations involving the logarithmic mean.* To appear in Z. Angew. Math. Mech.

[8] E.B. Leach and M.C. Sholander, *Extended Mean Values.* Amer. Math. Monthly **85** (1978), 84–90.

[9] E.B. Leach and M.C. Sholander, *Extended Mean Values II.* J. Math. Anal. Appl. **92** (1983), 207–223.

[10] T.-P. Lin, *The power mean and the logarithmic mean.* Amer. Math. Monthly **81** (1974), 879–883.

[11] D. S. Mitrinović, *Analytic inequalities.* Springer, Berlin-Heidelberg-New York, 1970.

[12] J. Nozička, *Mechanika a termomechanika pro elektroenergetica.* SNTL/ALFA, Praha, 1987.

[13] Zs. Páles, *Inequalities for sums of powers.* J. Math. Anal. Appl. **131** (1988,) 265–270.

[14] Zs. Páles, *Inequalities for differences of powers.* J. Math. Anal. Appl. **131** (1988), 271–281.

[15] H. Pichler, *Dynamik der Atmosphäre.* BI, Mannheim-Wien-Zürich, 1986.

[16] J. Sándor and Gh. Toader, *On some exponential means.* Babeş-Bolyai Univ. Sem. Math. Anal. Preprint No. **7** (1990), 35–40.

[17] K.B. Stolarsky, *Generalizations of the logarithmic mean.* Math. Mag. **48** (1975), 87–92.

[18] Gh. Toader, *Some remarks on means.* L'Anal. Numér. Théorie Approx. **20** (1991), 97–109.

Janusz Matkowski, Dept. of Mathematics, Technical University, Willowa 2, PL-43-309 Bielsko-Biała, Poland.

Jürg Rätz, Mathematisches Institut, Universität Bern, Sidlerstrasse 5, CH-3012 Bern, Switzerland.

International Series of Numerical Mathematics, Vol. 123, © 1997 Birkhäuser Verlag Basel

# Convex functions with respect to an arbitrary mean

Janusz Matkowski and Jürg Rätz

**Abstract**

For a mean $M$, a notion of $M$-convex function is introduced. A general criterion for the $M$-convexity of the sum of $M$-convex functions is given. As an application, we present conditions under which polynomials and the exponential functions are convex with respect to some of the Stolarsky means.

## Introduction

Let $J \subset \mathbb{R}$ be a fixed open interval and $M : J \times J \longrightarrow J$ a mean in $J$, i.e.,

$$\min\{x,y\} \le M(x,y) \le \max\{x,y\}, \qquad x,y \in J.$$

Let $I \subset J$ be an arbitrary open interval. A function $\phi : I \longrightarrow J$ is said to be *convex with respect to $M$ on $I$* (shortly *$M$-convex on $I$*) iff

$$\phi(M(x,y)) \le M(\phi(x),\phi(y)), \qquad x,y \in I.$$

Taking in the above definition $J = \mathbb{R}$ and $M$ the arithmetic mean, $M(x,y) = (x+y)/2$, we get the notion of *Jensen convex function.* The theory of Jensen convex functions, strictly related to the classical convexity, is important and well known (cf. for instance M. Kuczma [2], p. 122). This theory can be easily carried out to the $M$-convex functions if $M$ is a *quasi-arithmetic mean*, i.e., if $M$ is of the form

$$M(x,y) = f^{-1}\left(\frac{f(x)+f(y)}{2}\right), \qquad x,y \in J,$$

where $f : J \longrightarrow \mathbb{R}$ is a continuous and strictly monotonic function. Actually, it is easy to check that *if $f$ is increasing (decreasing) then $\phi$ is $M$-convex on $I \subset J$ iff the function $f \circ \phi \circ f^{-1}$ is Jensen convex (Jensen concave) on $f(I)$.* According to our best knowledge there is no theory of $M$-convex functions yet where $M$ is not the (weighted) arithmetic or a quasi-arithmetic mean.

The only exception is [8] where the problem of $M$-convexity of the family of power functions is considered. In [8] some earlier known inequalities are interpreted as

$M$-convexity of special power functions with respect to a suitable mean $M$. This approach permits to obtain natural generalizations as well as to look at some inequalities from a more systematic point of view.

In this paper we give a general criterion for $M$-convexity of the sum of $M$-convex functions (Theorem 1). Its version for a positively homogeneous mean has easy to verify assumptions. As an application we give conditions under which special polynomials and the exponential functions are convex with respect to some of the Stolarsky [10] means. The power means and the Gini means are also mentioned in this context.

## 1. Superadditivity of mean $M$ and $M$-convexity of sum of $M$-convex functions

Let $J \subset \mathbb{R}$ be a fixed open interval, $M : J \times J \longrightarrow J$ a mean in $J$, and $I \subset J$ an arbitrary open subinterval. A function $\phi : I \longrightarrow J$ is said to be *$M$-convex* on $I$ iff

$$\phi\left(M(x,y)\right) \leq M\left(\phi(x), \phi(y)\right), \qquad x,\, y \in I,$$

and *strictly $M$-convex* on $I$ iff

$$\phi\left(M(x,y)\right) < M\left(\phi(x), \phi(y)\right), \qquad x, y \in I,\ x \neq y.$$

If the inequalities are reversed, the function $\phi$ is said to be *$M$-concave* on $I$ or *strictly $M$-concave* on $I$, respectively. If $\phi$ is both $M$-convex and $M$-concave on $I$, i.e.,

$$\phi\left(M(x,y)\right) = M\left(\phi(x), \phi(y)\right), \qquad x, y \in I,$$

it is called *$M$-affine* on $I$.

These definitions are correct because $M(I,I) = I$ for all intervals $I \subseteq J$.

**Remark 1.** Let $M : J \times J \longrightarrow J$ be an arbitrary mean. Then the identity function $\phi(x) = x$, $x \in J$ and, for every $c \in J$, the constant function $\phi(x) = c$, $x \in J$, are $M$–affine on $J$.

Let us note the following obvious

**Remark 2.** Let $J \subset \mathbb{R}$ be open an interval, $M : J \times J \longrightarrow J$ a mean on $J$, and suppose that $I \subset J$ is an open subinterval.

**1°.** If $\phi : I \longrightarrow J$ and $\psi : J \longrightarrow J$ are $M$-convex and $\psi$ is increasing, then $\psi \circ \phi$ is $M$-convex.

**2°.** Let $\phi : I \longrightarrow J$ be bijective and (strictly) $M$-convex.

If $\phi$ is increasing, then $\phi^{-1}$ is (strictly) $M$-concave on $J$.

If $\phi$ is decreasing, then $\phi^{-1}$ is strictly $M$-convex on $J$.

In the sequel we put $\mathbb{R}_+^* := (0, \infty)$.

**Definition.** A function $M : \mathbb{R}_+^* \times \mathbb{R}_+^* \longrightarrow \mathbb{R}_+^*$ is said to be *superadditive* on $\mathbb{R}_+^* \times \mathbb{R}_+^*$ iff

$$M(x_1, y_1) + M(x_2, y_2) \leq M(x_1 + x_2, y_1 + y_2), \qquad x_1,\, x_2,\, y_1,\, y_2 > 0,$$

and *subadditive* iff the inequality is reversed.

**Theorem 1.** *Let $M : \mathbb{R}_+^* \times \mathbb{R}_+^* \longrightarrow \mathbb{R}_+^*$ be a superadditive mean and $I \subset \mathbb{R}_+^*$ an open interval. If $\phi$, $\psi : I \longrightarrow \mathbb{R}_+^*$ are $M$-convex then $\phi + \psi$ is $M$-convex; if moreover, $\phi$ or $\psi$ is strictly $M$-convex, then so is $\phi + \psi$.*

*Proof.* Applying in turn $M$-convexity of $\phi$ and $\psi$, and superadditivity of $M$, we obtain

$$\begin{aligned}
(\phi + \psi)\,(M(x, y)) &= \phi(M(x, y)) + \psi(M(x, y)) \\
&\leq M(\phi(x), \phi(y)) + M(\psi(x), \psi(y)) \\
&\leq M(\phi(x) + \psi(x), \phi(y) + \psi(y)) \\
&= M((\phi + \psi)(x), (\phi + \psi)(y))
\end{aligned}$$

for all $x$, $y \in I$, which means that $\phi + \psi$ is $M$-convex. In view of the definition of strict convexity, the remaining statement is obvious.

**Remark 3.** To get the counterpart of this result for $M$-concave functions it is enough to replace the superadditivity of $M$ by its subadditivity, and $M$-convexity of $\phi$ and $\psi$ by $M$-concavity.

In connection with Theorem 1, let us note the following

**Remark 4.** For a mean $M : \mathbb{R}_+^* \times \mathbb{R}_+^* \longrightarrow \mathbb{R}_+^*$ and an open interval $I \subset \mathbb{R}_+^*$, denote by $\mathrm{Conv}_M(I)$ the family of all $M$-convex functions $\phi : I \longrightarrow \mathbb{R}_+^*$. If the family $\mathrm{Conv}_M(I)$ has the two properties

1) for all $x$, $y \in \mathbb{R}_+^*$ and for all $s$, $t \in I$, $s \neq t$, there exists $\phi \in \mathrm{Conv}_M(I)$ such that

$$\phi(s) = x, \qquad \phi(t) = y, \qquad \phi(M(s, t)) = M(\phi(s), \phi(t));$$

2) for all $\phi$, $\psi \in \mathrm{Conv}_M(I)$, $\phi + \psi \in \mathrm{Conv}_M(I)$,

then $M$ is superadditive on $\mathbb{R}_+^* \times \mathbb{R}_+^*$.

*Proof.* Take arbitrary $x_1, x_2, y_1, y_2 > 0$ and $s, t \in I$, $s \neq t$. According to 1) there exist $\phi_1$, $\phi_2 \in \mathrm{Conv}_M(I)$ such that

$$\phi_i(s) = x_i, \quad \phi_i(t) = y_i, \quad \phi_i(M(s, t)) = M(\phi_i(s),\, \phi_i(t)) = M(x_i, y_i),$$

for $i = 1, 2$. Hence, making use of the relation $\phi_1 + \phi_2 \in \mathrm{Conv}_M(I)$, we get

$$\begin{aligned} M(x_1, y_1) + M(x_2, y_2) &= \phi_1\left(M(s,t)\right) + \phi_2\left(M(s,t)\right) \\ &= (\phi_1 + \phi_2)\left(M(s,t)\right) \\ &\leq M\left((\phi_1 + \phi_2)(s), (\phi_1 + \phi_2)(t)\right) \\ &= M\left(\phi_1(s) + \phi_2(s), \phi_1(t) + \phi_2(t)\right) \\ &= M\left(x_1 + x_2, y_1 + y_2\right), \end{aligned}$$

which was to be shown.

The superadditivity (resp. subadditivity) of a mean $M$ plays a key role in Theorem 1. Since there is no general criterion of superadditivity (subadditivity) of functions of two variables, it is in many cases a nontrivial question to decide if $M$ has this property. Now we shall show that if $M$ is positively homogeneous, the situation is not difficult.

For a two place function $M : \mathbb{R}_+^* \times \mathbb{R}_+^* \longrightarrow \mathbb{R}$ we denote the function $\mathbb{R}_+^* \ni x \longrightarrow M(x, 1)$ by $M(\cdot, 1)$. Similarly we define the function $M(1, \cdot)$.

Let us quote the following result (cf. [6], Theorem 9):

**Lemma 1.** *Let $M : \mathbb{R}_+^* \times \mathbb{R}_+^* \longrightarrow \mathbb{R}_+^*$ be a positively homogeneous function, i.e.,*

$$M(tx, ty) = tM(x, y), \qquad t,\, x,\, y > 0\,.$$

*Then $M$ is superadditive if, and only if, the function $M(\cdot, 1)$ (or $M(1, \cdot)$) is concave.*

*Proof.* Suppose that $M : \mathbb{R}_+^* \times \mathbb{R}_+^* \longrightarrow \mathbb{R}_+^*$ is positively homogeneous. Putting $h := M(\cdot, 1)$ we can write

$$M(x, y) = y\, h\left(\frac{x}{y}\right), \qquad x,\, y > 0,$$

Now it is easily seen that $h$ is concave if, and only if (cf. [4]),

$$(y_1 + y_2)h\left(\frac{x_1 + x_2}{y_1 + y_2}\right) \geq y_1 h\left(\frac{x_1}{y_1}\right) + y_2 h\left(\frac{x_2}{y_2}\right), \qquad x_1,\, x_2,\, y_1,\, y_2 > 0\,,$$

i.e., if, and only if, the function $M$ is superadditive.

A more general inequality as well as its integral counterpart are presented in [5], [9], and in [7] where also the equality case is considered.

Now we can prove

**Theorem 2.** *Let $M : \mathbb{R}_+^* \times \mathbb{R}_+^* \longrightarrow \mathbb{R}_+^*$ be a positively homogeneous mean. If the function $M(\cdot, 1)$ (or $M(1, \cdot)$) is concave, and $I \subset \mathbb{R}_+^*$ is an open interval, then*

*1) for every two $M$-convex functions $\phi$, $\psi : I \longrightarrow \mathbb{R}_+^*$, the function $\phi + \psi$ is $M$-convex, and it is strictly $M$-convex if $\phi$ or $\psi$ is strictly $M$-convex;*

*2) for every (strictly) $M$-convex $\phi : I \longrightarrow \mathbb{R}_+^*$ and $a > 0$, the function $a\phi$ is (strictly) $M$-convex.*

*Proof.* Part 1) is a consequence of Theorem 1 and Lemma 1. Part 2) follows immediately from the positive homogeneity of $M$.

Put $\mathbb{N}^0 := \mathbb{N} \cup \{0\}$.

**Corollary 1.** *Let $M : \mathbb{R}_+^* \times \mathbb{R}_+^* \longrightarrow \mathbb{R}_+^*$ be a positively homogeneous mean such that the function $M(\cdot, 1)$ (or $M(1, \cdot)$) is concave, and $I \subset \mathbb{R}_+^*$ an open interval. Suppose that for every $k \in \mathbb{N}^0$, $\phi_k : I \longrightarrow \mathbb{R}_+^*$ is $M$-convex on $I$, and $c_k > 0$. Then for every $n \in \mathbb{N}^0$, the function*

$$f_n := \sum_{k=0}^{n} c_k \phi_k$$

*is $M$-convex on $I$, and, if there is a $k \in \{0, 1, \ldots, n\}$ such that $\phi_k$ is strictly $M$-convex, then $f_n$ is strictly $M$-convex on $I$. Moreover, if the series*

$$f := \sum_{k=0}^{\infty} c_k \phi_k$$

*is pointwise convergent on $I$, then $f$ is $M$-convex on $I$, and it is strictly $M$-convex if at least one of the functions $\phi_k$ is strictly $M$-convex.*

*Proof.* By Theorem 2, the function $f_n$ is $M$-convex on $I$, i.e.,

$$f_n(M(x,y)) \le M(f_n(x), f_n(y)), \qquad x, y \in I,$$

for every $n \in \mathbb{N}^0$. The function $M(\cdot, 1)$, being concave, is continuous. Since $M(x, y) = y\, M(x/y, 1)$, $x, y > 0$, the function $M$ is continuous. Hence, letting $n \to \infty$ in the above inequality, we obtain

$$f(M(x,y)) \le M(f(x), f(y)), \qquad x, y \in I,$$

which completes the proof.

**Remark 5.** Obviously Theorem 1 and Lemma 1 remain valid on replacing superadditivity of $M$ by subadditivity of $M$, and $M$-convexity of functions $\phi$ and $\psi$ by $M$-concavity. Therefore the suitable counterparts of Theorem 2 and Corollary 1 also hold true.

## 2. Convex functions with respect to the Stolarsky means

An important class of means, strictly related to the Cauchy mean-value theorem, is the two parameter family of Stolarsky means $E_{r,s} : \mathbb{R}_+^* \times \mathbb{R}_+^* \longrightarrow \mathbb{R}_+^* (r, s \in \mathbb{R})$ defined by (cf. [10], also [1], p. 345)

$$E_{r,s}(x,y) := \begin{cases} \left(\frac{r}{s} \cdot \frac{x^s - y^s}{x^r - y^r}\right)^{1/(s-r)}, & r \neq s, \quad rs \neq 0; \quad x \neq y \\ \left(\frac{1}{r} \cdot \frac{x^r - y^r}{\log x - \log y}\right)^{\frac{1}{r}}, & r \neq 0; \quad s = 0; \quad x \neq y \\ \left(\frac{1}{s} \cdot \frac{x^s - y^s}{\log x - \log y}\right)^{\frac{1}{s}}, & r = 0; \quad s \neq 0; \quad x \neq y. \\ \left(e^{-\frac{1}{r}}\right) \cdot \left(\frac{y^{y^r}}{x^{x^r}}\right)^{1/(y^r - x^r)}, & r = s \neq 0; \quad x \neq y \\ \sqrt{xy}, & r = s = 0; \quad x \neq y \\ x, & r, s \in \mathbb{R}; \quad x = y \end{cases}$$

The family $(E_{r,s})_{r,s \in R}$ of positively homogeneous means contains the arithmetic $(E_{2,1})$, geometric $(E_{0,0})$, harmonic $(E_{-2,-1})$, and logarithmic $(E_{1,0})$ means as some special cases.

Recently Losonczi and Páles [3] gave conditions under which the means $E_{r,s}$ are superadditive or subadditive. We quote their result as

**Lemma 2.** *Let $r, s \in \mathbb{R}$, $r \neq s$, be fixed.*

**1°**. *$E_{r,s}$ is superadditive on $\mathbb{R}_+^* \times \mathbb{R}_+^*$ if $\min(r,s) \leq 1$ and $r + s \leq 3$.*

**2°**. *$E_{r,s}$ is subadditive on $\mathbb{R}_+^* \times \mathbb{R}_+^*$ if $\min(r,s) \geq 1$ and $r + s \geq 3$.*

In [8] (Theorem 13) we have proved the following result.

**Lemma 3.** *Let $r, s \in \mathbb{R}$.*

**1°**. *If $r + s > 0$, then*

$$\mathbb{R}_+^* \ni x \longrightarrow x^p \text{ is strictly } E_{r,s}\text{-convex on } \mathbb{R}_+^* \text{ for } p \in \mathbb{R} \setminus [0,1],$$

$$\mathbb{R}_+^* \ni x \longrightarrow x^p \text{ is strictly } E_{r,s}\text{-concave on } \mathbb{R}_+^* \text{ for } p \in (0,1).$$

**2°**. *If $r + s < 0$, then*

$$\mathbb{R}_+^* \ni x \longrightarrow x^p \text{ is strictly } E_{r,s}\text{-concave on } \mathbb{R}_+^* \text{ for } p \in \mathbb{R} \setminus [0,1],$$

$$\mathbb{R}_+^* \ni x \longrightarrow x^p \text{ is strictly } E_{r,s}\text{-convex on } \mathbb{R}_+^* \text{ for } p \in (0,1).$$

**3°**. *If $r + s = 0$, then $\mathbb{R}_+^* \ni x \longrightarrow x^p$ is $E_{r,s}$-affine on $\mathbb{R}_+^*$ for all $p \in \mathbb{R}$.*

Applying Lemma 1, Lemma 2, Lemma 3 and Corollary 1 gives

**Theorem 3.** *Let* $r,\ s \in \mathbb{R}$, $r \neq s$, *such that*

$$\min(r,s) \leq 1, \qquad r+s \leq 3,$$

*be fixed. Suppose that* $p_k \in \mathbb{R} \setminus (0,1)$ *and* $c_k > 0$, *for every* $k \in \mathbb{N}^0$. *Then for every* $n \in \mathbb{N}^0$, *the function* $f_n : \mathbb{R}_+^* \longrightarrow \mathbb{R}_+^*$,

$$f_n(x) := \sum_{k=0}^{n} c_k\, x^{p_k}$$

*is* $E_{r,s}$*-convex on* $\mathbb{R}_+^*$. *Moreover, if the series*

$$f(x) := \sum_{k=0}^{\infty} c_k\, x^{p_k}$$

*is pointwise convergent on an interval* $I \subset \mathbb{R}_+^*$, *then the function* $f$ *is* $E_{r,s}$*-convex on* $I$.

Hence, applying Remark **2.2°**, we get

**Corollary 2.** *Let* $r,\ s \in \mathbb{R}$, $r \neq s$, *such that*

$$\min(r,s) \leq 1, \qquad r+s \leq 3,$$

*be fixed. Then*

**1°**. *every polynomial*

$$f_n(x) := \sum_{k=0}^{n} c_k\, x^k, \qquad x \in \mathbb{R}_+^*,$$

*such that*

$$c_k \geq 0, \qquad k = 0,1,\ldots,n; \qquad \sum_{k=0}^{n} c_k > 0,$$

*is* $E_{r,s}$*-convex on* $\mathbb{R}_+^*$;

**2°**. *for every* $a > 1$ *the exponential function* $f : \mathbb{R}_+^* \longrightarrow \mathbb{R}_+^*$,

$$f(x) = a^x, \qquad x > 0,$$

*is strictly* $E_{r,s}$*-convex on* $\mathbb{R}_+^*$;

**3°**. *for every* $a > 1$ *the function* $f = \log_a$ *is strictly* $E_{r,s}$*-concave on the interval* $(1,\infty)$.

The assumption $a > 1$ is needed for applying Theorem 3 in Corollary **2.2°** since

$$a^x = \sum_{k=0}^{\infty} c_k x^k \qquad \textit{with } c_k = \frac{(\log a)^k}{k!} > 0 \qquad \textit{for all } k \in \mathbb{N}^0 ;$$

here log denotes the natural logarithm. We show now that $a > 1$ is really essential for part **2°** of Corollary 2.

**Remark 6.** Let $a \in (0,1)$ be fixed. Then the exponential function $f(x) = a^x$, $x > 0$, is neither $E_{1,0}$-convex nor $E_{1,0}$-concave on $\mathbb{R}_+^*$.

*Proof.* We first show that the function $h(x) := e^{-x}$, $x > 0$, is neither $E_{1,0}$-convex nor $E_{1,0}$-concave on $\mathbb{R}_+^*$, i.e., that neither the inequality

$$h\left(\frac{x-y}{\log(x)-\log(y)}\right) \le \frac{x-y}{\log h(x) - \log h(y)}, \qquad x,\, y > 0,$$

nor the reverse inequality holds true. For $y := e^2$ put

$$\alpha(x) := h\left(\frac{x-y}{\log(x)-\log(y)}\right), \quad \beta(x) := \frac{h(x)-h(y)}{\log h(x) - \log h(y)}, \quad x > 0,\ x \ne e^2 .$$

Then

$$\alpha(x) = \exp\left(-\frac{x-e^2}{\log(x)-2}\right), \qquad \beta(x) = \frac{e^{-x} - e^{-e^2}}{-x+e^2}, \qquad x > 0,\ x \ne e^2 .$$

Since

$$\alpha(0+) := \lim_{x \to 0+} \alpha(x) = 1, \qquad \beta(0) = \frac{1 - e^{-e^2}}{e^2},$$

we have $\alpha(0+) > \beta(0)$. This shows that $h$ cannot be $E_{1,0}$-convex on $\mathbb{R}_+^*$.

On the other hand we have

$$\alpha(1) = e^{\frac{1-e^2}{2}}, \qquad \beta(1) = \frac{e^{-1} - e^{-e^2}}{e^2 - 1}.$$

We shall show that

$$\alpha(1) < \beta(1).$$

It is easy to see that this is equivalent to the following inequality

$$e^{\frac{5-e^2}{2}} - e^{\frac{1-e^2}{2}} < e^{-1} - e^{-e^2}.$$

This inequality holds true because

$$e^{\frac{5-e^2}{2}} < e^{-1}, \qquad e^{\frac{1-e^2}{2}} > e^{-e^2}.$$

In fact, the first of these inequalities is equivalent to $e^2 > 7$, and the second one is obvious.

Suppose now that there is an $a \in (0,1)$ such that the function $f(x) = a^x$, $x > 0$, is $E_{1,0}$-convex (or $E_{1,0}$-concave) on $\mathbb{R}_+^*$. Choose $c > 0$ such that $a^c = e^{-1}$. Then, putting $g(x) := cx$, $x > 0$, we would have $h = f \circ g$. Since $g$ is $E_{1,0}$-affine on $\mathbb{R}_+^*$, the function $h$ is $E_{1,0}$-convex (or $E_{1,0}$-concave) on $\mathbb{R}_+^*$ by Remark 2.1$^0$. This contradiction completes the proof.

Corollary 2 allows to produce a lot of new inequalities. Consider for instance the following

**Example.** Taking $r = 1$, $s = 0$, and $f(x) = e^x$, $x > 0$, in Corollary **2.2°** gives

$$f\left(E_{1,0}(x,y)\right) \leq E_{1,0}\left(f(x), f(y)\right), \qquad x,\, y > 0\,.$$

By the definition of the logarithmic mean $E_{1,0}$ we hence get the inequality

$$e^{\frac{x-y}{\log(x)-\log(y)}} \leq \frac{e^x - e^y}{x-y}, \qquad x,\, y > 0,\ x \neq y\,.$$

## 3. Remarks about some other means

Let us fix $p \in \mathbb{R}$. The function $A_p : \mathbb{R}_+^* \times \mathbb{R}_+^* \longrightarrow \mathbb{R}_+^*$ defined by

$$A_p(x,y) := \begin{cases} \left(\frac{x^p+y^p}{2}\right)^{\frac{1}{p}}, & p \neq 0 \\ \sqrt{xy}\,, & p = 0 \end{cases}.$$

is a symmetric positively homogeneous mean, and it is called the *power* mean. Moreover, $A_0(\cdot,1)$ is concave, and for $p \neq 0$

$$\frac{d^2}{dx^2} A_p(x,1) = x^{p-2}\left(x^p+1\right)^{-2} A_p(x,1)(p-1)\,, \qquad x > 0\,,$$

so the function $A_p(\cdot,1)$ is concave for every $p \leq 1$ and convex for every $p \geq 1$.

Applying Theorem 1 and Corollary 1, it is easy to get the counterparts of Theorem 3 and Corollary 2 for the $A_p$-convex functions.

Let us also mention that, using some results of Losonczi and Páles [3], one can get more general results about $M_{r,s}$-convex functions, where $M_{r,s} : \mathbb{R}_+^* \times \mathbb{R}_+^* \longrightarrow \mathbb{R}_+^*$, is the two parameter family of Gini means, which for $r \neq s$ is defined by

$$M_{r,s}(x,y) := \left(\frac{x^r + y^r}{x^s + y^s}\right)^{1/(r-s)}, \qquad x,\, y > 0\,.$$

## References

1. P.S. Bullen, D.S. Mitrinović and P.M. Vasić, *Means and their inequalities.* Reidel, Dordrecht, 1988.

2. M. Kuczma, *An introduction to the theory of functional equations and inequalities. Cauchy's equation and Jensen's inequality.* Prace Nauk. Uniw. Śląsk. 489, Polish Scientific Publishers, Warsaw–Cracow–Katowice, 1985.

3. L. Losonczi and Zs. Páles, *Minkowski type inequalities for two variable homogeneous mean.* Private communication.

4. J. Matkowski, *A functional inequality characterizing convex functions, conjugacy and a generalization of Hölder's and Minkowski's inequalities.* Aequationes Math. **40** (1990), 168–180.

5. J. Matkowski, *A generalization of Hölder's and Minkowski's inequalities and conjugate functions.* In "Constantin Carathéodory: International Tribute", edited by Th.M. Rassias, 1991, World Scientific Publ. Co., 819–827.

6. J. Matkowski, *Subadditive functions and a relaxation of the homogeneity condition of seminorms.* Proc. Amer. Math. Soc. **117** (1993), 991–1001.

7. J. Matkowski and J. Rätz, *On the equality condition in Hölder's and Minkowski's inequalities.* Aequationes Math. **46** (1993), 220–228.

8. J. Matkowski and J. Rätz, *Convexity of power functions with respect to symmetric homogeneous means,* General Inequalities 7, Oberwolfach 1995 (this volume).

9. J. Matkowski, *The converse of the Hölder inequality and its generalizations.* Studia Math. **109** (1994), 171–182.

10. K.B. Stolarsky, *Generalizations of the logarithmic mean.* Math. Mag. **48** (1975), 87–92.

Janusz Matkowski, Department of Mathematics, Technical University, Willowa 2, PL 43-309 Bielsko-Biała, Poland.

Jürg Rätz, Institute of Mathematics, University of Bern, Sidlerstrasse 5, CH-3012 Bern, Switzerland.

International Series of Numerical Mathematics, Vol. 123, © 1997 Birkhäuser Verlag Basel

# Separation by semidefinite bilinear forms

Zsolt Páles*

**Abstract**

In this paper we offer necessary and sufficient conditions on the separability of two given functions by nonnegative definite symmetric bilinear forms. As a consequence, we characterize those functions that are pointwise suprema of symmetric nonnegative definite bilinear functions and also those set-valued functions that are generated by nonnegative selfadjoint operators.

## 1. Introduction

Our subject is to study the separability of two given functions by nonnegative definite (or positive semidefinite) symmetric bilinear forms. The methods used and the results obtained are parallel to those found in [1], [2] concerning the separability with bilinear and with symmetric bilinear functions. The theory of separation by quadratic forms has also been investigated in [4].

Now we briefly indicate the motivation of this research. The results for the symmetric bilinear separation ([2]) were used in [3] to prove that the second order generalized directional derivative of $C^{1,1}$ functions at a given point can be obtained as the pointwise maximum of a family of symmetric bilinear forms. This family of bilinear forms is then called the *generalized Hessian of the function at the given point*. In order to use this Hessian in optimality conditions, the representability by *semidefinite* bilinear forms has to be investigated as well. This problem then leads to the investigation of separability by semidefinite bilinear functions.

Throughout this paper, let $X$ be a locally convex Hausdorff topological vector space over the field of reals and denote by $X^*$ its dual space. Let $P, R : X \times X \to \mathbb{R}$ be given functions with $R \leq P$ on $X \times X$. The *separability* of $P$ and $R$ by a *nonnegative definite bilinear* function means that there exists a nonnegative definite symmetric bilinear function $Q$ satisfying $R \leq Q \leq P$ on $X \times X$. The analogous problem with bilinear and with symmetric bilinear forms has been considered in [1], [2]. The results for the (nonsymmetric) bilinear separation can be formulated as follows (see [1]).

*Research supported by the Hungarian National Foundation for Scientific Research (OTKA), Grant No. T-016846.

**Theorem 1** *Let $P : X \times X \to \mathbb{R}$ and $R : X \times X \to [-\infty, \infty[$ be two positively bihomogeneous functions. Then there exists a bilinear function $Q : X \times X \to \mathbb{R}$ such that $R \le Q \le P$ if and only if*

$$\sum_{i=1}^{n} R(u_i, v_i) \le \sum_{j=1}^{k} P(x_j, y_j) \tag{1}$$

*holds for all $n, k \in \mathbb{N}$, $u_1, \ldots, u_n, v_1, \ldots, v_n, x_1, \ldots, x_k, y_1, \ldots, y_k \in X$, whenever the following bilinear identity is satisfied*

$$\sum_{i=1}^{n} \langle x^*, u_i \rangle \langle y^*, v_i \rangle = \sum_{j=1}^{k} \langle x^*, x_j \rangle \langle y^*, y_j \rangle \qquad \forall\, x^*, y^* \in X^*. \tag{2}$$

For the problem of symmetric bilinear separation, the following statement is proved in [2].

**Theorem 2** *Let $P : X \times X \to \mathbb{R}$ and $R : X \times X \to [-\infty, \infty[$ be two positively bihomogeneous functions. Then there exists a symmetric bilinear function $Q : X \times X \to \mathbb{R}$ such that $R \le Q \le P$ if and only if* (1) *holds for all $n, k \in \mathbb{N}$, $u_1, \ldots, u_n, v_1, \ldots, v_n, x_1, \ldots, x_k, y_1, \ldots, y_k \in X$, whenever the following quadratic identity is satisfied:*

$$\sum_{i=1}^{n} \langle x^*, u_i \rangle \langle x^*, v_i \rangle = \sum_{j=1}^{k} \langle x^*, x_j \rangle \langle x^*, y_j \rangle \qquad \forall\, x^* \in X^*. \tag{3}$$

The next theorem, which is the main result of this paper, completely answers the problem of separation by nonnegative definite bilinear forms.

**Theorem 3** *Let $P : X \times X \to \mathbb{R}$ and $R : X \times X \to [-\infty, \infty[$ be two positively bihomogeneous functions. Then there exists a nonnegative definite symmetric bilinear function $Q : X \times X \to \mathbb{R}$ such that $R \le Q \le P$ if and only if* (1) *holds for all $n, k \in \mathbb{N}$, $u_1, \ldots, u_n, v_1, \ldots, v_n, x_1, \ldots, x_k, y_1, \ldots, y_k \in X$, whenever the following quadratic inequality is satisfied*

$$\sum_{i=1}^{n} \langle x^*, u_i \rangle \langle x^*, v_i \rangle \le \sum_{j=1}^{k} \langle x^*, x_j \rangle \langle x^*, y_j \rangle \qquad \forall\, x^* \in X^*. \tag{4}$$

In each case, inequality (1) must hold on the domains described by (2), (3), and (4), respectively. Clearly, among these domains, (1) is the smallest and (4) is the largest. We also note that the necessary and sufficient condition for the nonpositive definite separability of $R$ and $P$ is that (1) has to be satisfied on a domain which is described by the reverse inequality of (4). Obviously, all the results with nonnegative definiteness of this paper can be transformed into analogous results with nonpositive definiteness by changing the direction of a corresponding inequality.

Comparing the above results, the novelty in the proof of Theorem 3 is that here the principal axis theorem has to be used to obtain the necessity of the conditions and the Hahn Banach separation theorem on ordered vector spaces to obtain the sufficiency part.

The proof of the above Theorem 3 is given the next section. There, we also characterize those functions that can be obtained as the pointwise maximum of families of nonnegative definite bilinear forms. In the last section, we deal with the linear operator selection problem of set-valued functions in the case when the required selection has to be a positive semidefinite selfadjoint operator of a Hilbert space.

## 2. Main Results

We prove a special case of Theorem 3 when $R \equiv -\infty$ first. However, this particular case will imply Theorem 3 very easily.

**Theorem 4** *Let $P : X \times X \to \mathbb{R}$ be a positively bihomogeneous function. Then there exists a nonnegative definite symmetric bilinear function $Q : X \times X \to \mathbb{R}$ such that $Q \leq P$ if and only if*

$$0 \leq P(x_1, y_1) + \cdots + P(x_k, y_k) \tag{5}$$

*holds whenever $k \in \mathbb{N}$, $x_1, \ldots, x_k, y_1, \ldots, y_k \in X$ and*

$$0 \leq \langle x^*, x_1 \rangle \langle x^*, y_1 \rangle + \cdots + \langle x^*, x_k \rangle \langle x^*, y_k \rangle \qquad \forall x^* \in X^*. \tag{6}$$

*Proof.* To prove the necessity, assume that there exists a nonnegative definite bilinear function $Q$ such that $Q \leq P$. Let $x_1, \ldots, x_k, y_1, \ldots, y_k \in X$ and assume that (6) is satisfied. These vectors span a finite dimensional subspace $S \subset X$. By the principal axis theorem applied to the finite dimensional subspace $S$ and to the symmetric bilinear form $Q$, there exists a basis $e_1, \ldots, e_n$ of $S$ such that $Q(e_i, e_j) = 0$ for $i \neq j$. The nonnegative definiteness of $Q$ yields $Q(e_i, e_i) \geq 0$ for $i = 1, \ldots, n$.

On the other hand, as a consequence of the Hahn-Banach theorem, we can find a dual system $e_1^*, \ldots, e_n^* \in X^*$ such that

$$\langle e_i^*, e_j \rangle = \delta_{i,j},$$

where $\delta$ is the Kronecker function. Clearly,

$$Q(x,y) = \sum_{i=1}^{n} Q(e_i, e_i) \langle e_i^*, x\rangle \langle e_i^*, y\rangle \qquad \text{for} \quad x, y \in S,$$

since this equation is trivially satisfied for $x, y \in \{e_1, \ldots, e_n\}$. Applying this formula for $x = x_j$ and $y = y_j$, adding up the equations obtained for $j = 1, \ldots, k$ and using (6) for $x^* = e_1^*, \ldots, e_n^*$, we get

$$\sum_{j=1}^{k} P(x_j, y_j) \geq \sum_{j=1}^{k} Q(x_j, y_j) = \sum_{i=1}^{n} Q(e_i, e_i) \Big( \sum_{j=1}^{k} \langle e_i^*, x_j\rangle \langle e_i^*, y_j\rangle \Big) \geq 0.$$

Thus (5) must be valid.

To prove the sufficiency of the condition, assume that (5) holds on the domain indicated by (6). Let $Z$ be the set of functions $z : X^* \to \mathbb{R}$ that have the representation

$$z(x^*) = \sum_{i=1}^{m} \langle x^*, u_i\rangle \langle x^*, v_i\rangle \qquad \text{for} \quad x^* \in X^* \tag{7}$$

with some elements $u_1, \ldots, u_m, v_1, \ldots, v_m \in X$. Clearly, $Z$ is an ordered linear space. Define $\Phi, \Psi : Z \to [-\infty, \infty[$ by

$$\Phi(z) := \inf\Big\{ \sum_{j=1}^{k} P(x_j, y_j) \,\Big|\, \text{for } x_1, \ldots, x_k, y_1, \ldots, y_k$$
$$\text{satisfying } z(x^*) \leq \sum_{j=1}^{k} \langle x^*, x_j\rangle \langle x^*, y_j\rangle \ (\forall\, x^* \in X^*) \Big\},$$

$$\Psi(z) := \begin{cases} 0, & \text{if } z \geq 0 \\ -\infty, & \text{otherwise.} \end{cases}$$

Now we show that $\Phi$ is a monotonic sublinear function and that $\Psi$ is a superlinear function with $\Psi \leq \Phi$ on $Z$.

To prove the subadditivity of $\Phi$, let $z', z'' \in Z$ and $\zeta', \zeta'' \in \mathbb{R}$ such that $\Phi(z') < \zeta'$, $\Phi(z'') < \zeta''$. Then there exist $k' \in \mathbb{N}$, $x_1', \ldots, x_{k'}', y_1', \ldots, y_{k'}' \in X$ and $k'' \in \mathbb{N}$, $x_1'', \ldots, x_{k''}'', y_1', \ldots, y_{k''}' \in X$ such that

$$\sum_{j=1}^{k'} P(x_j', y_j') < \zeta', \qquad \sum_{j=1}^{k''} P(x_j'', y_j'') < \zeta''$$

and, for all $x^* \in X^*$,

$$z'(x^*) \leq \sum_{j=1}^{k'} \langle x^*, x_j'\rangle \langle x^*, y_j'\rangle, \qquad z''(x^*) \leq \sum_{j=1}^{k''} \langle x^*, x_j''\rangle \langle x^*, y_j''\rangle .$$

Thus,

$$(z' + z'')(x^*) \le \sum_{j=1}^{k'} \langle x^*, x_j' \rangle \langle x^*, y_j' \rangle + \sum_{j=1}^{k''} \langle x^*, x_j'' \rangle \langle x^*, y_j'' \rangle,$$

whence

$$\Phi(z' + z'') \le \sum_{j=1}^{k'} P(x_j', y_j') + \sum_{j=1}^{k''} P(x_j'', y_j'') < \zeta' + \zeta''.$$

Being $\zeta' > \Phi(z')$ and $\zeta'' > \Phi(z'')$ arbitrary, we get

$$\Phi(z' + z'') \le \Phi(z') + \Phi(z''),$$

which proves the subadditivity of $\Phi$.

The positive homogeneity of $\Phi$ follows from the positive bihomogeneity of $P$. To check the monotonicity, let $z' \le z''$, that is, $z'(x^*) \le z''(x^*)$ for all $x^* \in X^*$. Then

$$z''(x^*) \le \sum_{j=1}^{k} \langle x^*, x_j \rangle \langle x^*, y_j \rangle \quad \text{implies} \quad z'(x^*) \le \sum_{j=1}^{k} \langle x^*, x_j \rangle \langle x^*, y_j \rangle .$$

Thus, the set whose infimum is $\Phi(z')$ is larger than the set whose infimum is $\Phi(z'')$. This yields $\Phi(z') \le \Phi(z'')$. By (5), we have $\Phi(0) = 0$, therefore $\Phi$ must be real valued.

The superlinearity of $\Psi$ is obvious, since the set of nonnegative functions $z \in Z$ forms a convex cone. The monotonicity of $\Phi$ yields $\Phi(z) \ge \Phi(0) = 0 = \Psi(z)$ for $z \ge 0$. Thus, $\Psi \le \Phi$.

Applying the sandwich version of the Hahn-Banach theorem, we can find a linear function $\phi : Z \to \mathbb{R}$ such that $\Psi \le \phi \le \Phi$ holds on $Z$. Define the required function $Q : X \times X \to \mathbb{R}$ by

$$Q(x, y) = \phi(z) \qquad \text{if} \quad z(x^*) \equiv \langle x^*, x \rangle \langle x^*, y \rangle .$$

Then the symmetry of $Q$ is obvious. The functional $\phi$ is linear on $Z$, hence $Q$ is bilinear on $X$. If $z(x^*) \equiv \langle x^*, x \rangle \langle x^*, y \rangle$, then by the definition of $\Phi$, we have $\Phi(z) \le P(x, y)$. Thus, $\phi \le \Phi$ yields $Q \le P$. On the other hand, $\phi(z) \ge \Psi(z) = 0$ for $z \ge 0$. Therefore, with $z(x^*) \equiv \langle x^*, x \rangle^2$, we obtain $Q(x, x) = \phi(z) \ge 0$. This proves that $Q$ is nonnegative definite.

The proof of the theorem is complete.

**Remark.** If a bilinear $Q$ exists, then the inequality $Q \le P$ yields

$$-P(-x, y) \le Q(x, y) \le P(x, y) \qquad \text{for} \quad x, y \in X.$$

Thus, assuming the upper semicontinuity property of $P$, we always obtain that $Q$ is bounded in a neighborhood of the origin. Therefore it must be continuous as well.

Now we are ready to prove Theorem 3 on the nonnegative definite separability of two given functions.

*Proof* of Theorem 3. A bilinear functional $Q$ separates $R$ and $P$ if and only if

$$Q(x,y) \le \min\{P(x,y), -R(-x,y)\} =: P^*(x,y) \qquad \text{for} \quad x,y \in X.$$

Thus a separating $Q$ exists if and only if there is $Q \le P^*$. Now the statement of Theorem 3 follows directly from Theorem 4.

In the following theorem we characterize those functions that are pointwise suprema of nonnegative definite symmetric bilinear functions. This result and also its proof are analogous to that of Corollary 2 in [2] (cf. also Corollary 2 in [1]), therefore we may omit the proof.

**Theorem 5** *Let $P : X \times X \to \mathbb{R}$ be an arbitrary function. Then the following three assertions are equivalent:*

(i) *There exists a family $\mathcal{Q}$ of nonnegative definite symmetric bilinear forms such that*

$$P(x,y) = \max_{Q \in \mathcal{Q}} Q(x,y) \qquad \textit{for all} \quad x,y \in X;$$

(ii) *There exists a family $\mathcal{Q}$ of nonnegative definite symmetric bilinear forms such that*

$$P(x,y) = \sup_{Q \in \mathcal{Q}} Q(x,y) \qquad \textit{for all} \quad x,y \in X;$$

(iii) *The function $P$ is positively bihomogeneous and whenever the elements $x_0$, $x_1, \ldots, x_k$, $y_0, y_1, \ldots, y_k \in X$ satisfy*

$$\langle x^*, x_0\rangle \langle x^*, y_0\rangle \le \langle x^*, x_1\rangle \langle x^*, y_1\rangle + \cdots + \langle x^*, x_k\rangle \langle x^*, y_k\rangle \qquad \forall\, x^* \in X^*, \tag{8}$$

*then*

$$P(x_0,y_0) \le P(x_1,y_1) + \cdots + P(x_k,y_k).$$

In what follows, we give examples for the domain described in (8) and we list some interesting consequences of (iii). For, assume that $P$ satisfies (iii) of the previous theorem.

**Examples. 1.** For all $x_1, x_2, y \in X$ and $x^* \in X^*$, we have

$$\langle x^*, x_1 + x_2\rangle \langle x^*, y\rangle = \langle x^*, x_1\rangle \langle x^*, y\rangle + \langle x^*, x_2\rangle \langle x^*, y\rangle .$$

Hence, by (iii),

$$P(x_1 + x_2, y) \leq P(x_1, y) + P(x_2, y),$$

which is the subadditivity of $P(\,.\,, y)$. Therefore $P$ must be bisublinear. Similarly, (iii) implies the symmetry of $P$, as well.

**2.** The inequality

$$\begin{aligned} P(x_1 + x_2 + x_3, y_1 + y_2 + y_3) \quad \leq \quad & P(x_1, y_1 + y_2) + P(x_1 + x_2, y_3) \\ & + P(x_3, y_2 + y_3) + P(x_2 + x_3, y_1) + P(x_2, y_2), \end{aligned}$$

can be deduced analogously, but it cannot be obtained from the bisublinearity of the function $P$.

**3.** Consider now the obvious identity

$$\langle x^*, x \rangle \langle x^*, y \rangle \equiv \frac{1}{4} \langle x^*, x + y \rangle^2 - \frac{1}{4} \langle x^*, x - y \rangle^2 .$$

This yields that

$$\langle x^*, x \rangle \langle x^*, y \rangle \leq \frac{1}{4} \langle x^*, x + y \rangle^2 \qquad \forall x^* \in X^*,$$

whence, by (iii), we get

$$P(x, y) \leq \frac{1}{4} P(x + y, x + y).$$

**4.** Given $x, y \in X$, one can easily establish that

$$\langle x^*, x + y \rangle^2 \equiv 2 \langle x^*, x \rangle^2 + 2 \langle x^*, y \rangle^2 - \langle x^*, x - y \rangle^2 .$$

Hence

$$\langle x^*, x + y \rangle^2 \leq 2 \langle x^*, x \rangle^2 + 2 \langle x^*, y \rangle^2 \qquad \forall x^* \in X^*.$$

Therefore, by (iii),

$$P(x + y, x + y) \leq 2P(x, x) + 2P(y, y).$$

## 3. Nonnegative Definite Selfadjoint Selections of Set-valued Functions

Throughout this section assume that $X$ is a Hilbert space and $F : X \to 2^X$ is a set-valued mapping with nonempty weak compact convex values in $X$. We assume that $F$ is upper semicontinuous at the origin, i.e., for any neighborhood $V$ of the set $F(0) \subset X$ there exists a neighborhood $U$ of $0 \in X$ such that $F(x) \subset V$ whenever $x \in U$.

The *operator selection problem* for $F$ means that we search for conditions on $F$ such that $F$ admits a continuous linear operator selection $A \in B(X)$ [i.e., $A(x) \in F(x)$ $(\forall x \in X)$] satisfying prescribed conditions. A sufficient condition for the existence of a linear operator selection was found by Rodrigues-Salinas and Bou in [5]. The necessary and sufficient condition was discovered in [1]. In [2], necessary and sufficient conditions were obtained for the existence of a selfadjoint selection. The next result offers a criterion for the existence of a positive semidefinite selfadjoint selection. (Cf. [1, Corollary 4], [2, Theorem 2].)

**Theorem 6** *Let $F$ be a positively homogeneous nonempty weak compact convex set-valued function which is usc at the origin. Then $F$ admits a continuous nonnegative definite selfadjoint operator selection $A \in B(X)$ if and only if the inclusion*

$$0 \in \langle F(x_1), y_1\rangle + \cdots + \langle F(x_k), y_k\rangle$$

*holds true whenever $k \in \mathbb{N}$, $x_1, \ldots, x_k, y_1, \ldots, y_k \in X$ and (6) is valid.*

*Proof.* This result can be deduced from Theorem 4 by defining the function $P : X \times X \to \mathbb{R}$ by

$$P(x,y) := \sup \langle F(x), y\rangle \, .$$

and following the argument of the proof of Theorem 2 in [2].

The next statement describes those set valued mappings that can be generated by families of nonnegative definite selfadjoint operators. This result is analogous to that of [1, Corollary 5], [2, Corollary 3].

**Theorem 7** *For a weak compact convex set-valued map $F$, which is usc at the origin, the following three properties are equivalent:*

(i) *There exists a family $\mathcal{A} \subset B(X)$ of nonnegative definite selfadjoint operators such that*

$$F(x) = \{Ax : A \in \mathcal{A}\} \qquad \forall x \in X.$$

(ii) *There exists a family $\mathcal{A} \subset B(X)$ of nonnegative definite selfadjoint operators such that*

$$F(x) = \overline{\{Ax : A \in \mathcal{A}\}} \qquad \forall x \in X.$$

(iii) *$F$ is homogeneous, symmetric (i.e., $\langle F(x), y\rangle \equiv \langle F(y), x\rangle$) and, whenever $x_0, x_1, \ldots, x_k, y_0, y_1, \ldots, y_k \in X$ satisfy (8), then*

$$\langle F(x_0), y_0\rangle \subset \langle F(x_1), y_1\rangle + \cdots + \langle F(x_k), y_k\rangle$$

*holds.*

*Proof.* The equivalence of (i),(ii) and (iii) can be deduced from the equivalence of the analogous statements of Theorem 5 following the argument used in the proof of [2, Corollary 3].

**References**

[1] Zs. Páles, *Linear selections for set-valued functions and extension of bilinear forms*, Arch. Math., **62** (1994), 427–432.

[2] Zs. Páles, *Separation with symmetric bilinear forms and symmetric selections of set-valued functions*, Publ. Math. Debrecen, **46** (1995), 321–331.

[3] Zs. Páles and V. Zeidan, *Generalized Hessian for $C^{1,1}$ functions in infinite dimensional normed spaces*, Math. Programming, to appear.

[4] Zs. Páles and V. Zeidan, *Separation via quadratic functions*, Aequationes Math., to appear.

[5] B. Rodrigues-Salinas and L. Bou, *A Hahn-Banach theorem for arbitrary vector spaces*, Boll. Un. Math. Ital. (4)**10** (1974), 390–393.

Zsolt Páles, Institute of Mathematics, L. Kossuth University, H-4010 Debrecen, Pf. 12, Hungary.

# Inequalities in Functional Analysis and Functional Equations

International Series of Numerical Mathematics, Vol. 123, 

# Inequalities for selection probabilities

János Aczél [1] and Gyula Maksa [2]

**Abstract**

We suppose that selection probabilities are functions of the (ratio) scale values of the options to choose from. On one hand the aggregated ("overall") selection probability of each option should be a function of all componentwise ("individual") selection probabilities of all options, on the other hand it should be a function of the aggregated scale values. Using functional equations and inequalities, we derive from these suppositions under quite weak technical conditions that the scale values are aggregated by products of powers, in particular by weighted geometric means, and both the individual and the overall selection probabilities are relatively simple functions of the scale values of the options. Inequalities also play a key role in satisfying the natural requirement that the probability of selecting an option should increase with the value (attractiveness) of that option and decrease if the value of other options increases. Finally, as a special case, we characterize Luce's choice model.

## 1. Introduction

We suppose that the "overall" probability $P(e_j : E)$ $(j = 1, \ldots, n)$ of selecting an option (element) $e_j$ from a finite set $E$ is aggregated from the individual probabilities $P_i(e_1 : E), \ldots, P_i(e_n : E)$ $(i = 1, \ldots, m)$. Both the individual and the overall probabilities depend upon the options $e_k$ through ratio scale values $v_1(e_k), \ldots, v_m(e_k)$, $v(e_k)$ $(k = 1, \ldots, n)$, where $v(e_k)$ is a function of $v_1(e_k), \ldots, v_m(e_k)$.

[1] Supported in part by the NSERC (Canada) grant no. CGP0164211.

[2] Partially supported by OTKA (Hungary) grant no. I/3-1650.

Under some further conditions we determine all these functions. Surprisingly the inequalities $0 < P(e_j : E) < 1$ turn out to be the most difficult to satisfy. Inequalities play also a key role in satisfying the natural requirement that the probability of selecting an option should increase with the value (attractiveness) of that option and decrease if the value of other options increases.

Finally, as an important special case, we characterize Luce's choice model.

For more details on aspects other than inequalities, see Aczél-Maksa-Marley-Moszner 1995. Both papers originate from questions posed by A.A.J. Marley.

## 2. Functional equations and inequalities

We formulate our definitions and suppositions (cf. Aczél-Maksa-Marley-Moszner 1995) and derive from them a system of functional equations and inequalities which will be the subject of the rest of the paper.

**Definition 1.** Let $R$ be a set and let $S$ be the set of its finite nonempty subsets. The set $R$ and a function $P : R \times S \to [0, 1]$ form a *structure of selection probabilities* $(R, P)$ (a *selection model* for short), while the number $P(e : E)$ $(e \in R, E \in S)$ is a *selection probability*. As selection probability, $P(e : E)$ satisfies

$$0 \leq P(e : E) \leq 1,$$

$$P(e : E) = 0 \ \text{ if } \ e \notin E, \tag{1}$$

$$\sum_{e \in E} P(e : E) = 1. \tag{2}$$

In what follows, we suppose that the first chain of *inequalities* is *strengthened to*

$$0 < P(e : E) < 1. \tag{3}$$

It is convenient to think of the $e$ as options, of $E$ as a subset of options, $R$ the set of all options and $P(e : E)$ the probability that the option $e$ is chosen from $E$. Since $E \in S$, *the subsets $E$ are finite.* Choosing $e \notin E$ from $E$ has probability 0 by (1) and is excluded by (3). So *we suppose $e \in E$* throughout this paper. Also by (3), we exclude impossible options as $e$ and singletons (one-element sets) as $E$.

Of course, there can be several structures of selection probabilities on the same set $R$ and so also on the same subset $E$. As described in the Introduction, we will be interested in individual (componentwise) and aggregated (overall) selection probability functions $P_1, \ldots, P_m$ and $P$, respectively, all satisfying the above Definition 1.

Among selection models, Luce's choice model (Luce 1959, 1977) is a prominent example. We define it here.

**Definition 2.** A structure of (nonzero) selection probabilities is a *Luce choice model* if there exists a function $v$ from $R$ into the set $\mathbb{R}_{++}$ of all positive numbers such that

$$P(e:E) = v(e) \Big/ \sum_{d\in E} v(d). \tag{4}$$

The positive number $v(e)$ is the *scale value* of $e$, the function $v$ is the *scale.*

Here too, we can have several, in particular individual and overall selection probabilities by the choice of scales $v_1, \ldots, v_m, v$. *If we aggregate the individual scale values $v_1(e), \ldots, v_m(e)$ into an overall scale value by a product of powers*

$$v(e) = \prod_{i=1}^{m} v_i(e)^{a_i} \qquad \left(\sum_{i=1}^{m} a_i \neq 0\right), \tag{5}$$

*then in Luce's choice model the overall selection probability will be aggregated from the individual ones as follows:*

$$P(e:E) = \prod_{i=1}^{m} P_i(e:E)^{a_i} \Big/ \sum_{d\in E} \prod_{i=1}^{m} P_i(d:E)^{a_i}, \tag{6}$$

We look now at possible generalizations of this situation.

**Definition 3.** An *n-nonatomic scale* is a triple $(R, v, n)$ where $R$ is a set, $v: R \to \mathbb{R}_{++}$ is a scale (function), and $n$ is a positive integer such that for every $n$-tuple of positive numbers $(z_1, \ldots, z_n) \in \mathbb{R}^n_{++}$ there exist $e_1, \ldots, e_n \in R$ with $z_j = v(e_j)$ $(j = 1, \ldots, n)$.

**Definition 4.** An *n-nonatomic scale family of selection probabilities* (a *scale model* for short) is a 4-tuple $(R, P, v, n)$ where $(R, P)$ is a selection model and $(R, v, n)$ is an $n$-nonatomic scale such that for each $E \subseteq R$ with $E = \{e_1, \ldots, e_n\}$ there exist functions $F_j : \mathbb{R}^n_{++} \to ]0,1[$ with the property that for $e_j \in E$ $(j = 1, \ldots, n)$

$$P(e_j : E) = F_j[v(e_1), \ldots, v(e_n)].$$

Thus we generalize (4), (5), and (6) to

$$P(e_j : E) = F_j[v(e_1), \ldots, v(e_n)] \tag{$7_1$}$$

$$P_i(e_j : E) = F_j[v_i(e_1), \ldots, v_i(e_n)], \tag{$7_2$}$$

$$v(e_j) = G[v_1(e_j), \ldots, v_m(e_j)], \tag{8}$$

$$P(e_j : E) = H_j[P_1(e_1 : E), \ldots, P_m(e_1 : E), \ldots, P_1(e_n : E), \ldots, P_m(e_n : E)], \tag{9}$$

respectively. Here $i = 1, \dots, m$ and $j = 1, \dots, n$. By the $n$-nonatomicity of the scale families, $v(e_j), v_1(e_j), \dots, v_m(e_j)$ are *arbitrary* positive numbers, so $G : \mathbb{R}^m_{++} \to \mathbb{R}_{++}$. For the same reason, the domain of $F_j$ in (7) is $\mathbb{R}^n_{++}$, the range is, see (3), $]0,1[$. So $F_j : \mathbb{R}^n_{++} \to ]0,1[$ $(j = 1, \dots, n)$. However, we should take also (2), that is,

$$\sum_{j=1}^{n} P_i(e_j : E) = 1 = \sum_{j=1}^{n} P(e_j : E)$$

into consideration. Therefore the range of the *n-tuple* $(F_1, \dots, F_n)$ *of functions* is

$$\Gamma_n := \left\{ (p_1, \dots, p_n) \mid \sum_{j=1}^{n} p_j = 1; p_j > 0;\ j = 1, \dots, n \right\} \tag{10}$$

(that is, by definition, $\Gamma_n$ is the set of all positive $n$-tuples with sum 1). So $(F_1, \dots, F_n) : \mathbb{R}^n_{++} \to \Gamma_n$. Similarly, $(H_1, \dots, H_n) : \Gamma^m_n \to \Gamma_n$. Notice that $H_j$ has $mn$ (not only $n$) positive variables, $n$-tuples of which add up to 1.

Denoting $v_i(e_j)$ by $x_{ij}$, comparison of (7), (8) and (9) gives us our first system of $n$ *functional equations*:

$$\begin{aligned} &F_j[G(x_{11}, \dots, x_{m1}), \dots, G(x_{1n}, \dots, x_{mn})] \\ &\quad = H_j[F_1(x_{11}, \dots, x_{1n}), \dots, F_1(x_{m1}, \dots, x_{mn}), \dots, \\ &\qquad F_n(x_{11}, \dots, x_{1n}), \dots, F_n(x_{m1}, \dots, x_{mn})] \quad (j = 1, \dots, n). \end{aligned} \tag{11}$$

As a reminder, we list also the consequences of (2) and (3):

$$\sum_{j=1}^{n} F_j(z_1, \dots, z_n) = 1, \tag{12}$$

$$0 < F_j(z_1, \dots, z_n) < 1 \quad (j = 1, \dots, n), \tag{13}$$

$$\sum_{j=1}^{n} H_j(z_{11}, \dots, z_{m1}, \dots, z_{1n}, \dots, z_{mn}) = 1 \tag{14}$$

$$0 < H_j(z_{11}, \dots, z_{m1}, \dots, z_{1n}, \dots, z_{mn}) < 1 \quad (j = 1, \dots, n) \tag{15}$$

The scales $v_1, \dots, v_m, v$ are *ratio scales* here, that is, the representations (7) and (8) of probabilities and scales are invariant under linear transformation $v_j \mapsto \alpha_j v_j$ $(j = 1, \dots, m)$, $v \mapsto Mv$, $P_j \to P_j$. So, by (8) and (7),

$$G(\alpha_1 y_1, \dots, \alpha_m y_m) = M(\alpha_1, \dots, \alpha_m) G(y_1, \dots, y_m) \tag{16}$$

$$F_j(\alpha z_1, \dots, \alpha z_n) = F_j(z_1, \dots, z_n) \tag{17}$$

Since we have *nonatomic* families of *ratio scales*, the equations and inequalities (11)–(17) should be satisfied for *all* $x_{ij}$, $y_i$, $z_j$, $\alpha_i$, $\alpha \in \mathbb{R}_{++}$, and *all* $(z_{i1}, \ldots, z_{in}) \in \Gamma_n$ $(i = 1, \ldots, m;\ j = 1, \ldots, n)$.

For Luce's choice model

$$G(y_1, \ldots, y_m) = b \prod_{i=1}^{m} y_i^{a_i} = bM(y_1, \ldots, y_m) \quad \left( \sum_{i=1}^{m} a_i \neq 0, b > 0 \right),$$

$$\begin{aligned} F_j(z_1, \ldots, z_n) &= z_j \Big/ \sum_{k=1}^{n} z_k, \\ H_j(z_{11}, \ldots, z_{m1}, \ldots, z_{1n}, \ldots, z_{mn}) &= \prod_{i=1}^{m} z_{ij}^{a_i} \Big/ \sum_{k=1}^{n} \prod_{i=1}^{m} z_{ik}^{a_i} \ (j = 1, \ldots, n) \end{aligned} \tag{18}$$

These functions clearly satisfy (11)–(17). The question arises *what further conditions characterize the Luce model.*

The system (11)–(17) has way too many solutions (see Aczél-Maksa-Marley-Moszner 1995). So we strengthen the restrictions. In particular, it looks reasonable to specify the functions $H_j$ of $mn$ variables as compositions of $m$- and $n$-variable functions. One way is the following. As we see in $(7_2)$, the individual probabilities $P_i(e_j : E)$ $(i = 1, \ldots, m; j = 1, \ldots, n)$ all depend upon the scale values. Conversely, we introduce new scale values which depend upon the individual probabilities $P_i(e_j : E)$ and thus also upon $E$ :

$$\overline{v}(e_j : E) = \Phi[P_1(e_j : E), \ldots, P_m(e_j : E)] \quad (j = 1, \ldots, n) \tag{19}$$

and suppose that $(7_1)$ applies also to $\overline{v}(e_1 : E) \ldots, \overline{v}(e_n : E)$:

$$P(e_j : E) = F_j[\overline{v}(e_1 : E), \ldots, \overline{v}(e_n : E)]. \tag{20}$$

By (9), $(7_1)$, (20) and (19), writing $z_{ij} = P_i(e_j)$, we get

$$H_j(z_{11}, \ldots, z_{m1}, \ldots, z_{1n}, \ldots, z_{mn}) = F_j(\Phi(z_{11}, \ldots, z_{m1}), \ldots, \Phi(z_{1n}, \ldots, z_{mn}))$$

and (11) becomes

$$\begin{aligned} &F_j[G(x_{11}, \ldots, x_{m1}), \ldots, G(x_{1n}, \ldots, x_{mn})] \\ &\quad = F_j(\Phi[F_1(x_{11}, \ldots, x_{1n}), \ldots, F_1(x_{m1}, \ldots, x_{mn})], \ldots, \\ &\qquad \Phi[F_n(x_{11}, \ldots, x_{1n}), \ldots, F_n(x_{m1}, \ldots, x_{mn})]) \quad (j = 1, \ldots, n). \end{aligned} \tag{21}$$

## 3. Results

Of course, (17) implies

$$\begin{aligned} F_j(z_1,\ldots,z_n) &= F_j\left(z_1\Big/\sum_{k=1}^n z_k,\ldots,z_n\Big/\sum_{k=1}^n z_k\right) \\ &= f_j\left(z_1\Big/\sum_{k=1}^n z_k,\ldots,z_n\Big/\sum_{k=1}^n z_k\right), \end{aligned} \tag{22}$$

Notice that $f_j$ has been obtained as restriction of $F_j$ from $\mathbb{R}^n_{++}$ to $\Gamma_n$. The function given by (22), however, satisfies (17) with *arbitrary* $(f_1,\ldots,f_n):\Gamma_n\to\Gamma_n$. So (22) *with arbitrary* $(f_1,\ldots,f_n):\Gamma_n\to\Gamma_n$ *is the general solution* of (17).

As to (16), put $y_1=\ldots=y_m=1$ and $b=G(1,\ldots,1)>0$ into

$$\begin{aligned} &G(\alpha_1 y_1,\ldots,\alpha_m y_m) \\ &\quad = M(\alpha_1,\ldots,\alpha_m)G(y_1,\ldots,y_m) \quad (\alpha_i, y_i\in\mathbb{R}_{++},\ i=1,\ldots,m) \end{aligned} \tag{16}$$

to get that $G(\alpha_1,\ldots,\alpha_m)=M(\alpha_1,\ldots,\alpha_m)G(1,\ldots,1)=bM(\alpha_1,\ldots,\alpha_m)$ and that $M$ is *multiplicative*:

$$M(\alpha_1 y_1,\ldots,\alpha_m y_n)=M(\alpha_1,\ldots,\alpha_m)M(y_1,\ldots,y_n). \tag{23}$$

Since they always satisfy (16), we see that *the general solution of* (16) *is given by* $M$ *arbitrary multiplicative* (satisfying (23)) *and* $G=bM$ ($b$ *an arbitrary positive constant*), that is,

$$G(t_1,\ldots,t_m)=bM(t_1,\ldots,t_m) \quad (t_1,\ldots,t_m\in\mathbb{R}_{++}) \tag{24}$$

As is known (see e.g. Aczél 1987), *if* $G$ and thus $M$ *is locally bounded* (on a neighbourhood; for instance monotonic in each variable or continuous at a point) *and they satisfy* (16) (thus also (23) and (24)), then there exist constants $a_1,\ldots,a_m$ such that

$$M(y_1,\ldots,y_m)=\prod_{i=1}^m y_i^{a_i}, \quad G(y_1,\ldots,y_m)=b\prod_{i=1}^m y_i^{a_i}. \tag{25}$$

If $G(y,\ldots,y)$ is not constant then $a=\sum_i a_i\neq 0$. If, in particular, $G$ is strictly increasing in each variable at a point then $a_i>0$ $(i=1,\ldots,n)$. If

$$G(y,\ldots,y)=y \quad (y\in\mathbb{R}^n_{++}) \tag{26}$$

(and $G$ is locally bounded) then in (25)

$$\sum_{i=1}^m a_i=1=b \tag{27}$$

and we get a weighted geometric mean. Here we continue with the general case, where we suppose only that

$$G(y,\dots,y) \quad \text{is nonconstant} \tag{28}$$

(thus in (25) $a=\sum_i a_i \neq 0$ and $b \neq 0$; actually, $b>0$).

In addition to the very weak regularity condition that $G$ is locally bounded, we will suppose the somewhat less intuitive condition of *injectivity* for $F_1,\dots,F_n$, or really (see (22)) for $(f_1,\dots,f_n)$. As we saw, the latter $n$-tuple of functions (or one vector-valued function) maps $\Gamma_n$ into $\Gamma_n$ (cf. (10); notice that $\Gamma_n$ is $(n-1)$-dimensional, since $\sum_k t_k = 1$). We *suppose* that this mapping is *injective*, that is,

$$\begin{gathered}[f_j(t_1,\dots,t_n)=f_j(u_1,\dots,u_n) \ \text{ for } \ j=1,\dots,n] \\ \text{imply } \ t_k=u_k \ \text{ for } \ k=1,\dots,n\end{gathered} \tag{29}$$

or, since $f_j$ can be considered as the restriction of $F_j$ to $\Gamma_n$, equivalently,

$$\begin{gathered}[F_j(t_1,\dots,t_n)=F_j(u_1,\dots,u_n) \ \text{ for } \ j=1,\dots,n \\ \text{and } \ (t_1,\dots,t_n),(u_1,\dots,u_n)\in\Gamma_n] \\ \text{imply } \ t_k=u_k \quad (k=1,\dots,n).\end{gathered} \tag{30}$$

Note that we did *not* suppose that $(F_1,\dots,F_n)$ or $(f_1,\dots,f_n)$ be surjective ("onto") from $\Gamma_n$ onto $\Gamma_n$ but it is important that the mapping is between two sets of the same dimension because, if the range would be of smaller dimension than the domain, then it would be unreasonable to suppose that each (vector) function value is assumed at most once.

In view of (22) the equations (21) "reduce" to

$$\begin{aligned}&f_j\left(\frac{G(x_{11},\dots,x_{m1})}{\sum_{l=1}^n F(x_{1l},\dots,x_{ml})},\dots,\frac{G(x_{1n},\dots,x_{mn})}{\sum_{l=1}^n F(x_{1l},\dots,x_{ml})}\right)\\ &\quad=f_j\left(\frac{\Phi[F_1(x_{11},\dots,x_{1n}),\dots,F_1(x_{m1},\dots,x_{mn})]}{\sum_{l=1}^n \Phi[F_l(x_{11},\dots,x_{1n}),\dots,F_l(x_{m1},\dots,x_{mn})]},\dots\right.\\ &\quad\left.\dots,\frac{\Phi[F_n(x_{11},\dots,x_{1n}),\dots,F_n(x_{m1},\dots,x_{mn})]}{\sum_{l=1}^n \Phi[F_l(x_{11},\dots,x_{1n}),\dots,F_l(x_{m1},\dots,x_{mn})]}\right)\ (j=1,\dots,n).\end{aligned} \tag{31}$$

Or, because of the injectivity (29),

$$\begin{aligned}&\frac{G(x_{1k},\dots,x_{mk})}{\sum_{l=1}^n G(x_{1l},\dots,x_{ml})}\\ &\quad=\frac{\Phi[F_k(x_{11},\dots,x_{1n}),\dots,F_k(x_{m1},\dots,x_{mn})]}{\sum_{l=1}^n \Phi[F_l(x_{11},\dots,x_{ln}),\dots,F_l(x_{m1},\dots,x_{mn})]} \quad (k=1,\dots,n).\end{aligned} \tag{32}$$

We choose $x_{1l} = \ldots = x_{ml} = z_l$ $(l = 1, \ldots, n)$ to get, with (25) and $a = \sum_i a_i$ $(\neq 0$ by supposition),

$$\frac{z_k^a}{\sum_{l=1}^n z_l^a} = \frac{\Phi[F_k(z_1, \ldots, z_n), \ldots, F_k(z_1, \ldots, z_n)]}{\sum_{l=1}^n \Phi[F_l(z_1, \ldots, z_n), \ldots, F_l(z_1, \ldots, z_n)]} \quad (k = 1, \ldots, n). \tag{33}$$

Surprisingly, from our system (12), (13), (16), (17) and (21) the *inequalities*

$$0 < f_j(z_1, \ldots, z_n) < 1 \quad (j = 1, \ldots, n) \tag{13}$$

turn out to be the most difficult to satisfy. For instance

$$\begin{aligned} F_j(x_1, \ldots, x_n) &= \log x_j + \left(1 - \sum_{k=1}^n \log x_k\right)/n, \\ \Phi(y_1, \ldots, y_m) &= \exp\left(\sum_{i=1}^m a_i y_i\right), \end{aligned} \tag{34}$$

$$G(y_1, \ldots, y_m) = \prod_{i=1}^m y_i^{a_i} = M(y_l, \ldots, y_m) \quad \left(\sum_{i=1}^m a_i \neq 0\right)$$

satisfies (12), (16), (17), (21), but *not* (13). A more methodical approach is the following.

*We suppose that* $\Phi(y, \ldots, y)$ *is continuous and strictly monotonic* because we need that $\varphi$, defined by

$$\varphi(t) = \Phi(t, \ldots, t)^{1/a}, \tag{35}$$

be *continuous and strictly monotonic* and thus *its inverse* $\psi = \varphi^{-1}$ exist. With (35) the equation (33) becomes

$$\begin{aligned} \frac{\varphi[F_k(z_1, \ldots, z_k)]}{z_k} &= \left(\frac{\sum_{l=1}^n \varphi[F_l(z_1, \ldots, z_n)]^a}{\sum_{l=1}^n z_l^a}\right)^{1/a} \\ &= L(z_1, \ldots, z_n) \qquad (k = 1, \ldots, n). \end{aligned} \tag{36}$$

On the right hand side $L$ carries no subscript because it *does not depend upon* $k$. So

$$\begin{aligned} F_k(z_1, \ldots, z_n) &= \varphi^{-1}[z_k L(z_1, \ldots, z_n)] \\ &= \psi[z_k L(z_1, \ldots, z_n)] \qquad (k = 1, \ldots, n), \end{aligned} \tag{37}$$

where $\psi = \varphi^{-1}$ is the inverse of $\varphi$, the existence of which we made sure above. In order to guarantee (13) we need only that *the codomain of* $\psi$ *be a subset of* $]0, 1[$. Furthermore also (12) has to be satisfied, so we have to have

$$\sum_{k=1}^n \psi[z_k L(z_1, \ldots, z_n)] = 1. \tag{38}$$

Thus $L$ *should be the solution* $x$ *of*

$$\sum_{k=1}^{n} \psi(z_k x) = 1, \tag{39}$$

*unique*, if it exists, since $\psi$ is strictly monotonic.

The remaining requirement on $F_j$, the *homogeneity of degree* 0, that is (17), is satisfied if, and only if, in (37) $L$ is *homogeneous of degree* $-1$:

$$L(\alpha z_1, \ldots, \alpha z_n) = \alpha^{-1} L(z_1, \ldots, z_n). \tag{40}$$

Fortunately $L$ as solution of (38) satisfies this automatically: if we replace each $z_k$ by $\alpha z_k$, (38) becomes

$$\sum_{k=1}^{n} \psi[z_k \alpha L(\alpha z_1, \ldots, \alpha z_n)] = 1.$$

Comparing this to (38), the *uniqueness* of the solution of (39) yields (40).

How do we assure the existence of an $x$ satisfying (39)? Let us define a function $h$ by

$$h(x) = \sum_{k=1}^{n} \psi(z_k x), \tag{41}$$

This too is *strictly monotonic.* We want it to assume the value 1 at a point in order that (39) be satisfied. This happens if, and only if, either it starts (at 0) with a value smaller than 1 and "finishes" (at $\infty$) with a value greater than 1, or the other way round. This is realized, see (41), for all $z_k$ exactly if

$$A = \inf \psi(t) < 1/n \quad \textit{and} \quad B = \sup \psi(t) > 1/n. \tag{42}$$

So, *the codomain of* $\psi$ (and, at the same time, *the domain of* $\varphi$), *is an interval* $]A, B[$ *where* $B > 1/n$ *and* $A < 1/n$. Also $B = \infty$ is permissible, but, as we saw above, $A \geq 0$ is needed:

$$0 \leq A < 1/n < B. \tag{43}$$

(Since $\psi$ is strictly monotonic, therefore (42) and (43) guarantee that all values of $\psi$ are greater than 0). For Luce's choice model (18) this is satisfied because $\varphi(y) = y$, $\psi(t) = t$, $A = 0$, $B = \infty$. For (34) and (35), however, $\varphi(y) = \exp y$, $\psi(t) = \log t$, $A = -\infty$, $B = \infty$, so (43) is not satisfied.

Still, there clearly is a great variety of $\psi$'s from which we can choose. We give a few examples:

(i) Take the function $\psi(t) = t + t^{1/2}$. Here

$$A = \lim_{t\to 0+} \psi(t) = 0, \quad B = \lim_{t\to\infty} \psi(t) = \infty,$$

so (43) is satisfied. We calculate $L$ from (38):

$$1 = \sum_{k=1}^{n} \psi(z_k L) = L \sum_{k=1}^{n} z_k + L^{1/2} \sum_{k=1}^{n} z_k^{1/2},$$

$$L = \left( \frac{[(\sum_k z_k^{1/2})^2 + 4\sum_k z_k]^{1/2} - \sum_k z_k^{1/2}}{2\sum_k z_k} \right)^2. \tag{44}$$

Finally from (37),

$$F_j(z_1, \dots, z_n) = z_j L + z_j^{1/2} L^{1/2}, \tag{44'}$$

where $L$ is the not-quite-simple expression (44).

(ii) The choice of $\psi(t) = 1/t$ gives a simpler example; (43) holds again:

$$A = \lim_{t\to\infty} \psi(t) = 0, \quad B = \lim_{t\to 0+} \psi(t) = \infty.$$

Here

$$1 = \sum_{k=1}^{n} \frac{1}{z_k L}, \quad \text{so} \quad L = \sum_{k=1}^{n} (1/z_k),$$

and we get

$$F_j(z_1, \dots, z_n) = \frac{1}{z_j \sum_k (1/z_k)} = \frac{z_1 z_2 \dots z_{j-1} z_{j+1} \dots z_n}{z_2 z_3 \dots z_n + \dots + z_1 z_2 \dots z_{n-1}}. \tag{45}$$

(iii) More generally, $\psi(t) = t^c \, (c \neq 0)$ also gives $A = 0$, $B = \infty$, so (43) is satisfied. Here

$$1 = \sum_{k=1}^{n} z_k^c L^c \quad \text{so} \quad L = (\sum_{k=1}^{n} z_k^c)^{-1/c}$$

and

$$F_j(z_1, \dots, z_n) = z_j^c \Big/ \sum_{k=1}^{n} z_k^c. \tag{46}$$

Luce's choice model (18) is the case $c = 1$ of (46).

We can check directly that (44′), (45), and (46) satisfy (12), (13), (16), (17), and (21) with

$$G(y_1, \dots, y_m) = \prod_{i=1}^{m} y_i^{a_i} \tag{47}$$

and

$$\Phi(y_1,\dots,y_n)=\prod_{i=1}^{m}\psi^{-1}(y_i)^{a_i}=\prod_{i=1}^{m}\varphi(y_i)^{a_i}\qquad\left(\sum_{i=1}^{m}a_i=a\neq 0\right).\tag{48}$$

We prove this, however, in one sweep for all $F_j$ obtained in (37) by the above construction: The equations (16) and (17) are satisfied by all $G, M$ of the form (25) and $F_j$ of the form (22) which is the basis of our construction. The equation (12) and inequalities (13) hold because of (38), (42) and (43), respectively (the latter guarantees only that $F_k=\psi(z_kL)>0$ always but, with (12) which followed from (38), we have also $F_k<1$). Finally, we check (21) which, as we saw, is equivalent to (32) under our suppositions, as follows (the equations we use stand on top of the equality signs):

$$\frac{\Phi[F_k(x_{11},\dots,x_{1n}),\dots,F_k(x_{m1},\dots,x_{mn})]}{\sum_{l=1}^{n}\Phi[F_l(x_{11},\dots,x_{1n}),\dots,F_l(x_{m1},\dots,x_{mn})]}\overset{48}{=}\frac{\prod_{i=1}^{m}\varphi[F_k(x_{i1},\dots,x_{in})]^{a_i}}{\sum_{l=1}^{n}\prod_{i-1}^{m}\varphi[F_l(x_{i1},\dots,x_{in})]^{a_i}}$$

$$\overset{37}{=}\frac{\prod_{i=1}^{m}[x_{ik}L(x_{i1},\dots,x_{in})]^{a_i}}{\sum_{l=1}^{n}\prod_{i=1}^{m}[x_{il}L(x_{i1},\dots,x_{in})]^{a_i}}\frac{\prod_{i=1}^{m}x_{ik}^{a_i}}{\sum_{l=1}^{n}\prod_{i=1}^{n}x_{il}^{a_i}}\overset{25}{=}\frac{G(x_{1k},\dots,x_{mk})}{\sum_{l=1}^{n}G(x_{1l},\dots,x_{ml})},$$

so (32) and with it (21) indeed hold. Thus we proved the following.

**Theorem 1.** *The general solution of the system* (12), (13), (16), (17) *and* (21) *of functional equations and inequalities, with $G$ locally bounded, $G(y,\dots,y)$ non-constant, $\Phi(y,\dots,y)$ continuous and strictly monotonic and $(F_1,\dots,F_n)$ injective on $\Gamma_n$ (see* (10)*) is constructed as follows. Take an arbitrary continuous and strictly monotonic $\psi:\mathbb{R}_{++}\to]A,B[$, where $A$ and $B$ satisfy* (43) *and determine $L(z_1,\dots,z_n)$ as the unique solution of* (38). *Then the general $F_1,\dots,F_n$ are given by* (37), *$G$ and $M$ by* (25) *with $b>0$ and $\sum_i a_i\neq 0$, and $\Phi$ may be chosen as* (48).

**Remark.** It is, of course, desirable that $F_k(z_1,\dots,z_k,\dots,z_n)$ be *increasing in $z_k$ and decreasing in the other $(n-1)$ variables.* Indeed, cf. (7), the probability that an option is chosen should increase if its value (attractiveness) increases in the eyes of an individual or collective and/or the value of one or more of the others decrease. We prove the following.

**Corollary 2.** *The functions $F_k$, constructed in Theorem* 1, *are strictly increasing in $z_k$ and strictly decreasing in $z_j$ ($j\neq k$; $j=1,\dots,n$; $k=1,\dots,n$) if, and only if, $\psi$ is strictly increasing.*

We *prove* first that, *if $\psi:\mathbb{R}_{++}\to]A,B[$ is strictly increasing* then $L:\mathbb{R}^n_{++}\to\mathbb{R}_{++}$ *is strictly decreasing* in each variable. For otherwise there would be a point

$(z_1^0, \ldots, z_n^0)$ and (say) a $z_1 > z_1^0$ such that $L(z_1, z_2^0, \ldots, z_n^0) \geq L(z_1^0, z_2^0, \ldots, z_n^0)$. Then $z_1 L(z_1, z_2^0, \ldots, z_n^0) > z_1^0 L(z_1^0, z_2^0, \ldots, z_n^0)$ and

$$\begin{aligned} 1 &= \psi[z_1 L(z_1, z_2^0, \ldots, z_n^0)] + \sum_{j=2}^{n} \psi[z_j^0 L(z_1, z_2^0, \ldots, z_n^0)] \\ &> \psi[z_1^0 L(z_1^0, z_2^0, \ldots, z_n^0)] + \sum_{j=2}^{n} \psi[z_j^0 L(z_1^0, z_2^0, \ldots, z_n^0)] = 1 \end{aligned} \tag{49}$$

which is impossible. But $L$ is homogeneous of degree $-1$, so

$$\begin{aligned} F_k(z_1, \ldots, z_n) &= \psi[z_k L(z_1, \ldots, z_{k-1}, z_k, z_{k+1}, \ldots, z_n)] \\ &= \psi\left[L\left(\tfrac{z_1}{z_k}, \ldots, \tfrac{z_{k-1}}{z_k}, 1, \tfrac{z_{k+1}}{z_k}, \ldots, \tfrac{z_k}{z_k}\right)\right]. \end{aligned} \tag{50}$$

Thus, since $L$ strictly decreases in each variable, therefore $F_k$ *strictly increases in* $z_k$ *and strictly decreases in the rest of ems variables*, as asserted.

Now we prove that $L$ *is strictly decreasing in each variable also if* $\psi$ *is strictly decreasing*. Indeed, otherwise there would be a $(z_1^0, \ldots, z_n^0)$ and a $z_j < z_j^0$, say $z_1 < z_1^0$ such that $L(z_1, z_2^0, \ldots, z_n^0) \leq L(z_1^0, z_2^0, \ldots, z_n^0)$. Then

$$z_1 L(z_1, z_2^0, \ldots, z_n^0) < z_1^0 L(z_1^0, z_2^0, \ldots, z_n^0)$$

and, since $\psi$ is now strictly *decreasing*, so the impossible (49) would follow again. On the other hand, (50) holds here too, so *if* $\psi$ *is strictly decreasing then* $F_k$ *strictly decreases in* $z_k$ *and strictly increases in* $z_j$ $(j \neq k;\ j = 1, \ldots, n;\ k = 1, \ldots, n)$.

There are no other possibilities for the strictly monotonic $\psi$, so Corollary 2 is proved.

We see that, among the above examples, (iii) for $c > 0$ (in particular, for $c = 1$, Luce's choice model (18)) and (i) have the property called "desirable" in the remark, while (iii) for $c < 0$, in particular (ii) which is the case $c = -1$ of (iii), has the opposite property.

As we see, in the case where $\Phi$ may be different from $G$ but $G(y, \ldots, y)$ is nonconstant, we *obtained* $G$ and $F_j$ from the given equations and inequalities and showed that there *exists* a $\Phi$, namely (48), with which the problem is solved. In Aczél-Maksa-Marley-Moszner 1995 it has been shown that (48) is the *only* possible choice for $n > 2$ and all $\Phi$ 's have been determined also for $n = 2$.

This concludes the part where inequalities have been used in a substantial way. Theorem 1 shows however that, even for (21) (and (12), (13), (16), (17)), the solutions are still rather general. So we make the ultimate restriction, $\Phi = G$ with $G$ locally bounded and $G(y, \ldots, y)$ nonconstant, as before. Thus

$$\Phi(y_1, \ldots, y_m) = G(y_1, \ldots, y_m) = b \prod_{i=1}^{m} y_i^{a_i} \quad \text{with} \quad a = \sum_{i=1}^{m} a_i \neq 0, \quad b > 0.$$

So, what used to be a *condition, that* $\Phi(y,\ldots,y)$ (here $= by^a$) *be continuous and strictly monotonic, is automatically satisfied.* By (35), $\varphi(t) = b^{1/a}t$, $\psi(z) = \varphi^{-1}(z) = b^{-1/a}z$, so that (37) becomes

$$F_k(z_1,\ldots,z_n) = b^{-1/a}L(z_1,\ldots,z_n)z_k \quad (k = 1,\ldots,n). \tag{51}$$

In view of (12), we have

$$1 = \sum_{l=1}^{n} F_l(z_1,\ldots,z_n) = b^{-1/a}L(z_1,\ldots,z_n)\sum_{l=1}^{n} z_l,$$

that is,

$$b^{-1/a}L(z_1,\ldots,z_n) = 1\Big/\sum_{l=1}^{n} z_l$$

and, from (51)

$$F_k(z_1,\ldots,z_n) = z_k\Big/\sum_{l=1}^{n} z_l. \tag{52}$$

We have already (25) with $\sum_i a_i \neq 0$. So we got both formulas (18) of Luce's choice model, which we have thus *characterized.* Indeed, we obtained the following results.

**Theorem 3.** *The general solution of the system* (12), (13), (16), (17) *and* (21), *with* $\Phi = G$ *locally bounded,* $G(y,\ldots,y)$ *nonconstant and with* $(F_1,\ldots,F_n)$ *injective on* $\Gamma_n$, *is given by* (18), *where* $b > 0, a_1,\ldots,a_n$ *are arbitrary but* $\sum_i a_i \neq 0$.

To our question of *characterizing Luce's choice model among families of selection probabilities* we thus obtained the following answer.

**Theorem 3′**. *The only scale family of selection probabilities satisfying* (7), (8), (20) *and* (19) *with* $\Phi = G$ locally bounded, $G(y,\ldots,y)$ *nonconstant and with* $(F_1,\ldots,F_n)$ *injective on* $\Gamma_n$, *is Luce's choice model, described by* (4) *and by*

$$P_i(e:E) = v_i(e)\Big/\sum_{d\in E} v_i(d) \quad (i = 1,\ldots,n),$$

*with the scale values aggregated as in* (5) (*up to a positive constant multiplier*).

**References**

1. J. Aczél, *A Short Course on Functional Equations Based Upon Recent Applications in the Social and Behavioral Sciences.* Reidel-Kluwer, Dordrecht/ Boston, 1987.

2. J. Aczél, Gy. Maksa, A. A. J. Marley, and Z. Moszner, *Consistent aggregation of scale families of selection probabilities.* Submitted 1995.

3. R.D. Luce, *Individual Choice Behavior.* Wiley, New York, 1959.

4. R.D. Luce, *The choice axiom after twenty years.* J. Math. Psychol. **15** (1977), 215–235.

János Aczél, Department of Mathematics, University of Waterloo, Waterloo, Ontario N2L 3G1, Canada.

Gyula Maksa, Institute of Mathematics and Informatics, L. Kossuth University, Pf. 12, H-4010 Debrecen, Hungary.

International Series of Numerical Mathematics, Vol. 123, © 1997 Birkhäuser Verlag Basel

# Delta-exponential mappings in Banach algebras

Roman Ger

**Abstract**

An intriguing interplay between the theory of delta-convex mappings (in the sense of Veselý and Zajiček) and the Hyers-Ulam stability problems is developed by studying a functional inequality

$$\|F(x+y) - F(x)F(y)\| \leq f(x)f(y) - f(x+y)\,. \qquad (*)$$

This is an "exponential version" of the inequality

$$\|F(x+y) - F(x) - F(y)\| \leq \|x\| + \|y\| - \|x+y\|\,,$$

proposed first by D. Yost and then generalized to

$$\|F(x+y) - F(x) - F(y)\| \leq f(x) + f(y) - f(x+y)\,.$$

Superstability phenomenon in connection with $(*)$ is examined.

## 1. Introduction

Z. Gajda [3] has proved that given two Banach spaces $(X, \|\cdot\|)$ and $(Y, \|\cdot\|)$ and a mapping $F : X \longrightarrow Y$ satisfying the functional inequality

$$\|F(x+y) - F(x) - F(y)\| \leq \varepsilon\,(\|x\|^p + \|y\|^p), \qquad x, y \in X,$$

with an $\varepsilon \geq 0$ and a $p \in \mathbb{R}\setminus\{1\}$, there exists a unique additive mapping $A : X \to Y$ such that

$$\|F(x) - A(x)\| \leq \frac{2\varepsilon\,\mathrm{sgn}\,(p-1)}{2^p - 2}\|x\|^p, \quad x \in X.$$

For $p = 0$ the result was known already in 1941 by D.H. Hyers [9] and for $p \in [0,1)$ it was proved in 1978 by Th.M. Rassias [11]. The value $p = 1$ seems to be critical and it actually is; in [3] (see also Th.M. Rassias and P. Šemrl [12]) Z. Gajda presents an example of a continuous function $F : \mathbb{R} \longrightarrow \mathbb{R}$ such that $|F(x+y) - F(x) - F(y)| \leq |x| + |y|$ for all $x, y \in \mathbb{R}$, but there exists no additive mapping $A : \mathbb{R} \longrightarrow \mathbb{R}$ for which the function

$$\mathbb{R} \setminus \{0\} \ni x \longrightarrow \frac{1}{|x|}\,|F(x) - A(x)|$$

were bounded. In other words, the inequality

$$\|F(x+y)-F(x)-F(y)\| \leq \|x\|+\|y\|, \qquad x,y \in X,$$

is too weak for $F$ to enforce its asymptotic additivity.

On the other hand D. Yost (see [14] and [15]), motivated by some geometrical problem on complex analysis, repeatedly asked whether a homogeneous mapping $F : X \longrightarrow Y$ fulfilling a stronger inequality

$$\|F(x+y)-F(x)-F(y)\| \leq \|x\|+\|y\|-\|x+y\|, \quad x,y \in X,$$

has to be linear. This led me to the study of a somewhat more general problem

$$\|F(x+y)-F(x)-F(y)\| \leq f(x)+f(y)-f(x+y), \quad x,y \in X, \tag{1}$$

(see [4], [5] and [7]) which, in turn, corresponds very well to the deeply developed theory of the so-called *delta-convex* mappings in the sense of L. Veselý and L. Zajiček (see [13]). These are continuous solutions to the functional inequality

$$\left\|F\left(\frac{x+y}{2}\right)-\frac{F(x)+F(y)}{2}\right\| \leq \frac{f(x)+f(y)}{2}-f\left(\frac{x+y}{2}\right), \quad x,y \in X. \tag{2}$$

Scalar-valued delta-convex mappings are representable as a difference of two convex functionals. In particular, this family contains all $C^2$-functions on the real line which proves that the class of delta-convex mappings is wide enough to be examined. Moreover, inequality (2) may obviously be investigated without any regularity assumptions upon $F$ and f which, additionally, considerably enlarges the class of solutions (see author's paper [7] in that connection).

In his talk during the 7-th International Conference on General Inequalities (Oberwolfach, Germany, November 12-18, 1995) Th. Rassias has emphasized that there are no results on *exponential* analogues of inequality (1), suggesting to study the functional inequality

$$\|F(x+y)-F(x)F(y)\| \leq f(x)f(y)-f(x+y), \quad x,y \in X, \tag{3}$$

under the assumption that the target space for $F$ is endowed with a suitable multiplicative structure. It so happened that two days later, during the same conference, I was presenting a talk meeting precisely the suggestion of Th. Rassias. In what follows the details of this presentation are given.

## 2. Scalar-valued solutions

Let us recall that even in the case where $f = \text{const} =: \eta \geq 1$ and $(Y, \|\cdot\|) = (\mathbb{C}, |\cdot|)$ – the complex field, the behaviour of solutions of (3) is drastically different from those of its (sub)additive analogue (1). Namely (see J. Baker – J. Lawrence –

F. Zorzitto [2] and J. Baker [1]), each function $F$ from an arbitrary semigroup $(S, +)$ into $\mathbb{C}$ such that

$$|F(x+y) - F(x)F(y)| \leq \varepsilon := \eta^2 - \eta, \quad x, y \in S,$$

is either bounded by $\eta$ in absolute value or $F$ is simply *exponential*, i.e.

$$F(x+y) = F(x)F(y) \qquad \text{for all} \quad x, y \in S,$$

(*superstability* phenomenon). Such an alternative fails to hold, however (see J. Lawrence [10], for instance), if the absolute value sign is replaced by a norm in a Banach algebra $Y$ unless that norm obeys the property $\|u\,v\| = \|u\|\,\|v\|$, $u, v \in Y$ which, as well known, is satisfied exclusively for $Y \in \{\mathbb{R}, \mathbb{C}\}$ or $Y$ being the field of quaternions or Cayley numbers. Consequently, one should expect that such a sharp contrast continues to exist while comparing the behaviour of solutions of (1) and (3). That is really the case.

To begin with, let us first prove a result which will justify the terminology that will be used later on.

**Proposition 1.** *Let $(S, +)$ be an arbitrary grupoid and let $F, f : S \longrightarrow [0, \infty)$ be two functions satisfying the functional inequality*

$$|F(x+y) - F(x)F(y)| \leq f(x)f(y) - f(x+y), \quad x, y \in S. \tag{4}$$

*Then $F$ is representable as a difference of two nonnegative subexponential functions; more precisely, there exists a function $P : S \longrightarrow [0, \infty)$ such that $F = P - f$ and the inequalities*

$$P(x+y) \leq P(x)P(y) \quad \textit{as well as} \quad f(x+y) \leq f(x)f(y)$$

*hold true for all $x, y \in S$.*

*Proof.* Obviously (4) forces $f$ to be subexponential. Put $P := F + f$; then (4) implies that

$$\begin{aligned}
P(x+y) &\leq f(x)f(y) + F(x)F(y) = f(x)f(y) + (P(x) - f(x))\,(P(y) - f(y)) \\
&= P(x)P(y) + f(y)\,(f(x) - P(x)) + f(x)\,(f(y) - P(y)) \\
&= P(x)P(y) - [f(y)F(x) + f(x)F(x)] \\
&\leq P(x)P(y),
\end{aligned}$$

for all $x, y \in S$, because of the nonnegativity of both $F$ and $f$. The latter property gives also the nonnegativity of $P$ and completes the proof.

Unfortunately, the converse fails to hold, in general. To see this, take the positive half-line with the usual addition in place of $(S,+)$ and

$$P(x) := 3\,e^{\sqrt{x}}\,, \qquad f(x) := e^{\sqrt{x}}, \qquad x,y \in [0,\infty)\,.$$

Then a nonnegative function $[0,\infty) \ni x \mapsto F(x) := (P-f)(x) = 2\,e^{\sqrt{x}}$, is a difference of two nonnegative subexponential functions but it is easily seen that inequality (4) is not fulfilled.

In contrast to that we have the following:

**Proposition 2.** *Let $(S,+)$ be an arbitrary grupoid. Then functions $F, f : S \longrightarrow \mathbb{R}$ satisfy the functional inequality*

$$|F(x+y) - F(x) - F(y)| \le f(x) + f(y) - f(x+y), \quad x,y \in S\,,$$

*if and only if both $F + f$ and $f - F$ are subadditive. If that is the case, then $F$ is representable as a difference of two subadditive functions: $F = \frac{1}{2}(F+f) - \frac{1}{2}(f-F)$.*

*Proof.* Straightforward.

In spite of that striking disharmony (a kind of explanation may be found in [6]), for the sake of brevity, in the sequel we shall be using the label *delta-exponential* mappings for solutions $F$ of inequality (3) though even the term delta-subexponential is not quite adequate.

Our main result on scalar delta-exponential mappings reads as follows.

**Theorem 1.** *Let $(G,+)$ be a group (not necessarily commutative) and let functions $F : G \longrightarrow \mathbb{C}$, $f : G \longrightarrow (0,\infty)$ satisfy the functional inequality*

$$|F(x+y) - F(x)F(y)| \le f(x)f(y) - f(x+y), \quad x,y \in G.$$

*Then either $F$ is exponential or*

$$|F(x)| \le (1+\sqrt{2})\,f(x) \quad \textit{for all} \quad x \in G\,. \tag{5}$$

To prove this theorem we need two lemmas. The first of them yields a generalization of the basic idea from [1].

**Lemma 1.** *Let $(S,+)$ be an arbitrary semigroup and let functions $F : S \longrightarrow \mathbb{C}$, $R : S \times S \longrightarrow \mathbb{R}$ satisfy the functional inequality*

$$|F(x+y) - F(x)F(y)| \le R(x,y), \quad x,y \in S.$$

*Then, for every $z \in S$ such that $F(z) \neq 0$ one has*

$$|F(x+y) - F(x)F(y)| \leq \frac{1}{|F(z)|}\left(R(x+y,z) + R(x,y+z) + |F(x)|\, R(y,z)\right),$$

*for all $x, y \in S$*

*Proof.* Fix arbitrarily $x, y \in S$ and a $z \in S$ such that $F(z) \neq 0$. We have

$$|F(x+y+z) - F(x+y)F(z)| \leq R(x+y,z)$$

as well as

$$|-F(x+y+z) + F(x)F(y+z)| \leq R(x,y+z),$$

whence

$$|F(x)F(y+z) - F(x+y)F(z)| \leq R(x+y,z) + R(x,y+z).$$

Consequently,

$$\begin{aligned} |F(x+y)F(z) - F(x)F(y)F(z)| &\leq |F(x+y)F(z) - F(x)F(y+z)| \\ &\quad + |F(x)F(y+z) - F(x)F(y)F(z)| \\ &\leq R(x+y,z) + R(x,y+z) + |F(x)|\, R(y,z), \end{aligned}$$

whence our assertion follows immediately.

**Lemma 2.** *Let $(G,+)$ be an arbitrary group and let functions $F : G \longrightarrow \mathbb{C}$, $f : G \longrightarrow (0,\infty)$ satisfy the functional inequality*

$$|F(x+y) - F(x)F(y)| \leq f(x)f(y) - f(x+y), \quad x, y \in G.$$

*Then, $F$ is exponential provided that the quotient $\frac{1}{f}F$ is unbounded.*

*Proof.* Since, obviously, $f$ is subexponential, for every $x, y \in G$ one has

$$f(y) = f(-x + (x+y)) \leq f(-x)f(x+y)$$

whence

$$\frac{f(x)f(y)}{f(x+y)} - 1 \leq \varphi(x) := f(x)f(-x) - 1.$$

Clearly, $\varphi(x) \geq 0$ for all $x \in G$. Setting

$$R(x,y) := f(x)f(y) - f(x+y), \qquad x, y \in G,$$

we infer that the inequality

$$R(x,y) = f(x+y)\left[\frac{f(x)f(y)}{f(x+y)} - 1\right] \leq f(x+y)\varphi(x)$$

remains valid for every $x, y \in G$. Now, for any $x, y, z \in G$ we obtain the estimation

$$\begin{aligned} & R(x+y,z) + R(x,y+z) + |F(x)|R(y,z) \\ & \quad \le f(x+y+z)\,\varphi(x+y) + f(x+y+z)\varphi(x) + |F(x)|f(y+z)\,\varphi(y) \\ & \quad \le f(z)\,[f(x+y)\,\varphi(x+y) + f(x+y)\varphi(x) + |F(x)|f(y)\,\varphi(y)] \\ & \quad = f(z)\,A(x,y)\,, \end{aligned}$$

where $A(x, y)$ denotes the expression within the square brackets; plainly, the function $A$ does not depend upon the $z$-variable. By means of Lemma 1 we get finally that

$$|F(x+y) - F(x)F(y)| \le \frac{f(z)}{|F(z)|} A(x,y) \tag{6}$$

holds true for all $x, y, z \in G$ with $F(z) \neq 0$. Since, by assumption, the map $\frac{1}{f}F$ is unbounded, there exists a sequence $(z_n)_{n\in\mathbb{N}}$ of elements of $(G, +)$ such that

$$0 < \frac{|F(z_n)|}{f(z_n)} \longrightarrow \infty \quad \text{as} \quad n \longrightarrow \infty\,.$$

Consequently, setting $z = z_n$, $n \in \mathbb{N}$, in (6) and passing to the limit, we derive the equality

$$F(x+y) = F(x)F(y)$$

valid for all $x, y \in G$, as claimed.

*Proof of Theorem* 1. Put

$$H := \frac{1}{f}F \qquad \text{and} \qquad c := \sup\{|H(x)| : \ x \in G\}\,.$$

We shall distinguish two cases:

(i) $c < \infty$. Then from inequality (4) it follows that

$$\left| \frac{f(x+y)}{f(x)f(y)} H(x+y) - H(x)H(y) \right| \le 1 - \frac{f(x+y)}{f(x)f(y)} \le 1$$

holds for all $x, y \in G$. Consequently,

$$\left| \left( \frac{f(x+y)}{f(x)f(y)} H(x+y) - H(x+y) \right) + (H(x+y) - H(x)H(y)) \right| \le 1$$

whence

$$\begin{aligned} |H(x+y) - H(x)H(y)| & \le 1 + |H(x+y)| \left| \frac{f(x+y)}{f(x)f(y)} - 1 \right| \\ & \le 1 + |H(x+y)| \le 1 + c =: \delta \end{aligned}$$

for all $x, y \in G$. On account of J. Baker's result [1] we get the alternative:

$$\text{(a)} \quad |H(x)| \leq \tfrac{1}{2}\left(1+\sqrt{1+4\delta}\right) = \tfrac{1}{2}\left(1+\sqrt{4c+5}\right), \; x \in G,$$

or

$$\text{(b)} \quad H \;\text{ is exponential.}$$

The case (a) implies that $0 \leq c \leq \frac{1}{2}\left(1+\sqrt{4c+5}\right)$, i.e., $c \in [0, 1+\sqrt{2}\,]$. In other words, if $c > 1+\sqrt{2}$ then (b) occurs.

(ii) $c = \infty$. Then $H = \frac{1}{f}F$ is unbounded and Lemma 2 states that $H$ is exponential.

Therefore, we have shown that either $|H(x)| \leq 1+\sqrt{2}$, $x \in G$, or $H$ is exponential. Since the first possibility says that $|F(x)| \leq (1+\sqrt{2})f(x)$ for all $x \in G$ (the first part of the assertion alternative), it remains to consider the case where $H$ is exponential. In what follows we are going to show that, in such a case, $|F(x)| = f(x)$ for all $x \in G$ (and henceforth, in particular, $|F(x)| \leq (1+\sqrt{2})f(x)$, $x \in G$) or $F$ itself is exponential. To this end, observe first that if $H(x_0) = 0$ for some $x_0 \in G$, then $H(x+x_0) = 0$ for all $x \in G$, i.e., $H = 0$ and, subsequently, $F = 0$; so, $F$ is (trivially) exponential, as claimed.

Assume that $H(x) \neq 0$ for all $x \in G$. We have

$$|H(x+y)\left(f(x+y) - f(x)f(y)\right)| = |F(x+y) - F(x)F(y)| \leq f(x)f(y) - f(x+y),$$

for all $x, y \in G$, getting the implications

$$|H(x+y)| > 1 \Longrightarrow f(x+y) = f(x)f(y) \Longrightarrow F(x+y) = F(x)F(y)$$

for every $x, y \in G$. Put $Z := \{t \in G : \; |H(t)| \leq 1\}$ and consider two cases:

$$(*) \qquad Z \cup (a+Z) = G \qquad \text{for} \quad \text{some} \qquad a \in G\;;$$

$$(**) \qquad \text{non } (*)\;.$$

In case $(*)$, for every $t \in G$ we have either $|H(t)| \leq 1$ or $|H(-a+t)| \leq 1$, whence $|H(t)| \leq \max(1, |H(a)|)$ for all $t \in G$. The only bounded nonzero exponential functions are those having values in the unit circle; hence $|F(t)| = f(t)$ for all $t \in G$.

In case $(**)$ we are faced to a conditional Cauchy type functional equation of the form:

$$x + y \notin Z \Longrightarrow F(x+y) = F(x)F(y)\,, \qquad x, y \in G\,,$$

with $F$ mapping the group $(G, +)$ into the multiplicative group $(\mathbb{C} \setminus \{0\}, \cdot)$ and with $Z \cup (a+Z) \neq G$ for all $a \in G$. Fix arbitrarily $x, y \in G$. We have

$$Z \cup (-x+Z) \neq G$$

which implies that

$$(-y+Z)\cup(-y-x+Z)\neq G,$$

i.e., there exists a $z\in G$ such that

$$y+z\notin Z \qquad \text{and} \qquad x+y+z\notin Z.$$

Consequently, we get the equalities:

$$F(x+y)F(z)=F(x+y+z)=F(x)F(y+z)=F(x)F(y)F(z)$$

which clearly force $F$ to be exponential. Thus the proof has been completed.

**Remark 1.** The constant $1+\sqrt{2}$ occurring in the statement of Theorem 1 does not seem to be sharp. Trying to replace inequality (5) by

$$|F(x)|\le cf(x) \quad \text{forall} \qquad x\in G,$$

with another *universal* constant $c<1+\sqrt{2}$ we realize that such a $c$ (if any) has to be minorized by 1; indeed, take $(G,+)=(\mathbb{R},+)$ and $F(x)=f(x):=e^{|x|}$, $x\in\mathbb{R}$, or, more generally, $F(x)=f(x):=e^{p(x)}$, $x\in\mathbb{R}$, where $p:\mathbb{R}\longrightarrow\mathbb{R}$ is an arbitrary nonadditive subadditive function. The conjecture that $c=1$ is the optimal constant appears to be both attractive and plausible.

On the other hand, in *special* cases $c$ may happen to be arbitrarily close to zero. In fact, take any $\varepsilon>0$ and an arbitrary nonadditive subadditive function $p:\mathbb{R}\longrightarrow\mathbb{R}$ and put

$$F(x):=e^{p(x)}, \qquad f(x):=e^{p(x)+\varepsilon}, \quad x\in\mathbb{R}.$$

It is a simple matter to check that

$$|F(x+y)-F(x)F(y)|\le f(x)f(y)-f(x+y), \quad x,y\in\mathbb{R},$$

$F$ is not exponential and $|F(x)|=F(x)=e^{-\varepsilon}f(x)$ for all $x\in\mathbb{R}$.

## 3. Vector-valued solutions

Our next objective is to establish two theorems on solutions of inequality (3), similar in spirit to some of the results presented in [8]. Theorem 1 will serve as the main tool in these considerations.

**Theorem 2.** *Let $(G,+)$ be a group (not necessarilly abelian) and let $(Y,\|\cdot\|)$ be a commutative semisimple Banach algebra with a unit. Suppose that $F:G\longrightarrow Y$ and $f:G\longrightarrow(0,\infty)$ satisfy the functional inequality*

$$\|F(x+y)-F(x)F(y)\|\le f(x)f(y)-f(x+y), \quad x,y\in G.$$

*If the image $\left(y^*\circ\frac{F}{f}\right)(G)$ is unbounded for every nonzero linear and multiplicative member $y^*$ of the dual space $Y^*$, then $F$ is exponential.*

*Proof.* Let $y^* \neq 0$ be an arbitrarily fixed linear multiplicative functional on algebra $Y$. Then $\|y^*\| = 1$ whence, for every $x, y \in G$, one obtains

$$|y^*\left(F(x+y) - F(x)F(y)\right)| \leq \|F(x+y) - F(x)F(y)\| \leq f(x)f(y) - f(x+y),$$

which states that the function $\varphi := y^* \circ F$ satisfies the inequality

$$|\varphi(x+y) - \varphi(x)\varphi(y)| \leq f(x)f(y) - f(x+y), \quad x, y \in G.$$

Thus, an appeal to Theorem 1 leads to the alternative

$$\frac{|\varphi|}{f} \leq 1 + \sqrt{2} \qquad \text{or} \qquad \varphi \text{ is exponential}.$$

But, by assumption, the function

$$G \ni x \longmapsto \frac{|\varphi(x)|}{f(x)} = \frac{|y^*(F(x))|}{f(x)} = |y^*\left(\frac{F(x)}{f(x)}\right)| = |\left(y^* \circ \frac{F}{f}\right)(x)|$$

is unbounded and so $\varphi$ has to be exponential. It follows that

$$y^*(F(x+y)) = y^*(F(x))\, y^*(F(y))$$

holds true for all $x, y \in G$. In other words, for every pair $(x, y) \in G^2$, the Cauchy difference $F(x+y) - F(x)F(y)$ falls into the kernel of the functional $y^*$ and therefore, due to the unrestricted choice of $y^*$, into the radical of the algebra $Y$. Since $Y$ was supposed to be semisimple, it proves that $F(x+y) - F(x)F(y) = 0$, which is the desired conclusion.

**Theorem 3.** *Let $(G, +)$ be a group (not necessarilly abelian) and let $(Y, \|\cdot\|)$ be a commutative $C^*$-algebra. Suppose that $F : G \longrightarrow Y$ and $f : G \longrightarrow (0, \infty)$ satisfy the functional inequality*

$$\|F(x+y) - F(x)F(y)\| \leq f(x)f(y) - f(x+y), \quad x, y \in G.$$

*Then there exists a commutative $C^*$-algebra $Z$ such that $Y$ is isomorphic to a $C^*$-subalgebra $\tilde{Y}$ of $Z$, $Z = I \oplus J$, where $I$ and $J$ are closed ideals and, if $\Phi$ stands for the corresponding isomorphism between $Y$ and $\tilde{Y}$, then $\mathrm{proj}_I \circ \Phi \circ F$ is exponential whereas $\|\mathrm{proj}_J \circ \Phi \circ F(x)\| \leq (1+\sqrt{2})f(x)$ for all $x \in G$.*

*Proof.* By means of the celebrated Gelfand-Naimark Theorem, the Gelfand homomorphism of the algebra $Y$ yields an involution preserving isometric isomorphism of $Y$ onto the algebra $C(\Omega)$ of all complex-valued continuous functions on a compact Hausdorff space $\Omega$ with the uniform convergence norm. Let $\Gamma : G \longrightarrow C(\Omega)$ be defined by the formula:

$$\Gamma(x) := \Phi(F(x)), \qquad x \in G,$$

and let

$$\Omega^2 := \{t \in \Omega : \; |\Gamma(x)(t)| \le (1+\sqrt{2})f(x), \quad x \in G\},$$

$$\Omega_x := \{t \in \Omega : \; |\Gamma(x)(t)| \le (1+\sqrt{2})f(x)\}, \quad x \in G.$$

Since, for every $x \in G$, the function $\Gamma(x)$ is continuous, all the $\Omega_x$'s are closed and, henceforth, so is also $\Omega^2$ as the intersection of all $\Omega_x$'s.

Fix arbitrarily a $t \in \Omega^1 := \mathrm{cl}\,(\Omega \setminus \Omega^2)$. There exists a net $(t_\alpha)$ of elements of $\Omega \setminus \Omega^2$ such that $t = \lim t_\alpha$. For every $\alpha$ the function

$$G \in x \longmapsto \Gamma(x)(t_\alpha) \in \mathbb{C}$$

satisfies the assumptions of Theorem 1. Indeed, since $\Phi$ is an isometric isomorphism, we have

$$\begin{aligned} |\Gamma(x+y)(t_\alpha) - \Gamma(x)(t_\alpha)\Gamma(y)(t_\alpha)| &= |\left(\Gamma(x+y) - \Gamma(x)\Gamma(y)\right)(t_\alpha)| \\ &\le \sup_{t\in\Omega} |\left(\Gamma(x+y) - \Gamma(x)\Gamma(y)\right)(t_\alpha)| = \|\Gamma(x+y) - \Gamma(x)\Gamma(y)\| \\ &= \|\Phi\left(F(x+y) - F(x)F(y)\right)\| = \|F(x+y) - F(x)F(y)\| \\ &\le f(x)f(y) - f(x+y), \end{aligned}$$

and since $t_\alpha \notin \Omega^2$, for some $x \in G$ the inequality $|\Gamma(x)(t_\alpha)| > (1+\sqrt{2})f(x)$ is fulfilled. By Theorem 1, the function $\Gamma(\cdot)(t_\alpha)$ is exponential. Therefore, for every $x, y \in G$ and for every $\alpha$ we get the equality

$$\Gamma(x+y)(t_\alpha) = \Gamma(x)(t_\alpha)\,\Gamma(y)(t_\alpha)$$

whence, by means of the continuity of the functions $\Gamma(x+y)$, $\Gamma(x)$ and $\Gamma(y)$, we infer that

$$\Gamma(x+y)(t) = \Gamma(x)(t)\,\Gamma(y)(t).$$

In other words, the function $G \ni x \longmapsto \Gamma(x)|_{\Omega^1}$ is exponential.

Equip the space $Z := C(\Omega^1) \oplus C(\Omega^2)$ with the norm $\|u \oplus v\| := \max(\|u\|, \|v\|)$. Then it is easy to check that the map

$$Y \ni y \longmapsto (\Phi(y)|_{\Omega^1}) \oplus (\Phi(y)|_{\Omega^2}) \in Z$$

establishes an isomorphic isometry of $Y$ onto a $C^*$-subalgebra $\tilde{Y}$ of $Z$, and that

$$I := C(\Omega^1) \oplus \{0\}, \qquad J := \{0\} \oplus C(\Omega^2),$$

yield two closed ideals whose direct sum coincides with $Z$. Moreover, $\mathrm{proj}_I \circ \Phi \circ F = \mathrm{proj}_I \circ \Gamma = \Gamma(\cdot)|_{\Omega^1}$ is exponential whereas for every $x \in G$ one has

$$\|\mathrm{proj}_J \circ \Phi \circ F(x)\| = \|\Gamma(\cdot)|_{\Omega^2}\| = \max_{t\in\Omega^2} |\Gamma(x)(t)| = |\Gamma(x)(t_0)| \le (1+\sqrt{2})f(x);$$

here $t_0 \in \Omega^2$ is a point at which the continuous function $|\Gamma(x)|_{\Omega^2}|$ attains its global maximum and the latter inequality results from the definition of the set $\Omega^2$. This finishes the proof.

**Remark 2.** Given an $\varepsilon > 0$, in the very particular case where

$$f(x) \equiv \text{const} := \frac{1}{2}\left(1 + \sqrt{1+4\varepsilon}\right)$$

in Theorem 2 (resp. Theorem 3) we obtain Theorem 3.1 (resp. Theorem 3.2) from [8]. However, we have to emphasize that in the present paper we deal with groups as the domain structure whereas in [8] semigroups were admissible. A natural problem whether the results just established carry over to the case of semigroups as domains of the functions in question remains open.

**Acknowledgement.** This research was supported by KBN grant No. 2 P03A 049 09 (Poland).

## References

1. J.A. Baker, *The stability of the cosine equation.* Proc. Amer. Math. Soc. **80** (1980), 411–416.

2. J.A. Baker, J. Lawrence and F. Zorzitto, *The stability of the equation* $f(x+y) = f(x)f(y)$. Proc. Amer. Math. Soc. **74** (1979), 242–246.

3. Z. Gajda, *On the stability of additive mappings.* Internat. J. Math. & Math. Sci. **14** (1991), 431–434.

4. R. Ger, *On functional inequalities stemming from stability questions.* International Series of Numerical Analysis, Birkhäuser Verlag Basel, vol. 103 (1992), 227–240.

5. R. Ger, *The singular case in the stability behaviour of linear mappings.* Grazer Mathematische Berichte **316** (1992), 59–70.

6. R. Ger, *Superstability is not natural.* Rocznik Naukowo-Dydaktyczny WSP w Krakowie **159** (1993), 109–123.

7. R. Ger, *Stability aspects of delta-convexity.* In "Stability of Hyers-Ulam type", A Volume dedicated to D.H. Hyers & S. Ulam, (ed. Th.M. Rassias & J. Tabor), Hadronic Press, Inc., Palm Harbor (1994), 99–109.

8. R. Ger & P. Šemrl, *The stability of the exponential equation.* Proc. Amer. Math. Soc. (in print).

9. D.H. Hyers, *On the stability of the linear functional equation.* Proc. Nat. Acad. Sci. U.S.A. **27** (1941), 222–224.

10. J. Lawrence, *The stability of multiplicative semi-group homomorphisms to real normed algebras, I.* Aequationes Math. **28** (1985), 94-101.

11. Th.M. Rassias, *On the stability of the linear mapping in Banach spaces.* Proc. Amer. Math. Soc. **72** (1978), 297–300.

12. Th.M. Rassias & P. Šemrl, *On the behaviour of mappings which do not satisfy Hyers-Ulam stability.* Proc. Amer. Math. Soc. To appear.

13. L. Veselý & L. Zajiček, *Delta-convex mappings between Banach spaces and applications.* Dissertationes Math. **289**, Polish Scientific Publishers, Warszawa, 1989.

14. D. Yost, Talk at the 17-th Winter School on Abstract Analysis (Section of Analysis), January, 1989, Šrni, Czech Republic.

15. D. Yost, Talk at the 18-th Winter School on Abstract Analysis (Section of Analysis), January, 1990, Šrni, Czech Republic.

Roman Ger, Institute of Mathematics, Silesian University, Bankowa 14, 40-007 Katowice, Poland.

International Series of Numerical Mathematics, Vol. 123, © 1997 Birkhäuser Verlag Basel

# On a problem of S.M. Ulam and the asymptotic stability of the Cauchy functional equation with applications

Themistocles M. Rassias

**Abstract**

In 1940, S.M. Ulam proposed the following problem. Given a group $G_1$, a metric group $G_2$ with metric $d(\cdot,\cdot)$ and a positive number $\varepsilon$, does there exist a $\delta > 0$ such that if $f : G_1 \to G_2$ satisfies

$$d(f(xy), f(x)f(y)) \leq \delta \text{ for all } x, y \in G_1,$$

then a homomorphism $h : G_1 \to G_2$ exists with

$$d(f(x), h(x)) \leq \varepsilon \text{ for all } x \in G_1\,?$$

In this paper some new results for Ulam's problem are studied. It is also explained how the Hyers-Ulam stability theory can be used to study the asymptotic derivative of some nonlinear operators especially in fixed point theory.

## 1. Introduction

The most famous functional equation is the Cauchy equation

$$f(x+y) = f(x) + f(y), \tag{1.1}$$

any solution of which is called *additive*. It is known that for real valued functions defined on the real line, every Lebesgue measurable solution of (1.1) is of the form $f(x) = cx$ for some constant $c$. Non-measurable solutions also exist but they are "wild", being discontinuous everywhere and unbounded on every interval. A discussion of these facts may be found in Chapter 2 of the book by J. Aczél and J. Dhombres (1989).

The concept of stability for a functional equation arises when we replace the functional equation by an inequality which acts as a perturbation of the equation.

The stability question is: *How do the solutions of the inequality differ from those of the given equation?*

We shall consider functions defined on a real normed vector space $E_1$ with values in a real Banach space $E_2$. Given a positive number $\varepsilon$, we call a function $f : E_1 \to E_2$ *$\varepsilon$-additive* if it satisfies the inequality

$$\|f(x+y) - f(x) - f(y)\| < \varepsilon \tag{1.2}$$

for all $x$ and $y$ in $E_1$. We seek an additive function $g$ which is close to the given $\varepsilon$-additive function $f$. In order to do this we start by putting $y = x$ in (1.2) and then dividing by 2 to obtain

$$\left\| \frac{f(2x)}{2} - f(x) \right\| < \frac{\varepsilon}{2}. \tag{1.3}$$

In (1.3) replace $x$ by $2x$ and then divide by 2:

$$\left\| \frac{f(2^2x)}{2^2} - \frac{f(2x)}{2} \right\| < \frac{\varepsilon}{2^2}. \tag{1.4}$$

Combine (1.3) and (1.4) by use of the triangle inequality to get

$$\left\| \frac{f(2^2x)}{2^2} - f(x) \right\| < \varepsilon \left( \frac{1}{2} + \frac{1}{2^2} \right) = \varepsilon(1 - 2^{-2}).$$

Now proceed in this way to prove by induction that

$$\|2^{-n} f(2^n x) - f(x)\| < \varepsilon(1 - 2^{-n}), \quad n = 1, 2, \dots . \tag{1.5}$$

In order to show that the functions $g_n(x) = 2^{-n} f(2^n x)$ form a convergent sequence we use the Cauchy convergence criterion. Indeed, replace $x$ by $2^m x$ in (1.5) and divide the result by $2^m$, where $m$ is an arbitrary positive integer. We find that

$$\|2^{-(m+n)} f(2^{m+n} x) - 2^{-m} f(2^m x)\| < 2^{-m} \varepsilon,$$

for all positive integers.

Hence by the Cauchy criterion the limit $g(x) = \lim_{n \to \infty} g_n(x)$ exists for each $x$ in $E_1$. By taking the limit a $n \to \infty$ in (1.5) we see that $\|g(x) - f(x)\| \le \varepsilon$ for all $x$ in $E_1$. In order to show that $g$ satisfies (1.1), replace $x$ by $2^n x$ and $y$ by $2^n y$ in (1.2) and divide by $2^n$, to get

$$\|2^{-n} f(2^n(x+y)) - 2^{-n} f(2^n x) - 2^{-n} f(2^n y)\| < 2^{-n} \varepsilon.$$

Taking the limit as $n \to \infty$ we find that $g$ is additive. The inequality $\|g(x) - f(x)\| \leq \varepsilon$ is sharp as is seen by the example $f(x) = c$, where $c$ is any constant in $E_2$ with $\|c\| < \varepsilon$ and $g(x) \equiv 0$. Moreover $g$ is the *only* additive function with $\|g(x) - f(x)\| \leq \varepsilon$ for all $x$ in $E_1$. Indeed, suppose there is another such function $h : E_1 \to E_2$. Then for some $y$ in $E_1$ we would have $h(y) \neq g(y)$. Since $\|f(x) - h(x)\| \leq \varepsilon$ for all $x$ we have $\|g(x) - h(x)\| = \|g(x) - f(x) + f(x) - h(x)\| < 2\varepsilon$ for all $x$ in $E_1$.

Choose a positive integer $k$ so that $k\|g(y) - h(y)\| > 2\varepsilon$. Then since both $g$ and $h$ are additive it follows that $\|g(ky) - h(ky)\| = k\|g(y) - h(y)\| > 2\varepsilon$, which is a contradiction. Thus $h(x) = g(x)$ for all $x$ in $E_1$.

We note that so far no regularity conditions on the $\varepsilon$-additive function $f$ were required. Both $f$ and $g$ could be everywhere discontinuous. However, $g$ inherits regularity from $f$. In particular, suppose that $f$ is continuous at a single point $y$. Then we shall prove that $g$ is continuous everywhere. Assume the contrary. Then $g$ fails to be continuous at the origin, so there is an integer $a > 0$ and a sequence of points $x_n$ in $E_1$ converging to zero such that $g(x_n) > 1/a$ for $n = 1, 2, 3, \ldots$. Let $m$ be an integer greater than $3a\varepsilon$. Thus we have

$$\begin{aligned} \|g(mx_n + y) - g(y)\| &\leq \|g(mx_n + y) - f(mx_n + y)\| \\ &\quad + \|f(mx_n + y) - f(y)\| + \|f(y) - g(y)\| < 3\varepsilon \end{aligned}$$

for sufficiently large $n$ since $\lim_{n\to\infty} f(mx_n + y) = f(y)$ by the continuity of $f$. On the other hand we have $\|g(mx_n + y) - g(y)\| = \|g(mx_n)\| = m\|g(x_n)\| > 3\varepsilon$, which is a contradiction. Hence $g$ is continuous everywhere.

Finally, let us assume that the mapping $t \to f(tx)$ from $\mathbb{R}$ to $E_2$ is continuous in $t$ for each fixed $x$ in $E_1$. Then we apply the previous result to find that $g(tx) : \mathbb{R} \to E_2$, with $x$ fixed, is additive and continous in $t$. Hence $g(tx) = tg(x)$ for all real $t$, i.e., $g$ is linear.

Thus we have proved the following theorem (D.H. Hyers (1941)):

**Theorem 1.1.** *Let $E_1$ be a normed vector space, $E_2$ a Banach space and suppose that the mapping $f : E_1 \to E_2$ satisfies the inequality*

$$\|f(x + y) - f(x) - f(y)\| < \varepsilon \tag{1.6}$$

*for all $x$ and $y$ in $E_1$. Then the limit*

$$g(x) = \lim_{n\to\infty} 2^{-n} f(2^n x) \tag{1.7}$$

*exists for each $x$ in $E_1$ and $g$ is the unique additive mapping satisfying*

$$\|f(x) - g(x)\| \leq \varepsilon \tag{1.8}$$

*for all $x$ in $E_1$. Also, if for each $x$ the function $t \to f(tx)$ from $\mathbb{R}$ to $E_2$ is continuous for each fixed $x$, then $g$ is linear. If $f$ is continuous at a single point of $E_1$ then $g$ is continuous everywhere in $E_1$.*

As was pointed out by G.L. Forti (1987), the proof of (1.7) and (1.8) can be used to prove analogous results for mappings from an arbitrary semigroup into a Banach space. The following corollary is essentially Proposition 1 of Forti (1987).

**Corollary 1.2.** *Let $S$ be an arbitrary semigroup, with its operation indicated by juxtaposition, and let $E$ be a Banach space. Suppose that $F : S \to E$ satisfies the inequality $\|f(xy) - f(x) - f(y)\| < \varepsilon$ for all $x$ and $y$ in $S$. Then the limit $g(x) = \lim_{n\to\infty} 2^{-n} f(x^{2^n})$ exists for each $x$ in $S$ and (1.8) holds. Also, $g(x^2) = 2g(x)$ for all $x$ in $S$ and $g$ is the unique function with this property which also satisfies (1.8). Finally, if the semigroup is commutative then $g(xy) = g(x) + g(y)$.*

*Proof.* In the above proof of Theorem 1.1 no use was made of the commutativity of addition in $E_1$ for demonstrating (1.7) and (1.8), so the analgous formulas are true here. For each $x$ in $S$ we have

$$g(x^2) = \lim_{n\to\infty} 2^{-n} f(x^{2^{n+1}}) = 2 \lim_{n\to\infty} 2^{-(n+1)} f(x^{2^{n+1}}) = 2g(x).$$

To prove the uniqueness property, suppose that $h : S \to E$ satisfies (1.8) and that $h(x^2) = 2h(x)$ for all $x$ in $S$. Then

$$\|f(x^{2^n}) - h(x^{2^n})\| = \|f(x^{2^n}) - 2^n h(x)\| \leq \varepsilon.$$

Dividing by $2^n$ and letting $n \to \infty$ we find that $h(x) = g(x)$. □

This method of proof, resulting in a formula like (1.7), will be called a *direct method.* Unless otherwise stated all vector spaces are assumed to be real.

## 2. Approximately additive mappings

We have mentioned the concept of stability and given an example for the Cauchy functional equation (1.1). A definition of stability in the case of homomorphisms between metric groups was suggested by a problem posed by S.M. Ulam in 1940 (see Ulam (1960), p. 64). Given a group $(G_1, \cdot)$, a metric group $(G_2, *)$ with metric $d$ and a positive number $\eta$, suppose that there exists a positive number $\varepsilon = \varepsilon(\eta)$ such that, if $d(f(x \cdot y), f(x) * f(y)) \leq \varepsilon$ for some $f : G_1 \to G_2$ and all $x$ and $y$ in $G_1$, then a homomorphism $h : G_1 \to G_2$ exists with $d(f(x), h(x)) \leq \eta$ for all $x$ in $G_1$. In this case the equation of homomorphism $h(x \cdot y) = h(x) * h(y)$ is called *stable.*

In a number of stability theorems that have appeared in the literature the relation between $\eta$ and $\varepsilon$ is of the form $\eta = k\varepsilon$ for some constant $k$. Several other definitions of stability of functional equations have been used or implied by various authors. For a discussion of this topic one may see for example: R. Ger (1983), Z. Moszner (1985), D.H. Hyers and Th.M. Rassias (1992), P. Gavruta (1994), and Th.M. Rassias and J. Tabor (1994).

A generalization of Theorem 1.1 given by Th.M. Rassias (1978) allows the Cauchy difference to be unbounded, as follows:

**Theorem 2.1.** *Let $f : E_1 \to E_2$ be a mapping from a normed vector space $E_1$ into a Banach space $E_2$ subject to the inequality*

$$\|f(x+y) - f(x) - f(y)\| \le \varepsilon(\|x\|^p + \|y\|^p) \tag{2.1}$$

*for all $x$ and $y$ in $E_1$, where $\varepsilon$ and $p$ are constants with $\varepsilon > 0$ and $p < 1$. Then the limit $g(x) = \lim_{n\to\infty} 2^{-n} f(2^n x)$ exists for all $x$ in $E_1$ and $g$ is the unique additive mapping which satisfies*

$$\|f(x) - g(x)\| \le k\varepsilon\|x\|^p \tag{2.2}$$

*for all $x$ in $E_1$, where $k = 2/(2-2^p)$.*

It was also shown that, if for each $x$ the function $t \to f(tx)$ from $\mathbb{R}$ to $E_2$ is continuous for eached fixed $x$ in $E_1$ then $g$ is continuous on $E_1$. Th.M. Rassias asked during the 27th International Symposium on Functional Equations (1990) whether such a theorem can also be proved for $p \ge 1$.

Z. Gajda (1991) answered the above question of Th.M. Rassias for the case $p > 1$ in the following way, using the clever idea of replacing $n$ by $-n$ in the representation of $g(x)$ given in Theorem 2.1.

**Theorem 2.2.** *Given a mapping $f$ from a normed vector space $E_1$ into a Banach space $E_2$ which satisfies (2.1) for all $x$ and $y$ in $E_1$, where now $\varepsilon > 0$ and $p > 1$. Then the limit $g(x) = \lim_{n\to\infty} 2^n f(2^{-n}x)$ exists for all $x$ in $E_1$ and $g$ is the unique additive mapping which satisfies (2.2), but with $k = 2/(2^p - 2)$.*

**Corollary 2.3.** *When $p \ne 1$ and if $f$ is continuous at a single point $y \ne 0$ of $E_1$ then $g$ is continuous everywhere. Also when $p \ne 1$ and the mapping $t \to f(tx)$ is continuous for all real $t$ and $x$ fixed in $E_1$ then $g$ is linear.*

The proof of this corollary is easily adapted from the corresponding proof in Theorem 1.1 above. The restriction $y \ne 0$ is important when $p < 0$. For a generalization of these results see Th.M. Rassias and P. Šemrl (1993). A different generalization of Theorem 2.1 was established by G. Isac and Th.M. Rassias (1993). Instead of the function $t^p$ which is used in (2.1) they introduced a function $\psi$ mapping the real interval $(0,\infty)$ into itself subject to the conditions:

$$\lim_{t\to\infty} \frac{\psi(t)}{t} = 0, \tag{2.3}$$

$$\psi(ts) \leq \psi(t)\psi(s) \text{ for all } t>0,\ s>0, \tag{2.4}$$

$$\psi(t) < t \text{ for all } t>1, \tag{2.5}$$

and proved the following result.

**Theorem 2.4.** *Let $E_1$ be a normed vector space, $E_2$ a Banach space, $\varepsilon$ a positive number and $\psi : \mathbb{R}_+ \to \mathbb{R}_+$ a function with properties (2.3), (2.4) and (2.5). Let $f : E_1 \to E_2$ be a mapping such that $f(tx)$ is continuous in $t$ for each fixed $x$ in $E_1$ and $f$ satisfies the inequality*

$$\|f(x+y) - f(x) - f(y)\| \leq \varepsilon\left[\psi(\|x\|) + \psi(\|y\|)\right].$$

*Then there exists a unique linear mapping $T : E_1 \to E_2$ such that*

$$\|f(x) - T(x)\| \leq k\varepsilon\psi(\|x\|) \text{ for all } x \text{ in } E_1, \text{ where } k = \frac{2}{2-\psi(2)}. \tag{2.6}$$

**Proposition 2.5.** *If the given mapping $f : E_1 \to E_2$ is continuous at a single point $y \neq 0$, then $T$ is continuous everywhere in $E_1$.*

*Proof.* Assume that the additive mapping $T$ is not continuous. Then there exists a sequence of points $x_n \to 0$ such that $\|T(x_n)\| \to \infty$. We have

$$\begin{aligned} \|T(x_n)\| &= \|T(y+x_n) - T(y)\| \\ &\leq \|T(y+x_n) - f(y+x_n)\| + \|f(y) - T(y)\| + \|f(y+x_n) - f(y)\|. \end{aligned}$$

As $y \neq 0$ is fixed and $f$ is continuous at $y$, the second and third terms on the right side of this inequality remain bounded as $x_n \to 0$. Consider the first term. We have $\|T(y+x_n) - f(y+x_n)\| \leq k\varepsilon\psi(\|y+x_n\|)$. Put $c = 2/\|y\|$, so that $c^{-1} = \|y\|/2 > 0$. Then by (2.4) we may write

$$\psi(\|y+x_n\|) = \psi(c\|y+x_n\|c^{-1}) \leq \psi(c\|y+x_n\|)\psi(c^{-1}).$$

Since $x_n \to 0$ as $n \to \infty$ and $c\|y\| = 2$, we may choose a positive integer $n_0$ such that $n > n_0$ implies that $c\|y+x_n\| > 1$. Hence by (2.5) $\psi(\|y+x_n\|) \leq c\|y+x_n\|\psi(c^{-1})$. Therfore the first term also remains bounded as $n \to \infty$, contrary to our assumption, and the proposition is demonstrated.

Now we may apply this proposition to show that when the mapping $t \to f(tx)$ is continuous in $t$ for fixed $x$ then $T(tx)$ is continuous in $t$ for fixed $x$. If $x = 0$ there is nothing to prove. When $x \neq 0$ the proposition for the case where $E_1 = \mathbb{R}$ shows that $t \to T(tx)$ is continuous in $t$ for $x$ fixed, and hence $T$ is linear. □

D.H. Hyers, G. Isac and Th.M. Rassias [11] have just very recently proved the following asymptotic analogues of Th.M. Rassias' theorem obtained in 1978 for the Hyers-Ulam stability of mappings.

**Theorem 2.6.** *Given a real normed vector space $E_1$ and a real Banach space $E_2$, let numbers $M > 0$, $\varepsilon > 0$ and $p$ with $0 < p < 1$ be chosen. Let the mapping $f : E_1 \to E_2$ satisfy the inequality*

$$\|f(x+y) - f(x) - f(y)\| \leq \varepsilon(\|x\|^p + \|y\|^p) \tag{2.7}$$

*for all $x, y \in E_1$ such that*

$$\|x\|^p + \|y\|^p > M^p. \tag{2.8}$$

*Then there exists an additive mapping $\varphi : E_1 \to E_2$ such that*

$$\|\varphi(x) - f(x)\| < \beta(p)\varepsilon\|x\|^p \tag{2.9}$$

*for all $x \in E_1$ with $\|x\| > M/2^{1/p}$, where $\beta(p) = 2/(2-2^p)$ and $\varphi(x) = \lim_{n\to\infty} \dfrac{f(2^n x)}{2^n}$.*

For convenience in applications we give the following modified version of Theorem 2.6.

**Theorem 2.7.** *Given a real normed vector space $E_1$ and a real Banach space $E_2$, let numbers $m > 0$, $\varepsilon > 0$ and $p$ with $0 \leq p < 1$ be chosen. Suppose that the mapping $f : E_1 \to E_2$ satisfies the inequality*

$$\|f(x+y) - f(x) - f(y)\| \leq \varepsilon(\|x\|^p + \|y\|^p)$$

*for all $x$ and $y$ in $E_1$ such that $\|x\| > m$, $\|y\| > m$ and $\|x+y\| > m$. Then there exists an additive mapping $\varphi : E_1 \to E_2$ which satisfies*

$$\|\varphi(x) - f(x)\| \leq 2\varepsilon(2 - 2^p)^{-1}\|x\|^p$$

*for all $x$ in $E_1$ such that $\|x\| > m$. Moreover, $\varphi$ is given by the formula*

$$\varphi(x) = \lim_{n\to\infty} 2^{-n} f(2^n x)$$

*for all $x$ in $E_1$.*

Isac and Rassias (1993) pointed out that the inequality (2.6) together with condition (2.3) for $\psi$ implies that the mapping $f$ is asymptotically linear with $T$ as its asymptotic derivative in the sense of M.A. Krasnosel'skii (1964).

The case in which $p = 1$ was excluded above in Theorems 2.1 and 2.2. These results fail when $p = 1$ as has been shown by counterexamples in the literature. See D.G. Bourgin (1951), p. 224, or B.E. Johnson (1988), p. 314, or Z. Gajda (1991), or Th.M. Rassias and P. Šemrl (1992).

## 3. An application to nonlinear equations in Banach spaces

The results of Isac and Rassias (1993) given above have been used by these authors in studying nonlinear equations in ordered Banach spaces. In order to present these applications of G. Isac and Th.M. Rassias (1996) some background is needed. Given a real Banach space $E$, a subset $K$ of $E$ is called a *cone* if it is closed and has the following properties:

(1) If $x$ and $y$ belong to $K$ then $\lambda x + \mu y \in K$ for all non-negative real numbers $\lambda$ and $\mu$.

(2) $K \cap (-K) = \{0\}$.

Each cone $K$ induces a (partial) ordering on $E$ by the definition: $x \leq y$ when $y - x \in K$, and we say that $E$ is ordered by the cone $K$. Also $x < y$ means $y - x \in K$ and $y \neq x$. A cone $K$ is called *normal* if there exists $\delta > 1$ such that $\|x\| \leq \delta \|x + y\|$ for all $x$ and $y$ in $K$. A *generating cone* $K$ for $E$ is such that $E = K - K$. A cone $K$ in $E$ is called *solid* if it has a non-empty interior (i.e., $K$ contains an open set).

The *dual* cone of $K$ is the cone $K^*$ in the space $E^*$ of linear continuous functionals on $E$, defined by $K^* = \{\phi \in E^* : \phi(x) \geq 0 \text{ for all } x \text{ in } K\}$. We call a point $x_0$ in $K$ a *quasi-interior point* of $K$ if $\phi(x_0) > 0$ for each non-zero $\phi$ in $K^*$.

The space of bounded linear operators from $E$ into $E$ is denoted by $L(E, E)$. Every $T$ in $L(E, E)$ has a well-defined *spectral radius* $r(T) = \sup\{|\lambda| : \lambda \in \sigma(T)\}$, where $\sigma(T)$ is the spectrum of $T$. Moreover, the spectral radius of $T$ is given by Gelfand's formula:

$$r(T) = \lim_{n \to \infty} \|T^n\|^{1/n}.$$

Given two Banach spaces $E$ and $F$, a mapping $f : E \to F$ is called *compact* if it maps bounded sets onto relatively compact sets and is said to be *completely continuous* if it is continuous and compact. During the last couple of decades newer concepts such as the *measure of non-compactness* together with mappings called *k-set-contractions* have been used in generalizations of theorems concerning completely continuous operators. Given a bounded set $D \subset E$, the measure of non-compactness $\gamma(D)$ is the minimum of all positive numbers $\delta$ such that $D$ can be covered by a finite number of sets of diameter less than $\delta$. A mapping $f : E \to E$

is called a *k-set-contraction* if $f$ is continuous and if there exists a $k > 0$ such that for each bounded set $D \subset E$, we have $\gamma(f(D)) < k\gamma(D)$. When $k < 1$, $f$ is said to be a *strict-set-contraction.* It turns out that every completely continuous mapping is a strict-set-contraction, but not conversely.

In what follows the function $\psi : \mathbb{R}_+ \to \mathbb{R}_+$, as in Theorem 2.4 above, will always satisfy the three conditions (2.3), (2.4) and (2.5). We recall a function $f : E \to F$ is $\psi$-additive if there is a number $\theta > 0$ such that

$$\|f(x+y) - f(x) - f(y)\| \leq \theta\,[\psi(\|x\|) + \psi(\|y\|)] \quad \text{for all } x,\, y \text{ in } E.$$

Also by Theorem 2.4 and the proposition following it, if $f$ is $\psi$-additive and continuous then there exists a unique linear continuous operator $T$ in $L(E, F)$ such that

$$\|f(x) - T(x)\| \leq c\psi(\|x\|), \quad x \text{ in } E, \tag{3.1}$$

where the number $c$ is given by

$$c = \frac{2\theta}{2 - \psi(2)}, \tag{3.2}$$

and where

$$T(x) = \lim_{n\to\infty} \frac{f(2^n x)}{2^n}. \tag{3.3}$$

**Definition.** Let $E$ and $F$ be Banach spaces and let $K \subset E$ be a generating cone. A mapping $f : K \to F$ is called *asymptotically differentiable along* $K$ if there exists $T$ in $L(E, F)$ such that

$$\lim_{\substack{x\in K \\ \|x\|\to\infty}} \frac{\|f(x) - T(x)\|}{\|x\|} = 0,$$

and $T$ is called the *derivative at infinity along* $K$ of $f$.

**Theorem 3.1.** *Given a Banach space $E$ ordered by a generating cone $K$, a mapping $h : E \to E$ such that $h(K) \subseteq K$ and a mapping $g : K \to K$ such that $f = g + h$ is a strict-set-contraction from $K$ into $K$. Suppose that the following conditions hold:*

(A) $\displaystyle \lim_{\substack{x\in K \\ \|x\|\to\infty}} \frac{\|g(x)\|}{\|x\|} = 0,$

(B) *$h$ is completely continuous and $\psi$-additive,*

(C) *there is a quasi-interior point $x_0$ in $K$ and a number $\lambda_0$ in $(0, 1)$ such that*

$$\lim_{n\to\infty} \frac{h(2^n x_0)}{2^n} \leq \lambda_0 x_0.$$

*Then $f$ has a fixed point in $K$.*

*Proof.* Since $h$ is $\psi$-additive, from (3.1) and property (2.3) of $\psi$ listed above we have

$$\lim_{\substack{x\in K\\ \|x\|\to\infty}} \frac{\|h(x)-T(x)\|}{\|x\|} = 0, \tag{3.4}$$

so that $T$ is the derivative at infinity along $K$ of $h$. Since $h$ is completely continuous it follows (see, e.g., Lemma 3.1, p. 102, of M.A. Krasonsel'skii (1964)) that $T$ is also completely continuous. By hypothesis (A) it follows that $T$ is also the derivative at infinity along $K$ of $f$. As $T$ is completely continuous its spectrum consists of discrete eigenvalues and zero. Excluding the trivial case assume that the spectral radius $r(T)$ is positive. By hypothesis (C) and formula (3.3) we have $T(x_0) \leq \lambda_0 x_0$, where $x_0$ is a quasi-interior point of $K$ and $0 < \lambda_0 < 1$. Now we use the Krein-Rutman theorem (see, e.g., E. Zeidler (1986), p. 290, Proposition 7.26) to find that $r(T)$ is an eigenvalue of $T$ and of its dual operator $T^*$, with eigenvectors in $K$ and in $K^*$, respectively. Thus there exists an element $\phi_0 \in K^* \setminus \{0\}$ such that $T^*(\phi_0) = r(T)\phi_0$. Also $\phi_0(x_0) > 0$ since $x_0$ is a quasi-interior point of $K$. Hence

$$r(T) = \frac{[T^*(\phi_0)](x_0)}{\phi_0(x_0)} = \frac{\phi_0(T(x_0))}{\phi_0(x_0)} \leq \frac{\phi_0(\lambda_0 x_0)}{\phi_0(x_0)} = \lambda_0 < 1.$$

We have shown that all of the hypotheses of Theorem 1 of H. Amann (1973) are satisfied. Therefore $f$ has a fixed point. □

**Definition.** Given a generating cone $K \subset E$, a function $f : K \to E$ is said to have a derivative $f'(x_0)$ at $x_0 \in K$ along $K$ if $f'(x_0) \in L(E, E)$ and if

$$\lim_{\substack{x\in K\\ \|x\|\to 0}} \frac{\|f(x_0+x) - f(x_0) - f'(x_0)x\|}{\|x\|} = 0.$$

**Theorem 3.2.** *Let $E$ be a Banach space ordered by a generating cone $K$ and let mappings $g : K \to K$ and $h : E \to E$ be given such that $h(K) \subseteq K$ and $f = g + h$ is a strict-set-contraction from $K$ into $K$. Suppose that the above conditions* (A) *and* (B) *hold together with the following:*

(D) *$f$ has a derivative $f'(0)$ and $f(0) = 0$,*

(E) *$f'(0)$ does not have an eigenvector in $K \setminus \{0\}$ with eigenvalue $\geq 1$,*

(F) *$\|h(x) - x\| > c\psi(\|x\|)$ for all $x \in K \setminus \{0\}$, with $c$ given in* (3.2),

(G) *there exist $x_0 \notin -K$ and $\lambda_0 > 1$ such that $\displaystyle\lim_{n\to\infty} \frac{h(2^n x)}{2^n} \geq \lambda_0 x_0$.*

*Then $f$ has a fixed point in $K$.*

*Proof.* As in the proof of Theorem 3.1, we have that $T(x) = \lim_{n\to\infty} \frac{h(2^n x)}{2^n}$ is the derivative at infinity along $K$ of $h$ and that inequality (3.1) is satisfied. From (F) it follows that 1 is not an eigenvalue whose eigenvector belongs to $K \setminus \{0\}$. From (G) we deduce that $r(T) \geq \lambda_0$. For suppose on the contrary that $r(T) < \lambda_0$. By the well-known formula of Gelfand, $r(T) = \lim_{n\to\infty} \|T^n\|^{1/n}$. Thus $\lambda_0^{-n}\|T^n\| \leq k^n$ for some $k < 1$ and all sufficiently large $n$. But (G) implies that $T(x_0) \geq \lambda_0 x_0$, so $\lambda_0^{-n}T^n(x_0) \geq x_0$. In the limit as $n \to \infty$, since $\lambda_0 > 1$, we have $0 \geq x_0$, or $x_0 \in -K$, in contradiction to condition (G). Hence $r(T) \geq \lambda_0 > 1$. By the Krein-Rutman theorem $r(T)$ is an eigenvalue of $T$ with its eigenvector in $K$, and we know that it is greater than 1. Thus all the assumptions of Theorem 1 of N. Các and J. Gatica (1979) are satisfied and we conclude that $f$ has a fixed point in $K \setminus \{0\}$. □

**Corollary 3.3.** *Given the nonlinear eigenvalue problem:*

$$L(x) + h(x) = x + \lambda M(x). \tag{3.5}$$

*Assume that the mappings $L : E \to E$, $M : E \to E$, $h : E \to E$ and the number $\lambda > 0$ are such that $g = L - \lambda M$, $h$ and $f = g + h$ satisfy the hypotheses of Theorem* 3.2. *Then the nonlinear eigenvalue problem* (3.5) *has a solution.*

Next we look at the so-called *Hammerstein equation*:

$$x = \lambda A f(x), \tag{3.6}$$

subject to the following conditions:

(H1) $(E, K)$ and $(F, P)$ are real Banach spaces ordered by the cones $K$ and $P$, respectively, and the cone $K$ is normal and solid.

(H2) The mapping $f : E \to F$ is continuous and the operator $A : F \to E$ is linear and compact.

(H3) $A$ is strongly positive, that is $y - x \in P \setminus \{0\}$ implies that $A(y) - A(x) \in$ *int* $(K)$.

**Definition.** $(\mu_\infty, \infty)$ is said to be a *bifurcation from infinity* of equation (3.6) if $\mu_\infty > 0$ and there is a sequence of solutions $(\mu_n, x_n)$ of (3.6) such that $\mu_n \to \mu_\infty$ and $\|x_n\| \to \infty$ as $n \to \infty$.

**Theorem 3.4.** *Let conditions* (H1), (H2), (H3) *hold and suppose also that the following are true:*

(1) $f(K) \subset P$,

(2) *f is $\psi$-additive, and*

(3) $\lim_{n\to\infty} \dfrac{f(2^n x)}{2^n} > 0$ *for all* $x \in K \setminus \{0\}$.

*Put* $T(x) = \lim_{n\to\infty} \dfrac{f(2^n x)}{2^n}$ *and* $\mu_\infty = \dfrac{1}{r(AT)}$. *Then* $\mu_\infty, \infty)$ *is the only bifurcation from infinity of the Hammerstein equation* (3.6).

*Proof.* As before we have that $T : E \to F$ is the derivative at infinity along $K$ of $f$ when $x \in K$. By (3) $T(x) > 0$ when $x \in K \setminus \{0\}$, that is, $T$ is strictly positive. Hence all the hypotheses of Corollary 7.48 (b), p. 315, of E. Zeidler (1986) are satisfied and the theorem follows. □

## References


1. J. Aczél and J. Dhombres, *Functional Equations in Several Variables.* Encyclopedia of Mathematics and its Applications, vol. 31, Cambridge University Press, Cambridge, 1989.

2. H. Amann, *Fixed points of asymptotically linear maps in ordered Banach spaces.* J. Functional Anal. **14** (1973), 162–171.

3. D.G. Bourgin, *Classes of transformations and bordering transformations.* Bull. Amer. Math. Soc. **57** (1951), 223–237.

4. N. Các and J. Gatica, *Fixed point theorems for mappings in ordered Banach spaces.* J. Math. Anal. Appl. **71** (1979), 547–557.

5. G.L. Forti, *The stability of homomorphisms and amenability with applications to functional equations.* Abh. Math. Sem. Univ. Hamburg **57** (1987), 215–226.

6. Z. Gajda, *On stability of additive mappings.* Internat. J. Math. and Math. Sci. **14** (1991), 431–434.

7. P. Gavruta, *A generalization of the Hyers-Ulam-Rassias stability of approximately additive mappings.* J. Math. Anal. Appl. **184** (1994), 431–436.

8. R. Ger, *Almost approximate additive mappings.* In: “Proc. 3rd Internat. Conference on General Inequalities” (E.F. Beckenbach and W. Walter, eds.), ISNM 64, Birkhäuser Verlag, Basel, 1983, pp. 263–276.

9. D.H. Hyers, *On the stability of the linear functional equation.* Proc. Nat. Acad. Sci. USA **27** (1941), 222–224.

10. D.H. Hyers and Th.M. Rassias, *Approximate homomorphisms.* Aeq. Math. **44** (1992), 125–153.

11. D.H. Hyers, G. Isac and Th.M. Rassias, *On the asymptoticity aspect of Hyers-Ulam stability mappings.* Proc. Amer. Math. Soc. (to appear).

12. G. Isac, *Opérateurs asymptotiquement linéaires sur des espaces localement convexes.* Colloq. Math. **46** (1982), 67–72.

13. G. Isac and Th.M. Rassias, *On the Hyers-Ulam stability of $\psi$-additive mappings.* J. Approx. Theory **72** (1993), 131–137.

14. G. Isac and Th.M. Rassias, *Stability of $\psi$-additive mappings: Applications to nonlinear analysis.* Internat. J. Math. and Math. Sci. **19**(2) (1996), 219–228.

15. B.E. Johnson, *Approximately multiplicative maps between Banach algebras.* J. London Math. Soc. (2)**37** (1988), 294–316.

16. M.A. Krasnosel'skii, *Positive Solutions of Operator Equations.* Nordhoff, Groningen, 1964.

17. Z. Moszner, *Sur la stabilité de l'équation d'homomorphisme.* Aeq. Math. **29** (1985), 290–306.

18. Th.M. Rassias, *On the stability of the linear mapping in Banach spaces.* Proc. Amer. Math. Soc. **72** (1978), 297–300.

19. Th.M. Rassias, *The stability of mappings and related topics* (Abstract). Report of the 27th Internat. Symp. on Functional Equations, Aeq. Math. **39** (1990), 292–293. — *Problem 16*, 2°. Same report, p. 309.

20. Th.M. Rassias and P. Šemrl, *On the behaviour of mappings which do not satisfy Hyers-Ulam stability.* Proc. Amer. Math. Soc. **114** (1992), 989–993.

21. Th.M. Rassias and P. Šemrl, *On the Hyers-Ulam stability of linear mappings.* J. Math. Anal. and Appl. **173** (1993), 325–338.

22. Th.M. Rassias and J. Tabor (eds.), *Stability of Mappings of Hyers-Ulam Type.* Hadronic Press, Inc., Palm Harbor, Florida, 1994.

23. S.M. Ulam, *A Collection of Mathematical Problems.* Interscience Publ., New York, 1960 = Problems in Modern Mathematics, Wiley, New York, 1964.

24. E. Zeidler, *Nonlinear Functional Analysis and its Applications*, Vol 1, Springer-Verlag, New York – Berlin – Heidelberg Tokyo, 1986.

Themistocles M. Rassias, Department of Mathematics, National Technical University of Athens, Zografou Campus, 15780 Athens, Greece

International Series of Numerical Mathematics, Vol. 123, 

# Die Funktionalgleichung $f(x)+\max\{f(y),f(-y)\}=\max\{f(x+y),f(x-y)\}$

Raymond M. Redheffer* und Peter Volkmann

**Abstract**

Let $G$ be an abelian group. A function $f : G \to \mathbb{R}$ solves the equation in the title iff it is additive or $f(x) = |a(x)+p| - p$, where $p \geq 0$ and $a : G \to \mathbb{R}$ is additive. As an application, the solution of the equation $f(x) + g(y) = \max\{h(x+y), h(x-y)\}$ with three unknown functions $f, g, h : G \to \mathbb{R}$ can be determined.

## 1. Einleitung und Resultate

Es bezeichne $\mathbb{R}$, $\mathbb{Z}$, $\mathbb{N}$ den Bereich der reellen, ganzen bzw. natürlichen Zahlen; $\mathbb{N} = \{1, 2, 3, \ldots\}$. Ferner sei $G$ eine abelsche Gruppe. Nach [1] löst eine Funktion $f : G \to \mathbb{R}$ die Funktionalgleichung

$$f(x) + f(y) = \max\{f(x+y), f(x-y)\} \quad (x, y \in G) \tag{1.1}$$

genau dann, wenn $f(x) = |a(x)|$ $(x \in G)$ mit einer additiven Funktion $a : G \to \mathbb{R}$ gilt. (Die Additivität von $a$ bedeutet $a(x+y) = a(x) + a(y)$ für $x, y \in G$.) J. Tabor [2] betrachtet eine interessante Version von (1.1), bei welcher $f$ Werte in einem normierten Raume besitzt; darauf wird hier nicht weiter eingegangen. Studiert man die Pexidersche Form

$$f(x) + g(y) = \max\{h(x+y), h(x-y)\} \quad (x, y \in G) \tag{1.2}$$

von (1.1) mit drei unbekannten Funktionen $f, g, h : G \to \mathbb{R}$, so wird man leicht auf die Gleichung

$$f(x) + \max\{f(y), f(-y)\} = \max\{f(x+y), f(x-y)\} \quad (x, y \in G) \tag{$\star$}$$

geführt. Für $(\star)$ und (1.2) beweisen wir folgende Resultate:

*Anläßlich eines Aufenthaltes an der Universität Karlsruhe im April 1995.

**Satz 1.** *Eine Funktion $f : G \to \mathbb{R}$ löst $(\star)$ genau dann, wenn sie eine der Formen*

$$\text{(I)}\quad f = a, \qquad \text{(II)}\quad f(x) = |a(x) + p| - p \quad (x \in G)$$

*besitzt; dabei bedeutet $a : G \to \mathbb{R}$ eine beliebige additive Funktion und $p$ eine beliebige nichtnegative Zahl.*

**Satz 2.** *Funktionen $f, g, h : G \to \mathbb{R}$ lösen (1.2) genau dann, wenn mit einem additiven $a : G \to \mathbb{R}$ und $c_0, c_1, c_2 \in \mathbb{R}$ gilt:*

$$f(x) = a(x) + c_1, \quad g(x) = |a(x)| + c_2, \quad h(x) = a(x) + c_1 + c_2 \quad (x \in G) \tag{1.3}$$

*oder*

$$\left\{ \begin{array}{l} f(x) = |a(x) + c_0| + c_1, \quad g(x) = |a(x)| + c_2 \\ h(x) = |a(x) + c_0| + c_1 + c_2 \end{array} \right. \quad (x \in G). \tag{1.4}$$

In der folgenden Nr. wird Satz 2 mit Hilfe von Satz 1 bewiesen; der Rest der Arbeit ist dem Beweise des Satzes 1 gewidmet.

## 2. Zurückführung des Satzes 2 auf Satz 1

Die Funktionen (1.3) bzw. (1.4) lösen (1.2); es geht also um die Umkehrung: Das Funktionentripel $f, g, h : G \to \mathbb{R}$ löse (1.2). Mit $x = y = 0$ folgt $f(0) + g(0) = h(0)$, und indem man von $f$, $g$, $h$ zu $f(x) - f(0)$, $g(x) - g(0)$, $h(x) - h(0)$ übergeht, kann außer (1.2) noch

$$f(0) = g(0) = h(0) = 0 \tag{2.1}$$

vorausgesetzt werden. Mit $y = 0$ bzw. $x = 0$ folgt dann aus (1.2)

$$f(x) = h(x) \quad (x \in G), \tag{2.2}$$

$$g(x) = \max\{h(x), h(-x)\} \quad (x \in G). \tag{2.3}$$

In (1.2) können somit $g$ und $h$ durch $f$ ausgedrückt werden, und es ergibt sich die Funktionalgleichung $(\star)$.

Nach Satz 1 hat $f$ eine der Formen (I), (II). Im Falle (I) folgt mit (2.2), (2.3)

$$f = h = a, \quad g(x) = |a(x)| \quad (x \in G) \tag{2.4}$$

und im Falle (II) entsprechend

$$f(x) = h(x) = |a(x) + p| - p, \quad g(x) = |a(x)| \quad (x \in G). \tag{2.5}$$

Durch Aufhebung der Spezialisierung (2.1) (d.h. durch Inversion der Transformation, die auf (2.1) geführt hat) ergeben sich aus (2.4), (2.5) die Formen (1.3) bzw. (1.4) der Lösung von (1.2).

## 3. Erster Schritt im Beweise des Satzes 1

Man sieht leicht, da durch (I), (II) Lösungen von ($\star$) gegeben werden. Es sei also ab jetzt umgekehrt $f : G \to \mathbb{R}$ eine Lösung von ($\star$). Mit $x = y = 0$ folgt zunächst

$$f(0) = 0. \tag{3.1}$$

Es werde $g : G \to \mathbb{R}$ definiert durch

$$g(x) = \max\{f(x), f(-x)\} \quad (x \in G). \tag{3.2}$$

Ersetzt man in ($\star$) $x$ durch $-x$ und kombiniert man die so entstehende Gleichung mit ($\star$), so ergibt sich

$$g(x) + g(y) = \max\{g(x+y), g(x-y)\} \quad (x, y \in G).$$

Die Funktion $g$ genügt also der Funktionalgleichung (1.1), und nach dem in Nr. 1 erwähnten Resultat aus [1] ist $g(x) = |a(x)|$ mit einer additiven Funktion $a : G \to \mathbb{R}$. Damit läßt sich ($\star$) bzw. (3.2) schreiben als

$$f(x) + |a(y)| = \max\{f(x+y), f(x-y)\} \quad (x, y \in G), \tag{3.3}$$

$$\max\{f(x), f(-x)\} = |a(x)| \quad (x \in G). \tag{3.4}$$

Es wird nun das Ziel sein, hieraus (I) und (II) als einzig mögliche Formen für $f$ herzuleiten (eventuell mit $a$ ersetzt durch die Funktion $a(-x) = -a(x)$; Ersetzung von $a$ durch $-a$ ändert nichts an (3.3), (3.4)). Der Fall $a(x) \equiv 0$ kann gleich erledigt werden: Nach (3.4) ist dann $f(x) \leq 0$ $(x \in G)$, und (3.3) mit $y = x$ liefert weiter (man beachte (3.1))

$$f(x) = \max\{f(2x), 0\} = 0 \quad (x \in G),$$

d.h. es ist $f(x) \equiv 0$. Analog ist

$$f(x) = 0 \tag{3.5}$$

für ein $x \in G$, falls nur $a(x) = 0$ ist (z.B. falls $x$ endliche Ordnung hat): Die additive Funktion $a$ verschwindet auf der Gruppe $\mathbb{Z}x$, und Betrachtung von (3.3), (3.4) auf dieser Gruppe liefert (3.5). Wegen $a(x) = 0$ bedeutet (3.5) sowohl das Bestehen von $f(x) = \pm a(x)$ als auch von $f(x) = |\pm a(x) + p| - p$ mit beliebigem $p \geq 0$.

## 4. Der Fall $G = \mathbb{Z}$

Es sei $G = \mathbb{Z}$ die additive Gruppe der ganzen Zahlen; wir beweisen Satz 1 für diese Situation: Die in (3.3), (3.4) vorkommende additive Funktion $a : \mathbb{Z} \to \mathbb{R}$ ist von der Form $a(n) = cn$ $(n \in \mathbb{Z})$ mit $c \in \mathbb{R}$. Der Fall $a(n) \equiv 0$ ist schon erledigt, also kann $c \neq 0$ vorausgesetzt werden. Übergang von $f$ zu $(1/|c|)f$ liefert dann (3.3), (3.4) mit $a(n) \equiv n$, d.h.

$$f(m) + |n| = \max\{f(m+n), f(m-n)\} \quad (m, n \in \mathbb{Z}), \tag{4.1}$$

$$\max\{f(n), f(-n)\} = |n| \quad (n \in \mathbb{Z}).$$

Es ist also $f(1) = 1$ oder $f(-1) = 1$. Geht man gegebenenfalls von $f$ zur Funktion $f(-n)$ über, so kann man

$$f(1) = 1 \tag{4.2}$$

voraussetzen. (4.1) mit $n = 1$ liefert

$$f(m) + 1 = \max\{f(m+1), f(m-1)\} \quad (m \in \mathbb{N}).$$

Wegen $f(0) = 0$ und (4.2) erhält man hieraus sukzessive $f(2) = 2, f(3) = 3, \ldots$, also gilt

$$f(n) = n \quad (n = 0, 1, 2, \ldots). \tag{4.3}$$

Ist noch

$$f(m) \leq 0 \quad (m = -1, -2, -3, \ldots), \tag{4.4}$$

so folgt aus (4.1) für diese $m$ mit $n = -m$

$$f(m) + |m| = \max\{f(0), f(2m)\} = 0,$$

$$f(m) = -|m| = m \quad (m = -1, -2, -3, \ldots). \tag{4.5}$$

Zusammen mit (4.3) ergibt sich $f(n) = n$ $(n \in \mathbb{Z})$, d.h. $f$ hat die Form (I). Liegt (4.4) nicht vor, so gibt es ein $k \in \mathbb{N}$ mit

$$f(-1) \leq 0,\ f(-2) \leq 0,\ \ldots,\ f(-k+1) \leq 0, \quad f(-k) > 0. \tag{4.6}$$

Fall 1: $k$ ungerade, d.h. $k = 2N + 1$ mit einem ganzen $N \geq 0$. (4.6) bedeutet hier

$$f(-1) \leq 0,\ f(-2) \leq 0,\ \ldots,\ f(-2N) \leq 0, \quad f(-2N-1) > 0. \tag{4.7}$$

Analog zum Schluß von (4.4) auf (4.5) ergibt sich

$$f(-1) = -1,\ f(-2) = -2,\ \ldots,\ f(-N) = -N. \tag{4.8}$$

Setzt man

$$\alpha = f(-2N-1)$$

($\alpha > 0$ nach (4.7)), so folgt aus (4.1) mit (4.7) im Falle $N \geq 1$

$$\begin{aligned}
f(-N-1) + N &= \max\{f(-1), f(-2N-1)\} = \alpha,\\
f(-N-2) + N - 1 &= \max\{f(-3), f(-2N-1)\} = \alpha,\\
&\ldots\\
f(-2N) + 1 &= \max\{f(-2N+1), f(-2N-1)\} = \alpha,
\end{aligned}$$

und hieraus lassen sich $f(-N-1), \ldots, f(-2N)$ berechnen.

Wegen $\alpha - 1 = f(-2N) \leq 0$ folgt insbesondere $\alpha \leq 1$, d.h.

$$\alpha \in (0,1].$$

Im Falle $N = 0$ ist $\alpha = f(-1)$, und $\alpha \leq 1$ ergibt sich aus $\max\{f(1), f(-1)\} = 1$. Nun kann man die Beziehungen

$$f(-2N-m)+1 = \max\{f(-2N-m+1), f(-2N-m-1)\} \quad (m \in \mathbb{N})$$

benutzen, um mit $m = 1,2,3,\ldots$ sukzessive $f(-2N-2), f(-2N-3),\ldots$ zu bestimmen. Insgesamt erhält man

$$f(-N-m) = \alpha - N + m - 1 \quad (m \in \mathbb{N}). \tag{4.9}$$

Zusammen mit (4.3), (4.8) ergibt sich

$$f(n) = |n + N + \tfrac{1}{2}(1-\alpha)| - N - \tfrac{1}{2}(1-\alpha) \quad (n \in \mathbb{Z}),$$

also ist $f$ von der Form (II), wobei $p = N + \frac{1}{2}(1-\alpha) \geq 0$.

Fall 2: $k$ gerade, d.h. $k = 2N$ mit einem $N \in \mathbb{N}$. (4.6) bedeutet

$$f(-1) \leq 0,\ f(-2) \leq 0,\ \ldots,\ f(-2N+1) \leq 0, \quad \alpha := f(-2N) > 0.$$

Hier ergibt sich

$$f(-1) = -1, \ldots, f(-N+1) = -N+1$$

(falls $N \geq 2$), und analog zu (4.9) erhält man

$$f(-N-m) = \alpha - N + m \quad (m = 0,1,2,\ldots).$$

Zusammen mit (4.3) folgt

$$f(n) = |n + N - \tfrac{1}{2}\alpha| - N + \tfrac{1}{2}\alpha \quad (n \in \mathbb{Z}),$$

also gilt wieder (II), und zwar mit $p = N - \frac{1}{2}\alpha$. (Wegen $N \geq 1$ und $\alpha \in (0,1]$ ist $p \geq 0$.)

## 5. Der Fall $G = \mathbb{Z} \times \mathbb{Z}$, erster Teil

1. Für die in (3.3) vorkommende Funktion $a$ gilt hier $a(m,n) = \alpha m + \beta n$ $(m,n \in \mathbb{Z})$ mit geeigneten $\alpha, \beta \in \mathbb{R}$, und wir können etwa $\alpha \neq 0$ voraussetzen. (3.3) erhält die Form

$$f(m,n) + |\alpha k + \beta l| = \max\{f(m+k, n+l), f(m-k, n-l)\} \quad (k,l,m,n \in \mathbb{Z}). \tag{5.1}$$

Die Untergruppe $\{(m,0)|\ m \in \mathbb{Z}\}$ von $\mathbb{Z} \times \mathbb{Z}$ ist zu $\mathbb{Z}$ isomorph, auf ihr gilt $a(m,0) = \alpha m$, und die Gestalt von $f$ auf ihr ist nach Nr. 4 bereits bekannt. Wir können das Vorzeichen von $\alpha$ so gewählt denken, daß

$$f(m,0) = \alpha m \quad (m \in \mathbb{Z}) \tag{5.2}$$

oder

$$f(m,0) = |\alpha m + p| - p \quad (m \in \mathbb{Z}) \tag{5.3}$$

mit einem $p \geq 0$ gilt. Zur Erledigung des Falles $G = \mathbb{Z} \times \mathbb{Z}$ genügt es daher, von (5.2) auf

$$f(m,n) = \alpha m + \beta n \quad (m,n \in \mathbb{Z}) \tag{5.4}$$

und von (5.3) auf

$$f(m,n) = |\alpha m + \beta n + p| - p \quad (m,n \in \mathbb{Z}) \tag{5.5}$$

zu schließen.

2. Es sei $\beta = 0$: (5.1) mit $k = 0$ liefert

$$f(m,n) = \max\{f(m,n+l), f(m,n-l)\} \quad (l,m,n \in \mathbb{Z}),$$

und damit ist $f(m,n)$ unabhängig von der Variablen $n$ aus $\mathbb{Z}$. (5.2), (5.3) bedeuten also $f(m,n) = \alpha m$ bzw. $f(m,n) = |\alpha m + p| - p$, und das sind die Formeln (5.4), (5.5) für $\beta = 0$.

3. Ab jetzt sind $\alpha \neq 0$ und $\beta \neq 0$. Wir erledigen den Fall (5.2): Aus (5.1) mit $n = 0$, $k = m$ und $l$ ersetzt durch $-n$ folgt

$$\alpha m + |\alpha m - \beta n| = \max\{f(2m,-n), f(0,n)\} \quad (m,n \in \mathbb{Z}). \tag{5.6}$$

Zu $m \in \mathbb{Z}$ kann $n \in \mathbb{Z}$ gewählt werden mit

$$(\alpha/\beta)m - n \leq 0 < (\alpha/\beta)m - n + 1.$$

Es folgt dann $|\alpha m - \beta n| \leq |\beta|$, und damit ist die linke Seite von (5.6) auf $\mathbb{Z} \times \mathbb{Z}$ nicht nach unten beschränkt. Folglich kann auch $f(0,n)$ auf $\mathbb{Z}$ nicht nach unten beschränkt sein. Wegen $a(0,n) = \beta n$ ergeben sich nach Nr. 4 nur die beiden Möglichkeiten

$$f(0,n) = \beta n \quad (n \in \mathbb{Z})$$

oder

$$f(0,n) = -\beta n \quad (n \in \mathbb{Z}).$$

Im zweiten Falle ist nach (5.6)

$$-\beta n \leq \alpha m + |\alpha m - \beta n| \quad (m,n \in \mathbb{Z}),$$

und $m = -\text{sgn}\ \alpha$, $n = -\text{sgn}\ \beta$ liefert einen Widerspruch zu $\alpha, \beta \neq 0$. Es bleibt der erste Fall übrig; (5.1) mit $k = -m$, $l = n$ liefert dann für $(m,n) \in \mathbb{Z} \times \mathbb{Z}$

$$f(m,n) + |\beta n - \alpha m| = \max\{f(0,2n), f(2m,0)\} = \max\{2\beta n, 2\alpha m\},$$

und Auflösen nach $f(m,n)$ ergibt die gewünschte Form (5.4).

## 6. Der Fall $G = \mathbb{Z} \times \mathbb{Z}$, zweiter Teil

1. Wie in Nr. 5, Punkt 3 sind wieder $\alpha, \beta \neq 0$, und wir erledigen (5.3): Wäre $f(0,n)$ auf $\mathbb{Z}$ nach unten unbeschränkt (also $f(0,n) = \beta n$ oder $f(0,n) = -\beta n$), so wäre analog zu den Überlegungen in Punkt 3 der Nr. 5 auch $f(m,0)$ auf $\mathbb{Z}$ nach unten unbeschränkt, und das steht zu (5.3) im Widerspruch. Nach Nr. 4 bleiben also mit einem $q \geq 0$ die beiden Fälle

$$f(0,n) = |\beta n + q| - q \quad (n \in \mathbb{Z}), \tag{6.1}$$

$$f(0,n) = |-\beta n + q| - q \quad (n \in \mathbb{Z}) \tag{6.2}$$

übrig; der zweite Fall soll im nächsten Punkte ausgeschlossen werden.

2. Wir schreiben (5.1) mit $n = 0$, $k = m$ und $l$ ersetzt durch $n$ auf (und wir benutzen (5.3)):

$$|\alpha m + p| - p + |\alpha m + \beta n| = \max\{f(2m,n), f(0,-n)\} \quad (m,n \in \mathbb{Z}). \tag{6.3}$$

Im Falle (6.2) hat man dann

$$|\beta n + q| - q \leq |\alpha m + p| - p + |\alpha m + \beta n| \quad (m,n \in \mathbb{Z}).$$

Wählt man $m$, $n$ so, daß $\beta n > 0$, $p + \alpha m < 0$, $\alpha m + \beta n > 0$ sind, so erhält man im Falle $p > 0$ einen Widerspruch. Zur Behandlung des Falles $p = 0$ wählen wir in (5.1) $m = 0$, $l = n$, und wir ersetzen $k$ durch $m$:

$$f(0,n) + |\alpha m + \beta n| = \max\{f(m,2n), f(-m,0)\} \quad (m,n \in \mathbb{Z}).$$

Es folgt

$$f(-m,0) \leq f(0,n) + |\alpha n + \beta n| \quad (m,n \in \mathbb{Z}), \tag{6.4}$$

und (5.3) mit $p = 0$ sowie (6.2) liefern

$$|\alpha m| \leq |-\beta n + q| - q + |\alpha m + \beta n| \quad (m,n \in \mathbb{Z}).$$

Wählt man $m$, $n$ so, daß $\alpha m, -\beta n + q, \alpha m + \beta n < 0$ sind, so folgt $q \leq 0$, d.h. es ist $q = 0$. Damit kann (6.2) dem Fall (6.1) untergeordnet werden, denn für $q = 0$ besagen beide Formeln dasselbe.

3. Wir haben nun (5.3), (6.1) mit $p, q \geq 0$ zur Verfügung. Einsetzen in (6.4) ergibt

$$|-\alpha m + p| - p \leq |\beta n + q| - q + |\alpha m + \beta n| \quad (m,n \in \mathbb{Z}). \tag{6.5}$$

Aus (6.3) erhält man die analoge Ungleichung

$$|-\beta n + q| - q \leq |\alpha m + p| - p + |\alpha m + \beta n| \quad (m,n \in \mathbb{Z}). \tag{6.6}$$

Durch passende Wahl von $m, n$ erhält man aus (6.5) $q \leq p$ (es genügt, $-\alpha m + p < 0$, $\beta n + q < 0$, $\alpha m + \beta n > 0$ zu nehmen). Entsprechend folgt aus (6.6) $p \leq q$, und somit ist $p = q$. (5.1) mit $k = -m$, $l = n$ liefert dann

$$f(m,n) + |\beta n - \alpha m| = \max\{f(0,2n), f(2m,0)\} = \max\{|2\beta n + p| - p, |2\alpha m + p| - p\},$$

und Auflösen nach $f(m,n)$ beweist schließlich (5.5).

### 7. Ende des Beweises zu Satz 1

Es sei $G$ eine beliebige abelsche Gruppe. Nach Nr. 3 kann $a(x) \not\equiv 0$ vorausgesetzt werden, und es sei $b$ ein ab jetzt festes Element von $G$ mit $a(b) \neq 0$. Nach Nr. 4 gilt (bei passender Wahl des Vorzeichens der in (3.3), (3.4) auftretenden Funktion $a$)

$$f(x) = a(x) \quad (x \in \mathbb{Z}b) \quad \text{oder} \quad f(x) = |a(x) - p| + p \quad (x \in \mathbb{Z}b) \tag{7.1}$$

mit einem $p \geq 0$. Es genügt,

$$f(z) = a(z) \quad \text{bzw.} \quad f(z) = |a(z) - p| + p \tag{7.2}$$

für alle $z \in G$ zu beweisen. Dazu sei auch $z$ im folgenden fixiert. Hat $z$ endliche Ordnung, so gilt nach Nr. 3 $f(z) = a(z) = 0$, also ist (7.2) richtig. Hat $z$ unendliche Ordnung, so unterscheiden wir zwei Fälle:

$$mb + nz \neq 0 \quad \text{für alle} \quad m, n \in \mathbb{Z}, \quad (m,n) \neq (0,0), \tag{7.3}$$

$$Mb + Nz = 0 \quad \text{für ein} \quad (M,N) \in \mathbb{Z} \times \mathbb{Z} \setminus \{(0,0)\}. \tag{7.4}$$

Im Falle (7.3) ist die Untergruppe $\mathbb{Z}b + \mathbb{Z}z$ von $G$ zu $\mathbb{Z} \times \mathbb{Z}$ isomorph, und (7.2) ergibt sich aus Nr. 5 und Nr. 6. Liegt (7.4) vor, so sind $M, N \neq 0$, da $b$ und $z$ beide unendliche Ordnung haben. Aus (7.1) folgt

$$f(nMb) = a(nMb) \quad \text{bzw.} \quad f(nMb) = |a(nMb) - p| + p \quad (n \in \mathbb{Z}),$$

mit (7.4) ergibt sich dann

$$f(nNz) = a(nNz) \quad \text{bzw.} \quad f(nNz) = |a(nNz) - p| + p \quad (n \in \mathbb{Z}),$$

und hieraus folgt leicht

$$f(nz) = a(nz) \quad \text{bzw.} \quad f(nz) = |a(nz) - p| + p \quad (n \in \mathbb{Z}) \tag{7.5}$$

(da man die Struktur von $f$ auf der Gruppe $\mathbb{Z}z$ nach Nr. 4 bereits kennt). (7.5) mit $n = 1$ liefert (7.2).

### Literatur

1. Alice Simon (Chaljub-Simon) und Peter Volkmann, *Caractérisation du module d'une fonction additive à l'aide d'une équation fonctionnelle.* Aequationes Math. **47** (1994), 60–68.

2. Józef Tabor, *On some characterization of the absolute value of an additive function.* Ann. Math. Silesianae **8** (1994), 69–77.

Raymond M. Redheffer, Department of Mathematics, University of California, Los Angeles, CA 90024, U.S.A.

Peter Volkmann, Mathematisches Institut I, Universität Karlsruhe, 76128 Karlsruhe, Deutschland.

# Applications

International Series of Numerical Mathematics, Vol. 123, © 1997 Birkhäuser Verlag Basel

# Asymptotic analysis of nonlinear thin layers

Alain Brillard

**Résumé**

On étudie le comportement asymptotique, lorsque $\varepsilon$ tend vers 0, de la solution $u^\varepsilon$ d'un problème couplant une équation linéaire elliptique du second ordre posée dans deux domaines de $\mathbb{R}^3$, séparés par un cylindre d'épaisseur $2\varepsilon$ dans lequel une équation quasi-linéaire est posée. Ce comportement asymptotique sera établi à l'aide des méthodes d'épi-convergence. On montre en effet que $u^\varepsilon$ est solution d'un problème de minimisation. L'épi-convergence des fonctionnelles est alors prouvée en construisant des fonctions-test adaptées et en utilisant des inégalités sous-différentielles.

**Abstract**

We study the asymptotic behaviour, when $\varepsilon$ goes to 0, of the solution $u^\varepsilon$ of a problem coupling a linear second order elliptic equation posed in two domains of $\mathbb{R}^3$, bonded together by a cylindrical film of thickness $2\varepsilon$, in which a quasilinear equation is posed. This asymptotic behaviour is obtained by means of epi-convergence arguments. Indeed, we prove that $u^\varepsilon$ is the solution of a minimization problem. The epi-convergence of the linked functionals is proved building appropriate test-functions and using subdifferential inequalities.

## 1. Description of the problem

Let $\Omega_1$ and $\Omega_2$ be two bounded and smooth open subsets of $\mathbb{R}^3$ connected on an open subset $\omega$ in the plane $\{x_3 = 0\}$. We build the subsets $\Omega_1^\varepsilon$ and $\Omega_2^\varepsilon$ as

$$\begin{aligned} (x_1, x_2, x_3) \in \Omega_1 &\iff (x_1, x_2, x_3 + \varepsilon) \in \Omega_1^\varepsilon, \\ (x_1, x_2, x_3) \in \Omega_2 &\iff (x_1, x_2, x_3 - \varepsilon) \in \Omega_2^\varepsilon, \end{aligned} \tag{1}$$

where $\varepsilon$ is positive. This means that $\Omega_1^\varepsilon$ and $\Omega_2^\varepsilon$ are separated by the cylinder $\Omega_0^\varepsilon$ equal to $\omega \times ]-\varepsilon, \varepsilon[$. We denote by $\Omega^\varepsilon$ the union: $\Omega^\varepsilon = \Omega_1^\varepsilon \cup \Omega_2^\varepsilon \cup \Omega_0^\varepsilon$. The contact surfaces between the three domains are $S_1^\varepsilon = \omega \times \{\varepsilon\}$ and $S_2^\varepsilon = \omega \times \{-\varepsilon\}$. The boundary $\partial\Omega^\varepsilon$ of $\Omega^\varepsilon$ is supposed to be Lipschitz continuous and divided in two parts: $\partial\Omega^\varepsilon = (S_{1t}^\varepsilon \cup S_{2t}^\varepsilon \cup S_0^\varepsilon) \cup (S_{1u}^\varepsilon \cup S_{2u}^\varepsilon)$. Non homogeneous Neumann boundary conditions are applied on $S_t^\varepsilon = S_{1t}^\varepsilon \cup S_{2t}^\varepsilon \cup S_0^\varepsilon$, with $S_0^\varepsilon = \partial\omega \times ]\varepsilon, \varepsilon[$, while homogeneous Dirichlet boundary conditions are imposed on the remaining and non void part $S_{1u}^\varepsilon \cup S_{2u}^\varepsilon$ of $\partial\Omega_\varepsilon$.

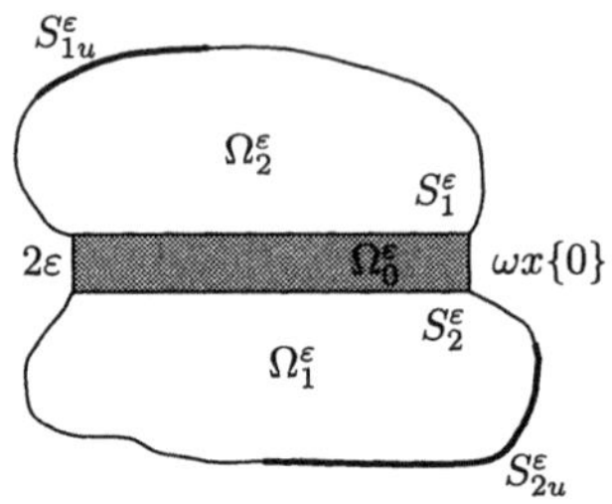


Figure 1: The domain

## 2. Existence and uniqueness of a solution

Let us introduce the problem

$$\begin{aligned}\int_{\Omega_1^\varepsilon\cup\Omega_2^\varepsilon} A_{ij}(x)\frac{\partial u^\varepsilon}{\partial x_i}\frac{\partial v}{\partial x_j}\,dx + \frac{1}{\lambda_\varepsilon^{p-1}}\int_{\Omega_0^\varepsilon}|\nabla u^\varepsilon|^{p-2}\nabla u^\varepsilon\nabla v\,dx \\ = \int_{\Omega^\varepsilon} f\,v\,dx + \int_{S_t^\varepsilon} t\,v\,d\sigma(x), \qquad \forall v\in V^\varepsilon,\end{aligned} \tag{2}$$

where the coefficients $A_{ij}$ satisfy the usual hypotheses of symmetry, coercivity, smoothness and boundedness on $\Omega^\varepsilon$, $\lambda_\varepsilon$ is positive, $p$ is strictly greater than 1, $f$ belongs to $L^\infty(\Omega^\varepsilon)$, $t$ belongs to $L^\infty(\partial\Omega^\varepsilon)$ and $V^\varepsilon$ is the subspace of $H^1(\Omega^\varepsilon)$ defined by

$$\begin{aligned}V^\varepsilon = \{v|\quad & v|_{\Omega_1^\varepsilon\cup\Omega_2^\varepsilon}\in H^1(\Omega_1^\varepsilon\cup\Omega_2^\varepsilon);\ v|_{\Omega_0^\varepsilon}\in W^{1,p}(\Omega_0^\varepsilon); \\ & [v]|_{\omega\times\{\pm\varepsilon\}} = 0;\ v=0 \ \text{ on } \ S_{1u}^\varepsilon\cup S_{2u}^\varepsilon\},\end{aligned} \tag{3}$$

where $[v]|_{\omega\times\{\pm\varepsilon\}}$ denotes the jump of $v$ on the surfaces $\omega\times\{\pm\varepsilon\}$, that is the difference between the two traces of $v$ on these surfaces $S_1^\varepsilon$ and $S_2^\varepsilon$. We will adopt the summation convention with respect to repeated indices. To this variational formulation (2), we associate the minimization problem

$$\min_{v\in V^\varepsilon}\left\{F^\varepsilon(v) - \int_{\Omega^\varepsilon} f\,v\,dx - \int_{S_t^\varepsilon} t\,v\,d\sigma(x)\right\}, \tag{4}$$

$F^\varepsilon$ being defined on $H^1(\Omega^\varepsilon)$ by

$$F^\varepsilon(v) = \begin{cases}\dfrac{1}{2}\displaystyle\int_{\Omega_1^\varepsilon\cup\Omega_2^\varepsilon} A_{ij}(x)\frac{\partial v}{\partial x_i}\frac{\partial v}{\partial x_j}\,dx + \frac{1}{p\lambda_\varepsilon^{p-1}}\int_{\Omega_0^\varepsilon}|\nabla v|^p\,dx & \text{if } v\in V^\varepsilon \\ +\infty & \text{otherwise.}\end{cases} \tag{5}$$

**Lemma 1.** *There exists a unique solution $u^\varepsilon$ of (4), which verifies (2). $u^\varepsilon$ is a weak solution of the problem*

$$
\begin{aligned}
-\frac{\partial}{\partial x_j}\left\{A_{ij}(x)\frac{\partial u^\varepsilon}{\partial x_i}\right\} &= f \quad in \ \Omega_1^\varepsilon \cup \Omega_2^\varepsilon,\\
-\frac{1}{\lambda_\varepsilon^{p-1}}\ \mathrm{div}\ (|\nabla u^\varepsilon|^{p-2}\,\nabla u^\varepsilon) &= f \quad in \ \Omega_0^\varepsilon,\\
-A_{ij}(x)\frac{\partial u^{\varepsilon\,-}}{\partial x_i}\ \nu_j = \frac{1}{\lambda_\varepsilon^{p-1}}\,|\nabla u^\varepsilon|^{p-2}\,\frac{\partial u^{\varepsilon\,+}}{\partial \nu} &\quad on \ \omega\times\{\pm\varepsilon\},\\
u^\varepsilon = 0 \quad on \ S_{1u}^\varepsilon \cup S_{2u}^\varepsilon,&\\
\frac{1}{\lambda_\varepsilon^{p-1}}\,|\nabla u^\varepsilon|^{p-2}\,\frac{\partial u^\varepsilon}{\partial\nu} &= t \quad on \ S_t^\varepsilon \cap \partial\Omega_0^\varepsilon,\\
A_{ij}(x)\frac{\partial u^\varepsilon}{\partial x_i}\nu_j &= t \quad on \ S_t^\varepsilon \cap \partial(\Omega_1^\varepsilon \cup \Omega_2^\varepsilon),
\end{aligned}
\tag{6}
$$

*the normal $\nu$ being the outward one with respect to $\Omega_0^\varepsilon$ and $\frac{\partial u^{\varepsilon}}{\partial\nu}^{+}$ (resp. $\frac{\partial u^{\varepsilon}}{\partial x_i}^{-}$) denoting the normal derivative of $u^\varepsilon$ (resp. the partial derivative of $u^\varepsilon$) "seen" from the domain $\Omega_0^\varepsilon$ (resp. from $\Omega_1^\varepsilon$ or $\Omega_2^\varepsilon$).*

*Proof.* $V^\varepsilon$ can be endowed with the norm

$$
\|v\|_{V^\varepsilon} = \|v_{|\Omega_1^\varepsilon\cup\Omega_2^\varepsilon}\|_{H^1(\Omega_1^\varepsilon\cup\Omega_2^\varepsilon)} + \varepsilon^{(p-1)/p}\|\nabla v_{|\Omega_0^\varepsilon}\|_{(L^p(\Omega_0^\varepsilon))^3}.
$$

The functional

$$
v \longmapsto F^\varepsilon(v) - \int_{\Omega^\varepsilon} f\,v\,dx - \int_{S_t^\varepsilon} t\,v\,d\sigma(x)
$$

is lower semi-continuous for the weak topology of $V^\varepsilon$. Because of the homogeneous Dirichlet boundary conditions imposed on the non void part $S_{1u}^\varepsilon \cup S_{2u}^\varepsilon$ of the boundary and thanks to the coercivity of the matrix $(A_{ij})_{ij}$, $F^\varepsilon$ is strictly coercive. Hence, (4) has a unique solution. We prove in the usual way that this solution satisfies the variational formulation (2) and the Euler equation (6). □

## A. Estimates

Let us assume that

$$
\lambda_\varepsilon = \frac{\lambda_0}{\varepsilon}. \tag{7}
$$

Then we will describe the asymptotic analysis of the problem in this situation. The purpose of this section is to establish estimates for the solution $u^\varepsilon$ of (4) as $\varepsilon$ goes to 0, in order to determine the topology which has to be considered when we will consider the asymptotic behaviour of the functional $F^\varepsilon$, defined in (5).

**Lemma 2.** *Assuming* (7), *the solution* $u^\varepsilon$ *of* (4) *satisfies*

$$\|u^\varepsilon\|_{H^1(\Omega_1^\varepsilon \cup \Omega_2^\varepsilon)} \leq Cte; \quad \varepsilon^{(p-1)/p} \|\nabla u^\varepsilon\|_{(L^p(\Omega_0^\varepsilon))^3} \leq Cte,$$

*where* $Cte$ *denotes a constant independant of* $\varepsilon$, *but depending of* $f$ *and* $t$.

*Proof.* From the variational formulation (2), we deduce taking $v = u^\varepsilon$

$$\int_{\Omega_1^\varepsilon \cup \Omega_2^\varepsilon} A_{ij}(x) \frac{\partial u^\varepsilon}{\partial x_i} \frac{\partial u^\varepsilon}{\partial x_j}\, dx + \frac{\varepsilon^{p-1}}{\lambda_0^{p-1}} \int_{\Omega_0^\varepsilon} |\nabla u^\varepsilon|^p \, dx = \int_{\Omega^\varepsilon} f\, u^\varepsilon \, dx + \int_{S_t^\varepsilon} t\, u^\varepsilon \, d\sigma(x).$$

Then, we write

$$\left| \int_{\Omega^\varepsilon} f\, u^\varepsilon \, dx \right| \leq \|f\|_{L^\infty(\Omega^\varepsilon)} \|u^\varepsilon\|_{L^1(\Omega^\varepsilon)} \leq Cte \|\nabla u^\varepsilon\|_{(L^1(\Omega^\varepsilon))^3},$$

thanks to the homogeneous Dirichlet boundary conditions on the non void part $S_{1u}^\varepsilon \cup S_{2u}^\varepsilon$ of $\partial\Omega^\varepsilon$. Applying two Hölder's inequalities, we obtain

$$\left| \int_{\Omega^\varepsilon} f\, u^\varepsilon \, dx \right| \leq Cte \left\{ \|\nabla u^\varepsilon\|_{(L^2(\Omega_1^\varepsilon \cup \Omega_2^\varepsilon))^3} + (\mathrm{meas}(\Omega_0^\varepsilon))^{1/q} \|\nabla u^\varepsilon\|_{(L^p(\Omega_0^\varepsilon))^3} \right\}$$

$$\Longrightarrow \left| \int_{\Omega^\varepsilon} f\, u^\varepsilon \, dx \right| \leq Cte \left\{ \|\nabla u^\varepsilon\|_{(L^2(\Omega_1^\varepsilon \cup \Omega_2^\varepsilon))^3} + \varepsilon^{1/q} \|\nabla u^\varepsilon\|_{(L^p(\Omega_0^\varepsilon))^3} \right\}.$$

Similarly,

$$\left| \int_{S_t^\varepsilon} t\, u^\varepsilon \, d\sigma(x) \right| \leq \|t\|_{L^\infty(\partial\Omega^\varepsilon)} \|u^\varepsilon\|_{L^1(\partial\Omega^\varepsilon)} \leq Cte \|\nabla u^\varepsilon\|_{(L^1(\Omega^\varepsilon))^3},$$

thanks to the trace theorems. This leads to a similar majorization of this boundary integral. Since $\frac{p}{q}$ is equal to $p-1$, we obtain

$$c \int_{\Omega_1^\varepsilon \cup \Omega_2^\varepsilon} |\nabla u^\varepsilon|^2 \, dx + \frac{\varepsilon^{p-1}}{\lambda_0^{p-1}} \int_{\Omega_0^\varepsilon} |\nabla u^\varepsilon|^p \, dx$$

$$\leq Cte \left( \left( \int_{\Omega_1^\varepsilon \cup \Omega_2^\varepsilon} |\nabla u^\varepsilon|^2 \, dx \right)^{1/2} + \left( \varepsilon^{p/q} \int_{\Omega_0^\varepsilon} |\nabla u^\varepsilon|^p \, dx \right)^{1/p} \right),$$

where $c$ is the constant of coercivity of the matrix $(A_{ij})_{ij}$. This situation can be formulated as

$$a_\varepsilon + b_\varepsilon^p \leq C_1 a_\varepsilon + C_2 b_\varepsilon, \tag{8}$$

with

$$a_\varepsilon = c \left( \int_{\Omega_1^\varepsilon \cup \Omega_2^\varepsilon} |\nabla u^\varepsilon|^2 \, dx \right)^{1/2}, \quad b_\varepsilon = \left( \frac{\varepsilon^{p-1}}{\lambda_0^{p-1}} \int_{\Omega_0^\varepsilon} |\nabla u^\varepsilon|^p \, dx \right)^{1/p}.$$

(8) implies that the two quantities $a_\varepsilon$ and $b_\varepsilon$ are bounded with respect to $\varepsilon$. □

**Remark 3.** We have

$$\int_\omega |u^\varepsilon(x_1,x_2,\varepsilon) - u^\varepsilon(x_1,x_2,-\varepsilon)|^p \, dx_1 dx_2 \leq Cte,$$

which means that the "jump of $u^\varepsilon$ across $\omega \times \{0\}$ is bounded in $L^p(\omega \times \{0\})$". Indeed, we notice that

$$\begin{aligned} |u^\varepsilon(x_1,x_2,\varepsilon) - u^\varepsilon(x_1,x_2,\varepsilon)| &= \left| \int_{-\varepsilon}^{\varepsilon} \frac{\partial u^\varepsilon}{\partial x_3}(x_1,x_2,s) \, ds \right| \\ &\leq \left( \int_{-\varepsilon}^{\varepsilon} 1 \, ds \right)^{1/q} \left( \int_{-\varepsilon}^{\varepsilon} |\frac{\partial u^\varepsilon}{\partial x_3}(x_1,x_2,s)|^p \, ds \right)^{1/p} \end{aligned}$$

which implies

$$\begin{aligned} &\int_\omega |u^\varepsilon(x_1,x_2,\varepsilon) - u^\varepsilon(x_1,x_2,\varepsilon)|^p \, dx_1 \, dx_2 \\ &\quad \leq Cte \, \varepsilon^{p/q} \int_\omega \int_{-\varepsilon}^{\varepsilon} |\frac{\partial u^\varepsilon}{\partial x_3}(x_1,x_2,s)|^p ds \, dx_1 \, dx_2 \leq Cte \, \varepsilon^{p-1} \int_{\Omega_0^\varepsilon} |\nabla u^\varepsilon|^p \, dx. \end{aligned}$$

We then use the second assertion of Lemma 2 in order to conclude.

## B. Asymptotic behaviour of the solution.

We specify in this section the "limit" of the sequence $(u^\varepsilon)_\varepsilon$, when $\varepsilon$ goes to 0. We will apply epi-convergence methods, which are adapted to the convergence of minimization problems. For the reader's convenience, we have gathered in an Appendix the definition and the main properties of this variational convergence. Our main result reads as follows

**Theorem 4.** *The sequence $(F^\varepsilon)_\varepsilon$ epi-converges in the weak topology of $H^1(\Omega_1) \times H^1(\Omega_2)$ and the weak topology of $L^p(\omega \times \{0\})$ for the traces on $\omega \times \{0\}$ of the functions (through the transformation (1) in order to bring the moving boundaries $S_1^\varepsilon$ and $S_2^\varepsilon$ to $\omega \times \{0\}$), to the functional $F^0$ defined on $H^1(\Omega_1) \times H^1(\Omega_2)$ by*

$$F^0(v) = \begin{cases} \displaystyle \frac{1}{2} \int_{\Omega_1 \cup \Omega_2} A_{ij}(x) \frac{\partial v}{\partial x_i} \frac{\partial v}{\partial x_j} \, dx \\ \displaystyle + \frac{1}{p(2\lambda_0)^{p-1}} \int_{\omega \times \{0\}} |[v](x_1,x_2,0)|^p \, dx_1 \, dx_2 & \text{if } v \in V^0 \\ +\infty & \text{otherwise,} \end{cases}$$

*$V^0$ being the subspace of $H^1(\Omega_1) \times H^1(\Omega_2)$ defined by*

$$V^0 = \{ v \in H^1(\Omega_1) \times H^1(\Omega_2) | [v] \in L^p(\omega \times \{0\}); \quad v = 0 \text{ on } S_{1u} \cup S_{2u} \}.$$

*Proof.* We have to verify the two assertions (15) and (16) (see the Appendix) of the epi-convergence of $(F^\varepsilon)_\varepsilon$ to $F^0$, which read in the present case as

$$\forall v \in V^0,\ \exists\, v^{\varepsilon 0} \in V^\varepsilon,\ v^{\varepsilon 0} \stackrel{\varepsilon\to 0}{\longrightarrow} v : \limsup_{\varepsilon\to 0} F^\varepsilon(v^{\varepsilon 0}) \le F^0(v) \tag{9}$$

$$\forall v \in V^0,\ \forall\, v_\varepsilon \in V^\varepsilon,\ v_\varepsilon \stackrel{\varepsilon\to 0}{\longrightarrow} v : \liminf_{\varepsilon\to 0} F^\varepsilon(v_\varepsilon) \ge F^0(v) \tag{10}$$

where the convergence of the sequences $(v^{\varepsilon 0})_\varepsilon$ and $(v_\varepsilon)_\varepsilon$ to $v$ takes place in the weak topology of $H^1(\Omega_1)\times H^1(\Omega_2)$ and the convergence of the traces in the weak topology of $L^p(\omega\times\{0\})$.

*First step: verification of* (9). We first choose a smooth function $v$ in $C^2(\overline{\Omega_1})\times C^2(\overline{\Omega_2})$ and define

$$\begin{aligned} v^{\varepsilon 0}(x) &= v(x) && \text{if } |x_3|\ge\varepsilon \\ &= \tfrac{x_3}{2\varepsilon}(v(\varepsilon)-v(-\varepsilon))+\tfrac12(v(\varepsilon)+v(-\varepsilon)) && \text{if } |x_3|\le\varepsilon, \end{aligned} \tag{11}$$

where, for simplicity, we have omitted the variables $x_1$ and $x_2$ and the transformation (1). This means that, in the domains $\Omega_1^\varepsilon$ and $\Omega_2^\varepsilon$, $v^{\varepsilon 0}$ coincides with $v$ (up to the transformation (1)) and in the domain $\Omega_0^\varepsilon$, $v^{\varepsilon 0}$ connects the traces of $v$ in an affine way with respect to $x_3$. It is easy to verify that the sequence $(\|v^{\varepsilon 0}\|_{V^\varepsilon})_\varepsilon$ is bounded. Indeed

$$\int_{\Omega_1^\varepsilon\cup\Omega_2^\varepsilon} A_{ij}(x)\frac{\partial v^{\varepsilon 0}}{\partial x_i}\frac{\partial v^{\varepsilon 0}}{\partial x_j}\,dx = \int_{\Omega_1^\varepsilon\cup\Omega_2^\varepsilon} A_{ij}(x)\frac{\partial v}{\partial x_i}\frac{\partial v}{\partial x_j}\,dx.$$

In $\Omega_0^\varepsilon$, we have for $\alpha=1,2$

$$\frac{\partial v^{\varepsilon 0}}{\partial x_\alpha}(x) = \frac{x_3}{2\varepsilon}\left(\frac{\partial v}{\partial x_\alpha}(\varepsilon)-\frac{\partial v}{\partial x_\alpha}(-\varepsilon)\right)+\frac12\left(\frac{\partial v}{\partial x_\alpha}(\varepsilon)+\frac{\partial v}{\partial x_\alpha}(-\varepsilon)\right)$$

$$\begin{aligned}\Longrightarrow \frac{\partial v^{\varepsilon 0}}{\partial x_\alpha}(x) &= \frac{x_3}{2\varepsilon}\left(\frac{\partial v}{\partial x_\alpha}(0^+)-\frac{\partial v}{\partial x_\alpha}(0^-)\right)+\frac{x_3}{2}\left(\frac{\partial^2 v}{\partial x_3\partial x_\alpha}(\varepsilon^+)-\frac{\partial^2 v}{\partial x_3\partial x_\alpha}(\varepsilon^-)\right) \\ &\quad +\frac12\left(\frac{\partial v}{\partial x_\alpha}(0^+)+\frac{\partial v}{\partial x_\alpha}(0^-)\right)+\frac{\varepsilon}{2}\left(\frac{\partial^2 v}{\partial x_3\partial x_\alpha}(\varepsilon^+)+\frac{\partial^2 v}{\partial x_3\partial x_\alpha}(\varepsilon^-)\right),\end{aligned}$$

with $\varepsilon^+\in]0,\varepsilon[$ and $\varepsilon^-\in]-\varepsilon,0[$. This implies

$$\left|\frac{\partial v^{\varepsilon 0}}{\partial x_\alpha}(x)\right| \le \left|\frac{\partial v}{\partial x_\alpha}(x_1,x_2,0^+)\right| + \left|\frac{\partial v}{\partial x_\alpha}(x_1,x_2,0^-)\right| + 2\varepsilon\, C_\alpha,$$

with $C_\alpha = \max_{x\in\overline{\Omega_1\cup\Omega_2}}\left|\frac{\partial^2 v}{\partial x_\alpha\partial x_3}\right|(x)$. For the third partial derivative of $v^{\varepsilon 0}$ in $\Omega_0^\varepsilon$, one has

$$\frac{\partial v^{\varepsilon 0}}{\partial x_3}(x) = \frac{1}{2\varepsilon}\Big(v(\varepsilon)-v(-\varepsilon)\Big) = \frac{1}{2\varepsilon}\Big(v(0^+)-v(0^-)\Big)+\frac12\left(\frac{\partial v}{\partial x_3}(\varepsilon^+)+\frac{\partial v}{\partial x_3}(\varepsilon^-)\right)$$

$$\Longrightarrow \left|\frac{\partial v^{\varepsilon 0}}{\partial x_3}(x)\right| \le \frac{1}{2\varepsilon}\left|v(0^+)-v(0^-)\right| + C_3 = \frac{1}{2\varepsilon}\left|[v](x_1,x_2,0)\right| + C_3,$$

with $C_3 = \max_{x \in \overline{\Omega_1 \cup \Omega_2}} \left| \frac{\partial v}{\partial x_3}(x) \right|$. Thus, one has

$$\limsup_{\varepsilon \to 0} \varepsilon^{p-1} \int_{\Omega_0^\varepsilon} |\nabla v^{\varepsilon 0}|^p \, dx \leq \int_\omega |[v](x_1, x_2, 0)|^p \, dx_1 \, dx_2.$$

This proves that $(\|v^{\varepsilon 0}\|_{V_\varepsilon})_\varepsilon$ is bounded. Since $v^{\varepsilon 0}$ coincides with $v$ on $\Omega_1^\varepsilon \cup \Omega_2^\varepsilon$, we thus prove that $(v^{\varepsilon 0})_\varepsilon$ converges to $v$ in the weak topology of $H^1(\Omega_1) \times H^1(\Omega_2)$. We compute

$$F^\varepsilon(v^{\varepsilon 0}) = \frac{1}{2} \int_{\Omega_1^\varepsilon \cup \Omega_2^\varepsilon} A_{ij}(x) \frac{\partial v}{\partial x_i} \frac{\partial v}{\partial x_j} \, dx + \frac{\varepsilon^{p-1}}{p \lambda_0^{p-1}} \int_{\Omega_0^\varepsilon} \left| \left( \frac{\partial v^{\varepsilon 0}}{\partial x_1}, \frac{\partial v^{\varepsilon 0}}{\partial x_2}, \frac{\partial v^{\varepsilon 0}}{\partial x_3} \right) \right|^p \, dx.$$

The preceding computations on the partial derivatives of $v^{\varepsilon 0}$ with respect to $x_\alpha$, $\alpha = 1, 2$ imply that

$$\limsup_{\varepsilon \to 0} \frac{\varepsilon^{p-1}}{p \lambda_0^{p-1}} \int_{\Omega_0^\varepsilon} \left| \frac{\partial v^{\varepsilon 0}}{\partial x_1} \right|^p dx = 0; \quad \limsup_{\varepsilon \to 0} \frac{\varepsilon^{p-1}}{p \lambda_0^{p-1}} \int_{\Omega_0^\varepsilon} \left| \frac{\partial v^{\varepsilon 0}}{\partial x_2} \right|^p dx = 0.$$

It remains

$$\limsup_{\varepsilon \to 0} F^\varepsilon(v^{\varepsilon 0}) = \frac{1}{2} \int_{\Omega_1 \cup \Omega_2} A_{ij}(x) \frac{\partial v}{\partial x_i} \frac{\partial v}{\partial x_j} \, dx + \limsup_{\varepsilon \to 0} \frac{\varepsilon^{p-1}}{p \lambda_0^{p-1}} \int_{\Omega_0^\varepsilon} \left| \frac{\partial v^{\varepsilon 0}}{\partial x_3} \right|^p dx$$

$$\begin{aligned} \Longrightarrow \limsup_{\varepsilon \to 0} F^\varepsilon(v^{\varepsilon 0}) &= \frac{1}{2} \int_{\Omega_1 \cup \Omega_2} A_{ij}(x) \frac{\partial v}{\partial x_i} \frac{\partial v}{\partial x_j} \, dx \\ &\quad + \frac{1}{p(2\lambda_0)^{p-1}} \int_\omega |[v]|^p \, dx_1 \, dx_2 = F^0(v). \end{aligned}$$

This proves (9) in the case of a smooth function $v$ in $C^2(\overline{\Omega_1}) \times C^2(\overline{\Omega_2})$. In the general case, we use the density of $C^2(\overline{\Omega_1}) \times C^2(\overline{\Omega_2})$ in $H^1(\Omega_1) \times H^1(\Omega_2)$ and the diagonalization argument exposed in Lemma A.4. Indeed, let $v$ be any element of $V^0$. There exists a sequence $(v_n)_n$ which converges to $v$ in the strong topology of $H^1(\Omega_1) \times H^1(\Omega_2)$ and $([v_n])_n$ converges to $[v]$ in the strong topology of $L^p(\omega \times \{0\})$. For every $n$, the preceding computations prove that

$$\limsup_{\varepsilon \to 0} F^\varepsilon((v_n)^{\varepsilon 0}) = F^0(v_n),$$

where $(v_n)^{\varepsilon 0}$ is associated to $v_n$ by (11). Hence

$$\limsup_{n \to +\infty} \limsup_{\varepsilon \to 0} F^\varepsilon((v_n)^{\varepsilon 0}) = F^0(v),$$

since $F^0$ is continuous for the strong topology of $H^1(\Omega_1) \times H^1(\Omega_2)$ and the strong topology of $L^p(\omega \times \{0\})$, for the traces. From Lemma A.4, we deduce the existence of a sequence $(n(\varepsilon))_\varepsilon$ growing to $+\infty$ such that $((v_{n(\varepsilon)})^{\varepsilon 0})_\varepsilon$ converges to $v$ in the weak topology of $H^1(\Omega_1) \times H^1(\Omega_2)$, $([(v_{n(\varepsilon)})^{\varepsilon 0}])_\varepsilon$ converges to $[v]$ in the weak topology of $L^p(\omega \times \{0\}$ and

$$\limsup_{\varepsilon \to 0} F^\varepsilon((v_{n(\varepsilon)})^{\varepsilon 0}) \leq F^0(v),$$

which proves (9) in the general case.

*Second step: verification of* (10). Let $v$ be any element of $V^0$ and $(v_\varepsilon)_\varepsilon$ be any sequence converging to $v$ in the weak topology of $H^1(\Omega_1) \times H^1(\Omega_2)$. Let $(v_n)_n$ be any sequence of smooth functions ($v_n \in C^2(\overline{\Omega_1}) \times C^2(\overline{\Omega_2})$, $\forall n \geq 0$) which converges to $v$ in the strong topology of $H^1(\Omega_1) \times H^1(\Omega_2)$ and $([v_n])_n$ converges to $[v]$ in the strong topology of $L^p(\omega \times \{0\})$. We use the following subdifferential inequality

$$\begin{aligned} F^\varepsilon(v_\varepsilon) \;\; &\geq \;\; F^\varepsilon((v_n)^{\varepsilon 0}) + \int_{\Omega_1^\varepsilon \cup \Omega_2^\varepsilon} A_{ij}(x) \frac{\partial (v_n)^{\varepsilon 0}}{\partial x_i} \frac{\partial (v_\varepsilon - (v_n)^{\varepsilon 0})}{\partial x_j}\, dx \\ &\quad + \frac{\varepsilon^{p-1}}{\lambda_0^{p-1}} \int_{\Omega_0^\varepsilon} |\nabla((v_n)^{\varepsilon 0})|^{p-2} \nabla((v_n)^{\varepsilon 0}) \nabla(v_\varepsilon - (v_n)^{\varepsilon 0})\, dx, \end{aligned}$$

where $(v_n)^{\varepsilon 0}$ is the function associated to $v_n$ by (11). Thus

$$\begin{aligned} &\liminf_{\varepsilon \to 0} F^\varepsilon(v_\varepsilon) \\ &\quad \geq F^0(v_n) + \liminf_{\varepsilon \to 0} \int_{\Omega_1^\varepsilon \cup \Omega_2^\varepsilon} A_{ij}(x) \frac{\partial (v_n)^{\varepsilon 0}}{\partial x_i} \frac{\partial (v_\varepsilon - (v_n)^{\varepsilon 0})}{\partial x_j}\, dx \\ &\quad + \liminf_{\varepsilon \to 0} \frac{\varepsilon^{p-1}}{\lambda_0^{p-1}} \int_{\Omega_0^\varepsilon} |\nabla((v_n)^{\varepsilon 0})|^{p-2} \times \nabla((v_n)^{\varepsilon 0}) \nabla(v_\varepsilon - (v_n)^{\varepsilon 0})\, dx, \end{aligned}$$

using the first step. From the computations of this first step, we infer

$$\begin{aligned} \liminf_{\varepsilon \to 0} F^\varepsilon(v_\varepsilon) \;\; &\geq \;\; F^0(v_n) + \liminf_{\varepsilon \to 0} \frac{1}{(2\lambda_0)^{p-1}} \int_\omega |v_n(\varepsilon) - v_n(-\varepsilon)|^{p-2} \\ &\quad \times (v_n(\varepsilon) - v_n(-\varepsilon)) \int_{-\varepsilon}^{\varepsilon} \frac{\partial (v_\varepsilon - (v_n)^{\varepsilon 0})}{\partial x_3}\, dx \end{aligned}$$

$$\begin{aligned} \Longrightarrow \liminf_{\varepsilon \to 0} F^\varepsilon(v_\varepsilon) \;\; &\geq \;\; F^0(v_n) + \liminf_{\varepsilon \to 0} \frac{1}{(2\lambda_0)^{p-1}} \int_\omega |[v_n]|^{p-2}(0)[v_n](0) \\ &\quad \times (v_\varepsilon(\varepsilon) - v_\varepsilon(-\varepsilon) - [v_n](0))\, dx_1\, dx_2 \end{aligned}$$

Using the transformation (1), the convergence of the sequence of functions deduced from $(v_\varepsilon)_\varepsilon$ through this transformation to $v$ in the weak topology of $H^1(\Omega_1) \times$

$H^1(\Omega_2)$, we deduce

$$\liminf_{\varepsilon\to 0} F^\varepsilon(v_\varepsilon) \geq F^0(v_n) + \frac{1}{(2\lambda_0)^{p-1}} \int_\omega |[v_n]|^{p-2}(0)[v_n](0) \\ \times ([v](0) - [v_n](0))\, dx_1\, dx_2.$$

Finally, we let $n$ go to $+\infty$. Since $F^0$ is continuous for the strong topology of $H^1(\Omega_1) \times H^1(\Omega_2)$, we obtain

$$\liminf_{\varepsilon\to 0} F^\varepsilon(v_\varepsilon) \geq F^0(v) + 0,$$

which proves (10). □

**Remark 5.** Thanks to the result exposed in the Remark 3, the presence of the constraint "$[v] \in L^p(\omega \times \{0\})$" in the limit functional $F^0$ is not surprising. Notice that this constraint disappears if $p$ is less than 2, thanks to the trace properties of the involved spaces.

The properties of the epi-convergence imply

**Corollary 6.** 1) *The sequence $(u^\varepsilon)_\varepsilon$, where $u^\varepsilon$ denotes the solution of (2), converges in the weak topology of $H^1(\Omega_1) \times H^1(\Omega_2)$ to the solution $u^0$ of*

$$\min_{v\in V^0} \left\{ F^0(v) - \int_{\Omega_1\cup\Omega_2} f\, v\, dx - \int_{S_t^0} t\, v\, d\sigma(x) \right\}. \tag{12}$$

2) *Moreover, $(F^\varepsilon(u^\varepsilon))_\varepsilon$ converges to $F^0(u^0)$, which means that*

$$\lim_{\varepsilon\to 0} \left( \frac{1}{2} \int_{\Omega_1^\varepsilon\cup\Omega_2^\varepsilon} A_{ij}(x) \frac{\partial u^\varepsilon}{\partial x_i} \frac{\partial u^\varepsilon}{\partial x_j}\, dx + \frac{\varepsilon^{p-1}}{p\lambda_0^{p-1}} \int_{\Omega_0^\varepsilon} |\nabla u^\varepsilon|^p\, dx \right) \\ = \frac{1}{2} \int_{\Omega_1\cup\Omega_2} A_{ij}(x) \frac{\partial u^0}{\partial x_i} \frac{\partial u^0}{\partial x_j}\, dx + \frac{1}{p(2\lambda_0)^{p-1}} \int_\omega |[u^0]|^p\, dx_1\, dx_2.$$

The proof is a consequence of Theorem 4, through Theorem A.2. □

The solution $u^0$ of (12) satisfies

**Lemma 7.** 1) *For every $v$ in $V^0$, we have*

$$\int_{\Omega_1\cup\Omega_2} A_{ij}(x) \frac{\partial u^0}{\partial x_i} \frac{\partial v}{\partial x_j}\, dx + \frac{1}{(2\lambda_0)^{p-1}} \int_\omega |[u^0]|^{p-2}[u^0][v]\, dx_1\, dx_2 \\ = \int_\Omega f\, v\, dx + \int_{S_t^0} t\, v\, d\sigma(x). \tag{13}$$

2) $u^0$ *is a weak solution of*

$$\begin{aligned}
-\frac{\partial}{\partial x_j}\left(A_{ij}(x)\frac{\partial u^0}{\partial x_i}\right) &= f \quad in \ \Omega_1 \cup \Omega_2 \\
-A_{ij}(x)\frac{\partial u^0}{\partial x_i}\nu_j &= \frac{1}{(2\lambda_0)^{p-1}}\left|[u^0]\right|^{p-2}[u^0] \quad on \ \omega \times \{0\} \\
u^0 &= 0 \quad on \ S_{1u} \cup S_{2u} \\
A_{ij}(x)\frac{\partial u^0}{\partial x_i}\nu_j &= t \quad on \ S_{1t} \cup S_{2t} = S_t^0.
\end{aligned} \tag{14}$$

**Remark 8.** The boundary condition on $\omega \times \{0\}$

$$-A_{ij}(x)\frac{\partial u^0}{\partial x_i}\nu_j = \frac{1}{(2\lambda_0)^{p-1}}|[u^0]|^{p-2}[u^0],$$

is equivalent to

$$[u^0] = -2\lambda_0\left|A_{ij}(x)\frac{\partial u^0}{\partial x_i}\nu_j\right|^{q-2} A_{ij}(x)\frac{\partial u^0}{\partial x_i}\nu_j,$$

where $q$ is the conjugate exponent of $p$. This is the limit contact law on $\omega \times \{0\}$.

## C. Rate of convergence

We describe the rate of convergence of $(u^\varepsilon)_\varepsilon$ to $u^0$.

**Proposition 9.** *Suppose that the solution $u^0$ of (12) belongs to $C^2(\overline{\Omega_1}) \times C^2(\overline{\Omega_2})$. One has*

$$\lim_{\varepsilon\to 0}\int_{\Omega_1^\varepsilon\cup\Omega_2^\varepsilon}|\nabla(u^\varepsilon - (u^0)^{\varepsilon 0})|^2\,dx = 0,$$

*where $(u^0)^{\varepsilon 0}$ is associated to $u^0$ by (11).*

*Proof.* We write

$$\begin{aligned}
&\frac{1}{2}\int_{\Omega_1^\varepsilon\cup\Omega_2^\varepsilon} A_{ij}(x)\frac{\partial(u^\varepsilon - (u^0)^{\varepsilon 0})}{\partial x_i}\frac{\partial(u^\varepsilon - (u^0)^{\varepsilon 0})}{\partial x_j}\,dx \\
&= \frac{1}{2}\int_{\Omega_1^\varepsilon\cup\Omega_2^\varepsilon} A_{ij}(x)\frac{\partial((u^0)^{\varepsilon 0})}{\partial x_i}\frac{\partial((u^0)^{\varepsilon 0})}{\partial x_j}\,dx - \frac{1}{2}\int_{\Omega_1^\varepsilon\cup\Omega_2^\varepsilon} A_{ij}(x)\frac{\partial u^\varepsilon}{\partial x_i}\frac{\partial u^\varepsilon}{\partial x_j}\,dx \\
&\quad - \int_{\Omega_1^\varepsilon\cup\Omega_2^\varepsilon} A_{ij}(x)\frac{\partial u^\varepsilon}{\partial x_i}\frac{\partial((u^0)^{\varepsilon 0} - u^\varepsilon)}{\partial x_j}\,dx.
\end{aligned}$$

We evaluate the last integral, using an integration by parts

$$\int_{\Omega_1^\varepsilon\cup\Omega_2^\varepsilon} A_{ij}(x)\frac{\partial u^\varepsilon}{\partial x_i}\frac{\partial((u^0)^{\varepsilon 0}-u^\varepsilon)}{\partial x_j}\,dx$$

$$= \int_{\Omega_1^\varepsilon\cup\Omega_2^\varepsilon} f((u^0)^{\varepsilon 0}-u^\varepsilon)\,dx + \int_{\partial(\Omega_1^\varepsilon\cup\Omega_2^\varepsilon)\cap S_t^\varepsilon} A_{ij}(x)\frac{\partial u^\varepsilon}{\partial x_i}((u^0)^{\varepsilon 0}-u^\varepsilon)\nu_j\,d\sigma(x)$$

$$- \int_{S_1^\varepsilon\cup S_2^\varepsilon} A_{ij}(x)\frac{\partial u^\varepsilon}{\partial x_i}((u^0)^{\varepsilon 0}-u^\varepsilon)\nu_j\,d\sigma(x),$$

the $-$ sign before the last integral being justified by the orientation of the normal. Using the boundary condition ($6_3$), we obtain

$$\int_{\Omega_1^\varepsilon\cup\Omega_2^\varepsilon} A_{ij}(x)\frac{\partial u^\varepsilon}{\partial x_i}\frac{\partial((u^0)^{\varepsilon 0}-u^\varepsilon)}{\partial x_j}\,dx$$

$$= \int_{\Omega_1^\varepsilon\cup\Omega_2^\varepsilon} f((u^0)^{\varepsilon 0}-u^\varepsilon)\,dx + \int_{S_t^\varepsilon\cap\partial(\Omega_1^\varepsilon\cup\Omega_2^\varepsilon)} t((u^0)^{\varepsilon 0}-u^\varepsilon)\,d\sigma(x)$$

$$-\frac{\varepsilon^{p-1}}{\lambda_0^{p-1}}\int_{\omega\times\{\pm\varepsilon\}} |\nabla u^\varepsilon|^{p-2}\times\frac{\partial u^\varepsilon}{\partial\nu}((u^0)^{\varepsilon 0}-u^\varepsilon)\,dx_1\,dx_2.$$

Let us interpret this last integral as the result of an integration by parts in $\Omega_0^\varepsilon$. We have

$$\frac{\varepsilon^{p-1}}{\lambda_0^{p-1}}\int_{\Omega_0^\varepsilon} |\nabla u^\varepsilon|^{p-2}\nabla u^\varepsilon\nabla((u^0)^{\varepsilon 0}-u^\varepsilon)\,dx$$

$$= -\frac{\varepsilon^{p-1}}{\lambda_0^{p-1}}\int_{\Omega_0^\varepsilon} \operatorname{div}(|\nabla u^\varepsilon|^{p-2}\nabla u^\varepsilon)((u^0)^{\varepsilon 0}-u^\varepsilon)\,dx$$

$$+\frac{\varepsilon^{p-1}}{\lambda_0^{p-1}}\int_{S_0^\varepsilon} |\nabla u^\varepsilon|^{p-2}\frac{\partial u^\varepsilon}{\partial\nu}((u^0)^{\varepsilon 0}-u^\varepsilon)\,d\sigma(x)$$

$$+\frac{\varepsilon^{p-1}}{\lambda_0^{p-1}}\int_{\omega\times\{\pm\varepsilon\}} |\nabla u^\varepsilon|^{p-2}\frac{\partial u^\varepsilon}{\partial\nu}((u^0)^{\varepsilon 0}-u^\varepsilon)\,dx_1\,dx_2$$

$$= \int_{\Omega_0^\varepsilon} f((u^0)^{\varepsilon 0}-u^\varepsilon)\,dx + \int_{S_0^\varepsilon} t((u^0)^{\varepsilon 0}-u^\varepsilon)\,d\sigma(x)$$

$$+\frac{\varepsilon^{p-1}}{\lambda_0^{p-1}}\int_{\omega\times\{\pm\varepsilon\}} |\nabla u^\varepsilon|^{p-2}\frac{\partial u^\varepsilon}{\partial\nu}((u^0)^{\varepsilon 0}-u^\varepsilon)\,dx_1\,dx_2$$

$$\Longrightarrow \int_{\Omega_1^\varepsilon \cup \Omega_2^\varepsilon} A_{ij}(x) \frac{\partial u^\varepsilon}{\partial x_i} \frac{\partial((u^0)^{\varepsilon 0} - u^\varepsilon)}{\partial x_j}\, dx$$

$$= \int_{\Omega^\varepsilon} f((u^0)^{\varepsilon 0} - u^\varepsilon)\, dx + \int_{S_t^\varepsilon} t((u^0)^{\varepsilon 0} - u^\varepsilon)\, d\sigma(x)$$

$$- \frac{\varepsilon^{p-1}}{\lambda_0^{p-1}} \int_{\Omega_0^\varepsilon} |\nabla u^\varepsilon|^{p-2} \nabla u^\varepsilon \nabla((u^0)^{\varepsilon 0} - u^\varepsilon)\, dx.$$

Moreover, the subdifferential inequality

$$\frac{\varepsilon^{p-1}}{p\lambda_0^{p-1}} \int_{\Omega_0^\varepsilon} |\nabla (u^0)^{\varepsilon 0}|^p\, dx$$

$$\geq \frac{\varepsilon^{p-1}}{p\lambda_0^{p-1}} \int_{\Omega_0^\varepsilon} |\nabla u^\varepsilon|^p\, dx + \frac{\varepsilon^{p-1}}{\lambda_0^{p-1}} \int_{\Omega_0^\varepsilon} |\nabla u^\varepsilon|^{p-2} \nabla u^\varepsilon \nabla((u^0)^{\varepsilon 0} - u^\varepsilon)\, dx$$

implies

$$\frac{1}{2} \int_{\Omega_1^\varepsilon \cup \Omega_2^\varepsilon} A_{ij}(x) \frac{\partial(u^\varepsilon - (u^0)^{\varepsilon 0})}{\partial x_j} \frac{\partial(u^\varepsilon - (u^0)^{\varepsilon 0})}{\partial x_j}\, dx$$

$$\leq \frac{1}{2} \int_{\Omega_1^\varepsilon \cup \Omega_2^\varepsilon} A_{ij}(x) \frac{\partial (u^0)^{\varepsilon 0}}{\partial x_i} \frac{\partial (u^0)^{\varepsilon 0}}{\partial x_j}\, dx + \frac{\varepsilon^{p-1}}{p\lambda_0^{p-1}} \int_{\Omega_0^\varepsilon} |\nabla (u^0)^{\varepsilon 0}|^p\, dx$$

$$- \frac{1}{2} \int_{\Omega_1^\varepsilon \cup \Omega_2^\varepsilon} A_{ij}(x) \frac{\partial u^\varepsilon}{\partial x_i} \frac{\partial u^\varepsilon}{\partial x_j}\, dx - \frac{\varepsilon^{p-1}}{p\lambda_0^{p-1}} \int_{\Omega_0^\varepsilon} |\nabla u^\varepsilon|^p dx$$

$$- \int_{\Omega^\varepsilon} f((u^0)^{\varepsilon 0} - u^\varepsilon)\, dx - \int_{S_t^\varepsilon} t((u^0)^{\varepsilon 0} - u^\varepsilon)\, d\sigma(x)$$

$$\leq F^\varepsilon((u^0)^{\varepsilon 0}) - F^\varepsilon(u^\varepsilon) - \int_{\Omega^\varepsilon} f((u^0)^{\varepsilon 0} - u^\varepsilon)\, dx$$

$$- \int_{S_t^\varepsilon} t((u^0)^{\varepsilon 0} - u^\varepsilon)\, d\sigma(x).$$

We then conclude using Corollary 6 and the compact embeddings from $H^1(\Omega_1) \times H^1(\Omega_2)$ into $L^2(\Omega_1 \cup \Omega_2)$ and into $L^2(\partial(\Omega_1 \cup \Omega_2))$. □

**Remark 10.** Since $(u^0)^{\varepsilon 0}$ coincides with $u^0$ on $\Omega_1^\varepsilon \cup \Omega_2^\varepsilon$, Proposition 9 implies

$$\lim_{\varepsilon \to 0} \int_{\Omega_1^\varepsilon \cup \Omega_2^\varepsilon} |\nabla(u^\varepsilon - u^0)|^2 dx = 0.$$

## 3. Study of the limit problem for other values of the coefficient $\lambda_\varepsilon$

We want to describe the behaviour of $(u^\varepsilon)_\varepsilon$, when the hypothesis (7) is not satisfied. We suppose that $\lambda_\varepsilon$ depends on $\varepsilon$ as

$$\lambda_\varepsilon = \lambda_0 \varepsilon^\alpha,$$

where $\alpha$ is a real constant. Then we denote by $F^{\varepsilon\alpha}$ the functional corresponding to this situation

$$F^{\varepsilon\alpha}(v) = \begin{cases} \dfrac{1}{2}\displaystyle\int_{\Omega_1^\varepsilon \cup \Omega_2^\varepsilon} A_{ij}(x)\frac{\partial v}{\partial x_i}\frac{\partial v}{\partial x_j}\,dx + \frac{\varepsilon^{-\alpha(p-1)}}{p\lambda_0^{p-1}}\int_{\Omega_0^\varepsilon} |\nabla v|^p\,dx & \text{if } v \in V^\varepsilon \\ +\infty & \text{otherwise,} \end{cases}$$

where $V^\varepsilon$ is defined in (3). Our main result is

**Theorem 11.** 1) *Suppose that $\alpha < -1$. Then $(F^{\varepsilon\alpha})_\varepsilon$ epi-converges in the weak topology of $H^1(\Omega_1) \times H^1(\Omega_2)$ to the functional $F_1^0$ defined on this space by*

$$F_1^0(v) = \begin{cases} \dfrac{1}{2}\displaystyle\int_{\Omega_1 \cup \Omega_2} A_{ij}(x)\frac{\partial v}{\partial x_i}\frac{\partial v}{\partial x_j}\,dx & \textit{if } v \in V_1^0 \\ +\infty & \textit{otherwise,} \end{cases}$$

*where $V_1^0$ is the subspace*

$$V_1^0 = \{v \in H^1(\Omega_1) \times H^1(\Omega_2) |\; v = 0 \;\; \textit{on} \;\; S_{1u} \cup S_{2u}\}.$$

2) *Suppose that $-1 < \alpha < \frac{1}{p-1}$. Then $(F^{\varepsilon\alpha})_\varepsilon$ epi-converges in the weak topology of $H^1(\Omega_1) \times H^1(\Omega_2)$ to the functional $F_2^0$ defined on this space by*

$$F_2^0(v) = \begin{cases} \dfrac{1}{2}\displaystyle\int_{\Omega_1 \cup \Omega_2} A_{ij}(x)\frac{\partial v}{\partial x_i}\frac{\partial v}{\partial x_j}\,dx & \textit{if } v \in V_2^0 \\ +\infty & \textit{otherwise,} \end{cases}$$

*where $V_2^0$ is the subspace*

$$V_2^0 = \{v \in H^1(\Omega_1 \cup \Omega_2 \cup (\omega \times \{0\})) |\; v = 0 \;\; \textit{on} \;\; S_{1u} \cup S_{2u}\}.$$

3) *Suppose that $\alpha = \frac{1}{p-1}$. Then $(F^{\varepsilon\alpha})_\varepsilon$ epi-converges in the weak topology of $H^1(\Omega_1) \times H^1(\Omega_2)$ to the functional $F_3^0$ defined on this space by*

$$F_3^0(v) = \begin{cases} \dfrac{1}{2}\displaystyle\int_{\Omega_1 \cup \Omega_2} A_{ij}(x)\frac{\partial v}{\partial x_i}\frac{\partial v}{\partial x_j}dx + \frac{1}{p(2\lambda_0)^{p-1}} I_p(v) & \textit{if } v \in V_3^0 \\ +\infty & \textit{otherwise,} \end{cases}$$

*where*

$$I_p(v)=\int_{\omega\times\{0\}}\int_{-1}^{1}\left\{\begin{array}{l}\left(\frac{y}{2}\left(\frac{\partial v}{\partial x_1}(0^+)-\frac{\partial v}{\partial x_1}(0^-)\right)+\frac{1}{2}\left(\frac{\partial v}{\partial x_1}(0^+)+\frac{\partial v}{\partial x_1}(0^-)\right)\right)^2\\ +\left(\frac{y}{2}\left(\frac{\partial v}{\partial x_2}(0^+)-\frac{\partial v}{\partial x_2}(0^-)\right)^2+\frac{1}{2}\left(\frac{\partial v}{\partial x_2}(0^+)\frac{\partial v}{\partial x_2}(0^-)\right)\right)^2\\ \quad+\left(\frac{\partial v}{\partial x_3}(0^+)+\frac{\partial v}{\partial x_3}(0^-)\right)^2\end{array}\right\}^{p/2}$$

*and $V_3^0$ is the subspace*

$$V_3^0=\{v\in H^1(\Omega_1\cup\Omega_2\cup(\omega\times\{0\}))|\ v=0 \text{ on } S_{1u}\cup S_{2u};\ I_p(v) \text{ exists}\}.$$

4) *Suppose that $\alpha>\frac{1}{p-1}$. Then $(F^{\varepsilon\alpha})_\varepsilon$ epi-converges in the weak topology of $H^1(\Omega_1)\times H^1(\Omega_2)$ to the functional $F_4^0$ defined on this space by*

$$F_4^0(v)=\begin{cases}\displaystyle\frac{1}{2}\int_{\Omega_1\cup\Omega_2}A_{ij}(x)\frac{\partial v}{\partial x_i}\frac{\partial v}{\partial x_j}\,dx & \textit{if } v\in V_4^0\\ +\infty & \textit{otherwise,}\end{cases}$$

*where $V_4^0$ is the subspace*

$$V_4^0 = \Big\{v\in H^1(\Omega_1\cup\Omega_2\cup(\omega\times\{0\}))|\ v=0 \text{ on } S_{1u}\cup S_{2u};$$
$$v=Cte \text{ on } \omega\times\{0\};\quad \frac{\partial v}{\partial x_3}(0^+)+\frac{\partial v}{\partial x_3}(0^-)=0\Big\}.$$

*Proof.* 1) We notice that in this case

$$\lim_{\varepsilon\to 0}\frac{\lambda_0\varepsilon^\alpha}{\frac{1}{\varepsilon}}=\lim_{\varepsilon\to 0}\lambda_0\varepsilon^{1+\alpha}=+\infty.$$

Hence, for every positive $\beta$, one has

$$\lambda_0\varepsilon^\alpha\geq\frac{\beta}{\varepsilon}\Longrightarrow\frac{\varepsilon}{\beta}\geq\frac{\varepsilon^{-\alpha}}{\lambda_0},$$

for $\varepsilon$ sufficiently small. Then, for every $v$ in $V^\varepsilon$

$$F^{\varepsilon\alpha}(v)\leq\frac{1}{2}\int_{\Omega_1^\varepsilon\cup\Omega_2^\varepsilon}A_{ij}(x)\frac{\partial v}{\partial x_i}\frac{\partial v}{\partial x_j}\,dx+\frac{\varepsilon^{p-1}}{p\beta^{p-1}}\int_{\Omega_0^\varepsilon}|\nabla v|^p dx.$$

Thanks to Theorem 4 and Proposition A.5, one derives from the preceding inequality that for every $v$ in $V^0$

$$F_1^0(v)\leq\frac{1}{2}\int_{\Omega_1^\varepsilon\cup\Omega_2^\varepsilon}A_{ij}(x)\frac{\partial v}{\partial x_i}\frac{\partial v}{\partial x_j}\,dx+\frac{1}{p(2\beta)^{p-1}}\int_{\omega\times\{0\}}|[v]|^p dx_1dx_2.$$

One can take the infimum over all positive $\beta$'s in this inequality. This proves that the constraint "$[v] \in L^p(\omega \times \{0\})$" disappears in the limit functional $F_1^0$. Hence this limit functional $F_1^0$ "lives" on $V_1^0$ (that is, is not equal to $+\infty$ for every $v$ in $V_1^0$) and

$$F_1^0(v) \leq \frac{1}{2} \int_{\Omega_1 \cup \Omega_2} A_{ij}(x) \frac{\partial v}{\partial x_i} \frac{\partial v}{\partial x_j}\, dx, \forall v \in V_1^0.$$

The reverse inequality is proved, using the test-function $v^{\varepsilon 0}$ associated to every smooth function $v$ in $C^2(\overline{\Omega_1}) \times C^2(\overline{\Omega_2}))$ (for which the jump $[v]$ belongs to $L^p(\omega \times \{0\})$ !). Indeed, for every $v$ in $V_1^0$ and for every sequence $(v_\varepsilon)_\varepsilon$ converging to $v$ in the weak topology of $H^1(\Omega_1) \times H^1(\Omega_2)$ and every sequence $(v_n)_n$ converging to $v$ in the strong topology of this space, one has using the subdifferential inequality

$$\begin{aligned}\liminf_{\varepsilon\to 0} F^{\varepsilon\alpha}(v_\varepsilon) \geq & \frac{1}{2}\int_{\Omega_1\cup\Omega_2} A_{ij}(x)\frac{\partial v_n}{\partial x_i}\frac{\partial v_n}{\partial x_j}\,dx + \int_{\Omega_1\cup\Omega_2} A_{ij}(x)\frac{\partial v_n}{\partial x_i}\frac{\partial(v-v_n)}{\partial x_j}\,dx \\ & + \liminf_{\varepsilon\to 0}\frac{\varepsilon^{-\alpha(p-1)}}{\lambda_0^{p-1}}\int_{\Omega_0^\varepsilon} |\nabla((v_n)^{\varepsilon 0})|^{p-2}\nabla(v_n)^{\varepsilon 0})\nabla(v_\varepsilon-(v_n)^{\varepsilon 0})\,dx.\end{aligned}$$

Thanks to the computations of the second step of the proof of Theorem 4, it remains

$$\begin{aligned}\liminf_{\varepsilon\to 0} F^{\varepsilon\alpha}(v_\varepsilon) \geq & \frac{1}{2}\int_{\Omega_1\cup\Omega_2} A_{ij}(x)\frac{\partial v_n}{\partial x_i}\frac{\partial v_n}{\partial x_j}\,dx + \int_{\Omega_1\cup\Omega_2} A_{ij}(x)\frac{\partial v_n}{\partial x_i}\frac{\partial(v-v_n)}{\partial x_j}\,dx \\ & + \liminf_{\varepsilon\to 0}\frac{\varepsilon^{-\alpha(p-1)}}{\varepsilon^{p-1}\lambda_0^{p-1}}\int_{\omega\times\{\pm\varepsilon\}} |[v_n]|^{p-2}\times[v_n][v_\varepsilon - v_n]\,d\sigma(x).\end{aligned}$$

Finally, we let $n$ increase to $+\infty$ and obtain

$$\liminf_{\varepsilon\to 0} F^{\varepsilon\alpha}(v_\varepsilon) \geq \frac{1}{2}\int_{\Omega_1\cup\Omega_2} A_{ij}(x)\frac{\partial v}{\partial x_i}\frac{\partial v}{\partial x_j}\,dx = F_1^0(v).$$

2) We proceed in a quite similar way. Notice that in the present case the constraint "$[v] \in L^p(\omega \times \{0\})$" also disappears, because the jump of $v$ is equal to 0.

3) In this case, we reproduce the preceding step and we notice that, since $v$ is equal to 0 on $\omega \times \{0\}$ and is smooth enough (in a first step)

$$\begin{aligned}\frac{\partial v^{\varepsilon 0}}{\partial x_\alpha} &= \frac{x_3}{2\varepsilon}\left(\frac{\partial v}{\partial x_\alpha}(\varepsilon) - \frac{\partial v}{\partial x_\alpha}(-\varepsilon)\right) + \frac{1}{2}\left(\frac{\partial v}{\partial x_\alpha}(\varepsilon) + \frac{\partial v}{\partial x_\alpha}(-\varepsilon)\right). \\ &\simeq \frac{x_3}{2\varepsilon}\left(\frac{\partial v}{\partial x_\alpha}(0^+) - \frac{\partial v}{\partial x_\alpha}(0^-)\right) + \frac{1}{2}\left(\frac{\partial v}{\partial x_\alpha}(0^+) + \frac{\partial v}{\partial x_\alpha}(0^-)\right)\end{aligned}$$

$$\frac{\partial v^{\varepsilon 0}}{\partial x_3} = \frac{1}{2\varepsilon}(v(x_1,x_2,\varepsilon) - v(x_1,x_2,-\varepsilon)) \simeq \frac{1}{2}\left(\frac{\partial v}{\partial x_3}(x_1,x_2,0^+) - \frac{\partial v}{\partial x_3}(x_1,x_2,0^-)\right).$$

In an easy way, we prove: $\limsup_{\varepsilon\to 0} F^{\varepsilon\alpha}(v^{\varepsilon 0}) = F_3^0(v)$, for $v$ smooth enough first and then for any $v$ in $V_3^0$. The second assertion of the epi-convergence of $(F^{\varepsilon\alpha})_\alpha$ to $F_3^0$ is proved like in Theorem 4.

4) We will partly deduce this case from the preceding one. We first observe that in the present case

$$\lim_{\varepsilon\to 0}\frac{\varepsilon^{-\alpha(p-1)}}{\varepsilon^{-1}}=\lim_{\varepsilon\to 0}\varepsilon^{1-\alpha(p-1)}=+\infty.$$

Hence, for every $v$ and for every $\beta$ positive

$$F^{\varepsilon\alpha}(v)\geq\frac{1}{2}\int_{\Omega_1^\varepsilon\cup\Omega_2^\varepsilon}A_{ij}(x)\frac{\partial v}{\partial x_i}\frac{\partial v}{\partial x_j}\,dx+\frac{\beta}{p\lambda_0^{p-1}}\frac{1}{\varepsilon}\int_{\Omega_0^\varepsilon}|\nabla v|^p dx.$$

Thus

$$\lim_{\varepsilon\to 0}F^{\varepsilon\alpha}(v)\geq\frac{1}{2}\int_{\Omega_1\cup\Omega_2}A_{ij}(x)\frac{\partial v}{\partial x_i}\frac{\partial v}{\partial x_j}\,dx+\frac{\beta}{p\lambda_0^{p-1}}I_p(v),$$

thanks to the preceding case. One takes the supremum of the right hand sided term with respect to $\beta$. Thus we are faced with the conditions

$$\forall y\in]-1,1[,\ \forall\alpha\in\{1,2\}:$$

$$\frac{y}{2}\left(\frac{\partial v}{\partial x_\alpha}(0^+)-\frac{\partial v}{\partial x_\alpha}(0^-)\right)+\frac{1}{2}\left(\frac{\partial v}{\partial x_\alpha}(0^+)+\frac{\partial v}{\partial x_\alpha}(0^-)\right)=0$$

$$\forall y\in]-1,1[:\ \frac{\partial v}{\partial x_3}(0^+)+\frac{\partial v}{\partial x_3}(0^-)=0,$$

which must be satisfied, otherwise $\lim_{\varepsilon\to 0}F^{\varepsilon\alpha}(v)$ is equal to $+\infty$. These conditions imply the ones written in this assertion 4. This ends the proof of Theorem 11. □

**Remark 12.** The asymptotic behaviour of thin elastic, adhesive, nonlinear and incompressible layers has been described in [3]. The asymptotic behaviour of thin viscoplastic, adhesive and incompressible layers has been presented in [4]. I would like to thank my colleagues J.F. Ganghoffer and J. Schultz from CNRS-ICSI in Mulhouse for the study of these corresponding elasticity problem.

## Appendix: Definition and properties of the epi-convergence

**Definition A.1** ([1], Definition 1.9). Let $(\mathbf{X},\tau)$ be a metric space, $(F_\varepsilon)_\varepsilon$ and $F_0$ be functionals from $\mathbf{X}$ with values in $\mathbb{R}\cup\{+\infty\}$. $(F_\varepsilon)_\varepsilon$ epi-converges to $F_0$ in the topology $\tau$ if and only if the two following properties are satisfied

$$\forall\, x\in\mathbf{X},\exists\, x_\varepsilon^0\in\mathbf{X},\quad \lim_{\varepsilon\to 0}x_\varepsilon^0=x:\limsup_{\varepsilon\to 0}F_\varepsilon(x_\varepsilon^0)\leq F_0(x),\tag{15}$$

$$\forall\, x\in\mathbf{X},\forall\, x_\varepsilon\in\mathbf{X},\quad \lim_{\varepsilon\to 0}x_\varepsilon=x:\liminf_{\varepsilon\to 0}F_\varepsilon(x_\varepsilon)\geq F_0(x)\tag{16}$$

the sequences $(x_\varepsilon^0)_\varepsilon$ and $(x_\varepsilon)_\varepsilon$ converging to $x$ in the topology $\tau$.

This epi-convergence is a special case of the Γ-convergence introduced by De Giorgi in [2]. It is well-fitted to the asymptotic analysis of sequences of minimization problems since one has the fundamental result

**Theorem A.2** ([1], Theorem 1.10). *Suppose that*

1) $F_\varepsilon$ *admits a minimizer* $\overline{x}_\varepsilon$ *on* $\mathbf{X}$,

2) *The sequence* $(\overline{x}_\varepsilon)_\varepsilon$ *is* $\tau$*-relatively compact,*

3) *The sequence* $(F_\varepsilon)_\varepsilon$ *epi-converges to* $F_0$ *in this topology* $\tau$.

*Then every limit point* $\overline{x}$ *of the sequence* $(\overline{x}_\varepsilon)_\varepsilon$ *minimizes* $F_0$ *on* $X$ *and*

$$\lim_{\varepsilon'\to 0} F_{\varepsilon'}(\overline{x}_{\varepsilon'}) = F_0(\overline{x}),$$

*if* $(\overline{x}_{\varepsilon'})_{\varepsilon'}$ *denotes the subsequence of* $(\overline{x}_\varepsilon)_\varepsilon$ *which converges to* $\overline{x}$.

Notice the following stability result of the epi-convergence

**Proposition A.3** ([1], page 40). *If* $(F_\varepsilon)_\varepsilon$ *epi-converges to* $F$ *in the topology* $\tau$ *and* $G$ *is* $\tau$*-continuous, then* $(F_\varepsilon + G)_\varepsilon$ *epi-converges to* $F + G$ *in this topology* $\tau$.

In the verification of (E1) or (E2), we usually need the following diagonalization result

**Lemma A. 4** ([1], Corollary 1.16). *Let* $(a_{\nu,\mu})_{\nu,\mu}$ *be a double indexed sequence of elements in* $\mathbb{R} \cup \{+\infty\}$. *There exists a function* $\nu \mapsto \mu(\nu)$ *increasing to* $+\infty$, *such that*

$$\limsup_{\nu\to+\infty} a_{\nu,\mu(\nu)} \leq \limsup_{\mu\to+\infty} \limsup_{\nu\to+\infty} a_{\nu,\mu}.$$

**Proposition A.5.** *Suppose that* $(F_\varepsilon)_\varepsilon$ *epi-converges to* $F$, $(G_\varepsilon)_\varepsilon$ *epi-converges to* $G$ *and:* $\forall x \in \mathbf{X}$, $\forall \varepsilon > 0 : F_\varepsilon(x) \leq G_\varepsilon(x)$. *Then:* $F \leq G$ *on* $\mathbf{X}$.

**References**

1. Attouch, H., *Variational convergence for functions and operators.* Pitman (London) 1984.

2. De Giorgi, E., *Convergence problems for functions and operators.* Proceedings of the International Congress on "Recent Methods in Nonlinear Analysis", Rome 1978. De Giorgi, Mosco, Spagnolo Eds. Pitagora Editrice (Bologna) 1979.

3. Ganghoffer, J.F., Brillard, A., Schultz, J., *Modelling of the mechanical behaviour of joints bonded by a nonlinear, incompressible elastic adhesive.* Submitted for publication.

4. Ganghoffer, J.F., Brillard, A., Schultz, J., *A justification of friction laws.* Actes of the 22$^{nd}$ Leeds-Lyon Symposium on Tribology. Lyon Sept. 1995 (to be published).

Alain Brillard, Laboratoire de Mathématiques, Faculté des Sciences et Techniques, 4 Rue des Frères Lumière F-68093, Mulhouse, Cedex France

International Series of Numerical Mathematics, Vol. 123, © 1997 Birkhäuser Verlag Basel

# The opaque square and the opaque circle

Bernd Kawohl

Imagine a square plot of ground $D$ of size 1 and suppose you want to build a fence on it that can block any line of sight across it. One way to build such a fence would look like Figure 1 and have length $\ell = 4$,

Figure 1: $\ell = 4$

Figure 2: $\ell = 3$

but the fence in Figure 2 would serve the same purpose and have length 3 only.

Question:

What is the shortest length fence that can block any line of sight across $D$?

In mathematical terms this problem can be restated as follows:

Find a one-dimensional set $S$ of minimal (1-dim. Hausdorff) measure such that every straight line intersecting $D$ intersects $S$.

This problem is known as the "opaque square problem" [1]. It was brought to my attention by F. Almgren. If the fence is supposed to be simply connected, then the shortest fence is depicted in Figure 3. If the fence may consist of more than one connected component, however, this is not the optimal solution. A better one is depicted in Figure 4.

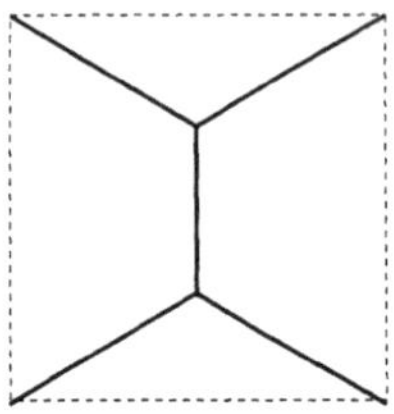

Figure 3: $\ell = 1 + \sqrt{3} \approx 2,732$

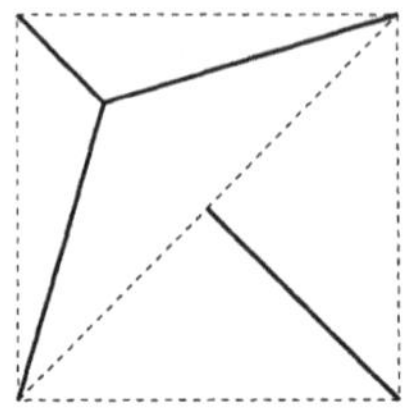

Figure 4: $\ell \approx 2,639$

At present it is unclear if a solution to the opaque square problem exists, but one has conjectured that it looks like the one in Figure 4.

The fence in Figure 4 has three straight segments meeting at angles of $2\pi/3$ and emanating from three corners, plus a segment from the fourth corner to the center. This fence is shorter than the shortest fence connecting all four corners [1, 2]. An intriguing feature of the "solution" in Figure 4 is its lack of symmetry.

There are heuristic reasons why one believes that Figure 4 is optimal.

**Reason 1.** The fence must contain all 4 corners, otherwise one could look across a corner.

**Reason 2.** If the optimal fence consists of a finite number of piecewise smooth connected components, no more than one should emanate from each corner.

**Observation 1.** Let $C_i$, $i = 1, \dots$ be connected components of a fence and suppose they are topologically closed sets. Then each $C_i$ blocks any line of sight from its closed convex hull $\operatorname{conv} C_i$. This observation proves conjecture 2 in [2, p. 263].

**Observation 2.** If $\operatorname{conv} C_i \cap \operatorname{conv} C_j \neq \emptyset$, then any line of sight is blocked from $\operatorname{conv}(\operatorname{conv} C_i \cap \operatorname{conv} C_j)$.

**Proposition 1.** *For any finite number n, there exists a solution to the problem of finding a "shortest" set S with at most n connected components, such that every line that intersects D has to intersect S. This was shown in* [3].

**Proposition 2.** *In particular we know that among all fences with at most two components there exists an optimal one. I claim that it is the one depicted in Figure* 4.

*Proof.* Since the best simply connected fence from Figure 3 is not optimal, we may assume that the optimal fence has exactly two connected components $C_1$ and $C_2$. Then $C_1$ may contain exactly 1 or exactly 2 corners. The second case cannot be optimal, since those two corners must be adjacent, otherwise $C_1$ and $C_2$ would not be disjoint. But if $C_1$ and $C_2$ form an optimal fence, then $\operatorname{conv} C_1 \cap \operatorname{conv} C_2 \neq \emptyset$, else by a separation theorem one could separate them by a line (of sight), see Figure 5. Let $z \in \operatorname{conv} C_1 \cap \operatorname{conv} C_2$ and replace $C_1$ by the two line segments connecting $z$ with the corners of $C_1$, and $C_2$ by the two line segments connecting $z$ to the corners of $C_2$, see Figure 6. One obtains a shorter and simply connected

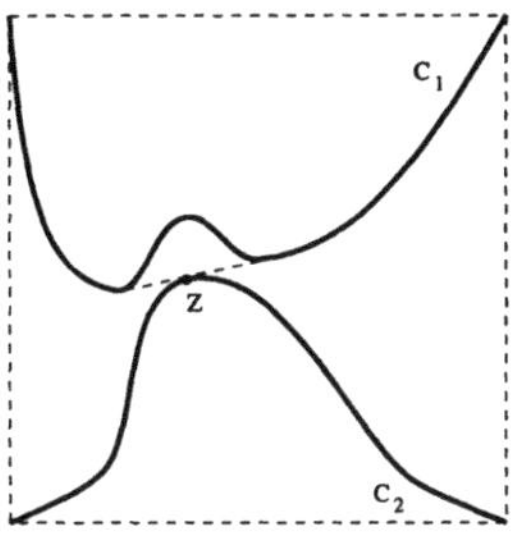


Figure 5

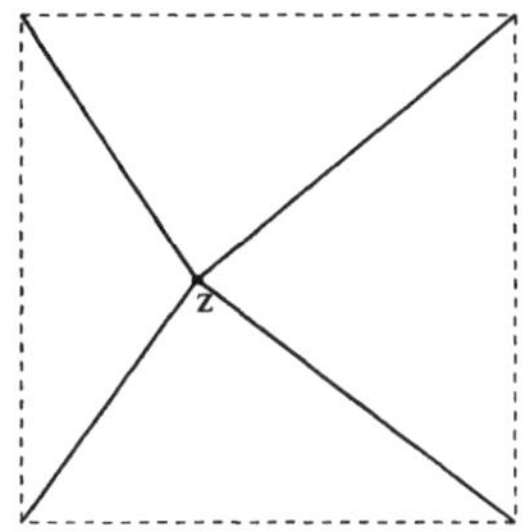


Figure 6

fence, a contradiction. Therefore $C_1$ contains exactly one and $C_2$ exactly three corners. The shortest path connecting three corners is depicted in Figure 4, and the shortest path connecting $\operatorname{conv} C_2$ to the fourth corner is the half diagonal. This argument shows that Figure 4 is optimal in the class of all fences with exactly two components.

**Observation 3.** Suppose the optimal fence has exactly three connected components $C_1$, $C_2$ and $C_3$, where $C_3$ contains two corners and both $C_1$ and $C_2$ contain exactly one of the remaining corners. Two cases can be distinguished.

In the case that $C_3$ contains two adjacent corners we may assume without loss of generality that $C_3$ and the line segment between the corners of $C_3$ bound a convex set. Now let $z_1$ and $z_2$ be points where $\operatorname{conv} C_1$ and $\operatorname{conv} C_2$ intersect $\operatorname{conv} C_3$, see Figure 7. Replacing $C_1$ and $C_2$ by straight line segments which connect $z_1$ and $z_2$ with the corners of $C_1$ and $C_2$, and replacing $C_3$ by a trapezoidal arc running through $z_1, z_2$ and the corners of $C_3$ (as in Figure 8) will yield a simply connected shorter fence, a contradiction.

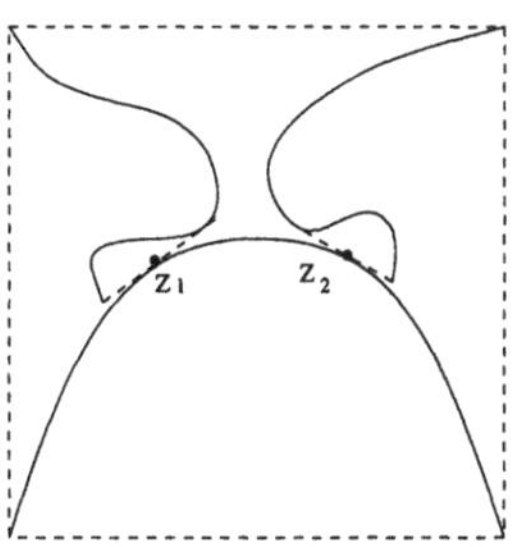


Figure 7

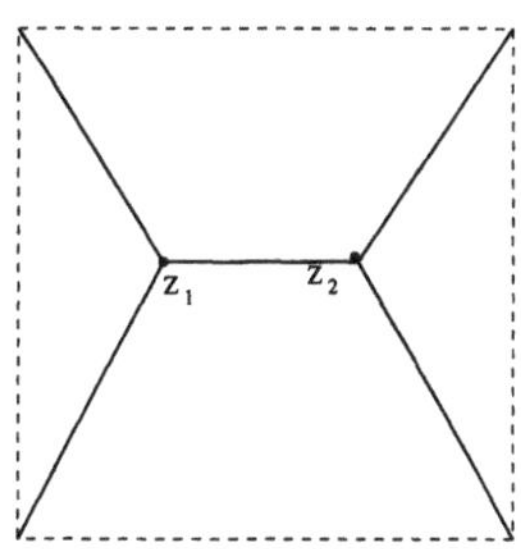


Figure 8

In the case that $C_3$ contains two diagonal corners, one can assume without loss of generality that $C_1$ and $C_2$ are two arcs ending in points $z_i \in \operatorname{conv} C_3 \backslash C_3$, $i = 1, 2$, see Figure 9.

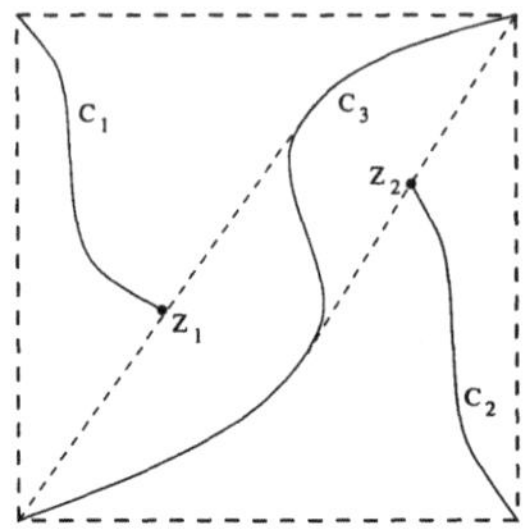


Figure 9

At present I cannot see how to rule out the situation in Figure 9 by reducing it to a shorter fence with fewer components.

When I presented this problem in Oberwolfach during the conference General Inequalities VII in November 1995, M. Plum raised the following question:

What if the unit square is replaced by a unit disc $D$? Then, since the fence that blocks any line of sight across $D$ must contain all its extremal points, the shortest such fence is the circle $\partial D$ of length $2\pi$? The analogous reasoning could be made about any strictly convex plane domain and is correct if the fence stays inside (the closure of) $D$.

As we shall see below, the answer depends heavily on whether the disc $D$ is open or closed, and on what we consider to be an admissible fence. In the meantime I found this question treated in the literature [3, 4] with different applications in mind. Suppose a telephone line is running straight through the ground, within a given radius, say 1, of a point that is marked by some point. One can locate the line by digging a circular trench of length $2\pi$ around this point. A shorter simply connected trench of length $\pi + 2$, running partly outside $D$ is known.

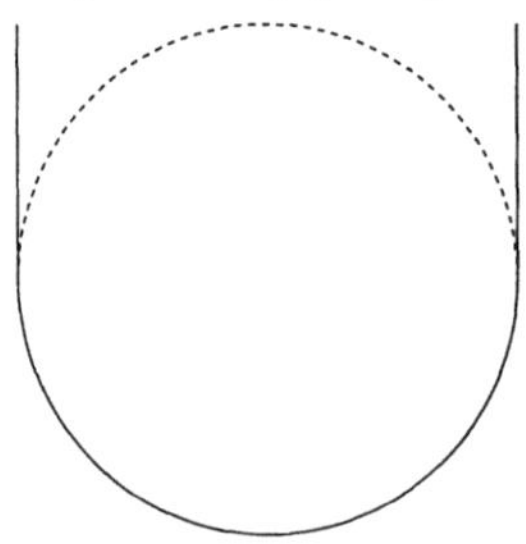

Figure 10: $\ell = \pi + 2$

In fact, one can show that the arc depicted in Figure 10 is the shortest simply connected arc that will serve this purpose, but multiply connected shorter ones are known (see [2, p. 264] or [4, p. 113] for doubly connected arcs with $\ell \approx 4,818$, and [3, p. 799] or [4, p. 113] for triply connected arcs with $\ell \approx 4,799$).

I now claim, that by chopping the arc up into small pieces and by moving them inside the disc, I can obtain the same effect.

**Proposition 3.** *The shortest trench inside the closed disc $D$ that intersects any transversal (i.e., nontangent) line through $D$ has length not exceeding* $(2+\pi)$.

For a proof let us study regular polygons. For the equilateral triangle the shortest trench that will uncover the telephone line is depicted in Figure 11,

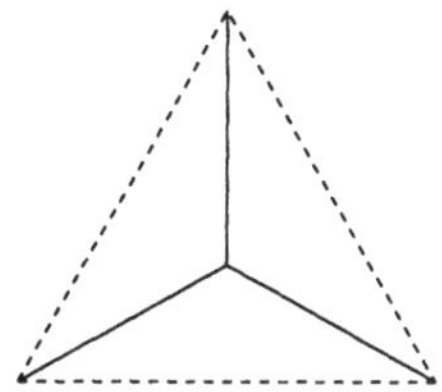

Figure 11: $\ell = \sqrt{3} \approx 1.732$

for the square the conjectured optimal solution is given in Figure 4, for the pentagon and hexagon one can find it in [2], see Figures 12 and 13.

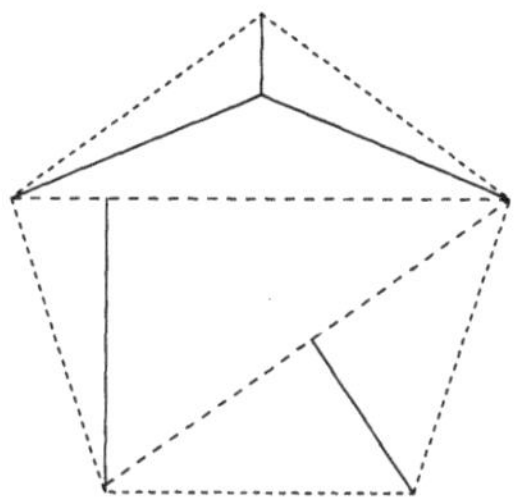

Figure 12: $\ell \approx 3.528$

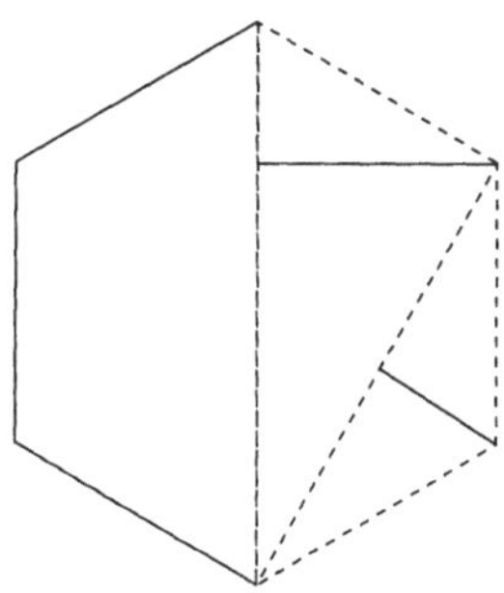

Figure 13: $\ell = (7+\sqrt{3})/2 \approx 4,366$

Extrapolating from the pentagon and hexagon one can now conjecture that for a regular $2N$-gon with $N \geq 3$ the optimal solution can be described as follows:

Set $C_0$ = half of the boundary of the polygon with $P_0$ and $P_N$ as endpoints, and suppose that the remaining $(N-1)$ corners are numbered $P_1, P_2, \ldots, P_{N-1}$ in consecutive order. For $i = 1, \ldots, N-1$ let $C_i$ be the straight line segment that connects $P_i$ to the closest point $Q_i$ on conv $\left(\bigcup_{j=1}^{i} C_j\right)$, see Figure 14.

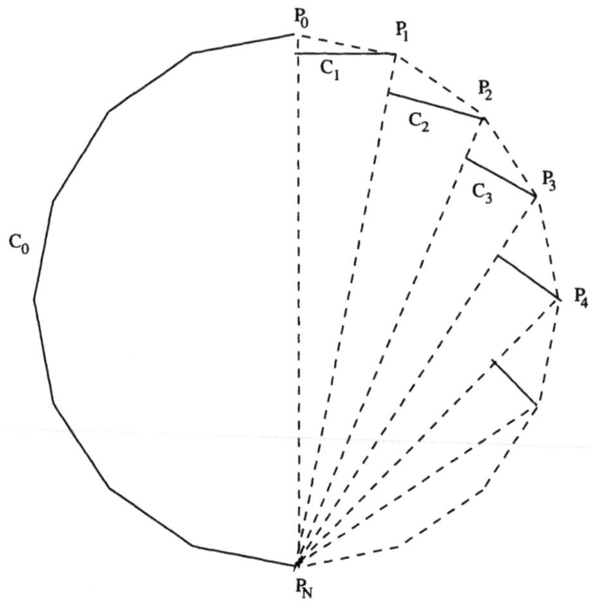


Figure 14

If the corners of the polygon lie on the unit circle, then as $N \to \infty$, this will provide a "trench" which uncovers the telephone line running through the disc $D$. So let us calculate its length $\ell_N$ by doing some trigonometry. According to [2, p. 265] this "looks extremely, hard to handle for $N$-gons with $N \geq 5$", but we shall see from analyzing Figure 15 that it is rather easy.

Let us denote the left endpoint of $C_{i+1}$ by $Q_{i+1}$. The angle $\beta$ between $P_iP_N$ and $P_{i+1}P_N$ is $\beta = \pi/2N$, since the angle between $P_0P_N$ and $P_iP_N$ is $i\pi/2N$. The length of the segment $\overline{P_iP_N}$ can be calculated using Thales' Theorem as $\overline{P_iP_N} = 2\cos(i\pi/2N)$, so $\sin\beta = \overline{Q_{i+1}P_{i+1}} \,/\, \overline{P_{i+1}P_N}$ or length of $C_{i+1} = \sin(\pi/2N) \cdot 2\cos((i+1)\pi/2N)$ for $i = 1, \ldots, N-1$.

To get the length of all the arcs we have to add them up

$$\ell_N = \text{ length of } C_0 + \sum_{i=1}^{N-1} \text{ length of } C_i \ .$$

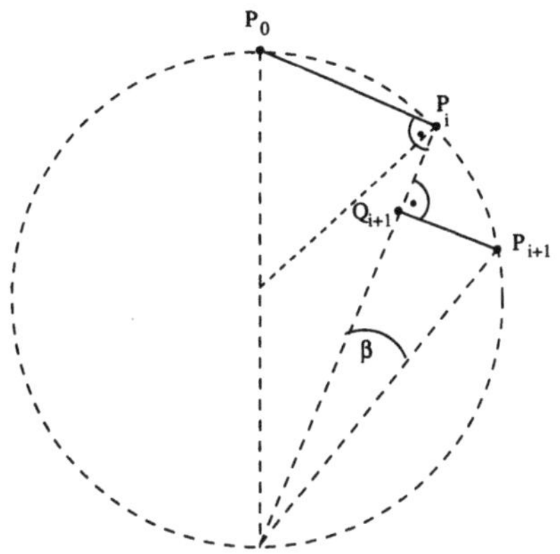


Figure 15: Computation of $\ell_N$

Now length of $C_0 = N \cdot \overline{P_0P_1} = N \cdot 2\sin(\pi/2N)$, and so

$$\ell_N = 2N \ \sin\left(\frac{\pi}{2N}\right) + \sum_{j=1}^{N-1} \sin\left(\frac{\pi}{2N}\right) \cdot 2\cos\left(\frac{j\pi}{2N}\right)$$

or, observing that $\cos\pi/2 = 0$,

$$l_N = 2N \ \sin\left(\frac{\pi}{2N}\right)\left(1 + \frac{1}{N}\sum_{j=1}^{N}\cos\left(\frac{j\pi}{2N}\right)\right) \quad .$$

Now $\frac{1}{N}\sum_{j=1}^{N}\cos\left(\frac{j}{N}\cdot\frac{\pi}{2}\right)$ is the Riemann sum for

$$\frac{2}{\pi}\int_0^{\frac{\pi}{2}} \cos x \ dx = \frac{2}{\pi} \quad ,$$

and for large $N$ we can approximate $\sin\pi/2N$ by $\pi/2N$, so that

$$\lim_{N\to\infty} \ell_N = \pi\left(1 + \frac{2}{\pi}\right) = \pi + 2 \quad ,$$

as claimed!

**Observation 4.** A slightly different version of Proposition 3 is due to Richard Laver and can be found in [3, p. 800].

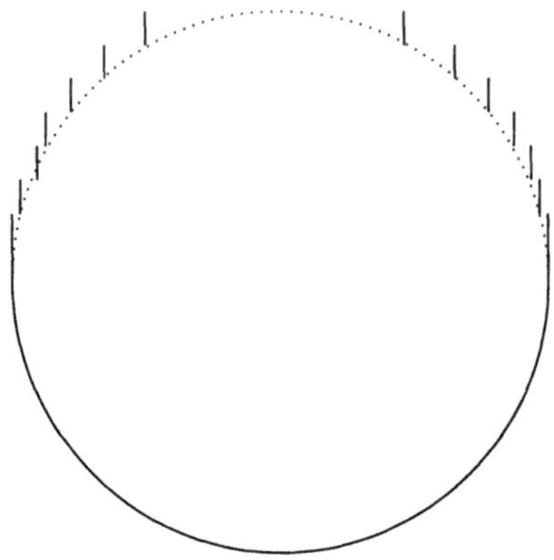

Figure 16: R. Laver's construction

For every $\varepsilon > 0$ there exists a set $S$ such that every straight line intersecting $D$ intersects $S$; and such that $S$ stays (partly outside of $D$, but) inside a strip of width $\varepsilon$ around $\partial D$.

**Acknowledgement.** It is my pleasure to thank F. Almgren for bringing the problem to my attention during a visit to Princeton in October 1995. This note was written up during a very pleasant stay at the Center for Dynamical Systems and Nonlinear Studies in Atlanta during February 1996. Finally I thank M. Mester for his help in drawing the figures.

## References

1. K.A. Brakke, *The Opaque Cube Problem.* Amer. Math. Monthly **99** (1992), p. 866–871.
2. V. Faber, J. Mycielski and P. Pedersen, *On the shortest curve which meets all the lines which meet a circle.* Ann. Polon. Math. **44** (1984), p. 249–266.
3. V. Faber and J. Mycielski, *The shortest curve that meets all the lines that meet a convex body.* Amer. Math. Monthly **93** (1986), p. 796–801.
4. I. Stewart, *The great drain robbery.* Scientific American (Sept. 1995), pp. 166–167 and (Dec. 1995), p. 87 and (Feb. 1996), p. 113.

Bernd Kawohl, Mathematisches Institut, Universität Köln, D-50923 Köln, Germany.

E-mail address: kawohl@mi.uni-koeln.de

International Series of Numerical Mathematics, Vol. 123, © 1997 Birkhäuser Verlag Basel

# Enclosure methods with existence proof for elliptic differential equations

Michael Plum

**Abstract**

For nonlinear second-order elliptic boundary value problems, a numerical method is proposed which may be used to obtain rigorous enclosures for solutions, and which proves the existence of a solution as an automatic by-product.

## 1. Introduction

We consider nonlinear elliptic boundary value problems of the form

$$-\Delta u + F(x, u, \nabla u) = 0 \text{ on } \Omega, \qquad B[u] = 0 \text{ on } \partial\Omega. \tag{1.1}$$

Here, $\Omega \subset \mathbb{R}^n$ is a bounded domain with Lipschitz continuous boundary $\partial\Omega$. The nonlinearity $F$ is defined on $\bar{\Omega} \times \mathbb{R} \times \mathbb{R}^n$, with values $F(x, y, z) \in \mathbb{R}$. $F$ and its derivatives $F_y := \partial F/\partial y$ and $F_z := (\partial F/\partial z_1, \ldots, \partial F/\partial z_n)^t$ are assumed to be continuous. In the case $n \geq 2$ we further assume that $F$ satisfies a quadratic growth condition (at $\infty$) with respect to $z$, i.e., that for any given $\alpha \geq 0$, some $C = C(\alpha) \geq 0$ exists such that

$$|F(x, y, z)| \leq C(1 + |z|^2) \text{ for } x \in \bar{\Omega},\ y \in \mathbb{R},\ |y| \leq \alpha,\ z \in \mathbb{R}^n. \tag{1.2}$$

Here, $|\cdot|$ denotes the modulus in $\mathbb{R}$ as well as the Euclidean norm in $\mathbb{R}^n$.

The boundary operator $B$ in (1.1) is of mixed type, i.e., there is some closed subset $\Gamma_0 \subset \partial\Omega$ such that $B[u](x) = u(x)$ on $\Gamma_0$ and $B[u](x) = (\partial u/\partial\nu)(x) := (\nabla u(x)) \cdot \nu(x)$ on $\Gamma_1 := \partial\Omega \setminus \Gamma_0$, where $\nu : \partial\Omega \to \mathbb{R}^n$ denotes the outer unit normal on $\partial\Omega$ and "$\cdot$" is the canonical inner product in $\mathbb{R}^n$. We assume that $\Gamma_0$ and $\Gamma_1$ are piecewise $C_2$-hypersurfaces. (This expression will be explained in more detail before Lemma 5.3.)

Finally, we want to make the assumption that the triple $(\Omega, \Gamma_0, \Gamma_1)$ is *regular* in the following sense: Let $H_2^B$ denote the closure of the set $\{u \in C_2(\bar{\Omega}) : B[u] = 0$ on $\partial\Omega\}$ in the Sobolev space $H_2(\Omega)$ consisting of all twice weakly differentiable functions with square-integrable derivatives up to second order. Thus, $H_2^B$ consists of functions in $H_2(\Omega)$ satisfying the boundary conditions in (1.1) in a suitably generalized sense. Our regularity condition for $(\Omega, \Gamma_0, \Gamma_1)$ now requires that some $\sigma \in \mathbb{R}$ exists such that the boundary value problem

$$u \in H_2^B, \quad -\Delta u + \sigma u = r \text{ on } \Omega$$

is uniquely solvable for all $r$ in the space $L_2(\Omega)$ consisting of all square-integrable measurable functions on $\Omega$.

Without aiming here at a complete description of all regular triples $(\Omega, \Gamma_0, \Gamma_1)$ we want to mention that, for instance, $(\Omega, \partial\Omega, \emptyset)$ and $(\Omega, \emptyset, \partial\Omega)$ are regular for *globally* $C_2$-smooth boundaries $\partial\Omega$, and that $(\Omega, \Gamma_0, \Gamma_1)$ is regular for *polygonal* domains $\Omega \subset \mathbb{R}^2$ if each edge of $\partial\Omega$ belongs completely either to $\Gamma_0$ or to $\overline{\Gamma}_1$, if the inner angles do not exceed $\pi$, and if in those particular corners where $\Gamma_0$ and $\overline{\Gamma}_1$ meet, the inner angles do even not exceed $\pi/2$. Further examples can be found in [24, Section 8]. It should be mentioned that reentrant corners are forbidden by the regularity condition.

In this article, we will propose a numerical method for computing an *enclosure* for a solution $u$ of problem (1.1), and proving its *existence* as an automatic by-product. The resulting statements are rigorous in a strict mathematical sense, and all kinds of discretization-, truncation-, and rounding errors generated in the numerical process are taken into account.

Rounding errors can nowadays be well controlled by *interval arithmetic*, which was advanced and realized in efficient software packages in particular by U. Kulisch and his co-workers (see [1, 3, 15] for an overview, and [6, 10, 11, 12, 13] as examples for concrete implementations). To pay regard also to discretization- and (if iterative procedures are used) to truncation errors, it has turned out not to be useful (or even impossible) to follow all single errors generated by the discretization and during runtime and to read off the overall error at the end. Instead, the following functional analytical approach, involving these errors in an indirect and much more structured way (*defect* of an approximate solution, *norm bounds* for some linear operator, etc.; see Section 3.), has proved to be successful.

In our terminology, an *enclosure* for $u$ will be a statement of the form

$$u \in U \tag{1.3}$$

with $U$ denoting a subset of a suitable Banach space $X$, which is described by a finite set of explicitly given numbers (so that $U$ is realizable on a computer).

A reasonable and frequently used general approach for obtaining such enclosure results consists in the transformation of the given problem into some equivalent *fixed point equation*

$$u \in X, \quad u = Tu \tag{1.4}$$

and the application of some fixed point theorem. Under suitable conditions on the space $X$ and the operator $T : X \to X$ (e.g., continuity, compactness, or contraction properties), which will usually be checked by theoretical means, the fixed point theorem provides the existence of a solution $u$ of (1.4) (and thus, of (1.1)) with the additional property (1.3), if the subset $U$ of $X$ satisfies, in addition to certain basic conditions (such as boundedness, closedness, convexity), the inclusion condition

$$TU \subset U. \tag{1.5}$$

As mentioned earlier, the set $U$ shall moreover be at hand in *explicit* form. Finally, the enclosure (1.3) should be "precise", i.e., $U$ should have a "small diameter" in an appropriate sense. The determination of a set $U$ having all these properties will usually need a lot of *numerical* work. The rigorous verification of condition (1.5) requires in addition the use of *interval analysis* on many levels between basic arithmetic and functional analysis.

A more detailed description of the existence and enclosure method requires now first the concrete specification of the space $X$, of the operator $T$, and possibly of the "class" of the sets $U$ which shall be taken into consideration for the verification of (1.5) (e.g., (norm-)balls, function intervals, direct sums of suitable lower dimensional sets, and the like). Here, different choices (providing different existence and enclosure methods) may be considered.

The oldest and most frequently used method which is applicable, under conditions to be discussed later, to boundary value problems of the form (1.1), is the method of *two-sided bounds* or *upper and lower solutions* which is closely related to the *maximum principle.* The method uses *function intervals* $U$, and the operator $T$ is constructed to possess certain *monotonicity* properties, which facilitates the inclusion condition (1.5) a lot. From the author's point of view, the most essential research in the field of such "monotonicity methods" is due to L. Collatz (see, e.g., [7]), to J. Schröder (see, e.g., [25, 26]) and to W. Walter (see, e.g., [27]). In the next section, we will describe some more details of this approach. Moreover, we want to explain there why the applicability of monotonicity methods to problem (1.1) is considerably restricted by certain assumptions which are indirectly required by condition (1.5).

In the subsequent sections, we will propose an existence and enclosure method developed more recently, which does not need the restrictive additional conditions of the monotonicity methods. The *basis* of this method is closely related to the Newton-Cantorovich Theorem. Essentially, it requires (for the definition of the operator $T$ and for the verification of condition (1.5)) the computation of an *approximate solution* $\omega \in H_2^B$ of problem (1.1), of an $L_2$-bound for its *defect* $d[\omega] := -\Delta\omega + F(\cdot, \omega, \nabla\omega)$ and of suitable *norm bounds* for the inverse of the differential operator $L$ generated by linearization of the left-hand side of (1.1) at the approximation $\omega$. We compute such bounds by means of enclosures for *eigenvalues* of $L$ (or of $L^*L$), in combination with analytical arguments providing, for instance, an *explicit* version of some Sobolev Embedding Theorem (i.e., explicit embedding constants).

Other existence and enclosure methods which are applicable to the problem class (1.1) and do not need the inherent restrictive assumptions of the monotonicity methods, have been proposed by M.T. Nakao [17] and by S. Oishi [18]. Their methods avoid the direct computation of bounds for $L^{-1}$ and use only the inverses of finite-dimensional *projections* of $L$. The (infinite-dimensional) "remainder terms" have to be bounded by other means, which requires additional assumptions resp. coarsens the computed bounds. In [24], Nakao's und Oishi's methods are discussed in more detail.

For boundary value problems with *ordinary* differential equations (which are of course contained in (1.1), as long as they are of second order), many existence and enclosure methods may be found in the literature. We will not address these approaches in the present article.

## 2. A "monotonicity"-approach

In this section, we will describe a particular construction of the fixed point operator $T$ in (1.4) and the enclosing set $U$ which is typical for the "monotonicity methods" mentioned in the previous section. For simplicity we assume here that the nonlinearity $F$ in (1.1) depends only *linearly* on $\nabla u$, i.e., that

$$F(x,y,z) = b(x) \cdot z + \tilde{F}(x,y) \tag{2.1}$$

for some continuous $b : \bar{\Omega} \to \mathbb{R}^n$ and some continuous $\tilde{F} : \bar{\Omega} \times \mathbb{R} \to \mathbb{R}$ which is continuously differentiable with respect to $y$. Moreover, we assume that $n \leq 3$ so that the Sobolev-Kondrachov Embedding Theorem [2, Theorem 6.2] provides the compactness of the embedding $E : H_2(\Omega) \hookrightarrow C(\bar{\Omega})$.

On $X := C(\bar{\Omega})$ (endowed with the maximum norm $\|\cdot\|_\infty$), we define the canonical order relation

$$u \leq v \quad :\Leftrightarrow \quad u(x) \leq v(x) \ \text{ for all } \ x \in \bar{\Omega}$$

and choose, as "candidates" for the enclosing set $U$ (see (1.3)), *function intervals*

$$U := [v,w] := \{u \in C(\bar{\Omega}) : v \leq u \leq w\}, \tag{2.2}$$

where the functions $v, w \in C(\bar{\Omega}), v \leq w$, are to be determined. The fixed point operator $T$ is constructed to be *monotone* on the set $U$ (with respect to the order relation defined above). For this purpose, we choose some constant $\tilde{c} > 0$ such that

$$\tilde{c} \geq \max\left\{\tilde{F}_y(x,y) : x \in \bar{\Omega},\ v(x) \leq y \leq w(x)\right\} \tag{2.3}$$

and define

$$\tilde{L}[u] := -\Delta u + b \cdot \nabla u + \tilde{c}u \qquad (u \in H_2^B). \tag{2.4}$$

Due to the positivity of $\tilde{c}$, the operator $\tilde{L}$ is *inverse-positive* on $H_2^B$, i.e., for all $u \in H_2^B$ satisfying $\tilde{L}[u](x) \geq 0$ for almost all $x \in \Omega$, the inequality $u(x) \geq 0$ holds for almost all $x \in \Omega$. (A proof of the inverse-positivity can be found, for instance, in [24, Corollary 3.2].) In particular, $\tilde{L} : H_2^B \to L_2(\Omega)$ is one-to-one. Furthermore, our general regularity condition on $(\Omega, \Gamma_0, \Gamma_1)$ implies the existence and the boundedness of $\tilde{L}^{-1} : L_2(\Omega) \to H_2^B$ (see [24, Theorem A.2]). Finally, the embedding $\hat{E} : C(\bar{\Omega}) \hookrightarrow L_2(\Omega)$ is bounded, and $E : H_2(\Omega) \hookrightarrow C(\bar{\Omega})$ is compact, as mentioned earlier. Altogether we obtain that $E\tilde{L}^{-1}\hat{E} : C(\bar{\Omega}) \to C(\bar{\Omega})$ is *compact* and *monotone*. Using (2.3) we obtain that

$$T : C(\bar{\Omega}) \to C(\bar{\Omega}), \quad Tu := E\tilde{L}^{-1}\hat{E}[\tilde{c}u - \tilde{F}(\cdot, u)], \tag{2.5}$$

is continuous and compact, and moreover, monotone on the function interval $U$ defined in (2.2). The equivalence of the fixed point equation (1.4) and the given problem (1.1) (with $F$ from (2.1)) is quite obvious. Schauder's Fixed Point Theorem therefore yields the existence of a solution $u^* \in H_2^B$ of problem (1.1) satisfying

$$v \leq u^* \leq w, \tag{2.6}$$

provided that the crucial condition (1.5) is satisfied. But since $T$ is monotone on $U$, the enclosure $Tv \leq Tu \leq Tw$ holds for all $u \in U = [v, w]$, so that (1.5) is equivalent to the following two inequalities:

$$v \leq Tv, \quad Tw \leq w. \tag{2.7}$$

If $v$ and $w$ are chosen in $H_2^B$, the inverse-positivity of $\tilde{L}$ on $H_2^B$ shows that the inequalities $\tilde{L}[v] \leq \tilde{L}[Tv]$, $\tilde{L}[Tw] \leq \tilde{L}[w]$ (to be understood pointwise almost everywhere in $\Omega$) are *sufficient* for (2.7). Due to (2.4), (2.5), (2.1), these *differential inequalities* read

$$-\Delta v + F(\cdot, v, \nabla v) \leq 0 \leq -\Delta w + F(\cdot, w, \nabla w) \ \ \text{(almost everywhere in } \Omega). \tag{2.8}$$

Conditions of the form (2.7) or (2.8) (and their consequence (2.6)) play a central role in the extensive research work done, in particular, by L. Collatz (see, e.g., [7]), by J. Schröder (see, e.g., [25, 26]), and by W. Walter (see, e.g., [27]) about "monotonicity methods" and their applications to differential equations. For many concrete cases, these and other authors could find solutions to the inequalities (2.7) resp. (2.8) (under the side condition $v \leq w$) using a suitable *ansatz* for $v$ and $w$. In this way, enclosure results were obtained for many problems with ordinary and partial differential equations already decades ago! Rounding errors were avoided by rational arithmetic, or they were neglected. Of course, this violated the rigorosity of the results, but the influence of rounding errors in the numerical treatment of boundary value problems may generally be regarded to be very small.

In case one is interested in a very *precise* enclosure, i.e., in upper and lower solutions $v$ und $w$ which are "closely together", the following approach, which was studied in all consequence first by J. Schröder, suggests itself. One chooses

$$v = \omega - z, \qquad w = \omega + z \quad , \tag{2.9}$$

with $\omega \in H_2^B$ denoting an *approximate* solution of problem (1.1), and with some "small" function $z \in H_2^B$, $z \geq 0$, to be determined. Inserting (2.9) into (2.8), we obtain the following two differential inequalities for $z$:

$$\begin{aligned} -\Delta z + F(\cdot,\omega+z,\nabla\omega+\nabla z) - F(\cdot,\omega,\nabla\omega) &\geq -d[\omega], \\ -\Delta z + F(\cdot,\omega,\nabla\omega) - F(\cdot,\omega-z,\nabla\omega-\nabla z) &\geq d[\omega] \end{aligned} \quad \text{(almost everywhere in } \Omega). \tag{2.10}$$

Here, $d[\omega] := -\Delta\omega + F(\cdot,\omega,\nabla\omega)$ denotes the *defect* of $\omega$. Using now (2.1) again and defining

$$\begin{aligned} c(x) := \tilde{F}_y(x,\omega(x)), \quad \tilde{f}(x,y) &:= \tilde{F}(x,\omega(x)+y) - \tilde{F}(x,\omega(x)) - c(x)y, \\ L[u] &:= -\Delta u + b\cdot\nabla u + cu, \end{aligned} \tag{2.11}$$

we find that (2.10) reads:

$$L[z] \geq \max\left\{-d[\omega] - \tilde{f}(\cdot,z), d[\omega] + \tilde{f}(\cdot,-z)\right\} \quad \text{(almost everywhere in } \Omega). \tag{2.12}$$

Since $\tilde{f}(x,y) = o(y)$ (for $y \to 0$), each "small" solution $z$ of (2.12) will, except in very special situations, necessarily have the property that

$$\operatorname*{ess\,inf}_{\Omega} L[z] > 0. \tag{2.13}$$

The existence of some function $z \in H_2^B$ with the properties $z \geq 0$ and (2.13), however, implies the *inverse-positivity* of $L$ on $H_2^B$ (see [24, Theorem 3.2] for a proof). Thus, the following statement follows from [24, Theorem 3.4]:

$$\textit{All real eigenvalues of } L \textit{ on } H_2^B \textit{ are positive.} \tag{2.14}$$

Using [14, Paragraph 6] one can show that (2.14) is equivalent to the statement that *all* eigenvalues of $L$ on $H_2^B$ have a *positive real part.* This condition on the *linearization* $L$ of the left-hand side of problem (1.1) in $\omega$, which is, due to the above considerations, an "almost" necessary consequence of the existence of "closely neighboring" upper and lower solutions, restricts of course the applicability of the "monotonicity methods" rather severely. To illustrate this fact, consider the following example taking its rise from semiconductor physics:

$$-u'' - (u\cdot sin(x))' + \lambda(u^2+u-1) = 0 \quad (0 < x < 2\pi), \quad u'(0) = u'(2\pi) = 0. \tag{2.15}$$

Applying a Newton-collocation procedure we computed the following bifurcation diagram of approximate solutions $\omega$. An additional branch is formed by the vertical line at $\lambda = 0$, where $u(x) := \mu \cdot \exp(cos(x))$ is a solution for each $\mu \in \mathbb{R}$.

On all branches (which were drawn after interpolation of a "grid" of computed approximate solutions $\omega$), one finds the respective numbers of *negative eigenvalues* of the linear operator $L$ defined in (2.11). Only two branches show a zero. Thus, only on these two branches all (real) eigenvalues of $L$ are positive, as required by the "almost" necessary condition (2.14) of the "monotonicity"-approach.

However, the new existence and enclosure method which we are going to propose in the following sections, was successful on *all* branches, except in direct neighborhoods of singular points (turning- and bifurcation points) where the inverse of $L$ does not exist or has at least a large norm.

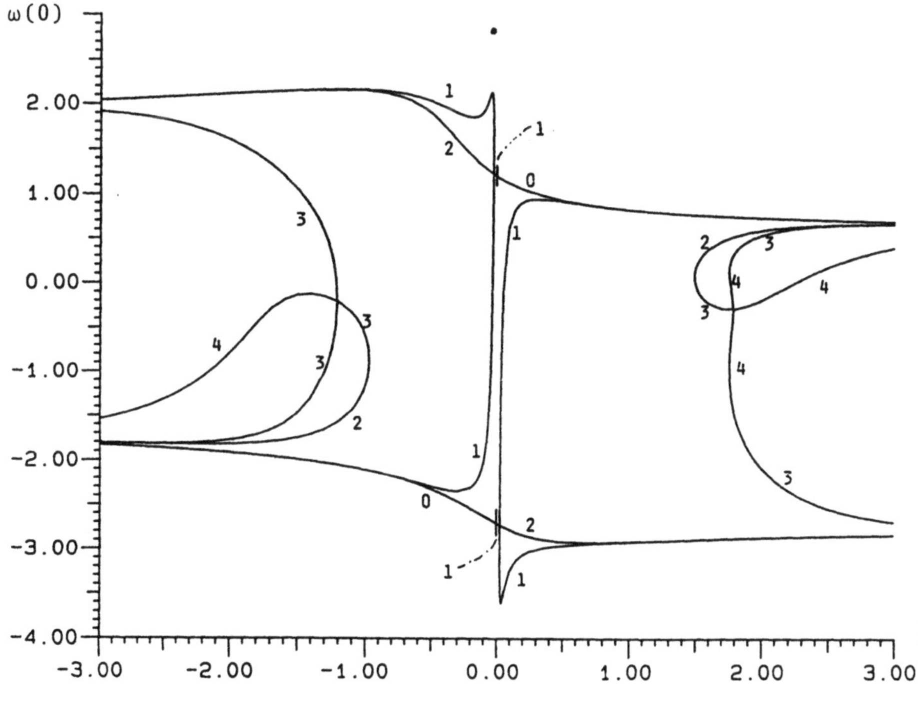


Figure 1: Bifurcation diagram for problem (2.15)

## 3. Enclosures by norm bounds

An important reason for the efficiency of the "monotonicity methods", which was demonstrated in numerous examples, is certainly the fact that the difficult inclusion condition (1.5) can be replaced by two inequalities (2.7) resp. (2.8), which are much better suited for constructive approaches, in particular because (2.8) does no longer contain any *inverse* of a differential operator. However, the applicability of the "monotonicity methods" is severely restricted by the "almost" necessary condition (2.14), so that a demand for new approaches arises. Due to the lack of the possibility to work with differential inequalities if (2.14) is not satisfied, we propose the following way of proceeding.

Starting with an *approximate* solution $\omega \in H_2^B$ of problem (1.1) (which usually will have to be computed by numerical means), we define (compare (2.11))

$$\begin{aligned} c(x) &:= F_y(x,\omega(x),\nabla\omega(x)), \quad b(x) := F_z(x,\omega(x),\nabla\omega(x)) \quad , \\ L[u] &:= -\Delta u + b\cdot\nabla u + cu \quad . \end{aligned} \tag{3.1}$$

We make the assumption that $L$ is *invertible* (one-to-one) on $H_2^B$, which is obviously a significantly weaker condition than (2.14). In Section 5.1 we will show how this condition can be checked numerically.

Moreover, we assume that $\omega$ and $\nabla\omega$ are essentially *bounded* on $\Omega$; this condition is satisfied for all "usual" approximations (such as conformal finite element approximations). Consequently, the coefficients $b$ and $c$ of $L$ are essentially bounded, too. The invertibility of $L$ therefore implies that is is *onto* (as a mapping from $H_2^B$ to $L_2(\Omega)$); see [24, Theoem A.2] for a proof. Thus, the inverse operator $L^{-1} : L_2(\Omega) \to H_2^B$ is bounded due to the Open Mapping Theorem.

Now, we define the fixed point operator $T : X \to X$ according to

$$Tu := L^{-1}\left[b\cdot\nabla u + cu - F(\cdot,u,\nabla u)\right], \tag{3.2}$$

with a Banach space $X$ which we require to have the following properties:

(X1) $\Phi : \left\{ \begin{array}{l} X \to L_2(\Omega) \\ u \mapsto b\cdot\nabla u + cu - F(\cdot,u,\nabla u) \end{array} \right\}$ is bounded on bounded sets and continuous,

(X2) the embedding $H_2^B \hookrightarrow X$ exists and is compact.

Therefore, $T : X \to X$ is continuous and compact. Thus, if some bounded, closed, convex subset $U$ of $X$ satisfying (1.5) can be found, Schauder's Fixed Point Theorem provides the existence of a fixed point $u^* \in U$ of $T$. Due to (3.2), $u^* \in H_2^B$ is then a solution of problem (1.1).

For the set $U$, we try the "midpoint form"

$$U = \omega + V := \{\omega + v : v \in V\} \tag{3.3}$$

with some "small" bounded, closed, convex set $V \subset X$. Due to (3.2), (3.3), condition (1.5) now reads

$$L^{-1}\left[-d[\omega] + \varphi(V)\right] \subset V \tag{3.4}$$

where

$$d[\omega] := -\Delta\omega + F(\cdot, \omega, \nabla\omega) \ \ (\textit{defect} \text{ of } \ \omega), \tag{3.5}$$

$$\varphi(v) := -\left[F(\cdot, \omega + v, \nabla\omega + \nabla v) - F(\cdot, \omega, \nabla\omega) - b \cdot \nabla v - cv\right]. \tag{3.6}$$

Since differential inequalities which are sufficient for (3.4) are not available without the restrictive condition (2.14), and since the inverse operator $L^{-1}$ can usually neither be computed in closed form nor be enclosed precisely, we propose *norm bounds* for the terms $L^{-1}$, $d[\omega]$, and $\varphi(V)$ in (3.4). Suppose that constants $K > 0$, $\delta \geq 0$ and some monotonically nondecreasing function $G : [0, \infty) \to [0, \infty)$ are known such that

$$\|L^{-1}[r]\|_X \leq K\|r\|_2 \ \text{ for all } \ r \in L_2(\Omega) \tag{3.7}$$

$$\|d[\omega]\|_2 \leq \delta, \tag{3.8}$$

$$\|\varphi(v)\|_2 \leq G(\|v\|_X) \ \text{ for all } \ v \in X, \tag{3.9}$$

where here and in the following, $\|\cdot\|_X$ denotes the norm in $X$ and $\|u\|_p := \left[\text{meas}\,(\Omega)^{-1} \int_\Omega |u|^p\,dx\right]^{\frac{1}{p}}$. In the subsequent sections, we will describe methods for computing the terms $K$, $\delta$, and $G$ in detail. In Section 4, we will show in particular that for the function $G$, the additional condition

$$G(t) = o(t) \ \text{ for } \ t \to 0 \tag{3.10}$$

may be required.

For the verification of the inclusion (3.4), we now commit ourselves to *norm balls* $V$: With $\alpha \geq 0$ to be chosen later, let

$$V := \{v \in X : \|v\|_X \leq \alpha\} \quad . \tag{3.11}$$

For all $v \in V$, (3.7), (3.8), (3.9) then provide

$$\begin{aligned} \|L^{-1}[-d[\omega] + \varphi(v)]\|_X &\leq K\| - d[\omega] + \varphi(v)\|_2 \leq K[\delta + \|\varphi(v)\|_2] \\ &\leq K[\delta + G(\|v\|_X)] \leq K[\delta + G(\alpha)], \end{aligned}$$

so that (3.4) is satisfied if $K[\delta + G(\alpha)] \leq \alpha$. We summarize the result in the following central Theorem:

**Theorem 3.1.** *Suppose that, for some constant* $\alpha \geq 0$,

$$\delta \leq \frac{\alpha}{K} - G(\alpha). \tag{3.12}$$

*Then, there exists a solution* $u^* \in H_2^B$ *of problem* (1.1) *satisfying*

$$\|u^* - \omega\|_X \leq \alpha. \tag{3.13}$$

((3.13) is equivalent to the statement $u^* \in U$, due to (3.3), (3.11).) (3.10) shows that the right-hand side of (3.12) is *positive* for "small" $\alpha \geq 0$. Consequently, (3.12) is satisfied for some "small" $\alpha$ (and (3.13) therefore yields a *sharp* enclosure), if the constant $\delta \geq 0$ is sufficiently small. Due to (3.5), (3.8), this requires that the approximate solution $\omega$ is computed with *sufficient accuracy*. The difficulties linked to the crucial requirement (3.12) are therefore largely transferred to the *approximate* computation, for which a large variety of well-established numerical methods is available.

Before we discuss the calculation of the terms required in (3.7), (3.8), (3.9), we first want to put the Banach space $X$ in concrete form. We assume again that

$$n \leq 3. \tag{3.14}$$

On one hand, $X$ should be "large" and be endowed with a "weak" norm, in order to facilitate conditions (X2) und (3.7). On the other hand, (X1) and (3.9) are facilitated by a "strong" norm in $X$, which moreover provides a stronger enclosure result (3.13). Thus, we choose the Banach space $X$ with the strongest norm still allowing the explicit computation of a constant $K$ having the property (3.7), and still satisfying condition (X2). In this way, we obtain

$$X := H_{1,4}(\Omega), \quad \|u\|_X := \max\{\|u\|_\infty, \gamma\|\nabla u\|_4\}, \tag{3.15}$$

where $\gamma > 0$ is a parameter which is free for the present. Condition (3.14) and Sobolev's Embedding Theorem [2, Theorem 5.4] show that the embedding $H_{1,4}(\Omega) \hookrightarrow C(\bar{\Omega})$ is bounded. The norm given in (3.15) is therefore equivalent to the usual $H_{1,4}$-norm (with $\|u\|_4$ instead of $\|u\|_\infty$). Moreover, the Sobolev-Kondrachov Embedding Theorem [2, Theorem 6.2] ensures condition (X2). (X1) will be addressed in the next section.

We wish to remark that the more special nonlinearity (2.1) also allows the choice

$$X := C(\bar{\Omega}), \quad \|u\|_X := \|u\|_\infty$$

(see Section 2), since then the terms $\Phi(u)$ in (X1) and $\varphi(v)$ in (3.6) do no longer contain any gradients.

Finally, it should be noted that for *ordinary* differential equations ($n = 1$, $\Omega = (\alpha, \beta)$), also the choice

$$X := C_1[\alpha, \beta], \quad \|u\|_X := \max\{\|u\|_\infty, \gamma\|u'\|_\infty\} \tag{3.16}$$

is possible, since the embedding $H_2(\alpha, \beta) \hookrightarrow C_1[\alpha, \beta]$ is compact. Obviously, (3.16) satisfies our requirement for a strong norm even better than (3.15), but since it cannot be used in the elliptic case $n \in \{2, 3\}$, we will not consider it further in this article.

## 4. Computation of $G$ and proof of (X1)

To abtain a monotonically nondecreasing function $G$ satisfying (3.9) and (3.10), we first calculate a function $\tilde{G} : [0,\infty) \times [0,\infty) \to [0,\infty)$ having the following properties:

(G1) $|F(x,\omega(x)+y,\nabla\omega(x)+z) - F(x,\omega(x),\nabla\omega(x)) - b(x)\cdot z - c(x)y| \leq \tilde{G}(|y|,|z|)$ for all $x \in \Omega$, $y \in \mathbb{R}$, $z \in \mathbb{R}^n$;

(G2) $\tilde{G}(\mu t, t) = o(t)$ $(t \to 0+)$ for each fixed $\mu > 0$;

(G3) $\tilde{G}$ is monotonically nondecreasing in both variables;

(G4) for each fixed $\alpha \geq 0$, $\left[\tilde{G}(\alpha, t^{\frac{1}{4}})\right]^2$ is a continuous and *concave* function of $t \in [0,\infty)$.

Obviously, (G4) requires in particular that $\tilde{G}$ grows, as a function of its second variable, at most quadratically at infinity. Due to (G1), this requires the quadratic growth restriction (1.2) for the given nonlinearity $F$. We wish to remark that, in the case $n = 1$, condition (G4) and thus, assumption (1.2) may be dropped if the setting (3.16) is under consideration.

The possibly simplest way to find a function $\tilde{G}$ having the properties (G1) to (G4) is to start from the ansatz

$$\tilde{G}(s,t) = \sum_{k=1}^{M} g_k(s) t^{\mu_k}$$

(with exponents $\mu_k \in [0,2]$ and monotonically nondecreasing functions $g_k$ satisfying $g_k(s) = o(s^{1-\mu_k})$ for $s \to 0+$), which ensures (G2) to (G4), and to adjust the terms $M$, $g_k$, and $\mu_k$ in order to fulfill (G1), too. It should be noted that the left-hand side of the inequality in (G1) is a term of the order $o(|y| + |z|)$ $(|y|, |z| \to 0)$, due to (3.1) and to the continuity of $F_y$ and $F_z$.

Using such a function $\tilde{G}$ we can bound the operator $\varphi$ defined in (3.6) as follows:

**Lemma 4.1.** *For all $u \in H_{1,4}(\Omega)$, the estimate $\|\varphi(u)\|_2 \leq \tilde{G}(\|u\|_\infty, \|\nabla u\|_4)$ holds.*

*Proof.* Let $\Psi(t) := \left[\tilde{G}(\|u\|_\infty, t^{\frac{1}{4}})\right]^2$ and $v := |\nabla u|^4 \in L_1(\Omega)$. (G1) and (G3) provide, for almost all $x \in \Omega$,

$$|\varphi(u)(x)|^2 \leq \left[\tilde{G}(|u(x)|, |\nabla u(x)|)\right]^2 \leq \left[\tilde{G}(\|u\|_\infty, v(x)^{\frac{1}{4}})\right]^2 = \Psi(v(x)),$$

so that (G4) and Jensen's Inequality [4, Theorem 56.6] imply:

$$\|\varphi(u)\|_2^2 \le \frac{1}{\text{meas}\,(\Omega)} \int_\Omega \Psi(v(x))\,dx$$
$$\le \Psi\left(\frac{1}{\text{meas}\,(\Omega)} \int_\Omega v(x)\,dx\right) = \left[\tilde{G}(\|u\|_\infty, \|\nabla u\|_4)\right]^2. \qquad \square$$

Using the norm defined in (3.15) we obtain from Lemma 4.1 und (G3) that

$$\|\varphi(u)\|_2 \le \tilde{G}(\|u\|_X, \frac{1}{\gamma}\|u\|_X) \quad \text{for all} \quad u \in H_{1,4}(\Omega) \tag{4.1}$$

so that (3.9) holds true for

$$G(t) := \tilde{G}(t, \frac{1}{\gamma}t) \quad (t \in [0,\infty)). \tag{4.2}$$

Moreover, this function $G$ is monotonically nondecreasing due to (G3), and (3.10) holds according to (G2).

Concerning the proof of property (X1) we first note that the identity $\Phi(u) = \Phi(\omega)+\varphi(u-\omega) \quad (u \in X)$ holds true for the operator $\Phi$ defined in (X1). Therefore, (4.1) shows that $\Phi : X \to L_2(\Omega)$ is bounded on bounded sets. The continuity of $\Phi$ (resp. of $\varphi$) is proved in [21, Lemma 2] with the aid of elementary measure theory and of estimates similar to those used in Lemma 4.1.

## 5. Bounds for $L^{-1}$

To derive (3.7) (with some *known* constant K) we will show in this section how constants $K$ and $K'$ with the properties

$$\|u\|_\infty \le K\,\|L[u]\|_2 \quad (\text{for all} \quad u \in H_2^B), \tag{5.1}$$

$$\|\nabla u\|_4 \le K'\|L[u]\|_2 \quad (\text{for all} \quad u \in H_2^B) \tag{5.2}$$

can be computed; here, $L$ is the linear operator defined in (3.1). According to (3.15) we then obtain $\|u\|_X \le K\|L[u]\|_2$ (for all $u \in H_2^B$), if we choose

$$\gamma := \frac{K}{K'}. \tag{5.3}$$

Therefore, (3.7) holds since, due to (5.1), $L : H_2^B \to L_2(\Omega)$ is one-to-one and thus, also onto, according to our general regularity assumptions.

To compute $K$ and $K'$ we first calculate constants $C_0$, $C_1$, $C_2$, $K_0$, $K_1$, $\kappa$, $K_2$ satisfying

$$\|u\|_\infty \le C_0\|u\|_2 + C_1\|\nabla u\|_2 + C_2\|u_{xx}\|_2 \quad (u \in H_2(\Omega)), \tag{5.4}$$

$$\|u\|_2 \le K_0\|L[u]\|_2 \quad (u \in H_2^B), \tag{5.5}$$

$$\|\nabla u\|_2 \le K_1\|L[u]\|_2,\ \|\Delta u\|_2 \le \kappa\|L[u]\|_2,\ \|u_{xx}\|_2 \le K_2\|L[u]\|_2 (u \in H_2^B). \tag{5.6}$$

Here, $u_{xx}$ denotes the Hessian matrix of $u$ and $\|u_{xx}\|_2$ its $L_2$-Frobenius-norm. Using these constants we immediately obtain (5.1) with

$$K := C_0K_0 + C_1K_1 + C_2K_2. \tag{5.7}$$

With the aid of some partial integration we also obtain (5.2) (see [21, Lemma 4]) with

$$K' := \sqrt{K(\kappa + 2K_2)}. \tag{5.8}$$

In the following, we will give a detailed description of the computation of the constants required in (5.4), (5.5), (5.6). Significant *numerical* work is necessary only for the computation of $K_0$, while all other constants can be calculated by *analytical* means.

### 5.1 Computation of $K_0$

Consider the following eigenvalue problem (in weak formulation):

$$u \in H_2^B, \quad \langle L[u], L[\psi]\rangle = \lambda\langle u, \psi\rangle \text{ for all } \psi \in H_2^B, \tag{5.9}$$

with $\langle\cdot,\cdot\rangle$ denoting the canonical inner product in $L_2(\Omega)$. The variational characterization of the smallest eigenvalue $\lambda_1$ of problem (5.9) reads

$$\lambda_1 = \min\left\{\frac{\langle L[u], L[u]\rangle}{\langle u, u\rangle} : u \in H_2^B,\ u \not\equiv 0\right\},$$

so that $L$ is one-to-one on $H_2^B$ – as required – if and only if $\lambda_1 > 0$. In the latter case, the above characterization of $\lambda_1$ shows furthermore that (5.5) holds for each

$$K_0 \geq \frac{1}{\sqrt{\lambda_1}}. \tag{5.10}$$

We therefore need a *positive lower bound* for $\lambda_1$. In general, one will have to use *numerical* methods for its computation. It has turned out that the most efficient methods for computing lower bounds to eigenvalues (for a very general class of eigenvalue problems) are the approaches by Lehmann [16] and by Goerisch [8]; the latter generalizes the former in a sense which is of great practical importance. These methods reduce the computation of the required eigenvalue bounds to *matrix*-eigenvalue problems (of small dimension), for which simpler and more direct enclosure methods are available. Both Lehmann's and Goerisch's methods require, as a priori knowledge, a (rough) lower bound for some higher eigenvalue. Such a bound can often be read off from a "comparison problem" which is solvable in closed form. In more complicated cases, the application of a homotopy method (developed independently in [9] and in [22]), which uses a "chain" of comparison problems, has turned out to be helpful. A detailed description of Lehmann's and Goerisch's methods is given in [5].

Since a *rigorous* lower bound for $\lambda_1$ is required by (5.10), so that also *rounding errors* have to be taken into account, some parts of the procedures have to be implemented in *interval arithmetic.*

If the coefficient function $b$ of $L$ is a gradient, $b \equiv \nabla\varphi$, one may use the following alternative to (5.10). $L$ is now *symmetric* on $H_2^B$ with respect to the inner product $\langle\cdot,\cdot\rangle_\varphi$ on $L_2(\Omega)$ weighted by $e^{-\varphi}$, so that a system of eigenfunctions of the problem

$$u \in H_2^B, \quad L[u] = \lambda u; \text{ auf } \Omega \tag{5.11}$$

exists which is orthonormal and complete with respect to $\langle\cdot,\cdot\rangle_\varphi$ . Supposing now that $\sigma$ is a *positive* number satisfying

$$\sigma \le |\lambda| \text{ for all eigenvalues } \lambda \text{ of problem (5.11)}, \tag{5.12}$$

we obtain, by eigenfunction expansion,

$$\|u\|_\varphi \le \frac{1}{\sigma}\|L[u]\|_\varphi \text{ for all } u \in H_2^B,$$

where $\|\cdot\|_\varphi := \sqrt{\langle\cdot,\cdot\rangle_\varphi}$. If $\underline{\varphi} \le \varphi \le \overline{\varphi}$ (on $\Omega$) for some constants $\underline{\varphi}$ and $\overline{\varphi}$, we therefore obtain (5.5) with

$$K_0 := \frac{1}{\sigma}\exp\left[\frac{1}{2}(\overline{\varphi} - \underline{\varphi})\right]. \tag{5.13}$$

The eigenvalue bounds needed for (5.12) can again be computed using the general methods described above. The easiest way to obtain *upper* eigenvalue bounds is to use the well-known Rayleigh/Ritz method. Since problem (5.11) has a simpler structure than (5.9), the computation of $K_0$ via (5.12), (5.13) has advantages compared with (5.10). However, the bound obtained by (5.13) is in general less accurate due to the rough estimate $\underline{\varphi} \le \varphi \le \overline{\varphi}$. Of course, this disadvantage cancels in case of a constant function $\varphi$, i.e., $b \equiv 0$.

## 5.2 Calculation of the remaining constants

The four following lemmata, which are given without proof here, show how the constants required in (5.4) and (5.6) can be calculated explicitly. More general versions of these lemmata (and their proofs) can be found in [20, 21].

**Lemma 5.1** [21, Lemma 5]. *Suppose that* (5.5) *holds for some constant* $K_0$. *Moreover, let* $\underline{c}$ *denote a constant lower bound for* $c$, *and* $\overline{b}$ *a constant upper bound for* $|b|$. *Then, the first inequality in* (5.6) *holds for*

$$K_1 := \begin{cases} \frac{1}{2\sqrt{\underline{c}} - \overline{b}}, & \text{if } \underline{c} > 0 \text{ and } (2\sqrt{\underline{c}} - \overline{b}) \cdot \sqrt{\underline{c}}K_0 \ge 1 \\ \frac{1}{2}\overline{b}K_0 + \sqrt{(\frac{1}{4}\overline{b}^2 - \underline{c})K_0^2 + K_0} & \text{otherwise.} \end{cases}$$

**Lemma 5.2** [21, Lemma 6]. *Suppose that* (5.5) *and the first inequality in (5.6) are satisfied for some constants* $K_0$ *and* $K_1$. *Then, the second inequality in* (5.6) *holds for*

$$\kappa := 1 + \|b\|_\infty K_1 + \|c\|_\infty K_0.$$

For the following lemma, we recall our general assumption that the two subsets $\Gamma_0$ and $\Gamma_1$ of $\partial\Omega$ determining the boundary conditions are piecewise $C_2$-hypersurfaces, i.e., that some $\partial\Omega$-measure-zero subset $Z \subset \partial\Omega$ exists such that $\Gamma_0 \setminus Z$ and $\Gamma_1 \setminus Z$ are relatively open in $\partial\Omega$ and $C_2$-hypersurfaces in $\mathbb{R}^n$. Consequently, the *mean curvature* $H$ and the *maximal principal curvature* $P$ are defined almost everywhere on $\partial\Omega$.

**Lemma 5.3** [20, Section 4] and [21, Lemma 7]. *Suppose that* (5.5) *and the first two inequalities in* (5.6) *are satisfied for some constants* $K_0, K_1, \kappa$. *Moreover, suppose that some Lipschitz continuous function* $g : \bar{\Omega} \to \mathbb{R}^n$ *exists such that, with* $\nu$ *denoting the outer unit normal,*

$$g\cdot\nu \geq (n-1)H \quad \text{almost everywhere on } \Gamma_0, \qquad -g\cdot\nu \geq P \quad \text{almost everywhere on } \Gamma_1,$$

*and that nonnegative constants* $G_0, G_1$ *are known such that*

$$|g| \leq G_0, \quad -\operatorname{div} g + \lambda_{\max}\left[J[g] + J[g]^t\right] \leq G_1 \quad \text{almost everywhere on } \Omega.$$

*Here,* $J[g]$ *denotes the Jacobian matrix of* $g$ *and* $\lambda_{\max}$ *the largest eigenvalue. Then, the third inequality in (5.6) holds for*

$$K_2 := \left[\kappa^2 + 2\kappa G_0 K_1 + G_1 K_1^2\right]^{\frac{1}{2}}.$$

As a corollary we obtain that, for *convex* domains $\Omega$, we may choose

$$K_2 := \kappa \tag{5.14}$$

since then the curvatures $H$ and $P$ are nonpositive on $\partial\Omega$, so that the assumptions of Lemma 5.3 are satisfied for $g \equiv 0$, $G_0 = G_1 = 0$. It is trivial that the choice (5.14) is also correct in the case $n = 1$, since then we have $\Delta u = u_{xx} = u''$, so that the second and third inequality in (5.6) are identical.

The calculation of constants $C_0, C_1, C_2$ satisfying (5.4), which is the topic of the next lemma, constitutes an *explicit version* of the Sobolev Embedding $H_2(\Omega) \hookrightarrow C(\bar{\Omega})$ which is bounded for $n \leq 3$. We wish to remark that in the case $n = 1$, where even the embedding $H_1(\Omega) \hookrightarrow C(\bar{\Omega})$ is bounded, an explicit version can be derived much easier (see [19, Lemma 2]).

**Lemma 5.4** ([20, Corollary 1]). *Let* $n \in \{2, 3\}$, *and let* $Q \subset \mathbb{R}^n$ *denote some compact and convex set with nonempty interior and with the following property: For each*

*$x_0 \in \bar{\Omega}$, there exists a set $\tilde{Q}$ which is congruent to $Q$ (i.e., is obtained from $Q$ by an affine-orthogonal transformation) and which satisfies $x_0 \in \tilde{Q} \subset \bar{\Omega}$. Then, (5.4) holds for*

$$C_j := \gamma_j \left[ \frac{\text{meas}\,(\Omega)}{\text{meas}\,(Q)} \cdot \max_{x_0 \in Q} \left\{ \frac{1}{\text{meas}\,(Q)} \int_Q |x - x_0|^{2j}\, dx \right\} \right]^{\frac{1}{2}} \qquad (j = 0, 1, 2),$$

*where*

$$\gamma_0 = 1 \qquad \gamma_1 = 1.1548 \quad \gamma_2 = 0.22361 \quad \text{if } n = 2,$$
$$\gamma_0 = 1.0708 \quad \gamma_1 = 1.6549 \quad \gamma_2 = 0.41413 \quad \text{if } n = 3.$$

## 6. Numerical procedures for the computation of $\omega$ and $\delta$

In this section, we report on numerical procedures for computing an approximate solution $\omega$ of problem (1.1) and some bound $\delta$ (see (3.8)) for the defect of $\omega$ defined in (3.5).

A suitable method for the computation of an approximate solution of the nonlinear problem (1.1) is *Newton's iteration*: Starting with some rough approximation $\omega^{(0)} \in H_2^B$, we compute iteratively approximate solutions $u^{(k)}$ of the linear boundary value problems

$$\begin{aligned} &-\Delta u + F_z(\cdot, \omega^{(k)}, \nabla\omega^{(k)}) \cdot \nabla u + F_y(\cdot, \omega^{(k)}, \nabla\omega^{(k)}) u \\ &= \Delta\omega^{(k)} - F(\cdot, \omega^{(k)}, \nabla\omega^{(k)}), \quad u \in H_2^B \end{aligned} \tag{6.1}$$

and update $\omega^{(k+1)} := \omega^{(k)} + u^{(k)}$.

Since, for our purposes, the approximate solution has to be defined on the *whole* of $\Omega$, the linear problems (6.1) should be treated by *projective* methods, such as finite element approaches, rather than by difference methods. Moreover, only $C_1$-elements (with continuously differentiable global basis functions) are allowed to ensure $u^{(k)} \in H_2(\Omega)$. Examples for such finite elements (in two dimensions) are

- Argyris triangular elements with quintic local basis functions,
- Bell triangular elements with reduced quintic local basis functions,
- Bogner-Schmidt-Fox rectangular elements with bi-cubic local basis functions,
- rectangular elements with bi-quintic local basis functions and the knot variables $u$, $\partial u/\partial x_1$, $\partial u/\partial x_2$, $\partial^2 u/\partial x_1 \partial x_2$ in the corners, the midpoints of the sides, and the midpoint of the rectangle.

The latter elements were used in the numerical examples presented in the following section. To fulfill the boundary conditions, we take of course only those basis functions into consideration which satisfy them. To approximate the integrals needed in the usual Ritz-Galerkin-approach for problem (6.1) one will use *quadrature formulas* (on each element); in our examples, we chose the composite trapezoidal or Simpson (product-) formula. The remaining linear algebraic system was solved approximately by Gaussian elimination (regarding the banded structure). Alternatively, one might use iterative procedures; however, the used method should not require a sign condition on the real parts of the eigenvalues of the matrix, in order to preserve the advantages of our existence and enclosure method compared with the "monotonicity methods".

To compute a defect bound $\delta$ (see (3.5), (3.8)) we obviously need numbers $\eta_1, \ldots, \eta_K$ such that

$$\int_{\Omega_k} \left[-\Delta\omega + F(\cdot, \omega, \nabla\omega)\right]^2 \, dx \leq \eta_k \quad (k = 1, ..., K), \tag{6.2}$$

with $\Omega_1, \ldots, \Omega_K$ denoting the elements of the used finite element method. We may then choose $\delta := \left[\text{meas}\,(\Omega)^{-1} \sum_{k=1}^{K} \eta_k\right]^{\frac{1}{2}}$. The additive splitting of the defect integral according to (6.2) has two significant advantages:

i) The integrals in (6.2) are "simple" because $\Omega_k$ has a simple geometry (triangle or rectangle), and $\omega$ is a polynomial of low degree on $\Omega_k$, according to the finite element approach.

ii) The bounds $\eta_1, \ldots, \eta_K$ can be computed *independently*, so that this part of the procedure is perfectly suited for *parallel* computation. The parallelization is in fact worth while, because the numerical effort needed to compute the numbers $\eta_k$ can be rather large, as we will explain in the following. A parallel implementation of the defect computation on a T800 transputer system (with 32 processors) has already been performed successfully.

The computation of the numbers $\eta_1, \ldots, \eta_K$ according to (6.2) obviously requires *verified bounds for integrals* (on a "simple" domain of integration and with a "simple" integrand). In case of a polynomial nonlinearity $F$, the whole integrand is a polynomial (on each $\Omega_k$), so that the required bounds can be obtained without significant problems by use of a *polynomial arithmetic*, which is easy to program, e.g., in a C++ environment. In the general (non-polynomial) case, one will have to use *quadrature formulas* $Q[\Phi]$ with known remainder term bound

$$\left| \int_{\Omega_k} \Phi(x)dx - Q[\Phi] \right| \leq R[\Phi]. \tag{6.3}$$

For rectangular elements $\Omega_k$, it is possible to apply a composite Newton-Cotes product formula (with stepsizes $h_1$ in $x_1$- and $h_2$ in $x_2$-direction), e.g., the trapezoidal rule with remainder term bound

$$R[\Phi] = \frac{1}{12} \text{meas}\,(\Omega_k) \left[ h_1^2 \left\| \frac{\partial^2 \Phi}{\partial x_1^2} \right\|_{\infty, \Omega_k} + h_2^2 \left\| \frac{\partial^2 \Phi}{\partial x_2^2} \right\|_{\infty, \Omega_k} \right] \tag{6.4}$$

or the Simpson formula with remainder term bound

$$R[\Phi] = \frac{1}{180}\text{meas}\,(\Omega_k)\left[h_1^4\left\|\frac{\partial^4\Phi}{\partial x_1^4}\right\|_{\infty,\Omega_k} + h_2^4\left\|\frac{\partial^4\Phi}{\partial x_2^4}\right\|_{\infty,\Omega_k}\right]. \tag{6.5}$$

Also for many other quadrature formulas, $R[\Phi]$ is given via maximum norms of certain higher derivatives of $\Phi$. For the computation of $\eta_k$, we therefore need

i) an enclosure (resp. upper bound) for $Q[\Phi]$,

ii) rough bounds for some higher derivatives of $\Phi$,

where $\Phi := [-\Delta\omega + F(\cdot,\omega,\nabla\omega)]^2\,|_{\Omega_k}$. For i) as well as for ii), one has to use *interval arithmetic*, in order to guard against rounding errors. For this purpose, the programming languages described in [6, 10, 11, 12, 13] are well suited. For i), we use the representation of $\omega|_{\Omega_k}$ as a linear combination of the finite element basis functions to enclose $\omega$, $\nabla\omega$, $\Delta\omega$ in the quadrature points; then, the nonlinearity $F$ is (interval-) evaluated to provide enclosures for $\Phi$ (in the quadrature points); finally, the quadrature formula itself is (interval-) evaluated, so that an enclosure for $Q[\Phi]$ is obtained; here, the accurate dot product available in the programming environments mentioned above facilitates the computation.

For ii), rough bounds are in fact sufficient since the remainder term bound $R[\Phi]$ can be kept small by choice of sufficiently many quadrature points (see (6.4), (6.5)). The required derivatives of $\Phi$ can be obtained, e.g., by automatic differentiation techniques. Then, either interval evaluations of these derivatives are performed (possibly after further subdivision of $\Omega_k$), or one might apply approximation theoretic results to compute the desired bounds.

## 7. Numerical examples

As a test for the proposed existence and enclosure method, several examples with ordinary and elliptic differential equations have been considered. Here, we report on some of the results obtained for elliptic problems.

Our first example is the well-known Bratu problem

$$-\Delta u = \lambda e^u \text{ on } \Omega := (0,1)^2, \quad u = 0 \text{ on } \partial\Omega, \tag{7.1}$$

with origin in combustion theory. After a certain transformation (aiming at the reduction of the corner singularities), approximate solutions $\omega$ were computed for several values of the parameter $\lambda > 0$; we used $8 \times 8$ rectangular elements with biquintic local basis functions. In this way we obtained – after interpolation between the approximations – the bifurcation diagram in Figure 2.

Our existence and enclosure method was successful on the "lower" branch (up to the turning point), and partly on the "upper" branch. In the further course of the upper branch, an increased accuracy of the finite element approximations would be needed to satisfy still condition (3.12). The "monotonicity methods" discussed in Section 2 are applicable *only* on the lower branch since the linearization $L$ defined in (3.1) has one negative eigenvalue on the upper branch. In the turning point itself, $L$ is not one-to-one, so that there (and in a small neighborhood) both methods break down. However, this difficulty can be overcome, in the context of our method, by *change of parameter* techniques; see [23] for details.

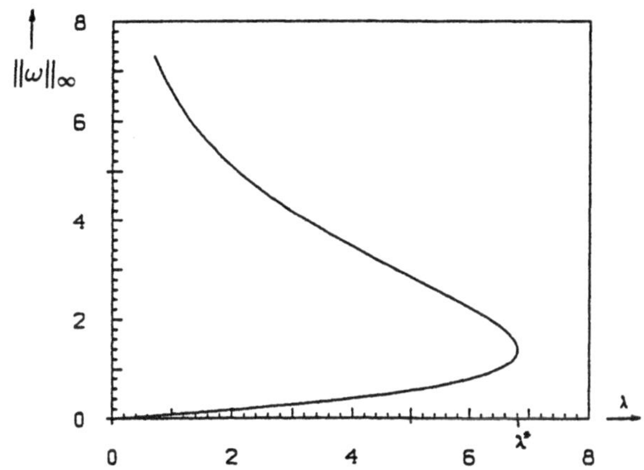


Figure 2: Bifurcation diagram for problem (7.1)

In our second example, we ask for nontrivial solutions of *Emden's* equation

$$-\Delta u = u^2 \text{ on } \Omega := (0,\ell) \times (0,\ell^{-1}), \quad u = 0 \text{ on } \partial\Omega \tag{7.2}$$

for several values $\ell > 0$. The solution $u$ represents the stationary temperature distribution in a plate with an interior heat source which is proportional to $u^2$. The problem of finding a starting approximation for the Newton-iteration was solved by embedding problem (7.1) into some appropriate *family* of problems; details can be found in [21].

Using 64 bi-quintic rectangular elements we computed approximate solutions for the three cases $\ell = 1, \ell = 2$, and $\ell = 2.5$. Our existence and enclosure method was successful in all three cases. Figure 3 shows plots of the three approximate solutions $\omega$. Table 1 contains approximate values for $\|\omega\|_\infty = \omega(\frac{1}{2}\ell, \frac{1}{2}\ell^{-1})$, the defect bounds $\delta$ (see (3.8)), the bounds $K$ for $L^{-1}$ (see (3.7), (5.1)), and the error bounds $\alpha$ for $\|u^* - \omega\|_\infty$ (see (3.13)).

We only looked for solutions which are *symmetric* with respect to reflection at the two symmetry axes (parallel to the $x-$ resp. $y-$ axis) of the rectangle $\Omega$. Thus, the differential equation was considered on $\tilde{\Omega} := (0, \frac{1}{2}\ell) \times (0, \frac{1}{2}\ell^{-1})$ only, with Neumann boundary conditions on $\partial\tilde{\Omega} \cap \Omega$. This reduces the numerical effort, and

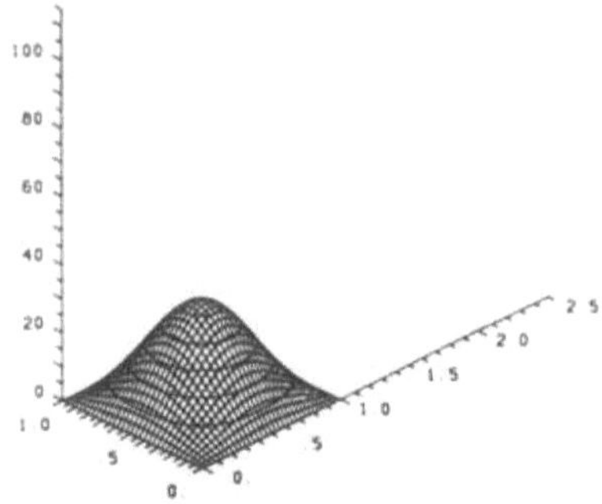

$\ell = 1$

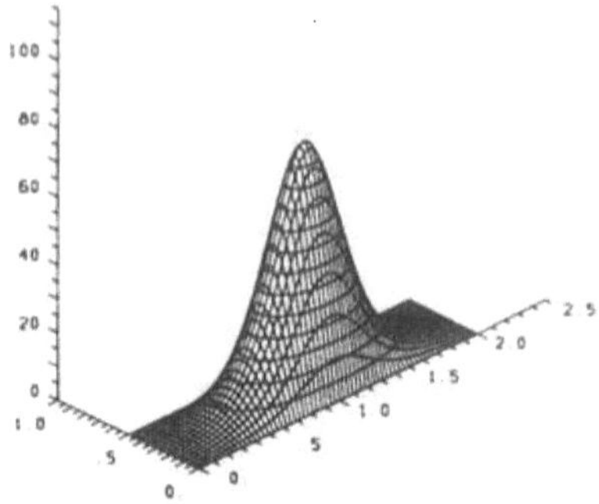

$\ell = 2$

| $\ell$ | $\|\omega\|_\infty \approx$ | $\delta$ | $K$ | $\alpha$ |
|---|---|---|---|---|
| 1 | 29.26 | 0.2207E–2 | 0.3832 | 0.8467E–3 |
| 2 | 71.52 | 0.8183E–1 | 0.3820 | 0.3167E–1 |
| 2.5 | 111.73 | 0.2231 | 0.3147 | 0.7188E–1 |

$\ell = 2.5$

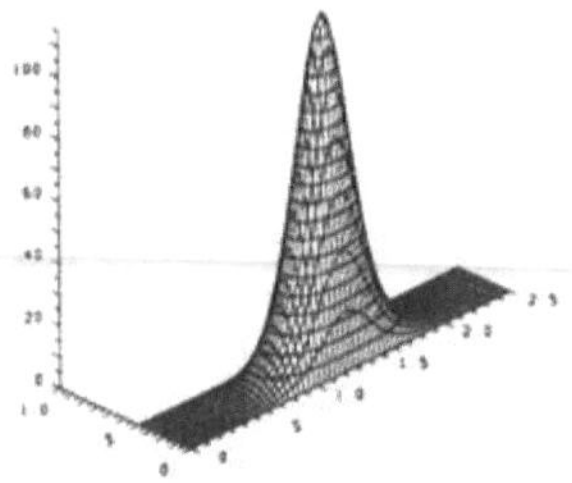

Figure 3 and Table 1: Emden's equation (7.2)

moreover, some of the eigenvalues of the linearization $L$ of the original problem are absent in this "symmetrized" problem, which may reduce the value of the constant $K_0$ considerably, according to (5.12), (5.13). The latter effect is very distinct for $\ell = 2$, and in particular, for $\ell = 2.5$, where the second eigenvalue of $L$ in the original problem is very small (in modulus) while this eigenvalue is missing in the "symmetrized" problem. We wish to remark that the *first* eigenvalue of $L$ is negative in all cases, so that the "monotonicity-methods" are not applicable.

## References


1. Adams, E., and Kulisch, U. (Eds.), *Scientific Computing with Automatic Result Verification. I. Language and Programming Support for Verified Scientific Computation*, II. *Enclosure Methods and Algorithms with Automatic Result Verification*, III. *Applications in the Engineering Sciences.* Academic Press, San Diego, 1993.

2. Adams, R. A., *Sobolev spaces.* Academic Press, New York, 1975.

3. Alefeld, G., and Herzberger, J., *Einführung in die Intervallrechnung*. Bibliographisches Institut (Reihe Informatik, Nr. 12), Mannheim Wien Zürich, 1974.
4. Bauer, H., *Wahrscheinlichkeitstheorie und Grundzüge der Maßtheorie*. 3rd Edition, de Gruyter, Berlin, 1978.
5. Behnke, H., and Goerisch, F., *Inclusions for Eigenvalues of Selfadjoint Problems*. In: Herzberger, J. (Ed.), Topics in Validated Computations. Studies in Computational Mathematics 5, North Holland (Elsevier), Amsterdam 1994, 277–322.
6. Bleher, J.H., Kulisch, U., Metzger, M., Rump, S.M., Ullrich, Ch., and Walter, W., *FORTRAN-SC: A Study of a FORTRAN Extension for Engineering/Scientific Computation with Access to ACRITH*. Computing 39 (1987), 93–110.
7. Collatz, L., *Aufgaben monotoner Art*. Arch. Math. 3 (1952), 366–376.
8. Goerisch, F., and Albrecht, J., *Eine einheitliche Herleitung von Einschließungssätzen für Eigenwerte*. In: Albrecht, J.; Collatz, L.; Velte, W. (Eds.), Numerical treatment of eigenvalue problems, Vol. 3, 58–88, ISNM 69, Birkhäuser, Basel, 1984.
9. Goerisch, F., *Ein Stufenverfahren zur Berechnung von Eigenwertschranken*. In: Albrecht, J.; Collatz, L.; Velte, W.; Wunderlich, W. (Eds.), Numerical Treatment of Eigenvalue Problems, Vol. 4, 104–114, ISNM 83, Birkhäuser, Basel, 1987.
10. IBM, *IBM High-Accuracy Arithmetic Subroutine Library (ACRITH)*. IBM Deutschland GmbH (Dept. 3282, Schönaicher Straße 220, 71032 Böblingen) 3rd Edition, 1986. 1. General Information Manual. GC 33–6163–02. 2. Program Description and User's Guide. SC 33–6164–02. 3. Reference Summary. GX 33–9009–02.
11. IBM, *ACRITH–XSC: IBM High Accuracy Arithmetic — Extended Scientific Computation*. Version 1, Release 1. IBM Deutschland GmbH (Schönaicher Straße 220, 71032 Böblingen), 1990. 1. General Information. GC 33–6461–01. 2. Reference. SC 33–6462–00. 3. Sample Programs. SC 33–6463–00. 4. How To Use. SC 33–6464–00. 5. Syntax Diagrams. SC 33-6466-00.
12. Klatte, R., Kulisch, U., Neaga, M., Ratz, D., and Ullrich, Ch., *PASCAL-XSC — Language Reference with Examples*. Springer-Verlag, Berlin Heidelberg New York, 1992.
13. Klatte, R., Kulisch, U., Lawo, C., Rauch, M., and Wiethoff, A., *C-XSC, A C++ Class Library for Extended Scientific Computing*. Springer-Verlag, Berlin Heidelberg New York, 1993.
14. Krein, M.G., and Rutman, M.A., *Linear operators leaving invariant a cone in a Banach space*. Amer. Math. Soc. Trans., Ser. 1, Nr. 10 (1962), 199–325.

15. Kulisch, U., *Numerik mit automatischer Ergebnisverifikation.* In: Jahrbuch Überblicke Mathematik, Vieweg Verlag, 1993.

16. Lehmann, N. J., *Optimale Eigenwerteinschließungen.* Numer. Math. 5 (1963), 246–272.

17. Nakao, M. T., *Solving Nonlinear Elliptic Problems with Result Verification Using an $H^{-1}$ Type Residual Iteration.* Computing, Suppl. 9 (1993), 161–173.

18. Oishi, S., *Numerical Verification of Existence and Inclusion of Solutions for Nonlinear Operator Equations.* Submitted for publication.

19. Plum, M., *Verified existence and inclusion results for two-point boundary value problems.* IMACS Annals on Computing and Applied Mathematics, Vol. 7 (1990), 341–355.

20. Plum, M., *Explicit $H_2$-estimates and pointwise bounds for solutions of second-order elliptic boundary value problems.* J. Math. Anal. Appl. 165 (1992), 36–61.

21. Plum, M., *Numerical Existence Proofs and Explicit Bounds for Solutions of Nonlinear Elliptic Boundary Value Problems.* Computing 49 (1992), 25–44.

22. Plum, M., *Bounds for eigenvalues of second-order elliptic differential operators.* Journal of Applied Mathematics and Physics (ZAMP) 42 (1991), 848–863.

23. Plum, M., *Enclosures for Solutions of Parameter-Dependent Nonlinear Elliptic Boundary Value Problems: Theory and Implementation on a Parallel Computer.* Interval Computations 3 (1994), 106–121.

24. Plum, M., *Inclusion Methods for Elliptic Boundary Value Problems.* In: Herzberger, J.(Ed.), Topics in Validated Computations. Studies in Computational Mathematics 5, North Holland (Elsevier), Amsterdam 1994, 323–379.

25. Schröder, J., *Operator Inequalities.* Academic Press, New York, 1980.

26. Schröder, J., *Operator Inequalities and Applications.* In: Everitt, W.N. (Ed.), Inequalities, Fifty Years On from Hardy, Littlewood and Pólya, 163–210. Marcel Dekker Inc., 1991.

27. Walter, W., *Differential and Integral Inequalities.* Springer, Berlin-Heidelberg, 1970.

Michael Plum, Mathematisches Institut I, Universität Karlsruhe, D-76128 Karlsruhe, Germany.

International Series of Numerical Mathematics, Vol. 123, © 1997 Birkhäuser Verlag Basel

# Weak persistence in Lotka-Volterra populations

Raymond M. Redheffer and Peter Volkmann

Let $u : [0, \infty) \to \mathbb{R}^n$ satisfy the Lotka-Volterra equations

$$\dot{u}_i = u_i\Big(a_i(t) - \sum_{j=1}^{n} b_{ij}(t)u_j\Big), \quad t > 0; \qquad u_i(0) > 0 \tag{1}$$

for $i = 1, \cdots, n$, where $a_i$ and $b_{ij}$ are piecewise continuous real-valued functions of the real variable $t$. Thus $u$ describes the behavior of a Lotka-Volterra population.

A particular solution $u$ is called *weakly persistent* if the population as a whole never dies out, though its individual members may get arbitrarily close to extinction. More precisely, *weak persistence* means

$$\liminf_{t\to\infty} m(t) > 0 \quad \text{where} \quad m(t) = \max_i u_i(t).$$

Hence at least one $u_i(t)$ must be somewhat large at each given $t$, but the question of which one is large can depend on $t$. The adjective *weak* distinguishes this type of persistence from the more familiar persistence in which a positive lower bound is required for each $u_i$. Weak persistence of the solution, or of the population, pertains to a particular $u$. If every solution is weakly persistent, the equation is called weakly persistent.

Our goal is to give conditions under which weak persistence holds and to discuss the behavior of $m(t)$ as $t \to \infty$. To simplify the presentation we do not strive for maximum generality and we postpone technical details.

## 1. Some useful inequalities

Since $u_i$ satisfies an equation of the form $\dot{u}_i = u_i f_i$, the inequality $u_i > 0$ holds on the interval of existence of $u$. The equation

$$\dot{m}(t) = m(t)\big(a_i(t) - b_{ii}(t)m(t) - \sum_{j\neq i} b_{ij}(t)u_j(t)\big) \tag{2}$$

holds on any open interval on which $m(t) = u_i(t)$. For a real-valued function $\phi$ we define

$$\phi^-(t) = \min\left(\phi(t),\ 0\right), \quad \phi^+(t) = \max\left(\phi(t),\ 0\right). \tag{3}$$

In this notation, the sum on the right of (2) can be assessed by

$$m(t)\sum_{j\neq i} b_{ij}{}^-(t) \le \sum_{j\neq i} b_{ij}(t)u_j(t) \le m(t)\sum_{j\neq i} b_{ij}{}^+(t).$$

Hence $m(t)$ satisfies a one-dimensional Lotka-Volterra equation,

$$\dot{m} = f(t)m - g(t)m^2, \qquad m(0) > 0, \tag{4}$$

where for some $i = i(t)$ we have $f(t) = a_i(t)$ and

$$b_{ii}(t) + \sum_{j\neq i} b_{ij}{}^-(t) \le g(t) \le b_{ii}(t) + \sum_{j\neq i} b_{ij}{}^+(t). \tag{5}$$

Since each $u_i(t) > 0$, both inequalities are strict unless the $b_{ij}(t)$ are all of one sign.

If $g(t) > 0$ for $t \ge 0$, we can factor out $mg(t)$ from the right side of (4) and conclude that

$$\inf \frac{f(t)}{g(t)} > 0 \Longrightarrow \inf m(t) > 0, \qquad \sup \frac{f(t)}{g(t)} < \infty \Longrightarrow \sup m(t) < \infty.$$

Since $u_i \le m$, the right-hand inequality gives a condition for $u$ to exist on $[0,\ \infty)$ and to be bounded there. The left-hand inequality gives a condition for weak persistence as defined above. Under the same hypotheses the condition

$$\int_0^\infty g(s)\,ds = \infty$$

leads to the sharper conclusion that

$$\liminf_{t\to\infty} \frac{f(t)}{g(t)} \le \liminf_{t\to\infty} m(t) \le \limsup_{t\to\infty} m(t) \le \limsup_{t\to\infty} \frac{f(t)}{g(t)}.$$

However, this is not needed here.

To get an explicit criterion, suppose $a_i(t) > 0$ and

$$b_{ii}(t) + \sum_{j\neq i} b_{ij}{}^-(t) > 0$$

for $i = 1, \cdots, n$. The fact that the same index $i = i(t)$ occurs both in $a_i$ and in the sum yields

$$\frac{a_i(t)}{b_{ii}(t) + \sum_{j\neq i} b_{ij}{}^+(t)} \le \frac{f(t)}{g(t)} \le \frac{a_i(t)}{b_{ii}(t) + \sum_{j\neq i} b_{ij}{}^-(t)}$$

for some value $i = i(t)$. If the left side has a lower bound $L(t)$ and the right side an upper bound $R(t)$, both independent of $i$, then the two conditions

$$\inf L(t) > 0, \qquad \sup R(t) < \infty \tag{6}$$

ensure weak persistence of every solution $u$. The second condition (6) is introduced only to guarantee existence of $u$ on $[0, \infty)$ and can be weakened as described next.

## 2. The linear equation

The familiar substitution $m = 1/y$ gives

$$\dot{y} + f(t)y = g(t), \qquad y(0) = y_0 > 0, \tag{7}$$

where $y_0 = 1/m(0)$. The solution is

$$y(t) = e^{-F(t)} \int_0^t e^{F(s)} g(s)\, ds + y_0 e^{-F(t)}, \qquad F(t) = \int_0^t f(s)\, ds. \tag{8}$$

Clearly

$$y_0 + \int_0^t e^{F(s)} g(s)\, ds > 0 \iff y(t) > 0 \iff u(t) < \infty.$$

Thus $m(t)$, and hence $u(t)$, exists on any interval $[0, T)$ on which the first of these inequalities holds.

As a special case we see that $u(t)$ exists on $[0, \infty)$ for all choices of the initial condition if $g(s) \geq 0$ for all $s > 0$. This in turn holds if

$$b_{ii}(t) + \sum_{j \neq i} b_{ij}{}^{-}(t) \geq 0, \qquad t > 0, \quad i = 1, \cdots, n. \tag{9}$$

The inequality (9) guarantees existence of $u(t)$ on $[0, \infty)$ and can replace the hypothesis $\sup R(t) < \infty$ in (6). In contrast to (6), however, (9) does not imply that the solution is bounded.

The difference $y = y_1 - y_2$ of two solutions $y_1$, $y_2$ satisfies (8) with $g = 0$, so

$$\lim_{t \to \infty} \big(y_1(t) - y_2(t)\big) = 0 \iff \int_0^\infty f(s)\, ds = \infty.$$

If the corresponding quantities $m_1$, $m_2$ are bounded away from 0 and also bounded above, the above result gives

$$\lim_{t \to \infty} \big(m_1(t) - m_2(t)\big) = 0 \iff \int_0^\infty f(s)\, ds = \infty.$$

This holds for all solutions if the equation is weakly persistent and all solutions are bounded above. In particular, it holds under the hypothesis (6). Since $f(t) = a_i(t)$ for some $i = i(t)$, a sufficient condition for divergence of the integral is

$$\int_0^\infty \min_i a_i(t)\, dt = \infty. \tag{10}$$

## 3. Two examples

When $n = 2$ the Lotka-Volterra system can be written in the form

$$a_1 = \frac{\dot{u}_1}{u_1} + b_{11}u_1 + b_{12}u_2, \quad a_2 = \frac{\dot{u}_2}{u_2} + b_{21}u_1 + b_{22}u_2.$$

This formulation makes it easy to choose the coefficients in such a way that any prescribed positive functions $u_1$, $u_2$ yield a solution. However, we want the resulting equation to satisfy both (6) and (10), so that the equation is weakly persistent and $m(t)$ has a unique limiting behavior as $t \to \infty$.

**Example 1.** As our first example let $u_1 = 1$, $u_2 = e^{-t}$. Then

$$a_1 = b_{11} + b_{12}e^{-t}, \quad a_2 = -1 + b_{21} + b_{22}e^{-t}.$$

For definiteness we take

$$b_{11} = 2, \quad b_{12} = 1, \quad b_{21} = 2, \quad b_{22} = 3.$$

Then $2 \leq a_1 \leq 3$, $1 \leq a_2 \leq 4$ and (5) gives $1 \leq f(t) \leq 4$, $1 \leq g(t) \leq 5$. It follows from (6) that the equation is weakly persistent and that all solutions are bounded on $[0, \infty)$. Since $m(t) = 1$ in this example, (10) shows that $\lim_{t\to\infty} m(t) = 1$ for all solutions. Nevertheless $\inf u_2(t) = 0$.

**Example 2.** We now give a more elaborate example to show that the lead solution (the one that is large at a given $t$) can change infinitely often as $t \to \infty$. Specifically, we construct a weakly persistent equation in which $\lim_{t\to\infty} m(t) = 1$ for all solutions and yet

$$\inf u_1(t) = \inf u_2(t) = 0 \tag{11}$$

for a particular solution $u$.

To this end let $0 < p < q < r = 2q - p$, the point $r$ being so chosen that $q - p = r - q$. We define $u_1 = 1$ on $(p, r)$. For $u_2$ we take the respective functions

$$e^{p-t}, \quad e^{p-2q+t}$$

on the intervals $(p, q)$ and $(q, r)$. Thus $u_2(t) = 1$ at the ends of the interval but the value $u_2(q)$ at the center can be small. We use the same coefficients $b_{ij}$ as those in Example 1 above. Then $a_1 = 2 + u_2$ and

$$a_2 = 1 + 3e^{p-t}, \quad a_2 = 3 + 3e^{p-2q+t}$$

on the two intervals respectively. This gives $1 \leq a_i \leq 6$ for $i = 1, 2$.

We do the above with $p = p_1$, $q = q_1$, $r = r_1$. We then proceed to intervals $(p_2, q_2)$ and $(q_2, r_2)$ where

$$p_2 = r_1, \quad r_2 = 2q_2 - p_2,$$

and make a similar construction with the roles of $u_1$ and $u_2$ interchanged. We proceed then to a third pair of intervals $(p_3, q_3)$, $(q_3, r_3)$ with $p_3 = r_2$, and so on. By making the lengths of these intervals tend to $\infty$ we can ensure (11). Equations (6) and (10) show nevertheless that the equation is weakly persistent and that $\lim_{t\to\infty} m(t) = 1$ for all solutions $u$.

## 4. Technical details

For simplicity, we have confined attention to the case in which $m(t) = m_i(t)$ holds on an open interval for some $i$. The general case is reduced to this one by [1], Lemmas 4 and 7.

We allowed piecewise continuous coefficients to facilitate the construction of counterexamples. The differential equation is interpreted by taking the last value of $u(t)$ on the preceding interval of continuity as initial value of $u(t)$ on the next; this is equivalent to various other interpretations of the solution depending on integration. At the points $t$ where intervals of continuity join, $\dot{u}(t)$ need not exist and the differential equation can fail. Nevertheless $m(t)$ is continuous and satisfies its differential equation except in a countable set, and this is sufficient.

If we define $m(t) = \sup u_i(t)/d_i$ as in [1], where $d_i$ are given positive constants, we get more general results containing multipliers. Since no new ideas are involved, the details are omitted.

## Reference

1. Ray Redheffer, *Nonautonomous Lotka-Volterra Systems* I. J. Differential Equations **127** (1996), 519–541. Further references are given here.

Ray Redheffer, University of California, Los Angeles, CA 90049, USA.

Peter Volkmann, University of Karlsruhe, D-76128 Karlsruhe, Germany.

International Series of Numerical Mathematics, Vol. 123, © 1997 Birkhäuser Verlag Basel

# Uniqueness for degenerate elliptic equations via Serrin's sweeping principle

Wolfgang Reichel

**Abstract**

For three different examples of degenerate elliptic equations uniqueness conditions for positive solutions are given. The main tool in the proofs is Serrin's sweeping principle.

## 1. Introduction

In this paper we obtain uniqueness theorems for three different degenerate elliptic equations, where, due to the degeneracy, comparison principles are not directly at hand. With suitable variants of Serrin's sweeping principle (its general form is given in Section 2) we shall prove the following results.

(I) For $a \geq b \geq 0$ there is at most one positive solution $u \in C^2(\Omega) \cap C(\overline{\Omega})$ of the anisotropic singular problem

$$u^a u_{xx} + u^b u_{yy} + p(x,y) = 0 \ \text{ in } \ \Omega \subset \mathbb{R}^2, \qquad u = 0 \ \text{ on } \ \partial\Omega, \tag{1}$$

if $p : \Omega \to \mathbb{R}$ satisfies

$$p(\tau_1 x, \tau_2 y) \geq p(x,y) \ \text{ for all } \ (x,y) \in \Omega, \ \tau_i \in [0,1] \tag{M}$$

and if the bounded domain $\Omega$ satisfies an *interior rectangle condition*, see Section 3, in order to make (M) meaningful. Existence for this problem on convex $C^{2,\alpha}$-domains is known from the work of Choi-Lazer-McKenna [4], whereas uniqueness was left as an open problem.

(II) Consider the nonlinear Keldysh-type equation

$$\left((u+x)u_x - \frac{u}{2}\right)_x + u_{yy} = 0 \ \text{ in } \ \Omega, \qquad u = g \ \text{ on } \ \partial\Omega, \tag{2}$$

where $\Omega = (0,X) \times (-Y,Y)$ is a rectangle in $\mathbb{R}^2$ and

$$g = 0 \ \text{ on } \ \Gamma_2 = \{0\} \times [-Y,Y]. \tag{3}$$

The boundary condition (3) makes the equation degenerate on $\Gamma_2 \subset \partial\Omega$. Čanić and Keyfitz [2], [3] consider this problem in the context of shock reflection. They prove the existence of elliptic solutions, i.e., solutions $u \geq -x$, under each of the following basic assumptions:

$$\frac{-x}{1+\sqrt{1-Cx}} \leq g(x,y) \leq 0 \tag{4$_-$}$$

or

$$A\sqrt{x} \leq g(x,y) \leq B\sqrt{x} \text{ on } \partial\Omega, \tag{5}$$

where $0 \leq C < 1/X$ and $0 \leq A \leq B$ (additional assumptions on $g$ are needed, see [2], [3]). The uniqueness question is of physical importance and due to a comparison principle (Theorem 2 in [2]) for sub-, supersolutions, uniqueness in the class of functions $u_{xx} \leq 0$ is known. However, solutions of (2) with $u_{xx} \geq 0$ are known, e.g., $u(x,y) = -x/(1+\sqrt{1+Cx})$, $C \geq 0$.

We shall prove a uniqueness theorem for positive solutions of (2), (3) under the assumptions that $g = g(x)$ depends on $x$ only, satisfies (5) and is (weakly) increasing in $x$; for a more general condition see Section 4.

(III) The positive weak solution $u = (u_1, \ldots, u_k) \in (W_0^{1,p}(\Omega))^k$ of the quasimonotone (cooperative) $p$-Laplace system (we use summation convention)

$$\Delta_p u_i + a_{ij}(x)u_j^{p-1} + f_i(x,u) = 0 \text{ in } \Omega \quad (i = 1, \ldots, k) \tag{6}$$

is unique, provided $(i,j = 1, \ldots, k)$

$$\partial\Omega \in C^{2,\alpha}, \quad a_{ij} \in C^\alpha(\overline{\Omega}), \quad a_{ij} \geq 0 \text{ for } i \neq j, \tag{H1}$$

$$f_i \in C^\alpha_{\text{loc}}(\overline{\Omega} \times [0,\infty)^k), \quad 0 \leq f_i(x,u) \leq K(1+u_i^{p-1}) \text{ for } u \geq 0, \tag{H2}$$

$$f_i(x,tu)/t^{p-1} \text{ strictly decreasing in } t > 0 \text{ for all } u \text{ with } u_1, \ldots, u_k > 0, \tag{H3}$$

$$f_i(x,u) \text{ increasing in } u \geq 0. \tag{H4}$$

For a single equation, this result is well known with less regularity of $f$, $\partial\Omega$ and without the sign restriction on $f$, see Huang [8]. It is based on a cleverly chosen test-function argument going back to Anane [1] and Lindqvist [13]. For the quasimonotone analogue we did not succeed in repeating this test-function argument. The same system with constant coefficients was studied by Fleckinger-Pellé and Takáč [5], where existence and uniqueness theorems for positive solutions are given. Our uniqueness theorem is independent of any existence theory.

Finally we note that the sweeping principle can also be applied to show the simplicity of the first eigenvalue of some homogeneous quasilinear (possibly degenerate) equations of divergence form of the type studied by Idogawa-Ôtani [9]. This will be elaborated elsewhere.

## 2. The sweeping principle

Let $\Omega \subset \mathbb{R}^N$ be a bounded domain and $Q$ be a second order quasilinear elliptic operator on $\Omega$. Serrin's sweeping principle in its general form can be stated as follows:

Find conditions on $Q$, $\Omega$ and $f : \Omega \times \mathbb{R} \times \mathbb{R}^N \to \mathbb{R}$ such that the following holds:

*Let $v \in C^2(\Omega) \cap C(\overline{\Omega})$ be a subsolution*

$$Qv + f(x, v, \nabla v) \geq 0 \quad in \;\; \Omega,$$

*and $w_\lambda \in C^2(\Omega) \cap C(\overline{\Omega})$ a family of supersolutions depending continuously on $\lambda \in [\lambda_1, \lambda_2]$*

$$Qw_\lambda + f(x, w_\lambda, \nabla w_\lambda) \leq 0 \quad in \;\; \Omega$$

*with $v \leq w_\lambda$ on $\partial\Omega$. If $v \leq w_{\lambda_2}$ in $\Omega$ and $v \not\equiv w_\lambda$ for $\lambda \in (\lambda_1, \lambda_2]$, then the conclusion $v \leq w_{\lambda_1}$ holds.*

The principle is known as *Serrin's sweeping principle*, see Sattinger [18], but goes back to McNabb [15]. It has been shown to hold for the Laplacian operator (and other uniformly elliptic operators) if $f(x, u, \nabla u) = f_1(x, u, \nabla u) + f_2(x, u)$, where $f_1$ is Lipschitz in $u$, $\nabla u$ and $f_2$ (weakly) increasing in $u$. Furthermore, one has to assume that either $w_\lambda > v$ on $\partial\Omega$ for $\lambda \in (\lambda_1, \lambda_2]$ or that $\Omega$ satisfies an interior ball condition, that $v, w \in C^1(\overline{\Omega})$ and that $\partial w_\lambda/\partial n\big|_{\partial\Omega}$ depends continuously on $\lambda \in (\lambda_1, \lambda_2]$. For a proof, see also McKenna and Walter [14], let $\lambda^*$ be the smallest $\lambda$ with $w_\lambda \geq v$ in $\Omega$. The difference $w_{\lambda^*} - v \geq 0$ satisfies a linear uniformly elliptic differential inequality

$$L\varphi + c\varphi \leq 0 \;\text{ in }\; \Omega \;\text{ with }\; -k \leq c \leq 0 \;\text{ in }\; \Omega.$$

If $\lambda^* > \lambda_1$, the strong maximum principle yields $\varphi = w_{\lambda^*} - v > 0$ in $\Omega$ and $\partial w_{\lambda^*}/\partial n < \partial v/\partial n$ at points $x \in \partial\Omega$, where $w_{\lambda^*}(x) = v(x)$. Both inequalities together contradict the minimality of $\lambda^*$.

Typical applications are:

(a) The uniqueness of positive solutions $u \in C^2(\Omega) \cap C^1(\overline{\Omega})$ of

$$\Delta u + f(x, u) = 0 \;\text{ in }\; \Omega, \qquad u = 0 \;\text{ on }\; \partial\Omega,$$

where $\Omega$ satisfies an interior ball condition, $f(x, 0) \geq 0$, $\partial f(x, u)/\partial u > -\infty$ (difference quotients are bounded below) and $f(x, u)/u$ is strictly decreasing in $u > 0$. For two solutions $v$, $w$ we may take $w_\lambda = \lambda w$ for $\lambda \in [1, \infty)$ and apply the sweeping principle.

(b) The radial symmetry result for positive solutions of

$$\Delta u + f(u) = 0 \text{ in a ball } B_R(0), \qquad u = 0 \text{ on } \partial B_R(0),$$

see Gidas, Ni and Nirenberg [6], where $f$ is Lipschitz in $u$. Usually the problem is reduced to showing symmetry in one direction, e.g., the $x_1$-direction. We may take $v(x) = u(x)$ and $w_\lambda(x) = u(x^\lambda)$, where $x^\lambda = (2\lambda - x_1, x')$ is the reflection of $x = (x_1, x')$ at the hyperplane $T_\lambda = \{x : x_1 = \lambda\}$, $\lambda \in [0, R]$. In this case the domain on which $w_\lambda$ lives is also parametrized, i.e., we take $\Omega_\lambda = B_R(0) \cap \{x : x_1 > \lambda\}$, the so-called right-hand cap. The sweeping principle (and an additional step showing that $\lambda^*$ is well defined and $< R$) then tells us $u(x^\lambda) \geq u(x)$ for $x \in \Omega_\lambda$ and all $\lambda \in [0, R]$, and in particular for $\lambda = 0$ we have

$$u(-x_1, x') \geq u(x) \text{ for } x \in B_R(0) \text{ with } x_1 > 0.$$

Repeating the argument with $\tilde{u}(x) = u(-x_1, x')$, the desired symmetry result follows.

For further applications of the sweeping principle see Guo-Webb [7], Korman-Leung [10] and Sweers [19]. In this paper we want to use variants of the sweeping principle to obtain uniqueness results for degenerate elliptic equations, where, due to the degeneracy, comparison principles are not directly at hand.

## 3. The anisotropic singular equation

For a convex bounded domain $\Omega \subset \mathbb{R}^2$, $\partial\Omega \in C^{2,\alpha}$ and $a \geq b \geq 0$ the existence of a positive solution of

$$u^a u_{xx} + u^b u_{yy} + p(x, y) = 0 \text{ in } \Omega, \qquad u = 0 \text{ on } \partial\Omega \tag{1}$$

was proved by Choi-Lazer-McKenna [4], provided $p$ is continuous with $P \geq p(x, y) \geq \varepsilon > 0$ in $\Omega$. They also prove a comparison principle for a positive sub-, super-solution pair $v$, $w$ under the condition that either $v_{xx} \leq 0$ or $w_{xx} \leq 0$. Hence uniqueness holds in the class of solutions $u$ with $u_{xx} \leq 0$. In Lair-Shaker [11] numerical evidence is given that $u_{xx} \leq 0$ is unrealistic for a solution $u$. Lair and Shaker give a uniqueness theorem under the condition that *one* solution satisfies $u_{xx} \leq K$, where $K$ depending on $\Omega$, $a$, $b$ and $p$ can be given explicitly.

For a more general uniqueness theorem we introduce the *interior rectangle condition.*

**Definition.** The bounded domain $\Omega \subset \mathbb{R}^2$ with $0 \in \Omega$ satisfies the interior rectangle condition, if for each $(x, y) \in \partial\Omega$ the rectangle $\{(\tau_1 x, \tau_2 y) : \tau_i \in [0, 1)\}$ is a subset of $\Omega$.

Obviously, balls centered at 0 and rectangles parallel to the axes with 0 as an interior point satisfy this condition, and the set of domains satisfying this condition is a subset of the sets star-shaped with respect to the origin. But also nonconvex, nonsmooth domains can satisfy the interior rectangle condition, see Figure 1 for an example.

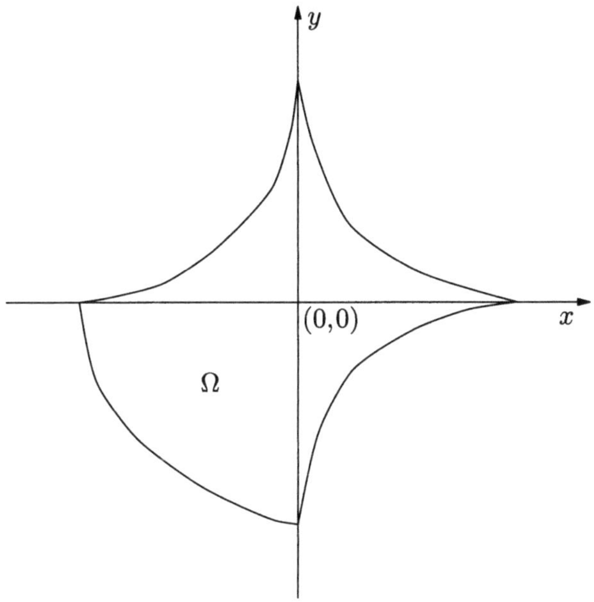


Figure 1: A domain with vertices on the axes

**Theorem 1.** *Suppose the bounded domain* $\Omega \subset \mathbb{R}^2$ *satisfies the interior rectangle condition and* $p : \Omega \to \mathbb{R}$ *satisfies*

$$p(\tau_1 x, \tau_2 y) \geq p(x,y) \quad \textit{for} \quad (x,y) \in \Omega, \; \tau_i \in [0,1]. \tag{M}$$

*Then there is at most one positive solution of* (1).

**Remarks.** 1) For $a = b \geq 0$, equation (1) is isotropic, and uniqueness of the positive solution is an immediate consequence of the weak comparison principle applied to $u_{xx} + u_{yy} + p(x,y)/u^a = 0$ in $\Omega$. For $a > b \geq 0$, the equation degenerates on the boundary.

2) (M) means that for every rectangle $Q$, defined by $(0,0)$ and $(x,y) \in \Omega$, the minimum of $p$ on $\overline{Q}$ is attained at the corner $(x,y)$.

*Proof.* Let $v$, $w$ be two solutions. We write $z = (x,y) \in \mathbb{R}^2$. For $\lambda > 1$ we define the scaling

$$w_\lambda(x,y) = \lambda w(x_\lambda, y_\lambda) = \lambda w(\lambda^{-(a+1)/2} x, \lambda^{-(b+1)/2} y)$$

for $(x,y) \in \Omega_\lambda = \{\lambda^{(a+1)/2}x, \lambda^{(b+1)/2}y \mid (x,y) \in \Omega\}$. For such $\lambda$ the set $\Omega_\lambda$ contains $\overline{\Omega}$ because of the interior rectangle condition. Furthermore, on $\Omega \subset \Omega_\lambda$ we have

$$\begin{aligned} w_\lambda^a w_{\lambda,xx} + w_\lambda^b w_{\lambda,yy} + p(x,y) &= (w^a w_{xx})(x_\lambda, y_\lambda) + (w^b w_{yy})(x_\lambda, y_\lambda) + p(x,y) \\ &\overset{(M)}{\leq} (w^a w_{xx})(x_\lambda, y_\lambda) + (w^b w_{yy})(x_\lambda, y_\lambda) + p(x_\lambda, y_\lambda) \\ &= 0, \end{aligned}$$

i.e., $w_\lambda$ is a supersolution. Again by $\overline{\Omega} \subset \Omega_\lambda$ there exists $\varepsilon_\lambda > 0$ with $w_\lambda \geq \varepsilon_\lambda > 0$ on $\overline{\Omega}$; for $\lambda$ sufficiently large we may take $\varepsilon_\lambda = \frac{\lambda}{2} w(0)$. Hence there exists a large $\bar{\lambda}$ such that $w_{\bar{\lambda}} \geq v$ on $\Omega$. Let $\lambda^* = \inf\{\lambda > 1 : w_\lambda \geq v \text{ in } \Omega\}$ and suppose for contradiction $\lambda^* > 1$. Then there exist sequences $(z_k) \subset \Omega$ and $\lambda_k \nearrow \lambda^*$, $\lambda_k > 1$ such that $w_{\lambda_k}(z_k) < v(z_k)$. Taking convergent subsequences we obtain $z_k \to \bar{z} \in \overline{\Omega}$, $w_{\lambda^*}(\bar{z}) \leq v(\bar{z})$ and, since $w_{\lambda^*} \geq v$ by definition, $w_{\lambda^*}(\bar{z}) = v(\bar{z})$. Clearly $\bar{z} \notin \partial\Omega$ since there $v = 0$ and $w_{\lambda^*} > 0$. Hence $w_{\lambda^*} - v$ has an interior zero-minimum at $\bar{z}$. Next we derive an elliptic differential inequality for $\varphi = w_{\lambda^*} - v \geq 0$ in $\Omega$. Let $\varphi_t = t w_{\lambda^*} + (1-t)v > 0$ on $\Omega$, $t \in [0,1]$, and define for $z \in \Omega$

$$a_{11}(z) = \int_0^1 \varphi_t^a(z)\,dt, \qquad a_{22}(z) = \int_0^1 \varphi_t^b(z)\,dt,$$

$$c(z) = \int_0^1 a\varphi_t^{a-1}(z)\varphi_{t,xx}(z) + b\varphi_t^{b-1}(z)\varphi_{t,yy}(z)\,dt.$$

If we subtract the differential inequalities

$$w_{\lambda^*}^a w_{\lambda^*,xx} + w_{\lambda^*}^b w_{\lambda^*,yy} + p(z) \leq 0 \text{ in } \Omega,$$
$$v^a v_{xx} + v^b v_{yy} + p(z) = 0 \text{ in } \Omega,$$

we obtain

$$0 \geq a_{11}\varphi_{xx} + a_{22}\varphi_{yy} + c\varphi \geq a_{11}\varphi_{xx} + a_{22}\varphi_{yy} + c^-\varphi \text{ in } \Omega \tag{7}$$

with $\varphi > 0$ on $\partial\Omega$ and $c = c^+ + c^-$. Inequality (7) is uniformly elliptic in $\Omega$, since the coefficients $a_{11} \geq \varepsilon_{\lambda^*}^a/(a+1)$, $a_{22} \geq \varepsilon_{\lambda^*}^b/(b+1)$ in $\Omega$ are bounded, and $c(z)$ is bounded in any compact subset of $\Omega$.

By the strong maximum principle (applied to every compact subset of $\Omega$) we get $\varphi \equiv 0$ or $\varphi > 0$ in $\Omega$, where the existence of the touching point $\bar{z} \in \Omega$ with $\varphi(\bar{z}) = 0$ forces $\varphi \equiv 0$ in contradiction to $\varphi > 0$ on $\partial\Omega$. Hence $\lambda^* = 1$ and $w \geq v$ for any two positive solutions, which is the desired uniqueness result. □

## 4. The nonlinear Keldysh equation

Čanić and Keyfitz proved in [2], [3] the existence of a solution of the nonlinear Keldysh-type equation

$$\left((u+x)u_x - \frac{u}{2}\right)_x + u_{yy} = 0 \text{ in } \Omega, \qquad u = g \text{ on } \Omega, \tag{2}$$

where $\Omega \subset \mathbb{R}^2$ is the rectangle $(0,X) \times (-Y,Y)$, and the boundary values $g$ are chosen such that

$$g = 0 \quad \text{on} \quad \Gamma_2 = \{0\} \times [-Y,Y] \tag{3}$$

and either

$$\frac{-x}{1+\sqrt{1-Cx}} \leq g(x,y) \leq 0 \quad \text{on} \quad \partial\Omega \tag{4$_-$}$$

or

$$A\sqrt{x} \leq g(x,y) \leq B\sqrt{x} \quad \text{on} \quad \partial\Omega, \tag{5}$$

where $0 \leq C < 1/X$, $0 \leq A \leq B$, respectively (additional assumptions on $g$ are needed, see [2], [3]). Because of the boundary condition (3), the operator in (2) degenerates on $\Gamma_2$.

Čanić and Keyfitz proved in [2] a comparison theorem for elliptic sub-, super-solutions $v$, $w$ of (2) under the hypothesis that either $v_{xx}$ or $w_{xx} \leq 0$ in $\Omega$, which implies uniqueness for (2) in the class of solutions with $u_{xx} \leq 0$. Three types of *explicit* solutions matching $g = 0$ on $\Gamma_2$ are known,

$$\frac{-x}{1+\sqrt{1-\gamma x}}, \quad \frac{-x}{1+\sqrt{1+\gamma x}} \quad \text{for} \quad \gamma \geq 0 \quad \text{and} \quad \alpha\sqrt{x} \quad \text{for} \quad \alpha \geq 0,$$

where only the first and the third satisfy $u_{xx} \leq 0$, whereas for the second $u_{xx} \geq 0$.

The following theorem gives a more general (but still only partial) answer to the uniqueness of *positive* elliptic solutions.

**Theorem 2.** *The boundary value problem* (2) *has at most one elliptic solution, if the boundary data $g$ is a function of $x$ only, satisfies* (5) *and*

$$g(X) = \max_{[0,X]} g(x), \quad \frac{g(x_2) - g(x_1)}{x_2 - x_1} > -1 \text{ for all } x_1, x_2 \in [0,X], x_1 \neq x_2. \tag{8}$$

**Remark.** Theorem 2 remains valid, if $g(x,y)$ satisfies (5) and if for all $y \in [-Y,Y]$ we have

$$g(X,y) = \max_{Q} g, \quad \frac{g(x_2,y) - g(x_1,y)}{x_2 - x_1} > -1, \quad x_1, x_2 \in [0,X], x_1 \neq x_2. \tag{8$'$}$$

A further consequence of the Čanić-Keyfitz comparison theorem is, that for solutions of (2) with boundary values satisfying either (4)$_-$ or (5) we have

$$\text{either} \quad \frac{-x}{1+\sqrt{1-Cx}} \leq u(x,y) \leq 0 \quad \text{or} \quad A\sqrt{x} \leq u(x,y) \leq B\sqrt{x} \quad \text{in} \quad \Omega. \tag{9}$$

For negative solutions we did not succeed in proving a uniqueness theorem. But we can prove an estimation theorem analogous to (9).

**Theorem 3.** *If the boundary data g satisfies*

$$\frac{-x}{1+\sqrt{1+Cx}} \leq g(x,y) \leq 0 \quad on \ \partial\Omega \tag{4$_+$}$$

*for some $C \geq 0$, then the following holds for every elliptic solution u of (2):*

$$\frac{-x}{1+\sqrt{1+Cx}} \leq u(x,y) \leq 0 \quad in \ \Omega. \tag{10}$$

**Remark.** Note that (10) does not follow from the comparison theorem since $\left(\frac{-x}{1+\sqrt{1+Cx}}\right)_{xx} \geq 0$ for $C \geq 0$.

*Proof of Theorem* 2. Let $v$, $w$ be two elliptic solutions of (2) with boundary data as in the theorem. We write $z = (x,y) \in \mathbb{R}^2$. Since $g(X)$ is a (constant) supersolution, the comparison theorem yields $v, w \leq g(X)$; here we have used $g(X) = \max_{[0,X]} g(x)$. We define

$$w_\lambda(x,y) = w(x+\lambda, y) + \lambda$$

for $(x,y) \in \Omega_\lambda = \{(\xi-\lambda,\eta) \mid (\xi,\eta) \in \Omega\}$. Note that $w_\lambda$ solves (2) on $\Omega_\lambda$ and attains the boundary values $g(x+\lambda)+\lambda$. We choose $\lambda \in (0,X)$ so close to $X$ that

$$A\sqrt{\lambda} + \lambda \geq B\sqrt{X-\lambda}$$

holds. For such $\lambda$ we find by (9) and the above equation

$$w_\lambda(x,y) \geq A(\sqrt{x+\lambda}+\lambda) \geq B\sqrt{X-\lambda} \geq B\sqrt{x} \geq v(x)$$

for all $(x,y) \in \overline{\Omega}_\lambda \cap \overline{\Omega} = \{(x,y) \in \overline{\Omega} : 0 \leq x \leq X-\lambda\}$. Now let $\lambda^* = \inf\{\lambda \in [0,X) : w_\lambda \geq v \text{ on } \overline{\Omega}_\lambda \cap \overline{\Omega}\}$ and suppose for contradiction $\lambda^* > 0$. Then there exist sequences $\lambda_n \nearrow \lambda^*$, $z_n \in \overline{\Omega}_{\lambda_n} \cap \overline{\Omega}$ with $z_n$ being the point where $w_{\lambda_n} - v$ attains its negative minimum on $\overline{\Omega}_{\lambda_n} \cap \overline{\Omega}$. Taking convergent subsequences $z_n \to \bar{z} \in \overline{\Omega}_{\lambda^*} \cap \overline{\Omega}$ we find $w_{\lambda^*}(\bar{z}) - v(\bar{z}) \leq 0$ and, due to the definition of $\lambda^*$, $w_{\lambda^*}(\bar{z}) - v(\bar{z}) = 0$.

We want to show $\bar{z} \notin \partial(\Omega_{\lambda^*} \cap \Omega)$. On the vertical part of that boundary $x=0$ we have $w_{\lambda^*}(0,y) \geq A\sqrt{\lambda^*} + \lambda^* > 0 = v(0,y)$ and at $x = X - \lambda^*$ we find $w_{\lambda^*}(x,y) = g(X) + \lambda^* > g(X) \geq v(x)$, as noted at the beginning of the proof. On the horizontal parts $y = \pm Y$ we observe $w_{\lambda^*}(x,y) = g(x+\lambda^*) + \lambda^* > g(x) = v(x,y)$; here we have used (8). Hence at $\bar{z}$ there is an interior negative minimum of $\varphi = w_{\lambda^*} - v$ on $\overline{\Omega}_{\lambda^*} \cap \overline{\Omega}$. Let $\varphi_t = tw_{\lambda^*} + (1-t)v$ on $\overline{\Omega}_{\lambda^*} \cap \overline{\Omega}$. Subtraction of the differential equations satisfied by $w_{\lambda^*}$ and $v$ results in the following linear equation for $\varphi$

$$a_{11}\varphi_{xx} + a_{22}\varphi_{yy} + b_1\varphi_x + c\varphi = 0 \quad \text{in} \ \Omega_{\lambda^*} \cap \Omega, \tag{11}$$

where the coefficients are defined for $z \in \Omega_{\lambda^*} \cap \Omega$ by

$$a_{11}(z) = \int_0^1 \varphi_t(z) + x\,dt, \qquad a_{22}(z) = 1,$$

$$b_1(z) = \int_0^1 \frac{1}{2} + 2\varphi_{t,x}(z)\,dt, \qquad c(z) = \int_0^1 \varphi_{t,xx}(z)\,dt.$$

Since $\varphi$ is nonnegative we find from (11)

$$a_{11}\varphi_{xx} + a_{22}\varphi_{yy} + b_1\varphi_x + c^-\varphi \leq 0 \ \text{ in } \ \Omega_{\lambda^*} \cap \Omega. \tag{12}$$

Since $a_{11} \geq \int_0^1 t\lambda + x\,dt \geq \frac{1}{2}\lambda$ in $\Omega_{\lambda^*} \cap \Omega$, the above equation is uniformly elliptic in $\Omega_{\lambda^*} \cap \Omega$. Since furthermore the coefficients $b_1, c, c^-$ are bounded in compact subsets of $\Omega_{\lambda^*} \cap \Omega$, the function $\varphi$ cannot have an interior zero minimum unless it vanishes identically (which is impossible since $\varphi > 0$ at $x = 0$). This contradiction shows that really $\lambda^* = 0$ and $w \geq v$ for any two positive solutions. Thus the uniqueness is proved. □

*Proof of Theorem* 3. Let $w$ be an elliptic solution of (2) with boundary data satisfying $(4)_+$, and let

$$v_\lambda(x) = \frac{-\lambda x}{1 + \sqrt{1 + Cx}}, \qquad \lambda > 1, \quad x \in [0, X]$$

(we use $v = v_1$). By the comparison lemma we get the estimate $-\frac{1}{2}x \leq w(x,y) \leq 0$ and therefore

$$\underline{D}_x^+ w(0,y) = \liminf_{h\to 0+} \frac{w(h,y)}{h} \geq -\frac{1}{2}, \quad y \in [-Y, Y].$$

Since $v_\lambda'(0) = -\lambda/2$ and since $v_\lambda < 0$ in $(0, X] \times [-Y, Y]$, there exists $\bar\lambda \gg 1$ with $v_\lambda \leq w$ in $\Omega$ for all $\lambda \geq \bar\lambda$. Let $\lambda^* = \inf\{\lambda > 1 : w \geq v_\lambda \text{ in } \Omega\}$ and suppose for contradiction $\lambda^* > 1$. Then there exist sequences $\lambda_n \nearrow \lambda^*$, $z_n \in \overline{\Omega}$ with $w - v_{\lambda_n}$ having a negative minimum at $z_n = (x_n, y_n)$. On $\partial\Omega$ we have $w(z) = g(z) \geq v(x) \geq v_\lambda(x)$ and therefore $z_n \in \Omega$, $\nabla w(z_n) = (\lambda_n v'(x_n), 0)$. Taking convergent subsequences and regarding the definition of $\lambda^*$, there is a $\bar z = (\bar x, \bar y) \in \overline{\Omega}$ with $w(\bar z) = v_{\lambda^*}(\bar x)$, $\nabla w(\bar z) = (\lambda^* v'(\bar x), 0)$. Since by assumption $\lambda^* > 1$ and hence $v_{\lambda^*}(x) < v(x) \leq g(x) = w(z)$ on $\partial\Omega \setminus \Gamma_2$, we see that $\bar z$ cannot be on the nondegenerate boundary $\partial\Omega \setminus \Gamma_2$. On $\Gamma_2 = \{0\} \times [-Y, Y]$ we have $\lambda^* v'(0) = -\frac{\lambda^*}{2} < -\frac{1}{2} \leq \underline{D}_x^+ w(0,y)$. Hence at $\bar z \in \Omega$ the function $w - v_{\lambda^*}$ attains its interior zero minimum.

Next we derive a differential inequality for $v_\lambda$. By some elementary calculus $(v^2)'' = 2(vv'' + v'^2) \geq 0$. If we consider $v$ as a function on $\Omega$, we calculate

$$\Big((v_\lambda + x)v_{\lambda,x} - \frac{v_\lambda}{2}\Big)_x + \underbrace{v_{\lambda,yy}}_{=0} = (\lambda v + x)\lambda v'' + \lambda^2 v'^2 + \frac{\lambda}{2}v' - \lambda\underbrace{\Big((v+x)v'' + v'^2 + \frac{1}{2}v'\Big)}_{=0}$$

$$= (\lambda^2 - \lambda)(vv'' + v'^2) \geq 0,$$

i.e., $v_\lambda$ is a subsolution on $\Omega$ for all $\lambda > 1$. With the same technique as in Theorem 2, we see that $w - v_{\lambda^*}$ satisfies a linear elliptic inequality, which is uniformly elliptic in compact subsets of $\Omega$. Therefore $w - v_{\lambda^*}$ cannot have an interior zero minimum unless it is identically zero (which is impossible since $\lambda^* v'(0) = -\frac{\lambda^*}{2} < -\frac{1}{2} \le \underline{D}_x^+ w(0,y)$). This contradiction shows $\lambda^* = 1$ and finishes the proof of the theorem. □

## 5. The quasimonotone $p$-Laplace system

We consider weak solutions $u \in (W_0^{1,p}(\Omega))^k$ of the quasimonotone (cooperative) $p$-Laplace system

$$\Delta_p u_i + a_{ij}(x) u_j^{p-1} + f_i(x,u) = 0 \text{ in } \Omega \qquad (i = 1, \ldots, k), \tag{6}$$

where for $i, j = 1, \ldots, k$ we require

$$\partial\Omega \in C^{2,\alpha}, \quad a_{ij} \in C^\alpha(\overline{\Omega}), \quad a_{ij} \ge 0 \text{ for } i \ne j, \quad (a_{ij})_{i,j=1}^k \text{ irreducible}, \tag{H1}$$

$$f_i \in C^\alpha_{\text{loc}}(\overline{\Omega} \times [0,\infty)^k), \quad 0 \le f_i(x,u) \le K(1 + u_i^{p-1}) \text{ for } u \ge 0, \tag{H2}$$

$$f_i(x,tu)/t^{p-1} \text{ strictly decreasing in } t > 0 \text{ for all } u \text{ with } u_1, \ldots, u_k > 0, \tag{H3}$$

$$f_i(x,u) \text{ increasing in } u \ge 0. \tag{H4}$$

**Theorem 4.** *Under condition* (H1)–(H4) *the system* (6) *has at most one nontrivial nonnegative solution $u \in (W_0^{1,p}(\Omega))^k$ on the bounded domain $\Omega$.*

**Remarks.** 1) Irreducibility for $(a_{ij})_{i,j=1}^k$ means that there is no splitting of indices $\{1, \ldots, k\}$ into disjoint subsets $\alpha$, $\beta$ such that $a_{ij} \equiv 0$ in $\Omega$ for $i \in \alpha$, $j \in \beta$, see Walter [21].

2) $C^\alpha_{\text{loc}}(\overline{\Omega} \times [0,\infty)^k)$ is the space of all locally $\alpha$-Hölder continuous functions on $\overline{\Omega} \times [0,\infty)^k$.

3) A particular difficulty for the $p$-Laplacian is the loss of uniform ellipticity at points $x$ with $\nabla u(x) = 0$. The maximal regularity of solutions is $C^{1,\alpha}$ and, furthermore, the strong comparison principle can fail for sub-, supersolutions of $p$-Laplacian inequalities, see Walter [22] and Reichel-Walter [16].

4) Existence and uniqueness theorems for systems like (6) are given in Fleckinger-Pellé and Takáč [5]. They are based on the theory of strongly subhomogeneous mappings. We prove our uniqueness theorem by maximum principles and regularity results only.

5) The sweeping principle in the $p$-Laplacian context for a single equation was used by Sakaguchi [17], Theorem A.1, to prove simplicity of the first eigenvalue.

*Proof.* Let $v$, $w$ be two nontrivial nonnegative solutions; we write $(av^{p-1})_i$ for $a_{ij}v_j^{p-1}$ and a corresponding expression for $w$. Because of the growth condition $0 \le f_i \le K(1+u_i^{p-1})$ the regularity result of Lieberman (Theorem 1 in [12]) applies and gives $v_i, w_i \in C^{1,\alpha}(\overline{\Omega})$ and then, furthermore, the Hölder continuity of $a_{ij}$ and $f_i$ allows us to apply the regularity result due to Barles, see Fleckinger-Pellé and Takáč, [5], Theorem 5.2, and we get $v, w \in C^2(\overline{\Omega}_\delta)$, where $\Omega_\delta = \{x \in \Omega : \text{dist}\,(x, \partial\Omega) < \delta\}$ is a sufficiently small neighbourhood of $\partial\Omega$. For small $\delta > 0$ the set $\Omega_\delta$ is connected due to the $C^{2,\alpha}$-regularity of $\partial\Omega$. By nonnegativity of $v$ we get

$$\Delta_p v_i + a_{ii}^- v_i^{p-1} \le 0 \;\text{ in }\; \Omega, \quad i = 1, \dots, k,$$

and by Hopf's maximum principle, see Tolksdorf [20], we obtain $v_i \equiv 0$ in $\overline{\Omega}$ or $v_i > 0$ in $\Omega$ and $\partial v_i/\partial n < 0$ on $\partial\Omega$. It is well known that the irreducibility of $(a_{ij})$ forces $v > 0$ or $v \equiv 0$ in $\Omega$, see Walter [21]. So in fact we have two positive solutions $v, w > 0$ in $\Omega$. Due to the boundary behaviour $\partial v_i/\partial n, \partial w_i/\partial n < 0$, there exists a large $\lambda \gg 1$ with $w_\lambda = \lambda w \ge v$ in $\Omega$. Let $\lambda^* = \inf\{\lambda \ge 1 : w_\lambda \ge v \text{ in } \Omega\}$ and suppose for contradiction $\lambda^* > 1$. We claim that there exists a touching point $x \in \Omega$ with $\lambda^* w_{i_0}(x) = v_{i_0}(x)$ for a certain index $i_0 \in \{1, \dots, k\}$. If this were not the case, then $\lambda^* w > v$ in $\Omega$. It then follows from

$$\begin{aligned}
&\Delta_p(\lambda^* w_i) + \lambda^{*p-1}(aw^{p-1})_i + f_i(x, \lambda^* w) \\
&\quad \le \lambda^{*p-1}(\Delta_p w_i + (aw^{p-1})_i + f_i(x, w)) \\
&\quad = 0 = \Delta_p v_i + (av^{p-1})_i + f_i(x, v))
\end{aligned} \tag{13}$$

(the first inequality follows from the subhomogeneity of $f$ and should be read in weak form after multiplication with a test function) and the structural assumptions of the system that

$$\Delta_p(\lambda^* w_i) + a_{ii}^-(\lambda^* w_i)^{p-1} \le \Delta_p v_i + a_{ii}^- v_i^{p-1}, \quad i = 1, \dots, k. \tag{14}$$

By Hopf's comparison principle (see Tolksdorf [20]) applied to $\overline{\Omega}_\delta$, it follows that $\lambda^* \partial w/\partial n > \partial v/\partial n$ on $\partial\Omega$. Together with $\lambda^* w > v$ in $\Omega$, this contradicts the minimality of $\lambda^*$. Hence a touching point $x \in \Omega$ exists. We claim that there must be a touching point at least in $\Omega_\delta$, the neighbourhood of $\partial\Omega$. If this were not the case, then there would exist $\varepsilon > 0$ such that $w_{\lambda^*} - v > \varepsilon(1, \dots, 1)$ on $\partial\Omega_{\delta/2} \cap \Omega$. By the strict subhomogeneity and continuity of $f$ we find $\tilde{\varepsilon} > 0$ with

$$f_i(x, \lambda^* w) + \tilde{\varepsilon} \le \lambda^{*p-1} f_i(x, w) \;\text{ on }\; \Omega \setminus \Omega_{\delta/2},\; i = 1, \dots, k,$$

and hence, choosing $\varepsilon > 0$ small enough, (13) and (14) read

$$\begin{aligned}
&\Delta_p v_i + a_{ii}^- v_i^{p-1} \ge \Delta_p(\lambda^* w_i) + a_{ii}^-(\lambda^* w_i)^{p-1} + \tilde{\varepsilon} \\
&\quad \ge \Delta_p(\lambda^* w_i - \varepsilon) + a_{ii}^-(\lambda^* w_i - \varepsilon)^{p-1} \qquad \text{in } \Omega \setminus \Omega_{\delta/2}, \quad i = 1, \dots, k.
\end{aligned} \tag{15}$$

By the weak comparison principle applied to (15), we find $\lambda^* w - v \ge \varepsilon(1, \dots, 1)$ in $\Omega \setminus \Omega_{\delta/2}$ in contradiction to the existence of a touching point in $\Omega \setminus \Omega_\delta$. Hence there is indeed a touching point in $\Omega_\delta$ and the strong comparison principle, see Tolksdorf [20], gives $\lambda^* w_{i_0} \equiv v_{i_0}$ in $\Omega_\delta$. Since (13) is valid pointwise in $\Omega_\delta$ and $\lambda^* > 1$, the strong subhomogeneity of $f$ results in a strict inequality in (13) and (14) on $\Omega_\delta$ in contradiction to $\lambda^* w_{i_0} \equiv v_{i_0}$. Hence the assumption $\lambda^* > 1$ must be dropped, and we find $w \geq v$ for any two nontrivial nonnegative solutions of (6). □

## References


[1] A. Anane, *Simplicité et isolation de la première valeur propre du p-laplacien avec poid.* C.R. Acad. Sci. Paris Sér. I Math. **305** (1987), 725–728.

[2] S. Čanić and B. Lee Keyfitz, *An elliptic problem arising from the unsteady transonic small disturbance equation.* J. Diff. Eqs. **125** (1996), 548–574.

[3] S. Čanić and B. Lee Keyfitz, *A smooth solution for a Keldysh type equation.* Commun. Part. Diff. Eqs. **21** (1996), 319–340.

[4] Y.S. Choi, A.C. Lazer and P.J. McKenna, *On a singular quasilinear anisotropic elliptic boundary value problem.* Trans. Amer. Math. Soc. **347** (1995), 2633–2641.

[5] J. Fleckinger-Pellé and P. Takáč, *Uniqueness of positive solutions for nonlinear cooperative systems with the p-Laplacian.* Indiana Univ. Math. J. **43** (1994), 1227–1253.

[6] B. Gidas, Wei-Ming Ni and L. Nirenberg, *Symmetry and related problems via the maximum principle.* Comm. Math. Phys. **68** (1979), 209–243.

[7] Z. Guo and J.R. Webb, *Uniqueness of positive solutions for quasilinear elliptic equations when a parameter is large.* Proc. Royal Soc. Edinburgh **124** A (1994), 189–198.

[8] Yin Xi Huang, *Existence of positive solutions for a class of the p-Laplace equations.* J. Austral. Math. Soc. Ser. B **36** (1994), 249–264.

[9] T. Idogawa and M. Ôtani, *The first eigenvalue of some abstract elliptic operators.* Funk. Ekvac. **38** (1995), 1–9.

[10] P. Korman and A. Leung, *On the existence and uniqueness of positive steady states in the Volterra-Lotka ecological models with diffusion.* Appl. Analysis **26** (1987), 145–160.

[11] A. Lair and A. Shaker, *Uniqueness of solutions to a singular quasilinear elliptic problem.* Nonlinear Analysis, TMA, to appear.

[12] G. M. Liebermann, *Boundary regularity of solutions of degenerate elliptic equations.* Nonlinear Analysis, TMA **12** (1988), 1203–1219.

[13] P. Lindqvist, *On the equation* div $(|\nabla u|^{p-2}\nabla u) + \lambda|u|^{p-2}u = 0$. Proc. Amer. Math. Soc. **109** (1990), 157–164.

[14] P.J. McKenna and W. Walter, *On the Dirichlet problem for elliptic systems.* Appl. Analysis **21** (1986), 207–224.

[15] A. McNabb, *Strong comparison theorems for elliptic equations of second order.* J. Math. Mech. **10** (1961), 431–440.

[16] W. Reichel and W. Walter, *Radial solutions of equations and inequalities involving the p-Laplacian*, in preparation.

[17] S. Sakaguchi, *Concavity properties of solutions to some degenerate quasilinear elliptic Dirichlet problems.* Ann. Sc. Norm. Sup. Pisa Cl. Sci. **14** (1987), 403–420.

[18] D.H. Sattinger, *Topics in Stability and Bifurcation Theory.* Lecture Notes in Math. **309**, Springer-Verlag, Berlin 1973.

[19] G. Sweers, *Some results for a semilinear elliptic problem with a large parameter.* Proc. of the first Int. Conf. on Ind. Appl. Math., Paris 1987, CWI Tracts **36** (1987), 109–116.

[20] P. Tolksdorf, *On the Dirichlet problem for quasilinear equations in domains with conical boundary points.* Comm. Part. Diff. Eqs. **8** (1983), 773–817.

[21] W. Walter, *The minimum principle for elliptic systems.* Appl. Analysis **47** (1992), 1–6.

[22] W. Walter, *A new approach to minimum and comparison principles for nonlinear ordinary differential operators of second order.* Nonlinear Analysis, TMA, **22** (1995), 1071-1078.

Wolfgang Reichel, Mathematisches Institut I, Universität Karlsruhe, 76128 Karlsruhe, Deutschland.

# Problems and Remarks

International Series of Numerical Mathematics, Vol. 123, © 1997 Birkhäuser Verlag Basel

# Overdetermined Hardy inequalities

Alois Kufner

Let $M_i \subset \{0, 1, \cdots, k-1\}$, $i = 0, 1$, and denote $AC^{k-1}(0,1; M_0, M_1) = \{u(t) \in AC^{k-1}(0,1), u^{(i)}(0) = 0$ for $i \in M_0$, $u^{(j)}(1) = 0$ for $j \in M_1\}$. Find necessary and sufficient conditions on the parameters $p, q$ $(0 < q < \infty, 1 < p < \infty)$ and on the weight functions $w_0, w_k$ under which the $k$-th order Hardy inequality

$$\left(\int_0^1 | u(t) |^q w_0(t) dt\right)^{1/q} \leq c \left(\int_0^1 | u^{(k)}(t) |^p w_k(t) dt\right)^{1/p}$$

holds for all $u \in AC^{k-1}(0,1; M_0, M_1)$, if $\#M_0 + \#M_1 > k$.

*Remark* The problem is solved for $k = 1$ (see [1]) and for $\#M_0 + \#M_1 = k$ (see [2] and [3]).

## References


[1] B. Opic and A. Kufner, *Hardy type inequalities.* Essex, 1990.

[2] A. Kufner, *Higher order Hardy type inequalities.* Bayreuth Math. Schriften **44** (1993), 105–146.

[3] G. Sinnamon, *On a conjecture of A. Kufner.* To appear in Real Anal. exchange.


Alois Kufner, Math. Inst. Acad. Sci., Žitná 25, 11567 Praha 1, Czech Republic.

International Series of Numerical Mathematics, Vol. 123, © 1997 Birkhäuser Verlag Basel

# A condition for monotony

E.R. Love

The following is a classical theorem, given (for instance) in Titchmarsh's "Theory of Functions" (2nd edition, 1938), p. 370 Lemma 2:

*If $f(x)$ is continuous in $(a,b)$ and $D^+f \geq 0$ almost everywhere therein, and $D^+f > -\infty$ everywhere, then $f$ is non-decreasing.*

The question is: Can the hypothesis $D^+f > -\infty$ be omitted?

*Answer by E. R. Love.* It turns out that this isn't the case because of the following counter-example: Let $-f(x)$ be the well-known singular, non-decreasing function defined on $[0,1]$ which is constant in each complementary interval of Cantor's ternary set in $[0,1]$. $f$ is continuous and $D^+f = 0$ almost everywhere, and $D^+f = -\infty$ at each point of Cantor's set. Thus $f$ satisfies the hypothesis of the above theorem except that $D^+f \not> -\infty$. But $f$ is not non-decreasing, in fact it is non-increasing.

*Remark by J. Rätz.* If $f$ is differentiable on $\mathbb{R}$, $f'(x) \geq 0$ a.e. on $\mathbb{R}$, then $f$ is non-decreasing [cf. A.Bruckner, *Differentiation of Real Functions*, CMR Monograph Series, Amer. Math. Soc., 5 (1994), p. 120].

E.R. Love, Department of Mathematics, University of Melbourne, Parkville, Victoria 3052, Australia.

International Series of Numerical Mathematics, Vol. 123, 

# A conjectured inequality of T.J. Lyons

E.R. Love

## 1. Introduction

The conjecture is

$$\alpha \sum_{j=0}^{n} \binom{\alpha n}{\alpha j} x^{\alpha j}(1-x)^{\alpha(n-j)} \leq 1 \tag{1}$$

where $0 < \alpha \leq 1$, $0 < x < 1$, $n$ is a positive integer and

$$\binom{\alpha n}{\alpha j} = \frac{(\alpha n)!}{(\alpha j)!(\alpha(n-j))!} = \frac{\Gamma(\alpha n + 1)}{\Gamma(\alpha j + 1)\Gamma(\alpha(n-j)+1)} \tag{2}$$

Notice that (1) holds with equality when $\alpha = 1$, by the binomial theorem. So if the left side were an increasing function of $\alpha$ the conjecture would follow. In a private communication, Lyons has proved the weaker inequality in which the right side of (1) is replaced by $1/\alpha$; but numerical trials suggest very strongly that (1) holds.

*Contribution by E.R. Love.* It will be shown that (1) holds when $(n+1)\alpha \leq 1$, and some approaches will be explored which might lead to a proof under less restrictive conditions.

## 2. A global maximum

Let $t = 1 - 2x$, so that $-1 < t < 1$, $x = (1-t)/2$ and $1 - x = (1+t)/2$. Let

$$F(t) = \sum_{j=0}^{n} \binom{\alpha n}{\alpha j} (1-t)^{\alpha j}(1+t)^{\alpha(n-j)} = 2^{\alpha n} \sum_{j=0}^{n} \binom{\alpha n}{\alpha j} x^{\alpha j}(1-x)^{\alpha(n-j)}. \tag{3}$$

The conjecture (1) is equivalent to

$$F(t) \leq \frac{2^{\alpha n}}{\alpha}, \text{ for } -1 < t < 1. \tag{4}$$

$F$ is an even function, because

$$\begin{aligned} F(-t) &= \sum_{j=0}^{n} \binom{\alpha n}{\alpha j} (1+t)^{\alpha j} (1-t)^{\alpha(n-j)} \\ &= \sum_{i=0}^{n} \binom{\alpha n}{\alpha(n-i)} (1+t)^{\alpha(n-i)} (t-1)^{\alpha i} = F(t). \end{aligned}$$

It follows that $F'(t)$ is odd and $F'(0) = 0$. Further

$$F'(t) = \sum_{j=0}^{n} \binom{\alpha n}{\alpha j} (1-t)^{\alpha j} (1+t)^{\alpha(n-j)} \left( \frac{\alpha(n-j)}{1+t} - \frac{\alpha j}{1-t} \right), \tag{5}$$

$$\begin{aligned} F''(t) &= \sum_{j=0}^{n} \binom{\alpha n}{\alpha j} (1-t)^{\alpha j} (1+t)^{\alpha(n-j)} \left\{ \left( \frac{\alpha(n-j)}{1+t} - \frac{\alpha j}{1-t} \right)^2 \right. \\ &\qquad \left. - \left( \frac{\alpha(n-j)}{(1+t)^2} - \frac{\alpha j}{(1-t)^2} \right) \right\} \\ &= \sum_{j=0}^{n} \binom{\alpha n}{\alpha j} (1-t)^{\alpha j - 2} (1+t)^{\alpha(n-j)-2} G_j(t) \end{aligned}$$

where $G_j(t)$ is quadratic in $t$; by completing the square

$$G_j(t) = \alpha n\{(\alpha n - 1)(t - 1 + 2j/n)^2 - 4(j/n)\,(1 - j/n)\}$$

for $0 \le j \le n$ and $-1 < t < 1$.

If $\alpha n < 1$, $G_j(t) = 0$ for all $j$ and $t$ involved; thus $F(t)$ is globally maximum at $t = 0$.

If $\alpha n = 1$, $G_j(t) = 0$ for $j = 0$ and $j = n$. If $n \ge 2$ there are values of $j$ in between 0 and $n$, and the corresponding terms, being negative, ensure that $G_j(t) < 0$. Thus $F(t)$ is again a maximum at $t = 0$. If $n = 1$ the same conclusion is obtained by elementary, though not entirely trivial methods.

If $\alpha n > 1$, (5) contains terms of both signs, and the sign of $F''(t)$ is not apparent.

To summarize,

$$F(t) \le F(0) \text{ if } \alpha n \le 1 \text{ and } -1 < t < 1. \tag{6}$$

## 3. A proof of (1) when $(n+1)\alpha \leq 1$

By (3), (4) and (6) it is sufficient to prove that

$$\sum_{j=0}^{n} \binom{\alpha n}{\alpha j} = F(0) \leq \frac{2^{\alpha n}}{\alpha}. \tag{7}$$

For this we have

$$\begin{aligned} \binom{\alpha n}{\alpha j} &= (\alpha n + 1)^{-1} B^{-1}(\alpha j + 1, \alpha(n-j) + 1) \\ &= (\alpha n + 1)^{-1} \left( \int_0^1 u^{\alpha j} (1-u)^{\alpha(n-j)} du \right)^{-1}. \end{aligned} \tag{8}$$

The integral here is equal to

$$\begin{aligned} &\int_0^{1/2} \left(\frac{1-u}{u}\right)^{\alpha(n-j)} u^{\alpha n} du + \int_{1/2}^1 \left(\frac{u}{1-u}\right)^{\alpha j} (1-u)^{\alpha n} du \\ &= \int_0^{1/2} \left(\frac{1-u}{u}\right)^{\alpha(n-j)} u^{\alpha n} du + \int_0^{1/2} \left(\frac{1-v}{v}\right)^{\alpha j} v^{\alpha n} dv \\ &> \int_0^{1/2} u^{\alpha n} du + \int_0^{1/2} v^{\alpha n} dv = \frac{1}{(\alpha n + 1) 2^{\alpha n}}; \end{aligned} \tag{9}$$

this inequality holds because, for

$$0 < u < \frac{1}{2}, \quad \frac{1-u}{u} = \frac{1}{u} - 1 > 2 - 1 = 1 \tag{10}$$

With (8) this gives

$$\binom{\alpha n}{\alpha j} < 2^{\alpha n} \tag{11}$$

$$F(0) = \sum_{j=0}^{n} \binom{\alpha n}{\alpha j} < (n+1) 2^{\alpha n},$$

$$F(0) < 2^{\alpha n} \alpha \text{ if } n + 1 \leq 1/\alpha. \tag{12}$$

By (7), this proves (1) if $(n+1)\alpha \leq 1$. Notice that this inequality (9) is quite crude, because $(1-u)/u$ is large when $u$ is small, whereas it has been replaced by 1 (as in (10)) even when $u$ is small. I have not been able to find an improvement on (10) which was useable. Observe further that the right side of (11) is independent of $j$, whereas for $j = 0$ and $j = n$ the left side is simply 1. Another proof of (1) when $(n+1)\alpha \leq 1$ is given in the next section.

## 4. Another approach

Formula 1.5 (30) in "Higher Transcendental Fuctions" by Erdély and others (Bateman Manuscript Project, McGraw-Hill, 1953) becomes, replacing $\alpha$ by $\alpha n$ and $\beta$ by $\alpha(n-2j)$,

$$\int_0^{\pi/2} \cos^{\alpha n}\theta \cos\alpha(n-2j)d\theta = \frac{\pi}{2^{\alpha n+1}}\frac{\Gamma(1+\alpha n)}{\Gamma(1+\alpha(n-j))\Gamma(1+\alpha j)} = \frac{\pi}{2^{\alpha n+1}}\binom{\alpha n}{\alpha j}$$

if $\mathrm{re}(\alpha n) > -1$. Writing $p = x/(1-x) = (1-t)/(1+t)$, and in particular $1+t = 2/(1+p)$

$$\binom{\alpha n}{\alpha j}(1-t)^{\alpha j}(1+t)^{\alpha(n-j)} = \left(\frac{2}{1+p}\right)^{\alpha n}\binom{\alpha n}{\alpha j}p^{\alpha j},$$

whence by (3)

$$\begin{aligned} F(t) &= \left(\frac{2}{1+p}\right)^{\alpha n}\frac{2^{\alpha n+1}}{\pi}\sum_{j=0}^{n} p^{\alpha j}\int_0^{\pi/2}\cos^{\alpha n}\theta\cos\alpha(n-2j)d\theta \\ &= \left(\frac{4}{1+p}\right)^{\alpha n}\frac{2}{\pi}\int_0^{\pi/2}\cos^{\alpha n}\theta\sum_{j=0}^{n}p^{\alpha j}\cos\alpha(n-2j)d\theta \qquad (13) \end{aligned}$$

$$=\left(\frac{4}{1+p}\right)^{\alpha n}\frac{2}{\pi}\int_0^{\pi/2}\cos^{\alpha n}\theta\frac{(1+p^{\alpha(n+2)})\cos\alpha n\theta - p^{\alpha}(1+p^{\alpha n})\cos(n+2)\alpha\theta}{1-2p^{\alpha}\cos 2\alpha\theta + p^{2\alpha}}d\theta$$

In particular, since $p=1$ when $t=0$, (13) gives

$$\begin{aligned} F(0) &= 2^{\alpha n}\frac{2}{\pi}\int_0^{\pi/2}\cos^{\alpha n}\theta\frac{\sin(n+1)\alpha\theta\sin\alpha\theta}{\sin^2\alpha\theta}d\theta \\ &= \frac{2^{\alpha n}}{\alpha}\frac{2}{\pi}\int_0^{\pi\alpha/2}\cos^{\alpha n}\left(\frac{\theta}{\alpha}\right)\frac{\sin(n+1)\theta}{\sin\theta}d\theta. \end{aligned} \qquad (14)$$

If $(n+1)\alpha \le 1$ the integrands in (14) are non-negative and so

$$\begin{aligned} 0 \le F(0) &< 2^{\alpha n}\frac{2}{\pi}\int_0^{\pi/2}\frac{\sin(n+1)\theta}{\sin\theta}d\theta \\ &< 2^{\alpha n}\frac{2}{\pi}\int_0^{\pi/2}(n+1)d\theta = 2^{\alpha n}(n+1) \le \frac{2^{\alpha n}}{\alpha} \end{aligned}$$

Taken with the first part of Section 2, this is a second proof of (12), and so of the conjecture (1) when $(n+1)\alpha \le 1$. If (13) could be shown to be less that $2^{\alpha n}/\alpha$ for $0 < p < 1$ as in (4), the truth of (1) would follow, taking into account the eveness of $F(t)$. But the complexity of (13) suggests that great delicacy would be needed; and this is highlighted by the numerical approximation that, when $\alpha = 7/8$ and $n = 8$, $F(0) = 146 \cdot 204$ and $2^{\alpha n}/\alpha = 146 \cdot 286$, or equivalently $\alpha F(0)/2^{\alpha n} = 0 \cdot 99944$.

## 5. Binary decimal values of $\alpha$

Let us write $N$ instead of $n$, and $F(\alpha, N, t)$ instead of $F(t)$. Then (14) gives

$$\frac{\alpha}{2^{\alpha N}} F(\alpha, N, 0) = \alpha \frac{2}{\pi} \int_0^{\pi/2} \cos^{\alpha N} \theta \frac{\sin(N+1)\alpha\theta}{\sin \alpha\theta} d\theta. \tag{15}$$

To satisfy (4) with $t = 0$, or equally (7), we would need to show that the right side of (15) is at most 1. Now let $N = 2^n$ and $\alpha = m/2^n$, where $m$ and $n$ are positive integers. The right side of (15) becomes

$$\begin{aligned}
\frac{2m}{N\pi} &\int_0^{\pi/2} \cos^m \theta \frac{\sin(1 + \frac{1}{N})\theta}{\sin(\frac{m\theta}{N})} d\theta \\
&= \frac{2m}{N\pi} \int_0^{\pi/2} \cos^m \theta \left( \sin m\theta \cot(\frac{m\theta}{N}) + \cos m\theta \right) d\theta \\
&= \frac{2m}{N\pi} \left( \int_0^{\pi/2} \int_0^{\pi/2} \cos^m \theta \sin m\theta \cot(m\theta/N) d\theta + \pi 2^{-(m+1)} \right) \\
&= \frac{2m}{N\pi} \left( \int_0^{\pi/2} \cos^m \theta \cot(m\theta/N) \sin m\theta d\theta + 2^{-m} \right).
\end{aligned} \tag{16}$$

If we could show that any of the expressions in (16) is less than one for $0 < m < N = 2^n$, (7) would be proved for binary decimal values of $\alpha$; and since these values are dense in $[0,1]$ and $F(\alpha, N, 0)$ is a continuous function of $\alpha$, (7) would follow for all $\alpha$ in $[0,1]$. This would be a substantial step towards proving (1), requiring in addition only that $F(\alpha, N, t)$ be maximum at $t = 0$ to complete the proof of (1).

Alternatively, if we could show that any expression of (16) is an increasing function of the integer $m$ in $0 < m \leq N = 2^n$, (7) would follow for binary decimal values of $\alpha$; because $\alpha = 1$ when $m = N$, and so

$$F(\alpha, N, 0) = F(1, N, 0) = \sum_{j=0}^{N} \binom{N}{j} = 2^N = 2^{\alpha N}/\alpha.$$

Then (7) would follow for all $\alpha$ in $[0,1]$, by the denseness of the binary decimals in the preceding paragraph, and the "substantial step towards proving (1)" in that paragraph would have been made.

E.R. Love, Department of Mathematics, University of Melbourne, Parkville, Victoria 3052, Australia.

International Series of Numerical Mathematics, Vol. 123, © 1997 Birkhäuser Verlag Basel

# A theorem of Pommerenke and a conjecture of Erdős

Themistocles M. Rassias

Ch. Pommerenke [4] proved the following theorem. Let

$$f(z) = z^n + a_1 z^{n-1} + a_2 z^{n-2} + \cdots + a_{n-1} z + a_n$$

be a polynomial of degree $n$ with some $a_j \neq 0$. Assume that the region $E_f = \{z \in \mathbb{C} : |f(z)| \leq 1\}$ is connected, where $\mathbb{C}$ denotes the field of complex numbers. Then

$$\max \{|f'(z)| : z \in E_f\} < \frac{1}{2} e n^2.$$

P. Erdős [1] reviewing Pommerenke's paper conjectured that

$$\max \{|f'(z)| = d : z \in E_f\} < \frac{1}{2} n^2.$$

is also true and it is best possible. Erdős reposed his conjecture as a problem in [2]. G.M. Rassias, J.M. Rassias and Th.M. Rassias [6] constructed a counterexample to Erdős conjecture.

**Problem.** Determine the least upper bound for $\max \{|f'(z)| = d : z \in E_f\}$.

The problem of analysing Erdős' conjecture is useful for questions dealing with some special lemniscates involving Green's function (cf. [3]). The problem was reposed in [5].

**References**

[1] P. Erdős, *Review 95.* Mathematical Reviews, Amer. Math. Soc. **22**, January 1961 (review of a paper by Ch. Pommerenke).

[2] P. Erdős, *Problem 564.* Colloquium Mathematicum **15** (1966), p.320.

[3] G.V. Milovanovic, D.S. Mitrinovic and Th.M. Rassias, *Topics in Polynomials: Extremal Problems, Inequalities, Zeros.* World Scientific Publ. Co., Singapore, New Jersey, London, 1994.

[4] Ch. Pommerenke, *On the derivative of a polynomial.* Michigan Math. J. **6** (1959), 373–375.

[5] Th.M. Rassias, *Research Problems.* In: Global Analysis-Analysis on Manifolds, Teubner-Texte zur Mathematik, Leipzig, **57** (1983), 365–376.

[6] G.M. Rassias, J.M. Rassias and Th.M. Rassias, *A counterexample to a conjecture of P.Erdős.* Proc. Japan Acad. Sci. **53** (1977), 119–121.

Themistocles M. Rassias, 4, Zagoras St., Paradissos, Amaroussion, 15125 Athens, Greece.

International Series of Numerical Mathematics, Vol. 123, © 1997 Birkhäuser Verlag Basel

# Problems on finite sums decompositions of functions

Themistocles M. Rassias

Let $f : X \times Y \to \mathbb{R}$ be a function and an integer $n \geq 1$ where $X$, $Y$ are arbitrary nonempty sets. Find the best approxmation $f(x,y) \approx \sum_{i=1}^{n} f_i(x) g_i(y)$ with respect to the supremum norm $\|f\| = \sup\{|f(x,y)| : x \in X \text{ and } y \in Y\}$. This is even an open problem for polynomial functions $f$ defined on the Cartesian product of two real intervals. If one changes the norm then the problem seem to be totally open. It will also be interested to study the best approximation problem

$$f(x_1, \cdots, x_k) \approx \sum_{i_1=1}^{m_1} \cdots \sum_{i_k=1}^{m_k} \alpha_{i_1 \cdots i_k} f_{i_1}^1(x_1) \cdots f_{i_k}^k(x_k)$$

for functions $f$ in more than two variables, even if the $L^2$-norm is considered. The corresponding problem of the best $L^2$-approximation (for functions of two variables only) has been solved in Chapter 7 of the book by Th.M. Rassias and J. Šimša [2] (see also [4]). The problem of finding a necessary and sufficient condition for a function $f$ in several variables to be represented by

$$f(x_1, x_2, \cdots, x_k) = \sum_{i=1}^{n} f_{i_1}(x_1) f_{i_2}(x_2) \cdots f_{i_k}(x_k)$$

was proposed in [3].

**Remark.** Functions of certain special forms have been investigated by several authors, including J. d'Alembert (1717–1783), who as early as 1747 proved that each sufficiently smooth scalar functions $f$ of the form $f(x,y) = f_1(x) f_2(y)$ has to satisfy the partial differential equation $\frac{\partial^2 \log f}{\partial x \partial y} = 0$ (see [1]).

## References

[1] J. D'Alembert, *Recherches sur la courbe que forme une corde tendue mise en vibration I-II.* Hist. Acad. Berlin (1747), 2114–249.

[2] Th.M. Rassias and J. Šimša J., *Finite Sums Decompositions in Mathematical Analysis.* John Wiley & Sons, 1995.

[3] Th.M. Rassias, *Problem P286.* Aequationes Math. **38** (1989), 11–112.

[4] J. Šimša, *The best $L^2$-approximation by finite sums of functions with separable variables.* Aequationes Math. **43** (1992), 248–263.

Themistocles M. Rassias, 4, Zagoras St., Paradissos, Amaroussion, 15125 Athens, Greece.

MIX
Papier aus verantwortungsvollen Quellen
Paper from responsible sources
FSC® C105338

If you have any concerns about our products,
you can contact us on
**ProductSafety@springernature.com**

In case Publisher is established outside the EU,
the EU authorized representative is:
**Springer Nature Customer Service Center GmbH**
**Europaplatz 3, 69115 Heidelberg, Germany**

Printed by Libri Plureos GmbH
in Hamburg, Germany